Coffee Pests, Diseases and their Management

Coffee Pests, Diseases and their Management

by

J.M. Waller

CAB International, Egham, Surrey, UK

M. Bigger

Lilac Cottage, Kingsland, Leominster, UK

and

R.J. Hillocks

Natural Resources Institute, University of Greenwich, Medway Campus, Chatham, UK

www.cabi.org

CABI is a trading name of CAB International

CABI Head Office
Nosworthy Way
Wallingford
Oxfordshire OX10 8DE
UK

Tel: +44 (0)1491 832111
Fax: +44 (0)1491 833508
E-mail: cabi@cabi.org
Website: www.cabi.org

CABI North American Office
875 Massachusetts Avenue
7th Floor
Cambridge, MA 02139
USA

Tel: +1 617 395 4056
Fax: +1 617 354 6875
E-mail: cabi-nao@cabi.org

A catalogue record for this book is available from the British Library, London, UK.

A catalogue record for this book is available from the Library of Congress, Washington, DC.

ISBN-13: 978 1 84593 129 2

Typeset by Columns Design Ltd, Reading, UK
Printed and bound in the UK by Biddles Ltd, King's Lynn, Norfolk

Contents

Colour plate section can be found following p. 216

Preface

There have been several texts dealing with general aspects of coffee cultivation and production during the last 20 years, but none has given a comprehensive coverage of pests and diseases. At a time when greater attention is being paid to environmental issues, both consumers and producers of coffee are becoming more concerned with the way that coffee is grown.

Over the last three decades there has been a marked shift away from the reliance on pesticides for management of pest and diseases, towards a more integrated approach using a variety of methods. An understanding of the biology and ecology of pest and pathogens is essential for this process. As a tropical perennial crop, coffee has a wider environmental effect as it is grown in some of the world's most ecologically sensitive regions. Not only can it act as a refuge for a large diversity of native fauna and flora, but it also helps to protect important watersheds of tropical montane areas, especially when grown under shade.

Coffee is a primary export of many developing countries that rely to a greater or lesser extent on its foreign exchange earnings for financing essential imports and services. Any decline in coffee export earnings can therefore have major economic and political repercussions. Coffee has been subjected to the rigorous discipline of market forces, with depressed prices resulting from excess of production over demand, interspersed with short periods of high prices stimulated by temporary setbacks in production.

Meanwhile the costs of inputs, such a transport, machinery, labour and materials, have continued to increase. The decreasing profit margins resulting from these opposing trends have forced coffee farmers to economize and this has often led to a reduction in the use of agricultural inputs necessary for optimal coffee production. The effects have been felt most by the millions of smallholder farmers who rely on coffee as their only or main cash crops and who lack financial resources and the possibilities of economy of scale. Resources allocated to crop protection have often been the first to be cut, and this has added further pressure for more sustainable methods of pest management.

Both economic forces and environmental pressures influence how coffee is grown. Pest and disease management forms an integral part of all coffee production systems, and a sustainable management system that can be integrated into the wider aspect of crop management cannot be achieved without knowledge of the biology and ecological interactions of pests, pathogens and their associated biota.

The increasingly evident climatic changes that are now occurring will also affect the incidence and severity of pests and diseases, and a closer understanding of their biology should facilitate any changes that may be required in their management. Currently, there is a marked tendency for scientists to examine the intimate details of the biochemical and molecular aspects of organisms, and by those concerned with crop production to focus on socio-economic aspects. This book goes some way toward addressing the resulting imbalance by presenting a comprehensive account of the biology, effects and management of coffee pest and pathogens.

The book will be useful to students of tropical agriculture, agriculturalists and extensionists, and those with an interest in crop protection and the incorporation of sustainable pest management into integrated crop management systems.

We are indebted to the following: Professor Dr Michael S. Engel, Natural History Museum, University of Kansas, USA, for advice and help on the taxonomy of bee species involved in the pollination of coffee; Dr D.J. Williams for reading the section on root mealybugs and for providing useful comments; Dr M. Shaffer, Natural History Museum, London, for information on the geographical distribution of *Prophantis* and *Thliptoceras* species; Dr Julio Ferrer, Swedish Museum of Natural History, for information on the African species of *Gonocephalus*; Dr Peter Baker, CABI Europe-UK, for initial review and provision of information; Dr Mike Rutherford, CABI Europe-UK, for help with photographic material; and the UK Department for International Development's Crop Protection Programme for covering the cost of the colour plates (under Project R8423).

I Coffee as a Crop and Commodity

The origins of the coffee crop can be traced back to the Ethiopian highlands for *Coffea arabica* (arabica), and the forests of West and Central Africa for *C. canephora* (robusta). The earliest records of its use as a beverage seem to be from the Yemen in the 14th century, from where it spread to other Middle Eastern countries in the 15th century and across the Arabian Sea to India.

European traders subsequently took coffee seeds to other parts of Asia, Africa and eventually to South America, where the crop flourished in the absence of many of its major pests and diseases. As a perennial plantation crop with a good export market and suitable for cultivation in cleared forest areas, coffee was well suited to the needs of colonial settlers. By the early 20th century, Brazil was the biggest producer and international trade in coffee grew to become a multimillion dollar industry.

Coffee is today grown in more than 60 tropical countries of the world and accounts for a significant part of the foreign exchange earnings of many. Coffee-growing is predominantly a smallholder enterprise throughout the world and the crop is grown on over 11 million ha worldwide. An estimated 25 million farmers depend on coffee for their livelihoods, even though smallholders often receive less than 5% of the retail value of a cup of coffee in Europe or North America.

The international coffee market has been subjected to large price fluctuation due to variations in supply and demand, and it is the poorest rural communities who are most affected by unstable world prices. The collapse of the quota system of the International Coffee Agreement in 1989 resulted in large price fluctuations and, with large-scale planting in Vietnam, led eventually to overproduction by the turn of the century, followed by a rapid fall in the world price to reach a 30-year low in 2003. This slump in the world coffee price has had an adverse effect on the livelihoods of millions of smallholders in many of the least developed countries. Many of the international aid organizations such as Oxfam and Christian Aid have developed action plans to assist poor farmers around the world who have been thrown back into poverty by the crisis.

The International Coffee Organisation (ICO) is addressing the issue of overproduction by promoting coffee consumption in emerging economies and through the Coffee Quality Improvement Programme (CQP). The CQP seeks the withdrawal from the market of poor-quality coffee that does not meet the minimum standards. Participation in the CQP is voluntary. From October 2002, the minimum export standard has been that there should be not more than 86 defective beans in a sample of 300 g, which must also meet required moisture content levels to prevent fungal growth that leads to mycotoxin contamination.

The effects of new plantings and investments in coffee during the period when prices were high are still expected to keep production above consumption for some time to come. However, if world consumption can continue to rise by 1.5–2% each year and the CQP is widely adhered to, the medium-term commodity outlook is more optimistic, a view supported by some recovery of prices since 2004.

1 The Basics of the Coffee Crop

The Origins of Coffee

Arabica coffee

The most important economic species of *Coffea* is *C. arabica* L., which is indigenous to the highland forests of Ethiopia at altitudes of 1370–1830 m above sea level and adjacent areas of the Boma Plateau in south-eastern Sudan, and to Marsabit in northern Kenya (see Fig. 1.1). About 400,000 ha of the ancient forest where coffee occurs as an understory shrub still remain in Ethiopia. The total extent of the area in which wild *C. arabica* exists is estimated at < 20,000 km^2, and with fragmented populations and continuing habitat degradation, Davis *et al.* (2007) consider that it is vulnerable to extinction.

Coffee was taken from Ethiopia to Arabia (Yemen) perhaps as early as AD 575 (Wellman, 1961). According to Haarer (1962), there was no reputable evidence of coffee as a beverage in Arabia until the 15th century. However, a manuscript in the Jacobs Suchard Museum in Zurich shows that coffee was well established as a drink in Yemen by the 14th century (Wrigley, 1988), and this agrees with evidence from other early Arabic texts quoted in the *Encyclopedia of Islam* (Bearman *et al.*, 2002). The subsequent spread and development of the coffee crop has been reviewed by several authors (Ukers, 1935; Wellman, 1961; Wrigley, 1988; Bigger, 2006).

The oldest commercial coffee-growing area of the world is the fertile area of Yemen, between 1350 and 2300 m above sea level, and the first European descriptions of coffee cultivation there date from the early 17th century. At this time, coffee was not cultivated in Ethiopia, but it was harvested from the forest. By then, coffee beans were already being traded from the port of Mocha in the Yemen by the Dutch, French and English. Demand for coffee in Europe expanded in the 17th century, and this encouraged new plantings. Java took

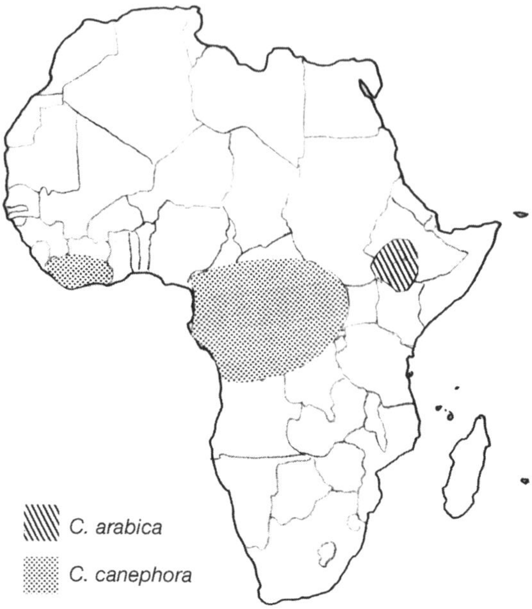

Fig. 1.1. Map of Africa showing indigenous centres of diversity of *Coffea arabica* and *Coffea canephora*.

over from Yemen as the main source of coffee before other French, and later, British, colonies entered the trade.

Coffee was introduced into India and Sri Lanka (then Ceylon) in the 17th century. Legend has it that Baba Budan, a Muslim pilgrim to Mecca, managed to smuggle some seeds out of Yemen, which were planted in Chickmagalur district in Western India. The plant then spread both naturally and by cultivation across the Western Ghats which, to this day, is still an important coffee-growing area of India. Arab trade across the Indian Ocean is likely to have resulted in other transfers to the Indian subcontinent and may have introduced coffee into Ceylon prior to the Portuguese invasion in 1517. The pre-European – or at least non-European – spread of coffee in Asia has been reviewed by Clarence-Smith (2001).

The Dutch were the first Europeans to take coffee seed from Yemen, first to the Malabar coast of Southern India and to Ceylon and, from there, material was transferred to Java during the 1690s (Ukers, 1935; Wrigley, 1988). A plant from Java was taken to the Amsterdam Botanic Gardens in 1706, where it flowered and produced berries: this was evidently the 'typica' variety of coffee (*C. arabica* L. var. *arabica*) which came to be planted throughout the New World.

Planting material from Amsterdam was sent to Surinam in 1718, from where its progeny were planted by the French in Cayenne, French Guyana. It was from there in 1727 that, despite tight security by both the French and Dutch authorities to prevent coffee-planting material being taken out of the Guyanas, coffee first came into the hands of the Portuguese in Brazil. The first plantings were made in Brazil in 1727 and the first 50 bags of coffee were shipped to Portugal in 1733.

Further introductions of 'typica' coffee were made by the French in the Caribbean from a plant in the Paris Botanic Garden donated from Amsterdam. The crop soon spread throughout the Caribbean islands from this source. Thus, much of the arabica coffee grown in the New World tropics was derived from the single tree in the Amsterdam Botanic Gardens, and is of very limited genetic variability (Purseglove, 1976). The Amsterdam tree also supplied the first planting material to the Philippines and to Hawaii.

The French took coffee seeds from Yemen to Bourbon (now Rèunion) in the early 16th Century; few trees survived, but from this source the bourbon variety (*C. arabica* L. var. *bourbon*) was derived and this material was introduced to East Africa by Catholic missionaries. The French also introduced bourbon coffee into the Caribbean and Latin America from the same source.

It was not until the early 19th century that the British began to develop coffee estates in India, but in Ceylon (now Sri Lanka) its cultivation on a large scale began around 1690, after the island was taken over by the Dutch (Heniger, 1986; quoted by Wrigley, 1988). Ceylon took over from Java as the main exporter of coffee, and the British continued to develop the trade when Ceylon fell into their hands in 1815. Coffee rust disease was first reported in Ceylon in 1869 and spread rapidly to affect all the coffee plantations. By 1890, rust had brought an end to the coffee industry in Ceylon and, instead, tea became the island's principal export.

The British introduced the coffee crop into their African territories. A single tree of var. *typica*, probably of West Indian origin, was taken from Edinburgh Botanic Gardens (UK) to Malawi (then Nyasaland) in 1878. Coffee-growing flourished on estates around Mulanje for around only 50 years before overbearing, fusarium bark disease and white stem borer combined to bring the industry into decline. From Malawi, arabica coffee was introduced into Uganda in 1900 under the name 'Nyasa'. Subsequently, and under the name 'Bugisha Local', it soon began to compete with the indigenous 'robusta' coffee.

The first introductions to Kenya came from both Aden and Rèunion via missionaries in Tanzania, and the resulting 'French Mission' coffee was quite heterogeneous, allowing some valuable selections to be made subsequently. Many other introductions to Kenya of 'bourbon' and 'typica' coffees, such as Blue Mountain from Jamaica, took place. The greatest expansion of the coffee industry in Kenya came with the 'Swynnerton plan', which encouraged smallholder production for the first time and this carried on when independence came in 1963.

Coffee production in Kenya expanded from 14,000 t in 1952, when the crop was grown on large plantations, to 129,000 t in 1984, of which 75,000 t were produced from 250,000 smallholdings (Swynnerton, 1985; cited by Wrigley, 1988). The Germans originally encouraged the cultivation of arabica in Cameroon in the early 1900s, which is the only West African country that still grows significant quantities of arabica.

Robusta coffee

The first European explorers found indigenous coffee (*C. canephora*) being semi-cultivated in central Africa. Burton reported its cultivation on the islands of Lake Victoria and to the north of the Kagera river in what is now Tanzania, and noted that the boiled berry was used for chewing. The expeditions of Emin Pasha also describe trade in coffee berries in the region south of Lake Albert, and that relatively large quantities were exported northwards from what is now Uganda (Wrigley, 1988).

French settlers found coffee growing on the banks between the Congo and Gabon rivers around the town of Kouillou. The plant was a small tree, similar to the 'nganda' form of robusta and became known as 'kouillou' ('conilon' in Brazil).

Coffee first described as *C. canephora* by the French botanist Pierre in 1879 was collected in Gabon. However, the name *C. robusta* was also used for similar specimens and it is for this reason that the term is still used to distinguish it from 'arabica'. Robusta coffee occurs in the wild in equatorial forest from West Africa to Lake Victoria, largely between 10° north and south of the equator, at altitudes between sea level and 1500 m.

There are two main centres of diversity (Charrier and Eskes, 2004): in forests of the Guinean countries and in the broader Congo basin area (see Fig. 1.1). It was introduced to the Dutch East Indies (now Indonesia) from Africa in the late 1800s, where there was some interest in its cultivation because it was resistant to coffee rust, which had arrived there in 1876. Arabica became confined to higher altitudes, while cultivation of robusta expanded at lower elevations. By the 1920s, Indonesia was producing 36,000 t of robusta coffee, becoming the world's largest producer by the 1980s, with some 350,000 t.

Although robusta is indigenous to the forests of West Africa, it was not used or cultivated there until the 20th century, and much of the material that was planted came from Indonesia rather than from the indigenous material in Africa. In Côte d'Ivoire, the first large-scale commercial plantations were established around 1927, when the French Government encouraged coffee-growing in their overseas territories.

By the middle of the 19th century, Côte d'Ivoire had become the largest producer of robusta coffee in the world. Uganda also exploited its indigenous coffee, and before the crop came under threat from coffee wilt disease in the latter part of the 20th century, the country had expanded its robusta production to become the second largest coffee producer in Africa.

Liberica coffee

Liberica coffee (*Coffea liberica*) is an indigenous tree of the humid tropical forests of Western and Central Africa, having a distribution similar to that of *C. canephora*, and is grown to a limited extent in parts of West Africa and South East Asia. It was grown most widely in Indonesia because of its early resistance to coffee rust, but to which it later succumbed. The liquoring quality is usually poor and it has a bitter taste, but has a market in South East Asia. It accounts for less than 1% of world coffee trade.

Botanical Features

The genus *Coffea* is a member of the family Rubiaceae. This is a large family of some 500 genera and over 6000 species, mostly trees and shrubs. Few are of commercial value but *Gardenia jasminoides* and some other shrubs are used as ornamentals. The best-known member of the family of economic importance – other than *Coffea* – is *Chichona* sp., from which quinine is extracted.

The taxonomy of the genus *Coffea* has recently been reviewed by Davis *et al.* (2007), who have enumerated 103 species of *Coffea* and seven infraspecific taxa (excluding autonyms), with 41 species in Africa, 59 in Madagascar and three in the Mascarenes. No naturally occurring species of *Coffea* are found outside these areas, and those species at one time thought to be present in Asia (section *Paracoffea*) were placed in the genus *Psilanthus* by Leroy (1980).

The three main centres of species diversity are Madagascar (mainly in the evergreen, humid forests of eastern Madagascar), Cameroon (14 spp.) and Tanzania (16 spp.). *Coffea* species in Africa inhabit a diversity of forest types, but generally most species occur in humid, evergreen forest.

The tropical African centre of diversity of the genus is also the centre of origin of its co-evolved pathogens, pests and their natural enemies.

Coffea spp. are evergreen, glabrous, glossy-leaved shrubs or trees 5–10 m high and most are adapted to a forest habitat. Leaves are elliptical, with pointed tips and occur in pairs. They have short petioles with small stipules, and domatia (small pits) are present on the undersides of leaves at the junction of the main veins. Flower clusters are produced in leaf axils. The fruit is a two-seeded drupe with a fleshy epicarp.

The stems exhibit dimorphic branching due to the different development of two buds that occur, one above the other in each leaf axil of the main stem. The upper bud develops to produce a lateral or primary branch. The primaries develop in succession from the base upwards and grow horizontally (plagiotropic) on opposite sides of each node, and they bear the flowers and fruits. The lower bud can only develop into a vertical (orthotropic) branch, and remains dormant until the main stem has been damaged or pruned, when it grows around the primary to produce a new vertical vegetative shoot.

In the leaf axils of the primaries, there are six buds that can develop into inflorescences or into secondary plagiotropic branches that remain vegetative. Under conditions conducive to flowering, usually the first three to four buds develop into inflorescences, each with typically four flowers. The buds can develop into vegetative branches by removal of the branch above their point of insertion or by damage to the terminal bud.

Damage by either *Antestiopsis* or boron deficiency can cause prolific development of vegetative shoots to produce 'fan' branching. Flowering occurs on plagiotropic stems (primaries) that have been produced during the previous season, so that conditions affecting vegetative growth influence the cropping level during the following season.

Only three species of *Coffea* are economically important: (i) *C. arabica* L., presently accounting for about 60% of world trade; (ii) *C. canephora* Pierre ex Froehner, accounting for most of the rest of the trade; with (iii) *C. liberica* Bull ex Hiern contributing less than 1%. The species may be distinguished using the following morphological key (Wrigley, 1988):

1. Stipules obtuse or occasionally acute, rarely apiculate; apex of leaves obtuse, rounded and shortly acuminate or rarely acute; domatia usually situated across the base of the lateral veins or occasionally in the vein axils .... *C. liberica.*
1. Stipules apiculate or aristate or occasionally acute; apices of leaves distinctly acuminate; domatia absent or situated in the vein axils .... 2.
2. Bracteoles bearing large subfoliaceous lobes (up to 2.2 cm long); pedicels usually very short, so that calyces do not exceed the bracteoles at anthesis; leaves 12–35(–40) cm long; lateral veins in (8–)11–15(–17) main pairs; domatia absent or pubescent; flowers 5–6(–7)-merous .... *C. canephora.*
2. Bracteoles bearing smaller subfoliaceous lobes (not exceeding 0.5 cm long); pedicels 1–2(–3) mm long, so that calyces exceed the bracteoles at anthesis; leaves 7–18 cm long; lateral veins in 7–10 main pairs; domatia glabrous or rarely ciliate, sometimes absent; flowers (4–)5(–6)-merous .... *C. arabica.*

The root systems of arabica and robusta differ somewhat. Arabica has a short tap root (45 cm) and extensive lateral roots going down to 2–3 m and extending 1–2 m horizontally. Robusta is more shallow-rooted, with a short tap root and the bulk of the feeding roots in the top 6 inches of soil. The leaves of robusta are usually larger than those of arabica and the lamina are more corrugated, while the leaves of liberica are much larger and leathery in texture. The primaries of robusta are longer than those of arabica, and flowering tends to be less confined to the seasonal flushes typical of arabica, whereas liberica flowers at irregular intervals.

Coffea arabica

The origin of arabica coffee has a great bearing on the subsequent incidence and spread of both its pests and diseases – and their natural enemies. Moreover, the ecological conditions under which it is often cultivated differ from those to which it is adapted in the wild state, and this has markedly influenced the severity of some pests and diseases. Shady, seasonally moist conditions and free-draining montane forest soils typify the conditions to which *C. arabica* has become adapted in its natural habitat in the highland forests of Ethiopia.

Coffea arabica is an allotetraploid ($2n = 4x = 44$), with most studies indicating that a species close to *C. eugenioides*, a small shrubby plant native to Easr African forests, and a species close to *C. canephora/congensis*, native to Central African forests, are its ancestral diploid parents (Charrier and Eskes, 2004). Neither of these occurs naturally in Ethiopia and it is presumed that *C.*

arabica arose sometime in the late Quaternary period, when much of the area bordering Kenya, Ethiopia, Uganda and southern Sudan was forested.

The coffee grown in Yemen, from which the early cultivated crop was derived, represents only a small proportion of the variability found in Ethiopia, and probably underwent some selection to produce types adapted to cultivation under the dry Yemeni conditions. The wider variability of the crop in its natural habitat has been exploited, firstly through collections made by individuals and secondly by organized botanical surveys (e.g. by FAO in 1964). This has led to the selection and subsequent use of valuable genotypes, providing sources of resistance or tolerance to a number of diseases and some pests.

These types were often named after the locations from which they were collected (Rume Sudan, Geisha, Kaffa, Tefarikela, Dilla), and many have useful resistance to disease. An FAO programme in the 1970s in Ethiopia selected material with resistance to coffee berry disease.

Coffea arabica is autogamous and mostly inbreeding. A degree of out-crossing (10–12%) occurs, and *C. arabica* can be crossed with most diploid species but, if this is prevented, it will breed true from seed. The flowers are heavily scented and visited by many insects that effect pollination. This is presumably largely self-pollination, but it has been shown to boost yields (see Appendix B for a detailed review of pollination). A small amount of natural variability occurs within the species due to natural hybridization and mutation. Two botanical 'varieties' (Harrar, 1962) were earlier recognized that gave rise to the traditional cultivars grown in the 19th and 20th centuries:

- *Coffea arabica* L. var. *arabica* (syn. var. *typica* Cramer and var. *abyssinica* Chev.) was the original type first cultivated in Yemen, from where most of the 'typica' stocks were derived. The primary fruiting branches are slender and grow horizontally; leaves are narrow, sometimes pendulous and bronze-tipped when young. Historically, the bulk of the world's commercial coffee production has been derived from 'typica'.
- *Coffea arabica* L. var. *bourbon* (B. Rodr.) Choussy arose as a spontaneous double-recessive mutant, and was first taken by the French to Bourbon. The tree is more slender than 'typica', with primary fruiting branches borne stiffly at an acute angle and bent down only at the tips during fruiting; the leaves are green-tipped when young.

A large number of other botanical varieties have been described, many informally but none of these are recognized by Davis *et al.* (2007) as valid, and all are regarded as synonyms of *Coffea arabica* L. Between 40 and 50 intraspecific taxa are recognized by the coffee industry as cultivars, several of which incorporate mutations.

Below are some of the more common mutant types:

- *Angustifolia*: elongated, narrow leaves and a poor producer.
- *Caturra*: dominant mutant from 'bourbon' but with semi-dwarfing mutation producing short internodes. It is heavy-bearing and is grown commercially in many parts of Latin America; crosses were made between

Caturra and Hybrido de Timor (HdT) (a spontaneous hybrid between arabica and robusta that was found on the island of Timor) to produce the highly successful 'Catimor' and related hybrids.

- *Columnaris*: tall cylindrical trees with short fruiting branches giving heavy yields under shade in Puerto Rico.
- *Erecta*: a dominant mutant with erect fruiting branches.
- *Semperflorens*: recessive mutant from 'bourbon' which flowers throughout the year.
- *Xanthocarpa*: recessive mutant from 'typica' with yellow fruits when ripe – grown on a small scale in Brazil as Amarelo de Botucatu.

Below are some of the best-known arabica cultivars:

- *Blue Mountain*: originated from 'typica' in Jamaica. Has been grown commercially in Kenya, where it shows some resistance to coffee berry disease.
- *SL28, SL34*: the SL selections were made in Kenya in the 1930s from the original French Mission and produce good yields of high-quality coffee, but are susceptible to coffee berry disease and rust.
- *Kent's (or Kent)*: originated as a mutation in Mysore, India in 1911 as a single tree selection that showed resistance to rust but became susceptible as new rust races appeared in the 1930s. It has the branching habit of 'bourbon', but leaves may be bronze-tipped and produces good yields of high-quality coffee. It was grown in India and East Africa, but has been replaced by more disease-resistant cultivars.
- *S795*: a cross between Kent's and a derivative of a *C. arabica* × *C. liberica* hybrid selected in India; this showed resistance to rust and remained 'durable' for about 20 years; still widely grown in India.
- *Mundo Novo*: originated in Brazil as a natural cross between 'typica' and 'bourbon'; it is vigorous and high-yielding and was grown commercially in Brazil, but has now been replaced by more modern cultivars.
- *Catuai*: selection from a cross between Caturra and Mundo Nova; grown widely in South America.
- *Catimor*: a family of hybrids between Caturra and HdT, selected for resistance to rust, but some Catimor populations are also a source of resistance to coffee berry disease.
- *Ruiru11*: an F_1 hybrid produced in Kenya and incorporating rust resistance from HdT and coffee berry disease resistance from Rume Sudan, as well as characteristics from other parental lines.

Coffea canephora

Coffea canephora (2n = 22) is self-sterile and cross-pollination has resulted in much more variability in the species than occurs in arabica. Growing wild in African equatorial forests, it has been widely distributed around the world and is more adaptable than *C. arabica*, and able to thrive in warmer, more lowland, conditions than those required for arabica.. Its resistance to rust made it more popular in the Old World Tropics than in the Americas. Less demanding than arabica, *C. canephora* is cheaper to grow and is popular in blends and for the production of instant coffee. It is generally larger that arabica and produces more yield. It can be grown successfully from sea level to 1400 m, with an optimum at between 300 and 800 m, but grows well around Lake Victoria in Uganda and near Bukoba in Tanzania, at 1150 m.

Two main forms of *C. canephora* were originally distinguished in Uganda: (i) upright 'erecta' forms known as *robusta* and considered to be the type of the species; and (ii) the more spreading *nganda* forms, which have a spreading canopy, forming domed-shaped shrubs. The leaves and berries are usually smaller than on the *robusta* forms. The 'kouillou' type was originally cultivated in many parts of West Africa, and is similar to the wild nganda type prevalent in Uganda.

The robusta type was first taken to Indonesia, where it became widely grown. Robusta and nganda were originally described as separate varieties of *C. canephora*, but although some traditional populations can be placed in these groups, there is much more variation within the species. Recent studies have shown that there are two main genetic groups of *C. canephora*, one in West Africa (Guinea area) and the other in Central Africa (Congo area) (Charrier and Eskes, 2004).

Coffea liberica

Coffea liberica (2n = 22) is a lowland species adapted to warm equatorial forests. It has been widely distributed in the lowland tropics from plants collected and distributed by Kew Gardens (London). It is an evergreen shrub or tree that can reach 17 m, has dimorphic branching and large leathery leaves. The large flowers are self-sterile and open at irregular intervals, rather than in flushes as in the other two cultivated species; it produces larger fruits than the other two species. It is tolerant of poor soils and thrives best in hot, wet conditions. *Coffea excelsa* A. Chev. ('excelsa' coffee) and *C. dewevrei* De Wild. & T. Durand were also grown commercially, but these are now considered by Davis *et al.* (2006) to be synonymous with *C. liberica var. dewevrei f. dewevrei* (De Wild. & T. Durand) Lebrun.

Hybrids

Spontaneuos hybrids between *C. arabica* and diploid species have been used in breeding programmes for selecting resistance to rust. Hybrids with *C. liberica* have occurred naturally in Java, where they showed resistance to coffee rust for a short period. They were also used in India, where selections crossed with Kent's produced the S795 family that had resistance to several races of rust. Hybrids with *C. canephora* have also been invaluable in producing hybrids with resistance to rust, such as Devamachy in India and Hybrido de Timor from Indonesia, a parent of the Catimor hybrids produced in Brazil (see above). Similar hybrid populations have been produced in Colombia.

Artificial hybrids between *C. canephora* and *C. arabica* have produced the 'arabusta' hybrids that are being used in robusta improvement programmes, and also 'Icatu', used as a source of rust resistance in Brazil. Hybrids between *C. canephora* and *C. congensis* have also occurred and are used in robusta breeding programmes (Charrier and Eskes, 2004).

Ecology and Cultivation

The conditions under which coffee is grown have a major influence on both the incidence of pests and diseases and the ability of the plant to resist or tolerate the damage that they cause. The impact of the environment is an integral part of the disease process of many pathogens, and hence coffee agronomy is of direct relevance in mitigating the effects of diseases. A brief review of the main features of coffee ecology and cultivation are given here, but more details of cultivation practices are presented in Part IV.

Ecological requirements

The evergreen nature of the coffee plant requires it to have access to water for transpiration throughout the year, but it also requires a dry period to initiate anthesis. Therefore, the moisture regimes under which coffee is grown can be critical. An annual rainfall of between 1100 and 2000 mm, with a 3–4-month dry season, is ideal for arabica, but longer dry seasons can be tolerated if the weather conditions are conducive to low evapo-transpiration (e.g. cool, cloudy with low wind speeds) or if supplementary irrigation is applied. The crop will also tolerate wetter conditions providing there is adequate drainage.

Coffee grows best in deep, friable soils with a pH < 7 typical of the volcanic soils in many parts of the montane tropics. Although *C. arabica* has a short tap root, axial roots growing down from laterals can penetrate to 3 m, and this allows it to reach moisture sources necessary for transpiration during dry seasons.

However, most feeder roots occur within the upper 30 cm or so of soil. Although it can be grown in shallower soils, drought stress is more likely to occur during dry weather unless irrigation is used. Coffee roots will not penetrate heavy clay subsoils and the plant is intolerant of 'wet feet'. The roots of robusta coffee are shallower, so it is less tolerant of dry conditions but grows well in soils that are wetter and shallower than those suited to arabica.

Arabica grows best in cool montane climates with average temperatures in the 15–25°C range, conditions usually found between 1000 and 2000 m in equatorial areas. However, it is grown at low altitudes at the edge of the tropics, where winters are cool and dry, providing there is no frost. Pest and disease incidence is greatly affected by the prevailing temperatures at different altitudes, with rust, leaf miner and stem borers being more severe at lower altitudes/higher temperatures and coffee berry disease worse under the cooler, wetter conditions at higher altitudes.

High temperatures (> 28°C) induce abnormalities such as star flowers and reduce yields, while cold temperatures (< 7°C), especially associated with wide diurnal fluctuations, can produce a malformation of shoots known as 'hot and cold' disease. Robusta grows best under warmer conditions typical of the lowland tropics. It is less tolerant of cool temperatures (< 10°C is damaging). Both species can be badly damaged by strong winds and by hail. Changes in the prevailing climatic conditions that are now beginning to occur will therefore have a significant affect on the condition of the crop and its pests and diseases.

The effects of pests and diseases on coffee are influenced by the physiology of the plant, which reflects its adaptation to a seasonally moist forest habitat. Most of the feeder roots of coffee are situated in the top layers of the soil and are most active during periods of vegetative and reproductive growth. Conditions hampering proper root function at this time, such as weed competition, drought, pest or disease damage, jeopardize the crop and can damage the whole plant. Flower initiation is greatly influenced by light levels, with higher radiation producing more flower initials.

Flowering occurs when moisture stress is released at the beginning of the wet season, and the time from anthesis to berry ripening is 6–9 months depending on temperatures. Hence, under conditions of a bimodal rainfall, crops from successive seasons overlap.

Vegetative growth also peaks during the wet seasons, and the competing effects of developing berries and vegetative growth place large demands on the tree, especially the root system.

The coffee plant has a poorly developed physiological mechanism for regulating fruiting levels so that, under conditions of heavy cropping, rather than excess berries being shed, premature ripening frequently occurs. This produces poor-quality beans and can lead to other secondary problems. Developing berries are a powerful physiologic sink for nutrients, and when the balance between cropping level, foliation and root function is disturbed, nutrients may be taken from other parts of the plant, resulting in root and shoot dieback and defoliation. The tendency for arabica coffee to overbear is a major factor influencing pest and disease damage. The physiology of the coffee tree has been reviewed by Cannell (1985).

Cultivation practices

Many of the husbandry practices used for coffee are concerned with environmental remediation necessary for the avoidance of stress under the prevailing agro-ecological conditions. Coffee cultivation practices are discussed in more detail in Part IV, but some that have particular relevance to pest and disease management are briefly mentioned below.

Site

The climatic and edaphic features of sites where coffee is grown or intended to be grown need to be suitable for the crop and, as far as possible, to meet its ecological requirements outlined above.

Shade

Traditionally, arabica coffee was grown under shade, sometimes from selected trees remaining after forest clearance, but more usually from planted species of genera such as *Grevillea*, *Erythrina*, *Albizia*, *Inga or Gliricidia.* Shade provides an environment similar to the natural habitat of coffee and is discussed in more detail in Chapter 13.

There are a number of implications for pest and disease management. Root pathogens such as *Armillaria* may spread from diseased shade trees and some can act as sources of inoculum of foliage pathogens such as *Mycena* and *Corticium* and of parasitic mistletoes, especially under heavy shade. Leguminous shade trees are attacked by a number of lepidopterous defoliators such as *Eurema hecabe*, which may move onto coffee when its food supply is exhausted. Shade trees can also be a source of infestation by some scale insects and mealybugs.

By contrast, shade reduces the incidence of coffee rust and the Asian white stem borer, and avoids damage by 'hot and cold' disease at higher altitudes. Shade also reduces cropping levels and much coffee is now grown without shade to maximize yields, but greater care is needed in the culture of unshaded coffee to avoid physiologic stress and overbearing.

Interplanting

Young coffee planted at wide spacing is often interplanted with arable crops such as beans but, as the trees mature, surface cultivation can damage feeder roots or stem bases and predispose plants to soilborne pathogens such as *Fusarium.*

However mixed perennial cropping is common in some areas, e.g. with banana in Uganda or fruit/spice trees in parts of Asia. This also provides shade, but can also enhance some soilborne pests and diseases, e.g. nematodes, basidiomycete root pathogens and scale insects.

Mulching

This is a desirable practice for unshaded coffee, but requires material for mulching and energy to transport and apply it. Mulching reduces weed growth, retains soil moisture and provides organic matter and nutrients to the rooting zone. The associated reduction in stress avoids predisposition to several diseases.

Pruning

There are two basic types of pruning, single stem and multiple stem, with several variations as described more fully in Chapter 12. In all cases, the objective is to encourage and control the production of new plagiotropic shoots that will bear the following season's crop and to control the density of the canopy. A more open canopy is less favourable to diseases that are dependent on high humidity such as web blights, but pruning wounds can encourage fusarium bark disease.

Most coffee is currently raised on multiple-stem systems, as this reduces pruning labour and the freer growth allows double cropping in equatorial climates that have bimodal rainy seasons. The downside from a disease perspective is that double cropping increases the incidence of berry diseases and there is less control over cropping levels, so that stress can be problematic under some situations.

Further Reading

There are a number of publications concerning coffee production, and many of these have chapters or sections dealing with pests and diseases: Wellman (1961); Harrer (1962); Clifford and Wilson (1985); Coste (1992); Carke and Macrae (1988); Wrigley (1988); Willson (1999); Wintgens (2004). Others deal more specifically with pests and/or diseases of coffee, but older publications are now out of print and those dealing with pests and disease of tropical crops also cover some of those affecting coffee. References to these publications are given in the introductions to Parts II and III.

References

Bearman, P.J., Bianquis, T., Bosworth, C.E., van Donzel, E. and Heinrichs, W.P. (eds) (2002) *Encyclopaedia of Islam* (online). Brill, Leiden, Netherlands.

Bigger, M. (2006) The dissemination of coffee cultivation throughout the world. *Tropical Agriculture Association Newsletter* 26, 15–19.

Cannell, M.G.R. (1985) The physiology of the coffee crop. In: Clifford, M.C. and Willson, K.C. (eds) *Coffee: Botany, Biochemistry and Production of Beans and Beverage*. Croom Helm, Sydney, Australia, pp. 108–134.

Charrier, A. and Eskes, A.B. (2004) Botany and genetics of coffee. In: Wintgens, J.N. (ed.) *Coffee: Growing, Processing and Sustainable Production*. Wiley-Verlach, Weinheim, Germany, pp. 25–56.

Clarence-Smith, W. (2001) The spread of coffee cultivation in Asia from the seventeenth to the early nineteenth century. In: *Le Commerce du Café avant L'Ère des Plantations Coloniales*. Institut Français d'Archéologie Orientale, Cairo, pp. 371–384.

Clarke, R. and Macrae, R. (1988) *Coffee*. Elsevier, London.

Clifford, M.C. and Willson, K.C. (1985) *Coffee: Botany, Biochemistry and Production of Beans and Beverage*. Croom Helm, Sydney, Australia.

Coste, R. (1992) *Coffee: The Plant and the Product*. Macmillan, London.

Davis, A.P., Govaerts, R., Bridson, D.M. and Stoffelen, P. (2007) An annotated taxonomic conspectus of the genus *Coffea* L. (Rubiaceae). *Botanical Journal of the Linnean Society*{In press}.

Haarer, A.E. (1962) *Modern Coffee Production*, 2nd edn. Leonard Hill, London.

Leroy, J.-F. (1980) *Les grandes lignées de Caféiers. Association Scientifique Internationale du Café* (ASIC) 9th Colloque, 473–477.

Purseglove, J.W. (ed.) (1976) Rubiacae. In: *Tropical Crops: Vol. I: Dicotyledons*. Longman, London, pp. 458–492.

Ukers, W.H. (1935) *All About Coffee*. Tea and Coffee Trade Journal Company, New York.
Wellman, F.L. (1961) *Coffee: Botany, Cultivation and Utilisation*. Leonard Hill, London.
Willson, K.C. (1999) *Coffee, Cocoa and Tea*. Crop Production Science in Horticulture No. 8. CAB International, Wallingford.
Wintgens, J.N. (ed.) (2004) *Coffee: Growing, Processing and Sustainable Production*. Wiley-Verlach, Weinheim, Germany.
Wrigley, G. (1988) *Coffee*. Longman, London.

2 World Coffee Production

Introduction

Coffee is one of the most valuable of traded commodities and comprises about 1% of the overall value of world trade. In the year 2004/5, total world production was 6.9 million t, valued at US$ 11.2 billion. An estimated 100 million people are employed in the industry through growing, processing and marketing, and the crop is grown by some 25–30 million coffee farmers, the majority of them smallholders, over 10.6 million ha in 68 countries.

Coffee production over the last 150 years has shown significant fluctuations, with the resulting imbalance of supply and demand causing marked price variation. Natural disasters in producing countries such as droughts and frosts have caused deficits, especially when these have occurred in Brazil, the world's largest producer, leading to rising prices. The response to this has been increased plantings, soon followed by overproduction, large, unsold stocks and falling prices. Wars and political problems have also left their mark on coffee prices and have, in turn, caused serious socio-economic upheavals.

The problems caused by these fluctuations led to the formation of a series of International Coffee Agreements between exporting and importing countries that set quotas and attempted to stablize prices. These agreements took shape in the late 1950s and have since been negotiated through the International Coffee Organisation (ICO), which was established under United Nations auspices in 1963 and represents producing and importing countries. However, the supply quota arrangements of the Agreement collapsed in 1989, and there have been marked price fluctuations again since then. Currently, 44 producing countries and 30 importing countries are members of the 2001 International Coffee Agreement.

The Global Market and Prices

In the early 1990s, earnings by coffee-producing countries were US$10–12 billion and the value of retail sales – largely in developed economies – was US$ 30 billion. The retail trade in 2003 was worth US$ 70 billion, but the earnings by coffee-producing countries had fallen to US$ 5.5 billion. This has had a major impact on countries such as Uganda and Ethiopia, where coffee provides a large proportion of their export revenue. The impact has been felt most keenly by millions of smallholders in Africa, Asia and Latin America, who depend on coffee for their livelihoods.

Different grades of coffee command different prices, but basically fall into three main groups, washed or unwashed arabica and robusta. Washed arabica refers to arabica coffee that is wet-processed, with the pericarp removed from the seed ('bean') by pulping, the bean then undergoing a short fermentation period and washing to remove remaining mucilage. It is then dried (to produce 'parchment' coffee) and hulled to remove the remaining dry integument ('silverskin').

Unwashed arabicas are hulled after the whole berry has dried to remove both the dried pericarp and the integument in one process. Robusta coffee is mostly hulled after the whole berry had been dried. Washed arabicas have the best quality and command the highest prices. Hulled coffee is referred to as green coffee.

Green coffee is traded in 60 kg bags, and production figures are usually quoted in numbers of bags, e.g. by the ICO, which also records annual production over crop years. Other production data such as those quoted by FAO are in tonnes, and annual data are calculated in calendar years to enable comparison with other agricultural commodities. The quality of statistical data from many developing countries may also lack accuracy and is collated or estimated in different ways. Hence, data from different sources may vary, especially those for cropped areas, which may be for the actual area of coffee harvested or for farms on which it is grown.

The data used here are those available (October 2006) on the ICO website (http://www.ICO.org), except for the tables of regional production, area and yield that are taken from the FAO database (http://www.FAO.org). Coffee prices also vary depending on type, quality and point of measurement. The ICO Composite Indicator Price in US cents/pound (lb) weight incorporates the different prices for arabica and robusta at the main centres of coffee-trading (New York, London, Bremen/Hamburg and Le Havre/Marseilles).

Total annual production since 1985 is shown in Fig 2.1. This shows an overall increase of about 17% over the last two decades as a result of new varieties, high-input technologies and, in some countries, new, large-scale planting. Although global consumption has increased – especially in the emerging economies – during the last 10 years, production has generally exceeded consumption; about 75% of production is consumed in importing countries.

Temporary falls in production after the early 1990s led to high prices during the period 1994–1997 that stimulated greater production, much of it achieved through more planting. When harvests from these new plantings

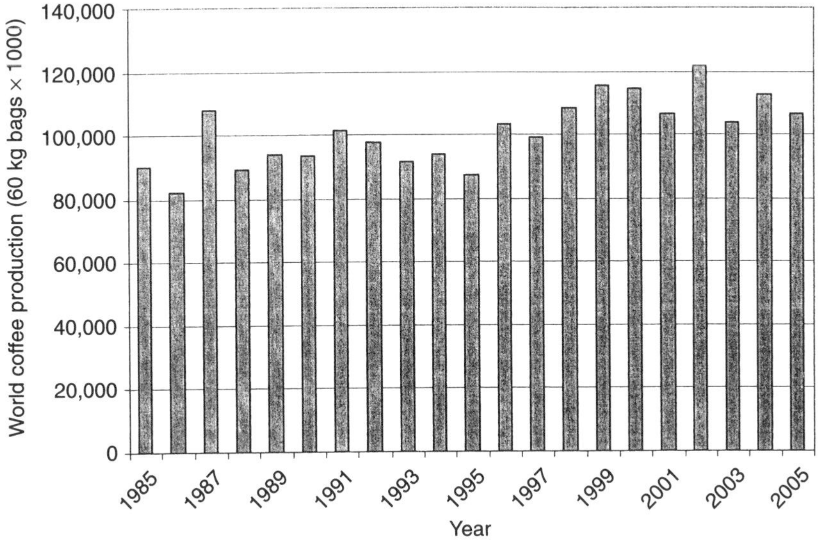

Fig. 2.1. Total world coffee production, 1985–2005 (60 kg bags × 1000).

began to reach the market at the end of the 1990s, production exceeded demand by an estimated 8% by 2002. Overproduction – already predicted in the late 1990s – resulted in a dramatic fall in price, from a peak in 1994 to a low of 45.6 US cents/lb in 2001 and caused a crisis in the industry.

The ICO has addressed the crisis through a series of measures taken following the 2001 International Coffee Agreement (Seudieu, 2003). The fluctuations in coffee prices since 1990 are shown in Fig. 2.2, which gives the quarterly average ICO Composite Indicator Price for the period.

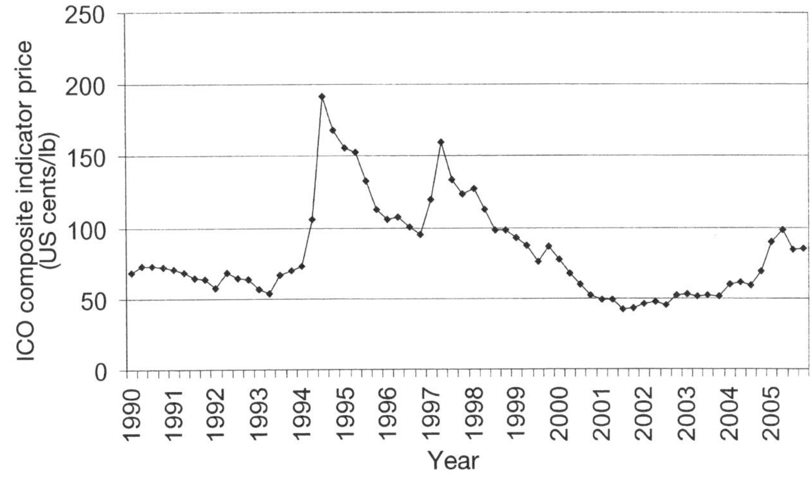

Fig. 2.2. ICO composite indicator price, 1990–2005; mean quarterly values.

Brazil has been for decades the largest coffee producer in the world, and increases in production efficiency during the 1990s resulted in more coffee from a production area that had decreased by 33%. Colombia is currently the second largest producer, but Vietnam's production has grown rapidly and had exceeded that of Indonesia (Table 2.1) and that of Colombia by the early 2000s. Vietnam was advised by international donors to embark on a campaign to persuade smallholders to grow robusta coffee. This campaign was so successful that Vietnam's production increased from 73,000 bags in 1980 to a peak of over 15 million bags in 2003. Latin America currently produces between 4.1 and 4.6 million t annually, compared to 1.9–2.1 million t from Asia and 1.0–1.2 million t from Africa. Average yields are highest in Asia, followed by Latin America, and lowest in Africa.

Table 2.1. World coffee production of the current top 25 producers in 1985 and 2005 (from ICO statistics, October 2006).

Country	Type of coffee	Production[a] (1985)	Production[a] (2005)	World production (2005, %)
1. Brazil	A/R	30,101	32,944	30.83
2. Colombia	A	11,764	11,900	10.81
3. Vietnam	R	466	11,000	10.29
4. Indonesia	R/A	5,624	7,654	6.33
5. India	A/R	1,571	4,630	4.32
6. Ethiopia	A	2,832	4,500	4.21
7. Mexico	A	4,941	4,000	3.93
8. Guatemala	A/R	2,633	3,675	3.44
9. Honduras	A	872	2,990	2.80
10. Uganda	R	2,758	2,366	2.57
11. Peru	A	1,243	2,420	2.57
12. Côte d'Ivoire	R	4,681	2,171	2.33
13. Costa Rica	A	1,325	2,157	2.02
14. Nicaragua	A	709	1,400	1.31
15. El Salvador	A	1,784	1,372	1.28
16. Papua New Guinea	A/R	860	1,267	1.15
17. Kenya	A	2,032	1,002	0.94
18. Cameroon	R/A	1,667	1,000	0.94
19. Venezuela	A	772	820	0.77
20. Thailand	R	527	764	0.72
21. Tanzania	A/R	832	750	0.70
22. Madagascar	R/A	879	733	0.69
23. Ecuador	A/R	1,989	720	0.67
24. Congo, Dem. Rep.	R	1,844	575	0.54
25. Dominican Rep.	A	431	500	0.47

[a] 60 kg bags × 1000.
A, arabica.
R, robusta.
A/R, mainly arabica, some robusta.
R/A, mainly robusta, some arabica.

Latin America

Coffee production in Latin America is dominated by the world's two largest producers, Brazil and Colombia. Statistics for Latin America as a whole, therefore, are affected disproportionately by what is happening in Brazil. Since 1985, in the region as a whole, FAO statistics show that there has been a small decrease in the area of coffee planted but a small increase in production, due to higher yields from improved varieties and technology (see Table 2.2). Production in Honduras and Peru has increased considerably over the last two decades, while it has changed little in Colombia, Mexico and Guatemala. Shortfalls in Brazilian production in 1995 stimulated the subsequent rise in coffee prices (see Table 2.3).

Brazil

Since the cultivation of coffee began there in the first part of the 18th century, production has grown to make Brazil the world's largest coffee producer (see Table 2.1). In 2005, Brazil produced 30% of the total world production and around 50% of all the coffee produced in Latin America.

Table 2.2. Area planted to coffee, mean yield and production in Latin America, 1985–2005 (from FAO database, 2006).

Year	Area (million ha)	Production (million t)	Yield (kg/ha)
1985	6.02	3.86	641
1990	6.48	3.88	599
1995	5.53	3.95	582
2000	5.79	4.26	736
2001	5.88	4.14	703
2002	5.88	4.59	780
2003	5.90	4.15	703
2004	5.71	4.60	805
2005	5.55	4.32	778

Table 2.3. Production trend in the top six coffee-producers in Latin America, 1985–2005 (60 kg bags × 1000).

Country	1985	1995	2000	2005
Brazil	30,101	18,003	34,100	32,944
Colombia	11,764	12,878	10,532	11,900
Mexico	4,941	5,300	4,815	4,000
Guatemala	2,633	4,002	4,940	3.675
Honduras	872	1,909	2,667	2,990
Peru	1,234	1,871	2,567	2,420

Coffee-growing began in the state of Rio de Janeiro and expanded into Minas Gerais and Espirito Santo, then south into the states of Sao Paulo and Parana, making the crop more vulnerable to frost. The effect of frost on production has discouraged further expansion of coffee-growing in Parana and Sao Paulo. Minas Gerais is now the largest coffee-producing state, with 50% of the country's coffee trees, while Espirito Santo accounts for a further 20%. More recently, Rondonia has become a coffee-producing state, with a large proportion of robusta.

Coffee-growing is one of Brazil's main agricultural activities, involving 220,000 farms and providing employment for over 3.5 million people. The average farm size is 9 ha, containing around 18,000 trees. Although many coffee producers are smallholders, there are also large estates, some of which are highly 'technified', using irrigation, mechanized harvesting and pest control. Brazil exports mainly unwashed arabica and some robusta. Export volumes vary between 20 and 35 million bags, depending on weather conditions – frost and drought having the greatest influence.

Leaf rust (*Hemileia vastatrix*) is the main disease problem and leaf miner (*Perileucoptera coffeella*) is one of the main insect pests. Leaf rust has been regarded as less of a problem in Brazil than elsewhere in the region, and the susceptible varieties Mundo Novo and Catuai continue to be widely grown. In some areas, fungicide spraying is not required but one to four sprays may be required in areas where climatic conditions favour the disease.

Colombia

Colombia has been for decades the second largest coffee producer in the world, but by the late 1990s Vietnam had become a rival for second place (see Table 2.1). Only arabica is exported, with close attention to quality control in harvesting and wet-processing. Although coffee has declined in importance in the national economy, it still employs 36% of the rural workforce. Coffee in Colombia is grown in the hilly north-west of the country at altitudes of between 800 and 2000 m. The main production centres are the departments of Caldas, Antioquia, Cauca, Cundinamarca, Tolima and Valle.

There are around 500,000 producers occupying more than 1,000,000 ha. Production is divided between the larger-scale technified sector (69%) and the traditional smallholder sector, but only 7% of coffee farms comprise more than 10 ha (Coste, 1992).

The main varieties grown are Catuai, Caturra and Colombia – a multi-line variety from Hybrido de Timor/Caturra crosses that was introduced from the late 1980s to combat leaf rust. The main disease problems are rust, South American blight (*Mycena citricolor*) and pink disease (*Erythricium salmonicolor*). The main insect pest is the berry borer (*Hypothenemus hampei*).

The world crisis in coffee prices has seen the export value of Colombia's coffee in 2001 decline by 50% from the average value from 1994 to 1999. The country's coffee organisation, FEDECAFE, has adopted a rigorous policy to promote sustainable coffee production. This involves new planting, which had reached an area of some 200,000 ha by 2001, increased tree density to 6000 trees/ha and implementation of control measures for coffee berry borer that has decreased infestation levels from 11 to 3% of harvested berries (Gemeil, 2002).

Mexico

Coffee production in Mexico is the third largest in the Americas and among the top ten in the world (see Table 2.1). Mexican coffees are of high quality and a significant proportion are of organic or gourmet status. Most of the crop is wet-processed, but dry processing is also carried out in some areas. The main coffee-growing areas are in the southern mountain zone, mainly in the states of Oaxaca, Chiapas, Puebla and Veracruz. There are around 268,000 coffee-growers in Mexico occupying 761,165 ha, and 90% of the producers are smallholders with farms less that 5 ha. Much of the coffee is shade-grown, with some in polycultures. The main pest problems are coffee leaf rust and berry borer.

Guatemala

Guatemala is Central America's largest coffee producer. Due partly to good extension and quality control provided by the producers' association, ANACAFE, Guatemala produces high-quality wet-processed arabica coffees that are mostly shade-grown. Eighty per cent of the 62,500 producers are smallholders, accounting for 20% of annual production. The crop is produced in many different areas of Guatemala, but mainly from the volcanic soils of San Marcos, Suchitepéque, Quezaltenango and Santa Rosa, and from the limestone regions close to the cities of Huetuetenango and Coban.

Labour involvement in coffee production is very high (221 days/ha), and low prices result in unemployment problems. The good coffee prices in the second half of the 1990s saw coffee production expand to produce a record crop of 4.5 million bags in 1997, accounting for 35% of the country's export earnings (Cuchet, 1997). Most of the coffee is grown under shade. The variety Bourbon is still widely grown and the other main variety is Caturra. Nematodes have caused problems and control is obtained through grafting arabica onto robusta root stocks. Rust and berry borer are the main pest problems.

Costa Rica

The main coffee-growing areas of Costa Rica cover an area of around 100,000 ha and are found on either side of the central plateau in the provinces of San José, Cartago, Alajuela and Heredia. There are about 40,000 registered coffee-growers, of which half are smallholders who farm less than 5 ha. Great care is taken in harvesting and processing, so that most of the coffee, especially from the higher altitudes, is of high-quality washed arabica. Introduction of high-density planting and modern farming methods have resulted in yields of 1200–1500 kg/ha. About 40% of the coffee is sun-grown. Rust, other foliage diseases and nematodes are the major pest problems.

Honduras

The area planted to coffee is about 250,000 ha, but much is on mixed holdings. Of about 89,000 producers, over 90% are smallholdings of less than 10 ha. Most coffee is shade-grown arabica. The main areas of production are Santa Barbara, El Paraiso, Comayague and Cortes.

Peru

Peru produces some high-quality coffee from the many small plantations at high altitude (1600–2000 m). The most famous of the coffee areas is Chanchamayo, in the centre of the country. Coffee is grown on a total area of 150,000 ha, and 60% of plantations are owned by smallholders. Only arabica is exported and most of this is washed.

El Salvador

Coffee-growing in El Salvador extends from the hills close to the Pacific coast to the high plateaux, covering about 180,000 ha. The main areas of production are Santa Ana, Libertad, Usulutan and Sonsomate. There are more than 50,000 plantations producing high-quality washed arabica and most of these are smallholdings. Coffee plays an important role in biodiversity conservation in El Salvador, as 80% of the country's natural or cultivated forest is shade cover for coffee.

Ecuador

Ecuador exports washed and unwashed arabica and there is also some robusta. There are a large number of coffee-growers, but most of the holdings are small and many are grouped into cooperatives. Coffee is cultivated over some 350,000 ha of land, over half of which is found in the hilly areas of Manabi Province, but about 90% is managed in the traditional system with native shade tree species or other perennial crops in polycultures.

Asia

In Asia, the two largest producers are Vietnam and Indonesia, with Vietnam rapidly having increased production during the 1990s as a result of large-scale planting, to overtake both India and Indonesia (see Table. 2.1). In the region as a whole, the area under coffee increased considerably during the 1980s and 1990s while, during the same period, average yields (kg/ha) have increased by around 30%, resulting in a level of production that has more than trebled since 1985 (see Table 2.4).

There has been a steady increase in production in India over the last 20 years, with some increased production in Papua New Guinea and Thailand. Only the Philippines among the top six producers in the region produced less coffee in 2005 than in 1985 (see Table 2.5).

Vietnam

Coffee production, mainly robusta, increased greatly in Vietnam during the last decade of the 20th century, making it the second largest producer in the world after Brazil by the early 2000s. Since independence and nationalization of private estates, the crop has been produced by collective state farms and cooperatives. The bulk of the crop is robusta, with some arabica. The main coffee regions are Dalat, Bam me Thuot, Darlac, Tuyen Quang and Co Nghia.

Table 2.4. Area planted to coffee, mean yield and production in Asia, 1985–2005 (from FAO database, 2006).

Year	Area (million ha)	Production (million t)	Yield (kg/ha)
1985	1.01	0.72	662
1990	1.34	0.86	643
1995	1.58	1.12	709
2000	1.99	2.00	1006
2001	1.98	2.07	1043
2002	1.99	1.91	961
2003	2.03	1.91	939
2004	2.24	2.09	933
2005	2.81	2.31	822

Table 2.5. Production trend in the top six coffee-producers in Asia, 1985–2005 (60 kg bags × 1000).

Country	1985	1995	2000	2005
Vietnam	466	3,938	14,775	11,000
Indonesia	5,624	5,865	6,978	7,654
India	1,571	3,560	4,516	4,630
Papua N.G.	860	1,002	1,041	1,267
Thailand	527	1,317	1,692	764
Philippines	940	850	775	500

By 2001, Vietnam had become a victim of its own success and falling coffee prices began to drive smallholders out of the crop, but production has continued to increase. One plan proposed was to decrease the area of robusta by 100,000 to 400,000 ha and to increase the planting of arabica to 100,000 ha (King, 2002). The Vietnamese government is encouraging conversion to other crops on the less productive coffee farms. The eventual target is to produce around 550,000 t from 365,000 ha (Anon, 2004), but production has decreased in recent years (see Table 2.5).

Indonesia

Indonesia is one of the oldest coffee-growing countries, being among the first where robusta was grown on a commercial scale and was the third largest coffee-producing country in the world, trebling its output since the 1960s, until overtaken by Vietnam. Around 5% of the crop is arabica, but most of the exported crop is dry-processed robusta. The main areas of production are Sumatra, Java and Sulawesi, with the crop grown on 1 million smallholdings averaging 1.4 ha in size, although many of these are mixed holdings producing coffee, along with other perennial crops, in polycultures.

India

Coffee in India is produced mainly in the hills of the Western Ghats in the southern states of Karnataka (72%), Kerala (20%) and Tamil Nadu (7%), with a small amount produced in the 'non-traditional' areas of Andhra Pradesh and Orissa (1%). Just over half the total production is of robusta, with the remainder arabica. Most robusta is produced in Kerala and most arabica in Tamil Nadu, with Karnataka producing both in roughly equal quantities. All but 2% of coffee holdings are classified as smallholder (< 10 ha) and are responsible for 60% of the country's production (Reddy, 2001).

Over the last 50 years, coffee production in India has increased some 15-fold, to make it the fifth largest producer in the world, with around 80% of the crop being exported. Much coffee in India has traditionally been grown in polycultures with other perennial cash crops such areca palm, pepper, citrus and cardamom. One study showed that the net return from polycultures with robusta coffee was four times that of the mono-crop and provided almost year-round income and employment (Korikanthimath *et al.*, 1998). This has given some economic protection to growers during the recent poor coffee prices on the world market.

Most coffee in India is grown under shade, and this is promoted as a sustainable system that maintains biodiversity and protects hillside soils from erosion (Indira and Santaram, 1999). The main diseases affecting coffee in India are leaf rust and black rot. The main insect pest is the coffee white stem borer, *Xylotrechus quadripes*, for which a pheromone is currently being developed for control by trapping. Mealy bugs have also been an epidemic pest recently, and some success has been achieved in biological control with the parasitoid *Leptomastix dectylopii*. Coffee berry borer, *Hypothenemus hampei*, was first recorded in India in 1990 and is currently the subject of an intensive control campaign.

Philippines

The Philippines is another country where coffee production increased at the end of last century with support from the government. There are some 150,000 coffee plantations. Before the coffee price fell drastically at the end of the 1990s, the area of arabica planted by smallholders was being extended in the mountainous region of Benguet, but today robusta still accounts for 90% of the crop.

Papua New Guinea (PNG)

Conditions in the mountainous central regions of PNG are suitable for arabica, which is grown mainly (70%) by smallholders on a total area of 150,000 ha. Most of the crop is grown at altitudes > 1000 m and is exported as washed arabica. It is the country's most valuable agricultural export and up to half the nation's population derive direct or indirect benefit from the coffee industry.

Thailand

Coffee is a popular beverage in Thailand, and the government has encouraged planting to decrease dependency on foreign imports. In 2003, only 66,000 ha of coffee were harvested and the robusta trees are commonly interplanted with rubber and food crops.

Africa

Ethiopia, Côte d'Ivoire and Uganda are the big coffee producers in Africa (see Table 2.7), although they produce much less than the largest producers in Asia and Latin America. Since the early 1990s, total coffee production in Africa has declined by around 20%. The average yield for the region is more or less unchanged since the 1980s. The figures (see Table 2.6) would suggest that declining production was due to a decrease in the area harvested but, over the last few years, poor world prices for coffee have resulted in falling yields in many African countries as inputs have become unaffordable.

While in Asia and Latin America, some of the leading producers have achieved their maximum yields in the years since 1999, production has declined over the same period for all the main African producers, who achieved their maximum outputs in the 1980s and 1990s. Côte d'Ivoire, Kenya

Table 2.6. Area planted to coffee, mean yield and production in Africa, 1985–2005 (from FAO database, 2006).

Year	Area (million ha)	Production (million t)	Yield (kg/ha)
1985	3.16	1.18	374
1990	3.42	1.25	367
1995	2.73	1.13	413
2000	2.64	1.18	446
2001	2.37	1.04	439
2002	2.02	0.99	490
2003	1.94	0.89	459
2004	2.08	1.02	489
2005	2.13	0.99	465

Table 2.7. Production in the top six coffee-producers in Africa, 1985–2005 (60 kg bags × 1000).

Country	1985	1995	2000	2005
Ethiopia	2832	2860	2768	4500
Uganda	2758	3244	3205	2366
Côte d'Ivoire	4681	2532	4846	2171
Kenya	2032	1664	1001	1002
Cameroon	1667	660	1113	1000
Tanzania	832	897	809	750

and Cameroon produced more coffee in 1985 than in 2005. In Côte d'Ivoire, production peaked at almost 3.5 million tonnes in 1998 but, by 2005, output had fallen by 40% and Ethiopia had become the largest coffee producer in Africa (see Table 2.7).

The decline in production in recent years has been greatest in Kenya. Although the poor world price is universally felt as a disincentive, in Kenya the coffee industry in has additionally been beset by internal marketing problems, and horticulture has provided a more lucrative enterprise. In Côte d'Ivoire, internal problems have also affected production and, in Uganda, production has been adversely affected by an aged coffee tree population and by coffee wilt disease.

Ethiopia

Ethiopia is the largest producer of arabica in Africa and vies with Côte d'Ivoire as the largest coffee exporter in Africa. The Ethiopian highlands are the centre of diversity for *C. arabica*, and the forests on the Kaffa plateau and around Lake Tana contain natural stands of 'wild coffee'. In addition to representing a valuable genetic resource, these are being exploited to some extent for the speciality market. The main areas of arabica cultivation are Kaffa, Sidamo,

Illubaor, Harrar and Wollega, with some 331,000 farms cultivating coffee on about 450,000 ha, and 94% of farmers being smallholders cultivating an average of 0.5 ha. Most of the plantations are at altitudes of 1300–2000 m.

Coffee is produced under four main systems: (i) forest and (ii) semi-forest coffee rely on natural or secondary forest cover; (iii) garden coffee is a low-input smallholder system; and (iv) more intensively managed smallholdings or former state farms account for 20% of production. Before 1991, agricultural policy in Ethiopia was centrally planned and prices were controlled. With the change of regime, coffee production and marketing were liberalized, bringing private exporters and intermediaries into the sector. This has improved grower prices as a proportion of the export price and provided greater incentive for investment in coffee production.

The quality of Ethiopian coffee is good, and forest coffee has a speciality market. However, deregulation caused some problems with quality control due to low standards of picking, drying, primary processing and storage. Ethiopia is one of the countries particularly badly affected by the slump in world coffee prices. In the late 1990s, coffee accounted for about 54% of Ethiopia's export earnings but, between 1998 and 2001, the value of coffee exports fell by 58%. However, more than 40% of Ethiopia's production is consumed locally, thus ensuring a substantial internal market.

There are around 15 million people in the country who are dependent on coffee for their livelihood, and Oxfam estimate that some 700,000 households suffered increased poverty or famine as a consequence of the low price they received for their coffee in 2003. The government's approach to the crisis is to promote the largely shade-grown and pesticide-free crop as organically grown.

Côte d'Ivoire

Côte d'Ivoire was formerly Africa's leading coffee producer, but production has declined significantly over the last decade, partly as a result of internal political problems. Both Ethiopia and Uganda now produce more coffee. The crop consists almost entirely of robusta grown in the low-altitude forest areas in the southern regions of Divo, Tiassale, Aboisso and Agboville and the central regions of Dimbokro, Oumé and Gagnoa. Coffee-growing is carried out predominately by smallholders over a total area of some 1,300,000 ha (Coste, 1992), although currently the harvested area is < 800,000 ha. Coffee berry borer is the major pest, but largely kept under control through sanitation and natural enemies.

Uganda

Uganda does not approach the world's top producers in terms of total coffee output, but is currently the top producer of robusta in Africa. The high plateau (1100–1300 m) and proximity to the equator – with adequate rainfall and good soils – is ideal for the cultivation of robusta. Most robusta is grown by smallholders in the central part of the country, extending south-west to the borders with Congo, Rwanda and Tanzania. Uganda also produces good-quality arabica, mainly from the slopes of Mount Elgon, close to the Kenya border, but also some from West Nile in the north-west of the country.

In total, about 300,000 ha of coffee are grown, mainly divided into 500,000 family-run plots of 0.5–3.0 ha. Coffee is grown in 31 of the 57 districts in Uganda, and up to 2.5 m people are engaged in its production and processing. It is estimated that a total of 5–6 m people may be directly or indirectly dependent on coffee for their livelihood, representing about 25% of the population. Coffee production in Uganda picked up after the political problems of the 1970s and early 1980s, but has declined again to levels similar to those of the mid-1980s.

Although incentives to farmers have improved with market liberalization, many of the coffee trees are over 40–50 years old and have reached the end of their economic life. Poor management during the 1970s, when incentives to grow coffee were poor, has also had an adverse effect on the productivity of the trees. Since 1994, around 10% of the robusta trees have been killed by coffee wilt disease. The present programme to address these issues involves the replacement of the ageing robusta trees with new clonal coffee, which is considered to be high-yielding and produces a crop of better quality. Since 1999, wilt resistance has become an important coffee-breeding objective.

Kenya

Kenya has a reputation for good-quality arabica, grown at altitudes of 1500–2100 m, that is carefully processed and has a characteristic acidic flavour. The total area of coffee cultivation is about 165,000 ha, mainly in the districts of Kiambu, Thika and Ruiru, producing average yields of 500 (smallholders)–1000 kg/ha (estates). Roughly 75% of the coffee area consists of 600,000 family-run smallholdings, and the remainder is under 1300 medium to large estates.

Traditionally, the Kenyan coffee growers were affiliated to cooperatives, grouped into unions that sold to the Coffee Board of Kenya that carefully controlled the quality. By 1998, growers were unhappy with the marketing monopoly of the Coffee Board due to poor prices and delays in payment for coffee delivered. The Kenya Coffee Growers Association urged growers to boycott the central auction trading system and market their crop through association-approved private brokers.

Discontent among coffee-growers has been reflected in the decline in production from a high of 140,000 tonnes to only 67,000 tonnes in 2000, as many found more lucrative enterprises. As the slump in coffee prices has deepened, production fell to 48,000 tonnes by 2003 but has since begun to pick up again. However, production can vary widely from year to year, due mainly to weather conditions; the main constraints to yield being periodic droughts, coffee berry disease and coffee leaf rust.

Since 1994, the Coffee Research Foundation has been promoting the F_1 hybrid 'Ruiru 11', which has resistance to both rust and coffee berry disease. Despite the downturn in the coffee industry, demand for Ruiru 11 at 15 million seedlings a year exceeds supply. The popularity of Ruiru 11 is probably because production costs are up to one-third less than for the traditional cultivars that require fungicide applications.

Madagascar

Coffee plantations cover about 200,000 ha in Madagascar, consisting almost entirely of robusta grown by some 400,000 smallholders with 20,000 ha under large estates. Coffee represents about 25% of the country's export trade. Most of the coffee comes from the alluvial plains on the east coast, where yields average 250 kg/ha but may reach 800 kg/ha on the better-maintained farms.

Cameroon

In Cameroon, arabica is grown on the high plateaux of Bamiléké and Bamoun, while robusta, which is the more important crop, is grown in the mid-altitude regions in the west of the country and to some extent in the east, in Abong Mbang. Coffee represents about 25% of the total export trade. With the assistance of the French government, careful crop management and processing, the superior quality of Cameroon arabicas has been maintained.

Tanzania

Tanzania has about 220,000 ha of coffee of which less than 10% is robusta, grown in the west of the country around Bukoba; 90% is arabica, grown mainly on the slopes of mount Kilimanjaro in the districts of Arusha, Oldeani and Moshi. Estates account for about 7000 ha of arabica and the remainder is under smallholdings. Official estimates put the number of families who derive at least part of their income from coffee at 270,000.

During the 1980s, considerable resources were devoted to coffee research and development, and two large processing factories were built. It was estimated that the country had the potential for production of 144,000 t of coffee, but production struggled to reach 55,000 t during the mid-1990s. Smallholder yields are very low, partly due to extension problems and earlier parastatal marketing arrangements.

Although the present Government has supported market liberalization, the downturn in the world coffee market has continued to discourage investment in coffee-growing and, despite investment by the EU in the coffee sector and early problems with liberalization having been overcome, production remains low. New, high-yielding and disease-resistant cultivars are now available for mass distribution, which should give a boost to the industry.

Some Economic Considerations

In contrast to falling coffee prices, the costs of the inputs required for coffee production have increased steadily. Improved cultural practices and marketing arrangements go some way toward enabling coffee farmers to be more efficient and to receive a greater share of the market price to offset increased costs but, as with many other agricultural enterprises, profit margins are being continually squeezed.

The costs incurred in coffee production can be attributed to various categories, as described by Rodriguez and Vasques (2004), and the 'break-even' point – in terms of yield at which production costs balance farm gate prices – varies widely between regions, countries and types of enterprise. In general, higher costs of the larger, more intensive, production systems are offset by greater yields, whereas the lower-input costs of smallholder coffee result in a lower break-even point but lower overall yields.

Harvesting and general labour costs form the greatest proportion of production costs, especially for smallholders, and the costs of crop protection measures – especially in Africa, where spraying against both rust and coffee berry disease may be required for arabica production – are also large.

A recent value chain analysis undertaken by the World Bank Group in Kenya (Ingram, 2005) found that spraying costs accounted for one-third of all production costs incurred by farmers, and over 90% of this was for imported agrochemicals. This analysis also reveals the substantial costs incurred between production and export, leaving a reduced proportion of the market price being paid to the farmer.

Labour costs vary greatly between regions, but are highest where steep, mountainous terrain prevents mechanization – as in many Andean countries. Even relatively small fluctuations in coffee prices can cause major economic problems for growers who are at the end of the value chain, and who tend to receive what is left after other costs have been taken out and at some period after the crop has been harvested.

Small growers are most vulnerable to price downturns, as their margins are often precarious and they lack economy of scale and resources. However, they may offset their higher labour costs by the use of unpaid family labour, and they can also be more flexible and revert to other crops, which has happened in several African countries such as Kenya, where horticulture has displaced coffee to some extent.

Low prices and delayed payments often result in agronomic or crop protection practices not being undertaken due to lack of cash or credit. This then leads to lower yields, and a downward spiral of poor yields and lower returns drives farmers away from coffee production. Those who grow coffee as part of a mixed enterprise or in a low-input sustainable system are best able to weather periods of low prices, as crop diversity acts as a both a biological and an economic buffer.

References

Anon (2004) Vietnam – VICOFA says it's cutting back. *Coffee and Cocoa International* 30, 46–48.

Coste, R. (1992) *Coffee: The Plant and the Product*. MacMillan, London, 328 pp.

Cuchet, R. (1997) A record crop at the right time. *Coffee and Cocoa International* November/December, 28–29.

Gemeil, J. (2002) Leading from the front. *Coffee and Cocoa International* July, 32–33.

Indira, M. and Santaram, A. (1999) Coffee for sustaining Indian Forests. *Indian Coffee* 63, 9–11.

Ingram, M. (2005) Summary of Kenya value chain analysis. *World Bank Group African Region, Private Sector Unit*. Note No. 8. World Bank, Washington, DC.

King, K. (2002) Changing for the better. *Coffee and Cocoa International* September, 34–36.

Korikanthimath, V.S., Kiresur, V., Hiremath, G.M., Rajendra, H., Ravindra, M. and Hosmani, M.M. (1998) Economics of mixed cropping of pepper, Coorg mandarin and cardamom in robusta coffee. *Journal of Plantation Crops* 26, 149–155.

Reddy, D.R.B. (2001) Trends in Indian coffee industry – prospects and challenges. *Journal of Plantation Crops* 29, 22 –26.

Rodriguez, B.P. and Vasquez, M.M. (2004) Economic aspects of coffee production. In: Wintgens, J.N. (ed.) *Coffee: Growing Processing and Sustainable Production*. Wiley-Verlach, Weinheim, Germany, pp. 823–830.

Seudieu, D.O. (2003) World coffee market. *Tropical Agriculture Association Newsletter* 24, 23–25.

II Insect Pests and their Management

Introduction

Le Pelley (1968) provided the most comprehensive review of the pests of coffee, but good general accounts of many of the pests are also to be found in Chevalier (1947), Coste (1955), Hill (1975), Kranz *et al.* (1978), Wrigley (1988), Wintgens (2004) and CABI (2005), and in regional accounts such as Tothill (1940) (Uganda), Risbec (1942) (New Caledonia), Kalshoven (1950–1951) (Indonesia), Wille (1952) (Peru), Lavabre (1961) (Cameroon), Robinson (1964) (Tanzania), Anon (1967), CRF (1978) (Kenya), Kalshoven and Van der Laan (1981), Khoo *et al.* (1991) (Malaysia) and Cárdenas and Posada (2001) (Colombia).

Geographical Distribution of Pests

The coffee plant (*Coffea* spp.) is host to a wide range of arthropod pests, many of which are found throughout the tropical and subtropical areas where coffee is grown. There are published records of over 3000 species of insects and mites associated with coffee worldwide, either feeding on the plant or associated with other species as parasites or predators, etc. There are several reasons for this. First, the crop is a long-lived tree, so it provides a stable and persistent environment that does not change significantly from season to season.

Secondly, the cultivated species of *Coffea* have been widely disseminated from their African centre of origin to most parts of the tropics and sub-tropics suitable for their cultivation, and are now found in over 100 countries worldwide. As a result, the crop has attracted new indigenous pests and diseases wherever it has been planted.

The Rubiaceae are a widespread and successful family, containing a large number of woody perennials which harbour organisms that have been able to translocate to coffee. In the past, but now to a lesser extent, coffee was planted under shade trees, and these add a further source of coffee pests.

Much of the dissemination of *C. arabica*, after its early introduction into Yemen from its origin in Ethiopia, took place during the period between the mid-17th to the late 19th century (Bigger, 2006), when quarantine was not considered an important issue. During this period many pests, no doubt, moved to new locations, along with their host plant. The introduction of quarantine restrictions during the 20th century was designed to halt the spread of pests and diseases, but sadly has not prevented the spread of serious insect pests and pathogens such as coffee berry borer (see Chapter 4) and coffee leaf rust (see Chapter 7).

Some coffee insects, such as scale insects and mealybugs, are polyphagous and found on other important crops such as citrus, cocoa, tea or mango, increasing the possibility that they will be spread to other countries on vegetative material. The result is that scales and mealybugs are found almost everywhere that coffee is grown, while others such as antestia bug (*Antestiopsis* spp.) and white stem borer (*Monochamus leuconotus*), which are not so easily distributed accidentally, are of more restricted distribution.

Ecology of Insect Pests

Coffee is a perennial crop that remains in the field for many years. This allows some insects to maintain an uninterrupted succession of generations without leaving the plant, unlike those on annual crops where the pest must move elsewhere after the plant dies. Others may be permanently associated with coffee and have a narrow host range, but their populations increase to damaging levels under certain favourable conditions, e.g. antestia and the Lyonetiid moths (*Leucoptera* spp.).

Those that feed upon the berries, such as the berry borer (*Hypothenemus hampei*), may be more easily controlled in areas with a defined flowering period rather than in those that experience intermittent rainfall throughout the year, with continuous availability of berries. Insect pests rarely kill the tree, but those that do, such as stem borers, may have a permanent effect on the plantation, as it can be difficult to re-establish bearing trees in the gap left by a dead tree. Leaf retention is essential for maximization of coffee yields, and low-level but continuous loss of leaves due to leaf-feeding insect pests and Homoptera – which literally drain the plant of nutrients – can contribute to physiological dieback if they are not controlled.

Most insect pests have a high intrinsic rate of increase and are kept in check from increasing exponentially by the action of parasites, predators, diseases and, ultimately, starvation when their food supply is consumed. Kirkpatrick (1935) gives the example of a hypothetical, parthenogenetic female mealybug that lays 100 eggs and requires 36 d for a complete life cycle. If the average daily mortality from all causes is 12.33%, the population will remain steady through succeeding generations. However, by increasing the egg-laying egg-laying capacity to 105 eggs and reducing the generation length by 1 day, the same daily rate of loss (12.33%) will lead to a 20% increase in each succeeding generation, and the population will have increased sixfold during the year.

It is clear that even a small change in ambient conditions can have a large effect on the growth of pest populations. The fecundity of the female will depend on the quality of her nutrition which, in turn, will depend on the well-being of the host plant, and the length of the life cycle of her offspring will depend on temperature and relative humidity. At the same time, natural enemies will be subject to similar climatic effects, and the success of entomopathogens may be affected by rainfall.

Kirkpatrick (1935) made a very detailed study of the climate within a coffee plantation in the highlands of Kenya at Kiambu during 1932 and 1933. This study compared the climate found in close-spaced, unshaded coffee on exposed level ground with that of coffee grown on slopes of various aspect, or shaded, heavily pruned or with windbreaks, etc. Slopes and hollows tend to raise the daytime air temperature within the crop, but valleys and hollows are colder than flat ground at night. Shade, as might be expected, lowers both air and soil temperature during the day, but tends to raise it during the night and, crucially, the temperature of the leaf surface is considerably lowered under shade and the relative humidity raised during the day, but this difference is reversed at night.

Part of his table which summarizes these data was reproduced by Le Pelley (1968, Table 2). Later, Kirkpatrick (1937) went on examine the effect of microclimate on the autecology of antestia. His general conclusion was that pest populations were regulated more by the cumulative effect on fertility and mortality of quite small climatic variations than by extreme weather conditions, and that these were more important than natural enemies in limiting increase in population.

This means that crop management practices will have a great influence on pest populations, and whether a field is shaded or not may determine the species of pests that are likely to be found and the severity of their attack. The use of soil mulches, for instance, is highly beneficial, especially in unshaded coffee, through the conservation of soil moisture and the restoration of organic matter and nutrients to the soil. However, the use of grass mulches is believed to lead to an increased infestation by the leaf miner, *Leucoptera meyricki*.

Integrated Pest Management

Because pest population increases can be set off by fairly minor perturbations, it is not surprising that natural enemies are often unable to contain them adequately at all times and that damaging outbreaks occur. At these times, a decision has to be made as to whether it is preferable to accept that some crop loss is inevitable and to wait until the pest/parasite balance is restored, or to intervene with chemicals or other forms of control.

The decision as to when and where to use chemicals has to be based on an intimate knowledge of the habits of the pest and a constant monitoring of population levels, so that an outbreak can be caught in its early stages and dealt with promptly. The particular chemical to be used has to be chosen with care so that it does not harm the pest's natural enemies or precipitate outbreaks of other pests. Spot treatment of smaller areas in the crop where the outbreak originates is preferable to blanket spraying of the entire plantation, and the number of applications should be kept to a minimum.

Prophylactic treatment is undesirable. This essentially is the philosophy of integrated pest management dealt with in more detail in Chapter 15. Integrated pest management is not a new concept and, prior to the Second World War, when the chemical arsenal was limited, there was little alternative to adopting alternative methods of control such as hand collection – particularly of larger pests – and a great effort was put into searching for natural enemies in countries of origin of the pest, a process which has continued to the present day with a good deal of success.

Pesticides

During the war, the development of DDT opened up new hope for chemical control, particularly for malarial mosquitos, and aerial spraying in the tropical war zones for malarial control became routine. After the war, and with the development of other powerful synthetic chemicals in the 1950s, blanket spraying of crops – both from the air and from the ground – became routine, often with unexpected short-term and long-term effects. Coffee was no exception, and the experience with coffee leaf miners in Tanzania, where changes in populations have been well documented, will illustrate the dangers.

Before the war, the ecology of leaf miners and their many natural enemies had been well studied by Notley (1948, 1956). At that time, leaf miners were considered to be significant but not serious pests. Between 1948 and 1960, DDT – and later, dieldrin – were used for the control of antestia and led to serious outbreaks of leaf miners.

In 1960, a 1 ha plot of coffee was set aside for monitoring and received no further insecticidal treatment. Tapley (1961) showed that the complex of parasites had changed since Notley's time and, indeed, the status of the leaf miners themselves, since *Leucoptera meyricki* had become the dominant species in both shaded and unshaded coffee, whereas *L. caffeina* had previously been considered dominant in shaded coffee.

Observations on the same plot were carried out, more or less continuously, for 10 years (Bigger and Tapley, 1969), and it was shown that the species complex of natural enemies was gradually reverting to its pre-war status over the whole time scale (Bigger, 1973). With the introduction of more benign pesticides and a general lessening of their use, the importance of leaf miner in Tanzania has since declined.

With the growing awareness during the second half of the 20th century of the environmental hazards of pesticides and their effect on human health, both of farmers and consumers, much greater care is now taken in the choice of pesticides. The Codex Alimentarius was set up jointly by FAO and WHO in

1963 to protect the health of consumers and currently (March, 2006) safety residue levels in green coffee have been established for 14 commonly used pesticides (http://www.faostat.fao.org).

'The Common Code for the Coffee Commodity' is a new initiative set up in 2003 by a consortium of international, regional and national coffee organizations, NGOs and some private firms to encourage more sustainable coffee production. A series of recommendations for pesticide use on coffee have been made (Jansen, 2005), with the aim of reducing the use of more toxic substances in favour of safer alternatives. These alternative recommendations will need to be tested in the field and some may prove to be impractical but, nevertheless, it is a move that is to be welcomed.

The biology and control of the main insect pests of coffee are considered in this section according to the part(s) of the plant that are affected. The cover of species is by no means exhaustive, but the main pests have been included. Pest control is discussed further under IPM in Chapter 15.

References

Anon (1967) *Coffee Pests and their Control.* Coffee Research Foundation, Nairobi, Kenya, 90 pp.

Bigger, M. (1973) An investigation by fourier analysis into the interaction between coffee leaf miners and their larval parasites. *Journal of Animal Ecology* 42, 417–434.

Bigger, M. (2006) The dissemination of coffee cultivation throughout the world. *Tropical Agriculture Association Newsletter* 26, 15–19.

Bigger, M. and Tapley, R.G. (1969) Prediction of outbreaks of coffee leaf miners on Kilimanjaro. *Bulletin of Entomological Research* 58, 601–617.

CABI (2005) *Crop Protection Compendium 2005 Edition.* CAB International, Wallingford, UK.

Cárdenas, M.R. and Posada, F.J. (2001) *Los Insectos y otres Habitantes de Cafetales y Platanales.* Centro Nacional de Investigaciones de Café, Colombia, 250 pp.

Chevalier, A. (1947) Les caféiers du globe, fascicle III. Systématique des caféiers et faux-caféiers, maladies et insectes nuisibles. *Encyclopédie Biologique XXVIII*, Lechevalier, Paris, 357 pp.

Coste, R. (1955) *Les Caféiers et les Cafés dans le Monde.* Larose, Paris, 381 pp.

CRF (1978) *An Atlas of Coffee Pests and Diseases.* Coffee Board of Kenya, Nairobi, Kenya, 150 pp.

Hill, D.S. (1975) *Agricultural Insect Pests of the Tropics and their Control.* Cambridge University Press, London, 516 pp.

Jansen, A.E. (2005) *Plant Protection in Coffee. Recommendations for the Common Code for the Coffee Community Initiative.* Deutsche Gesellschaft für Technische Zusammenarbeit, GmbH. 65 pp.

Kalshoven, L.G.E. (1950–1951) *De Plagen van de Cultuurgewassen in Indonesië.* Van Hoeve, Gravenhage, Netherlands/Bandoeng, Indonesia, 1065 pp.

Kalshoven, L.G.E. and Van der Laan, P.A. (1981) *Pests of Crops in Indonesia.* Van Hoeve, Jakarta, Indonesia, 701 pp.

Khoo, K.C., Ooi, P.A.C. and Ho, C.T. (1991) *Crop Pests and their Management in Malaysia.* Tropical Press, Kuala Lumpur, 242 pp.

Kirkpatrick, T.W. (1935) *The Climate and Eco-climates of Coffee Plantations.* Crown Agents, London, 66 pp.

Kirkpatrick, T.W. (1937) Studies on the ecology of coffee plantations in East Africa. II. The autecology of *Antestia* spp. with a particular account of a Strepsipterous parasite. *Transactions of the Royal Entomological Society 86*, London, 247–343.

Kranz, J., Schmutterer, H. and Koch, W. (1978) *Diseases, Pests and Weeds in Tropical Crops*. Wiley, Chichester, UK, 666 pp.
Lavabre, E.M. (1961) *Protection des Cultures de Caféiers, Cacaoyers et autres Plantes Pérennes Tropicales*. Institut Français du Café et du Cacao, Paris, 266 pp.
Le Pelley, R.H. (1968) *Pests of Coffee*. Longmans, London, 590 pp
Notley, F.B. (1948) The *Leucoptera* leaf miners of coffee on Kilimanjaro I. *Leucoptera coffeella* Guer. *Bulletin of Entomological Research* 39, 399–416.
Notley, F.B. (1956) The *Leucoptera* leaf miners of coffee on Kilimanjaro II. *Leucoptera caffeina* Wshbn. *Bulletin of Entomological Research* 46, 899–912.
Risbec, J. (1942) *Observations sur les Insectes des Plantations en Nouvelle-Calédonie*. Secrétariat d'État aux Colonies, Paris, 128 pp.
Robinson, J.B.D. (ed.) (1964) *A Handbook on Arabica Coffee in Tanganyika*. Tanganyika Coffee Board, 182 pp.
Tapley, R.G. (1961) Coffee Leaf Miner Epidemics in Relation to the Use of Persistent Insecticides. Research Report of the Coffee Research Station, Lyamungu, Tanzania, 1960, pp. 43–55.
Tothill, J.D. (1940) *Agriculture in Uganda*. Oxford University Press, Oxford, UK, 551 pp.
Wille, J.E. (1952) *Entomologia Agricola del Peru. Manual para Entomólogos, Ingenieros, Agronómos, Agricultores y Estudientes de Algricultura*. Junta de Sanidad Vegetal, Ministerio de Agricultura, Lima, 544 pp.
Wintgens, J.N. (2004) *Coffee: Growing, Processing, Sustainable Production. A Guidebook for Growers, Processors, Traders and Researchers*. Wiley, Chichester, UK, pp. 421–458.
Wrigley, G. (1988) *Coffee*. Longman (Tropical Agriculture Series), Harlow, UK, 639 pp.

3 Stem- and Branch-borers

Introduction

Insects in a number of families of the Coleoptera (beetles), Lepidoptera (moths), Isoptera (termites) and Hymenoptera (bees and wasps) can tunnel into and damage woody stems and branches of the coffee plant. The xylem tissue forming the woody part of the stem is composed mainly of cellulose and lignin, which cannot be digested directly by most insects, although some have developed enzymes that can break down these substances. Feeding, therefore, tends to be confined to the phloem and vascular cambium layers towards the exterior, which are more readily digested, and tunnels are excavated in the xylem by these species more as shelters than as a source of nutrients. The bark is also unattractive because of the presence of tannins.

Other groups, such as the Cossidae and Bostrichidae, make use of the minute quantities of sugars, proteins and starches to be found in xylem and reject the lignin and cellulose. This results in very extensive tunnels and the production of a large quantity of waste material. Still others, such as some Scolytidae, depend on fungi to break down the lignin and cellulose for them, and they then feed off the fungal spores and mycelia. The spores that initiate the fungal colonies are carried on the bodies of the beetles when they excavate their tunnel.

By far the most important order of wood borers is the Coleoptera. Around 90 species of long-horned beetles (Cerambycidae) have been recorded from coffee, the most important of which have been dealt with in detail below; other common species recorded from coffee are listed in Table 3.1. Many of these are large insects and burrow within the main stem, but others are purely branch borers.

The other important family of beetles is the Scolytidae. These are small insects, with about 40 species recorded from coffee. They are mainly borers in twigs, and two species of *Xylosandrus* are discussed below. Some *Xyleborus*

Table 3.1. The commoner species of Cerambycidae recorded from coffee (apart from those dealt with in the text).

Family/species	Recorded in/by
Cerambycinae	
Callichroma collare Jordan	Congo (Vayssiere, 1955); Gabon (Vayssiere, 1955); Democratic Republic of Congo (Vayssiere, 1955)
Plocaederus spinicornis (Fabricius)	Cameroon (Vayssiere, 1955); Congo (Vayssiere, 1955); Guinea (Mallamaire, 1955); Democratic Republic of Congo (Mallamaire, 1955)
Xylotrechus grayi (White)	China (Kung, 1977); Taiwan (Miwa, 1936)
Lamiinae	
Acalolepta rusticator (Fabricius)	India (Beeson, 1919); Java (Koningsberger, 1908)
Ancylonotus tribulus (Fabricius)	Angola (Vayssiere, 1955); Cameroon (Vayssiere, 1955); Congo (Morstatt, 1935); Guinea (Zacher, 1921); Côte d'Ivoire (Jover, 1954); Democratic Republic of Congo (Marchal, 1900)
Ceroplesis adusta Harold	Cameroon (Lepesme and Villiers, 1941); Democratic Republic of Congo (Ahrens and Vandenput, 1952)
Ceroplesis buettneri (Kolbe)	Ghana (Vayssiere, 1955); Togo (Vayssiere, 1955)
Ceroplesis calabarica Chevrolat	Cameroon (Vayssiere, 1955); Congo (Aulmann and La Baume, 1911); Togo (Mallamaire, 1955); Democratic Republic of Congo (Ahrens and Vandenput, 1952)
Ceroplesis molator (Fabricius)	Ghana (Vayssiere, 1955); Togo (Morstatt, 1935)
Coptops aedificator (Fabricius)	Congo (Aulmann and La Baume, 1911); Guinea (Zacher, 1921); India (Lepesme and Villiers, 1944); Kenya (Anderson, 1919); Madagascar (Lepesme and Villiers, 1944); Mali (Lepesme and Villiers, 1944); Mauritius (Lepesme and Villiers, 1944); Réunion (Lepesme and Villiers, 1944); São Tomé and Príncipe (Aulmann and La Baume, 1911); Sri Lanka (Lepesme and Villiers, 1944); Tanzania (Le Pelley, 1959); Democratic Republic of Congo (Bredo, 1939)
Frea maculiformis Thomson	Congo (Vayssiere, 1955); Gabon (Vayssiere, 1955); Sao Tome and Principe (Vayssiere, 1955)
Lasiopezus marmoratus (Olivier)	Democratic Republic of Congo (Aulmann and La Baume, 1911); Guinea (Zacher, 1921)
Lasiopezus nigromaculatus Quedenfeldt	Congo (Vayssiere, 1955); Guinea (Mallamaire, 1955); Togo (Mallamaire, 1955); Democratic Republic of Congo (Bredo, 1939)
Lasiopezus sordidus (Olivier)	Congo (Aulmann and La Baume, 1911); Guinea (Zacher, 1921)
Lasiopezus variegator Fabricius	Congo (Vayssiere, 1955); Democratic Republic of Congo (Vayssiere, 1955)
Monochamus ruspator (Fabricius)	Congo (Lepesme and Villiers, 1944); Guinea (Zacher, 1921)

Monochamus scabiosus Quedenfeldt	Benin (Mallamaire, 1955); Cameroon (Mallamaire, 1955); Gabon (Mallamaire, 1955); Tanzania (Harris, 1937); Democratic Republic of Congo (Mallamaire, 1955)
Plagiohammus maculosus (Bates)	El Salvador (Vayssiere, 1955); Honduras (Munoz, *c.*2000); Mexico (Vayssiere, 1955)
Sophronica calceata Chevrolat	Ghana (Forsyth, 1966); Côte d'Ivoire (Lepesme, 1953); Togo (Vega *et al.*, 1999)
Sophronica ventralis Aurivillius	Ghana (Forsyth, 1966); Kenya (Le Poer Trench and Anderson, 1930); Uganda (Tothill, 1940)
Sternotomis chrysopras (Voet)	Angola (Fonseca Ferrao, 1965); Cameroon (Lepesme and Villiers, 1944); Guinea (Zacher, 1921)
Sternotomis virescens (Westwood)	Angola (Fonseca Ferrao, 1965); Congo (Vayssiere, 1955)
Sthenias cylindrator (Fabricius)	Congo (Lepesme and Villiers, 1944); Democratic Republic of Congo (Vrijdagh, 1930)
Tragocephala guerini White	Côte d'Ivoire (Vayssiere, 1955); Democratic Republic of Congo (Mertens, 1920)
Tragocephala nobilis (Fabricius)	Côte d'Ivoire (Mallamaire, 1955); Democratic Republic of Congo (Mayne and Donis, 1962)
Tragocephala nobilis chloris Chevrolat	Côte d'Ivoire (Mallamaire, 1955); Democratic Republic of Congo (Mayne and Donis, 1962)
Prioninae	
Stenodontes downesii (Hope)	Guinea (Zacher, 1921); Madagascar (Vayssiere, 1955)

species bore into larger wood, whilst other species such as *Hypothenemus hampei*, dealt with in Chapter 4, are seed feeders.

A few Curculionidae (weevils) and Buprestidae have larvae that feed beneath the bark. With other families the damage is purely mechanical, and the beetles do not breed in the tree. These include the Bostrichidae – dealt with below – and the Cicindellidae. The latter is an interesting family, because the larvae are predaceous and live within burrows in branches, from which they prey on passing insects. They are not of importance as pests of coffee.

Important wood borers are found within the family Cossidae in the Lepidoptera. Around ten species have been recorded from coffee, of which the most important is *Zeuzera coffeae*, discussed below. Hymenoptera cause purely mechanical damage, hollowing out nesting burrows which may weaken branches. A few ants (Formicidae) and carpenter bees (Apidae) have this habit.

African White Stem Borer

Monochamus leuconotus (Pascoe) [Coleoptera: Cerambycidae] (Syn. *Anthores leuconotus* Pascoe)

Morphology

The adult beetle is 30–35 mm long, grey in colour with black head and thorax and black at the tip of the wing casings. The antennae are conspicuously long (20–30 mm). The egg is light cream in colour and 4–5 mm long by 1.5–2.0 mm diameter (Tapley, 1960; Schoeman *et al.*, 1998). The mature larvae are white or cream in colour, 3–5 cm long , 10 mm wide at the head , tapering to 5 mm at the tail. Female pupae (31 mm) are larger than the males (28 mm) and can be distinguished by two genital lobes on the last abdominal sternite (Schoeman *et al.*, 1998) (see Plate 1).

Pest status and distribution

Monochamus leuconotus is considered to be a serious pest of arabica coffee in Africa at altitudes below about 1700 m (Hill, 1975). It is confined to the southern half of Africa, with the equator roughly representing its northern boundary, an area that includes Kenya, Uganda, Tanzania, Republic of Congo, Cameroon, Angola, Malawi, Zimbabwe, Mozambique and South Africa (CAB International, 1965). The earliest reports of white stem borer came from South Africa in the late 1860s (Schoeman, 1995). In Tanzania and Kenya, the pest appears to have migrated from indigenous hosts with the introduction of coffee as a plantation crop and to have spread gradually from lower elevations (*c.*1000 m) to higher elevations (up to 1600 m), with the expansion of coffee cultivation between the two world wars (Tapley, 1960).

The lack of consistent documented observation in any one country during the 20th century makes it difficult to follow the pattern of decline and outbreak of white stem borer. It is clear from the report by Tapley (1960) that the problem became severe in Tanzania during the 1950s, due in part to the breakdown of control measures during the Second World War. In Malawi (Hillocks, 2000), Zimbabwe and South Africa (Schoeman, 1995), a resurgence of white stem borer may have occurred during the 1980s and 1990s due to a combination of an ageing coffee tree population and the withdrawal due to environmental concerns, of Dieldrin, which was widely used for control of the pest.

Damage

The adult beetle does little damage during its feeding on the bark of branches, and the main damage is caused by ring barking by the early instars that interferes with vascular transport. This causes trees to become unthrifty, with yellowing foliage, shoot die-back and varying degrees of defoliation. Ring barking is particularly destructive of younger trees with small trunk diameter, but rarely circles the trunk of larger trees. Young trees may therefore be killed quite rapidly but older trees can survive, although in an unproductive state.

Ring barking – which is the evidence of invasion – may sometimes be quite inconspicuous and initially go unnoticed, unless the soil is scraped away from the stem base. The most conspicuous evidence of attack is the exit holes that may be 8–10 mm in diameter, and the frass extruding from them. However, wood boring by later instars is not as damaging as ring barking, although extensive tunnelling does weaken the structure of the tree (see Plate 2).

Life cycle

The complete life cycle of *M. leuconotus* takes between 18 and 24 months, depending on latitude and altitude (see Fig. 3.1). The mature female beetle lays eggs just under the bark, usually on the lower 0.5 m of the main stem. The newly emerged beetles feed on bark and, in East Africa, the females lay on average 20–25 eggs, but the range may be between ten and 40 eggs, depending mainly on temperature (Moffat and Allan, 1934; Knight, 1939). In South Africa, a single female may lay more than 80 eggs (Shoeman *et al.*, 1998). The eggs hatch in about 21 days and the young larvae bore downwards under the bark towards the base of the trunk.

The larvae penetrate the wood of the tree in the crown area, often where a lateral root joins the taproot. There are seven instars and the later ones bore into the wood, making the characteristic longitudinal tunnels (see Plate 3).

At about 20 months after eggs hatch, the fully developed larva excavates a large pupation chamber within the trunk. Estimates vary for the length of the pupal stage – from 20 to 42 days – and the adult emerges from the trunk at the start of the rains by boring a circular hole to the exterior. The exact survival period of the adult is unknown.

Tapley (1960) states that female beetles lived for 40–60 days at Lyamungu in Tanzania, while at Mpumalanga in South Africa, the mean survival period was reported to be twice that long, at between 112 and 120 days (Schoeman *et al.*, 1998). In East Africa and northern Malawi, beetles can be found in coffee trees in greatest abundance from the onset of the rains in November through to December and January.

A second emergence occurs with the next rainfall peak in March/April. Beetle emergence is followed by egg-laying, peaking in mid-December to mid-January, and which gradually declines going into May and June. Ring-barking activity peaks in January and February, and wood boring larvae become active in April and May, throughout the next year, and begin to pupate in September through to the following February.

Host range

Although the damage they cause is more severe on *Coffea arabica*, African white stem borer breeds successfully on all commercial coffee species and a number of wild hosts, including *Canthium* sp., *Erythroxylon emarginatum*, *Gardenia urcelliformis*, *Oxyanthus speciosus*, *Pavetta oliveriana*, *Randia* sp., *Rytigynia schumanii* and *Vangueria linearisepala* (Duffy, 1957).

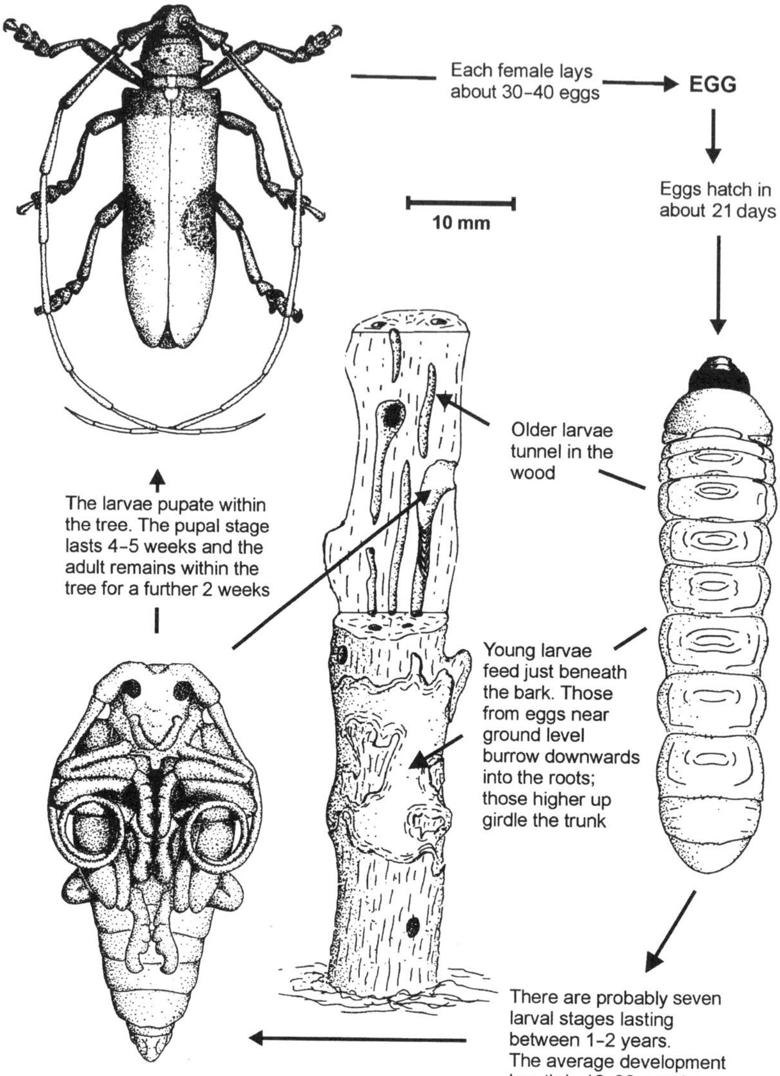

Fig. 3.1. Life cycle of *Monochamus leuconotus*.

Natural enemies

Both in Tanzania and South Africa, the eggs are parasitized by an undetermined species of *Aprostocetus*. Up to 12 individuals of this Braconid wasp may develop within a single egg but, as the parasites are not common, they have little effect. Larval mortality is high, Tapley (1960) having recorded 45% losses in Tanzania, much of which must be due to parasites and predators. Several parasitic wasps have been recorded (see Appendix A) but, due to difficulties in rearing, many have not been identified to species or even to genus.

Exceptions are the Braconid, *Hybogaster varipalpis* (Cameron), recorded from Kenya and Zimbabwe and the Ichneumonid, *Afrocoelichneumon didymatus* (Morley), from Tanzania. The former was reported to have killed 48% of larvae in a study in Kenya (Knight, 1939). A number of ant species have been recorded as predators in South Africa and Democratic Republic of Congo and a Clerid beetle, *Gyponyx retrocinctus* (Chevrolat), was frequently found in association with dead borer larvae (S. Schoeman, personal communication).

Control

Prior to its withdrawal, dieldrin was the recommended insecticide for white stem borer control, applied to the bark of the stem with a brush or with a sprayer. A single application just before the onset of the rains was sufficient to kill either adults during oviposition or the young larvae boring into the treated bark. By 1958, the widespread adoption in East Africa of stem banding with dieldrin had reduced white stem borer to minor pest status (Le Pelley, 1968). Well-maintained trees are less likely to be attacked and better able to withstand single invasions. In the absence of dieldrin, no other chemical has been found that is as cost-effective. The use of metal spokes poked into the entry holes, or other insecticides poured into the holes, is practised in some areas, but there is no evidence that this is effective.

Experiments were conducted in South Africa with stem paints containing 2% chlordane (Schoeman and Pasques, 1993), but it was considered uneconomic for large plantations, and its high mammalian toxicity makes its use undesirable. Field trials conducted in Malawi with fipronil and imidacloprid, applied as stem paints, showed the chemicals to be effective in preventing invasion (G. Odour, unpublished report). In Zimbabawe, chlorpyriphos is also reputed to be effective as a stem treatment (D. Kutywayo, personal communication). Biological control with *Beauvaria bassiana* has been investigated in South Africa. Preliminary laboratory tests suggest that some formulations are effective against adult beetles and sixth instar larvae (Schoeman and Schoeman, 1997).

The practice of smoothing the bark of the lower stem to remove crevices suitable for oviposition is well known by farmers in several African countries, including Malawi but, although it is used as a control measure, there is no published information on its efficacy. A maize cob or stick is used to carry out the smoothing, or sometimes a knife, but there is a risk of damaging the stem under the bark.

Asian White Stem Borer

Xylotrechus quadripes (Chevrolat) [Coleoptera: Cerambycidae]

Morphology

Smaller than the African white stem borer, the adult beetle of the Asian white stem borer is only 8–20 mm long and black in colour, with grey pubescence. The forewings are black with white bands and the head of the male is conspicuously ridged (see Fig. 3.2). Eggs are white to cream and 1.25×0.5 mm in size. The whitish larvae are up to 30 mm long when fully developed.

Pest status and distribution

Xylotrechus quadripes is the most serious insect pest of arabica coffee in India (Hall *et al.*, 2006). It was first reported in India in 1838 (Stokes, quoted by Duffy, 1968). As in Africa, the insect has become an increasing problem following the withdrawal of dieldrin and poor standards of crop management resulting from declining world coffee prices.

The traditional areas of coffee production in India are Karnataka, Tamil Nadu and Kerala. More recently, the crop has been introduced to Andhra Pradesh, Orissa, Madhya Pradesh and the north-eastern hill states, collectively known as the non-traditional coffee areas. In 2002, *X. quadripes* was particularly severe in the Kadoga district of Karnataka, where a 10% loss of production was attributed to the pest (ICB, 2002). *Xylotrechus quadripes* is

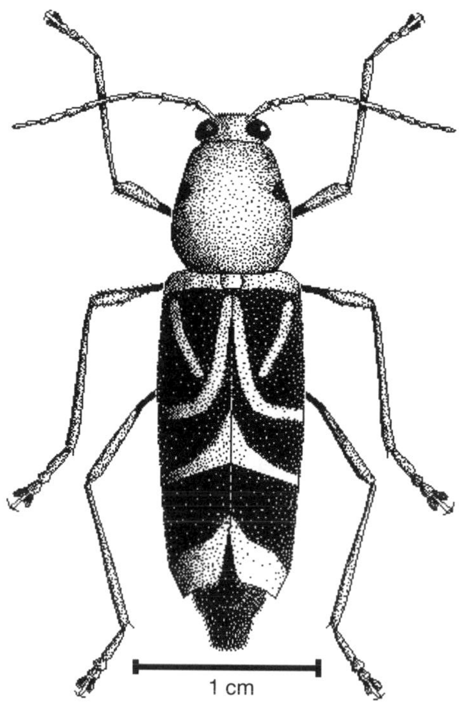

Fig. 3.2. Adult of *Xylotrechus quadripes*.

found also in Sri Lanka, China, Taiwan, Myanmar, Thailand, Laos, Vietnam, the Philippines and Indonesia.

Damage

Both ring barking, as the larvae bore under the bark, and extensive tunneling in the wood combine to kill young trees and make older trees unproductive. The foliage on affected trees appears yellow and wilted.

Host range

The main host of *X. quadripes* is *Coffea arabica*, but the pest is also found in *C. canephora, C. liberica* and a number of other hosts including *Albizia, Canarium, Cudrania javensis, Ixora coccinea, Jasminum dispersum, Olea dioica, Oroxylon, Premna pyramidata, Pterocarpus marsupium, Randia spinosa, Rhus semiciliata, Sponia whightii* and *Tectona grandis* (Duffy, 1968).

Life cycle

In Southern India, the adult beetle emerges and begins egg laying during the north-east monsoon in October–November. In Tamil Nadu, adults can be found between October and January (Reddy and Bhat, 1987). The eggs hatch in about 14 days and the first instar larvae tunnel just under the bark for a period of 2 months. Later instars tunnel for up to a further 9 months and then bore a chamber close to the exterior of the stem before pupating. Visitpanich (1994a) found that there were six larval instars in northern Thailand with a mean development time of 172 days. The larval stages are completed in 6 months, when the last instar reaches the pupal chamber. Pupation lasts for 30–40 days.

The adult beetle remains in the pupal chamber for 3–7 days before cutting an exit hole, and the emerging adult immediately searches for a mate. The adults live for 2–3 weeks but fly little, and this tends to restrict the infestations. They are, however, more active at night and can be attracted longer distances by light (Wrigley, 1988). The adult female survives for 9–30 days, during which time up to 100 eggs are laid in cracks and crevices of the bark on the main stem and thick primary branches (Le Pelley, 1968).The life cycle is completed in about 12 months. *Xylotrechus quadripes* has an annual cycle on coffee in India, but Kung (1977) reported that there were two generations per year in Guangxi Province of China.

Natural enemies

The natural enemies of *X. quadripes* are well known thanks to the work of Duport (1924) in Vietnam during and after the First World War, many of the species being described by Kieffer (1921). Duport found a surprisingly large number of parasites for a wood-boring beetle (see Appendix A), and he instituted a breeding programme for two of the more promising species, the Braconid, *Parallorhogas pallidiceps* (Perkins) (referred to as *Doryctes strioliger* Kieffer in contemporary literature), and the Bethylid, *Sclerodermus domesticus* Klug.

By 1924, some 4 million specimens of *P. pallidiceps* had been bred, 1.8 million of which were released in the vicinity of the experimental station at Cho-Gahn, and 2 million distributed to other estates, apparently without having any discernible effect on the beetle population (Duport, 1924). *Sclerodermus domesticus* was also bred and showed more promise in controlling the beetle, but was only effective if continuous releases were made (Du Pasquier, 1930). Du Pasquier concluded that the use of parasites indigenous to Vietnam was not a practical proposition.

Subramanyam (1934) mentioned that a Eupelmid, *Metapelma* sp., parsitizes *X. quadripes* larvae and was widely distributed in India and, more recently, a Bethylid parasite, *Apenesia* sp., has been reported from India (Seetharama *et al.*, 2002). In Thailand, Hymenopterous parasitoids of *X. quadripes*, *Pristaulacus* sp. and *Disatephanus* sp., were identified in groups located in cracks in the bark on the lower trunk (Visitpanich, 1994b).

Control

Xylotrechus quadripes can be controlled by a combination of well-managed shading (as the pest prefers to lay its eggs in sunshine), removal and burning of infected tree trunks before the period of beetle emergence and manual smoothing of the trunk by removing loose bark (Bhat, 1987). Painting the stem with an insecticide such as BHC or dieldrin was previously recommended, especially for unshaded coffee. During the late 1990s, the male pheromone of *X. quadripes* was synthesized and proved effective in trapping the females, with good prospects for commercial production (Hall *et al.*, 2006). *Beauveria bassiana* has been used successfully as a control measure in China (see under *Acalolepta cervina*, below).

West African Coffee Borer

Bixadus sierricola White [Coleoptera: Cerambicidae]

Morphology

The adult beetle of the West African coffee borer is 21–28 mm long. It is distinguished from *M. leuconotus* by its coloration, being chestnut brown, but concealed by a whitish tomentum. The markings on the back are brown and roughly in the shape of a W at the rear and a V in the middle. The antennae are longer than the body. The larva reaches 5 cm long × 8 mm wide (Le Pelley, 1968) (see Fig. 3.3).

Pest status, host range and distribution

Bixadus sierricola is the most damaging of the longicorn beetles affecting coffee in the region. All coffee species are attacked, but arabica suffers the most damage. The pest can develop on a wide range of woody species (Vayssiere, 1955). The West African Coffee Borer is confined to Africa, particularly the West and Central forest regions. It has been recorded from Angola, Benin, Cameroon, Central African Republic, Democratic Republic of Congo, Dahomey, Equatorial Guinea – including Bioko, Gabon, Gambia, Ghana, Guinea, Guinea Bissau, Côte d'Ivoire, Nigeria, Sierra Leone, Togo and Uganda.

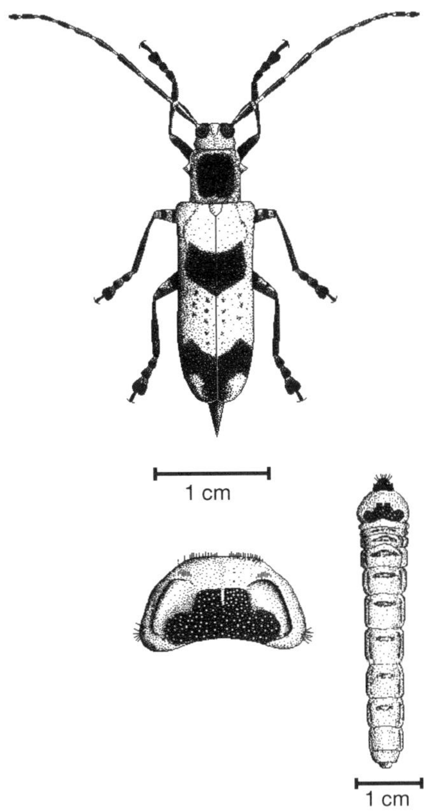

Fig. 3.3. Adult and larva of *Bixadus sierricola.*

Damage

It is fortunate that *B. sierricola* does not fly readily, as it is potentially very damaging to both healthy and old or weak coffee trees. Ring barking is the most damaging effect, leading to loss of branches or death of younger trees. Trees younger that 3 years are rarely attacked (Vayssiere, 1955). Older trees may not be killed, but internal tunnelling renders them susceptible to wind damage and termite attack. Trees of 4–5 years old are preferred by the insect, but usually only a few trees are attacked. However, occasionally severe infestations have occurred, with 60–80% of trees affected, resulting in the death of many trees (Lepesme and Villiers, 1944; Wrigley, 1988). Damage is often inconspicuous until the tree has begun to die.

Life cycle

The eggs are laid, usually singly, in cracks in the bark of the trunk 15–20 cm from the ground. They hatch in 6–7 d and the larva bores into the sapwood and moves downwards into the roots. Sometimes they bore upwards into the larger branches up to 2 m from the ground. Some young larvae feed in the sapwood just beneath the bark in a circular gallery that can girdle the trunk. The larval stage lasts for 5–6 months and pupation takes place in a chamber at the top of the burrow. The adult beetle emerges from the pupa after 30–40 d and searches for a mate. The mature adults fly little and this restricts the areas of infestation. The beetles survive for 2–3 weeks and are more active at night (Le Pelley, 1968; Wrigley, 1988).

Natural enemies

Records of natural enemies are sparse. The Ichneumonid, *Gabunia ruficoxis* Kreichbaumer attacks the larvae in Equatorial Guinea (Baguena Corella, 1942) and a Tachinid, *Phorastoma* sp. in Cameroon (Lepesme and Paulian, 1943), but neither is common nor an effective control agent.

Control

In the absence of affordable and safe contact insecticides with which to paint the base of the trunk, integrated control programmes are required in areas close to forests that are prone to attack by *B. sierricola*.

The recommendations for IPM include combinations of the following measures: (i) separate the coffee trees from the forest with a 30 m belt of food crops; (ii) uproot and burn affected coffee trees; and (iii) keep the base of the tree free of lichen, etc. to facilitate early detection of the pest. The use of light traps has also been proposed as a control measure (Wrigley, 1988). This would be of rather limited use because of the need for an electricity source close to infested trees, although Vayssiere (1955) reports that the pest can be attracted to light over 1.5 km away.

A fumigant paste of 57% aluminium phosphide (Gastoxin) has been developed in Brazil for use against wood borers in fruit trees such as citrus. This comes in tubes with a nozzle that can be inserted into the ventilation holes of the borer and a small amount of the fumigant injected, after which the hole is sealed. In Ghana, good control of larvae of *B. sierricola* was obtained experimentally using this paste (Padi and Adu-Acheampong, 2001), which has also been used in that country to control the Cossid borer of cocoa, *Eulophonotus myrmeleon* Felder.

Yellow-headed Borer

Dirphya nigricornis Olivier, *Neonitocris princeps* (Jordan) [Coleoptera: Cerambycidae]

Morphology

The mature adult of *D. nigricornis* is a slender beetle about 25–30 mm long with antennae as long as the body. The body is brownish black and the thorax and first quarter of the wing cases are yellow/orange. *Neonitocris princeps* is similar except that the elytra and most of the thorax are black (see Fig. 3.4). The eggs are brown, oval and 3.6 × 1.1 mm. The larva is red or brown at first, becoming yellow/orange when mature and about 50 mm long. The pupa is about 30 mm long (Hill, 1975).

Pest status, distribution and damage

Neonitocris princeps has a western distribution, from Côte d'Ivoire to Uganda and the western parts of Tanzania (on coffee: Cameroon, Congo, Gabon, Côte d'Ivoire, Tanzania, Uganda, Democratic Republic of Congo), whilst *D. nigricornis* is found in Eastern and Southern Africa (Kenya, Tanzania, Malawi, Zimbabwe). A record of *D. nigriconis* from Senegal (Lepesme and Villiers, 1944) probably refers to *N. princeps*, as does an unspecified report of a *Neonitocris* sp. from Ghana (Forsyth, 1966).

Neonitocris princeps is a pest of arabica, robusta and excelsa coffee, having been recorded in Uganda as early as 1910 (Gowdey, 1914), but *D. nigricornis* attacks only arabica. It is normally a minor pest, but severe attacks have been known in the Taita Hills of Kenya and in parts of northern Malawi. The larva of both species begin to tunnel in the primary branches, making a series of holes (flute-holes) along the branch, through which frass is extruded, causing the tips to wilt. Attacked branches are weakened and may break with the weight of developing berries (Crowe, 1962). As the larvae grow older, the tunnels extend down into the stem close to the ground and may even penetrate the main root.

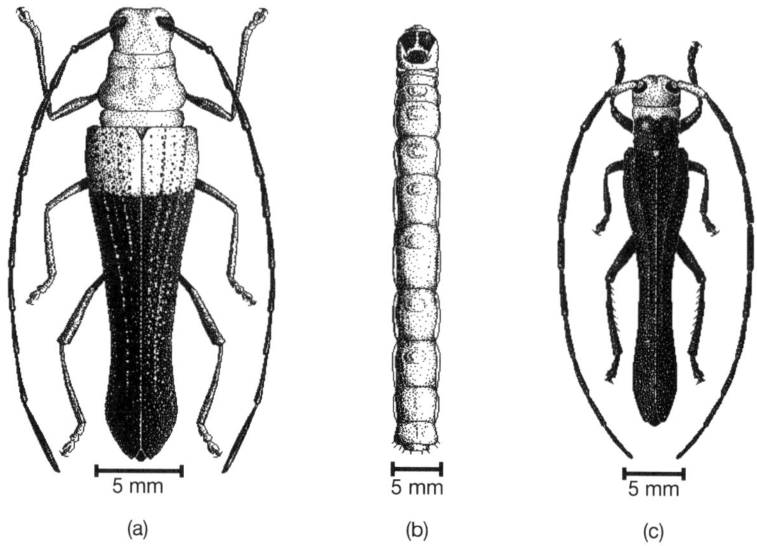

Fig. 3.4. (a) Adult of *Dirphya nigricornis*; (b) larva of *D. nigricornis*; (c) adult of *Neonitocris princeps*.

Host range

No alternative hosts for *N. princeps* are known, but *D. nigricornis* has been found to breed in *Rytigynia schumanii* in Tanzania (Kilimanjaro) (Harris, 1937) and on *Vangueria rotundata* in Kenya (Crowe, 1962). Crowe also observed adults resting on *Vangueria apiculata*, *Rytigynia eickii* and *R. schumannii* in Kenya, and speculated that these might also be host plants.

Life cycle

Eggs are laid singly near the tip of a branch under a small flap of green bark. On hatching, the larvae bore into the green shoots and move down the primary towards the main stem, making flute-like holes to the outside. The frass is pushed out through these holes. The burrowing continues down the branch and into the main stem for about 10 months in the case of *D. nigricornis* (Crowe, 1962), but that of *N. princeps* seems to be rather longer, lasting 2–3 years (Gowdey, 1914). Pupation occurs in a cell excavated by the larva in the main stem close to ground level. The pupal stage lasts for about 7–9 weeks (Gowdey, 1914; Crowe, 1962).

Natural enemies

The Braconid wasp, *Hybogaster varipalpis* (Cameron), is a larval parasite of both species, whilst in Uganda *N. princeps* has two other Braconid larval parasites, *Zaglyptogastra pulchricaudis* (Szepligeti) and *Bathyaulax* sp., as well as an unidentified Braconid pupal parasite. The Reduviid, *Margasus afzelii* (Stal), predates the adults of *N. princeps* in Uganda (Hancock, 1926).

Control

At the first sign of attack (tip-wilting and flute-holes at the top of a primary branch), the affected primaries should be removed and burned. If many trees are affected, a suitable insecticide can be poured into the stem through a flute-hole after widening one of them with a sharp blade (Crowe, 1959).

Brown Borer

Acalolepta cervina (Hope) [Coleoptera: Cerambycidae]

Morphology

The adult beetle is 15–22 mm long and greyish brown in colour, with antennae twice as long as the body in the male, shorter in the female but still much longer than the body. The larva can be up to 48 mm long and around 9 mm wide at the prothorax. The egg is 3.25 mm long and 0.75 mm wide (Duffy, 1968).

Pest status and distribution

The species has been known for many years as an important pest of teak and other forest trees in Pakistan, India, Myanmar and Thailand (Browne, 1968), but in recent years it has been recognized as an important pest of coffee in southern China and Thailand (Lan and Wintgens, 2004). There are no records of damage to coffee from India. Its range also extends to Sikkim, Nepal, Japan and Korea.

A survey of damage in eight plantations in South Yunnan by Rhainds *et al.* (2002) showed that *A. cervina* was much more prevalent than *Xylotrechus quadripes*, although the latter caused more damage. Some damage is caused to the bark by the adults during their maturation feed, but the main damage is caused by the larvae. These feed initially in the cambial layer just under the bark, but later penetrate more deeply.

Host range

Duffy (1968) lists 16 alternative host plants. Beeson (1941) considers *Clerodendron infortunatum* to be the commonest food plant of this species but lists a number of other hosts of which *Tectona grandis* (teak) and *Gmelina arborea* are the most important. He also includes tea as a host plant.

Life cycle

The female cuts a transverse incision in the bark into which the egg is inserted into the cambium layer. A female can lay up to 60 eggs during her lifetime. The egg hatches after 5–7 days and the larva burrows just below the bark in a downward direction in teak (Beeson, 1941), but upward in coffee (Lan and Wintgens, 2004). There are six larval stages lasting 288 days in China (Lan and Wintgens, 2004) and 320 days in India (Beeson, 1941), after which a pupal chamber is excavated in the wood and the insect pupates for 10–15 days.

There is a single generation per year and adults emerge during June in China and between April and September in different parts of India.

Control

Some success has been achieved in China (Wei and Kuang, 2002) by forcing 'fungal mud' containing spores of *Beauveria bassiana* into the larval tunnels, leading to over 90% mortality within 20 d. A similar treatment was used for the control of *Xylotrechus quadripes*, with equal success. Lan and Wintgens (2004) recommend swabbing the bark with 0.1% solution of dichlorvos 80 EC at oviposition time to prevent larvae from penetrating the stem.

Black Borer

Apate monachus Fabricius, *A. indistincta* Murray, *A. terebrans* (Pallas), *A. femoralis lignicolor* Fairmaire [Coleoptera: Bostrichidae]

Morphology

The adult beetle is black and up to 20 mm long, with head held under the large thorax (see Fig. 3.5).

Pest status, distribution and damage

The black borer is a minor pest of coffee and usually only a few trees in a plantation are attacked. The various species are widely distributed through Africa and have been introduced to South America and the Caribbean. *Apate monachus* is, in addition, found in some Mediterranean countries.

Records of attack on coffee are from the following countries:

Fig. 3.5. Adult of *Apate monachus*.

- Caribbean: Cuba (*A.m.*), Dominican Republic (*A.m.)*, Haiti (*A.m.*), Jamaica (*A.m., A.t.*), Martinique (*A.m.*), Puerto Rico (*A.m.*)
- South America: Brazil (*A.m., A.t.*)
- West Africa: Guinea (*A.m.*), Côte d'Ivoire (*A.m., A.t.*), Togo (*A.m., A.t.*)
- Central Africa: Congo (*A.m.*), Democratic Republic of Congo (*A.m., A.t.*)
- Eastern Africa: Eritrea (*A.m.*), Ethiopia (*A.m., A.i.*), Kenya (*A.m., A.i.*), Tanzania (*A.m., A.i.*), Uganda (*A.m.*)
- Southern Africa: Angola (*A.m., A.t.*), Malawi (*A.i.*), South Africa (*A.m.*), Zimbabwe (*A.m.*)
- Indian Ocean and Arabia: Madagascar (*A.m., A.f.*), Yemen (*A.m.*).

The beetle makes an entry hole on the sunny side of the trunk and excavates a straight tunnel about 6 mm in diameter, moving upwards in the main stem. Entry holes may be from ground level right to the tip of the main stem. Sawdust accumulates at the base of trees that are being actively bored. In Côte d'Ivoire, the damage can be more severe on *C. liberica* than on other *Coffea* spp. Serious damage by black borer has been reported in Puerto Rico (Wolcott, 1921).

Life cycle and host range

Only the adult stage is known on coffee, and the eggs, larvae and pupae occur in alternative hosts. Bostrichid larvae require wood with a high starch content, and this is most readily available in felled or moribund trees. Attacks on healthy trees result from the need of newly emerged adults to feed in order for their gonads to mature. *Apate monachus* and *A. terebrans* are well known as forestry pests and have been recorded in a wide range of hosts (Browne, 1968). In Puerto Rico, eggs were reported to be laid within the tunnel after the tree had been killed (Wolcott, 1921), but the larvae did not develop.

Control

Control is rarely required, but a cloth or cotton wool plug soaked in an insecticide such as fenitrothion can be pushed up the entrance hole of the tunnel. Alternatively, the beetle can be speared in its tunnel using a sharp length of wire such a bicycle spoke.

Twig Borer

Xylosandrus compactus (Eichhoff), *X. morigerus* (Blandford), *X. discolor* (Blandford) [Coleoptera: Scolytidae]

Morphology

The females are small, brown to black, cylindrical beetles with a body length of around 1.5–2.0 mm and a width of around 0.75 mm. The males are more rounded and smaller, with a length of about 1 mm × 0.5 mm wide. The creamy white larvae have a brown head, curved, legless body and are about 2 mm long. Females of the two main species are very similar in general appearance (see Fig. 3.6).

Pest status and distribution

Xylosandrus discolor is distributed through South East Asia from Pakistan, through India, to Indonesia and north to China and is a borer of green twigs, but is not so important as the other two species.

Xylosandrus compactus is a widely distributed species of ambrosia beetle, being found in most of tropical Africa, including Madagascar, Southern India and Sri Lanka, China, Japan, Thailand, Vietnam, West Malaysia, Sabah, Indonesia, New Guinea and many Pacific islands.

The centre of distribution of *X. morigerus* is from West Malaysia, through Indonesia and New Guinea, to Australia and through the Pacific. There are

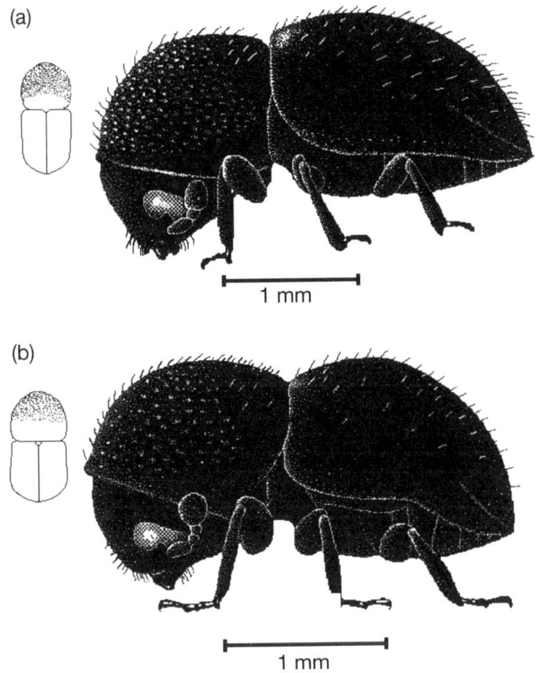

Fig. 3.6. Adults of (a) *Xylosandrus compactus* and (b) *X. morigerus*.

scattered reports from Madagascar, Réunion, Sri Lanka and Vietnam, and doubtful records from Taiwan and the Philippines.

Prior to 1971, distribution of *X. morigerus* in the New World had been confined to Brazil, where it was first recorded in 1940 (CABI, 1971) and Colombia, but it has since spread to other countries in Central and South America and the Caribbean, including Puerto Rico, Trinidad and Tobago, Mexico, Costa Rica, Honduras, Panama, Equador and Venezuela.

In 1982 (CABI, 1982), *X. compactus* was found in the western hemisphere only in the Southern USA, but has since been recorded from Cuba, Curaçao, Puerto Rico and Brazil.

On coffee the three species have been recorded from:

- Central America: Mexico (*X.m.*)
- Caribbean: Cuba (*X.c.*), Puerto Rico (*X.c.*)
- South America: Colombia (*X.m*), Ecuador (*X.m.*)
- West Africa: Guinea (*X.c.*), Liberia (*X.c.*), Sierra Leone (*X.c.*), Côte d'Ivoire (*X.c.*), Ghana (*X.c.*), Togo (*X.c.*), Nigeria (*X.c.*), Benin (*X.c.*), Cameroon (*X.c.*), Gabon (*X.c.*)
- Central Africa: Central African Republic (*X.c.*), Equatorial Guinea (*X.c.*), Democratic Republic of Congo (*X.c.*)
- Eastern Africa: Tanzania (*X.c.*)

- Indian Ocean: Madagascar (*X.c.*, *X.m.*), Seychelles (*X.c.*), Mauritius (*X.c.*), Réunion (*X.m.*)
- Asia: India (*X.c.*, *X.d.*), Sri Lanka (*X.c.*, *X.m.*, *X.d.*), Malaysia (*X.m.*), Myanmar (*X.d.*), Indonesia Java: (*X.c.*, *X.m.*, *X.d.*), Sulawesi: (*X.c.*); Sumatra: (*X.c.*, *X.d.*); Timor: (*X.m.*), Vietnam (*X.c.*, *X.m.*, *X.d.*)
- Pacific Ocean: Fiji (*X.c.*), Hawaii (*X.c.*).

The two main species are primarily pests of *C. canephora*, but *C. arabica* is also attacked. They are polyphagous insects and have been recorded from a wide range of woody plants including cocoa, avocado and tea. On coffee, the stems of seedling trees are susceptible to damage, but on older trees small branches with a diameter < 25 mm are also attacked. Mallamaire (1955) reported damage to upwards of 25% of nursery trees in Côte d'Ivoire.

Both species bore into small branches to form a nest, occupied by a male and female beetle. The nest consists of a short, radial entrance tunnel about 1mm in diameter and, if the branch is large enough, this curves round in a horizontal plane and follows the circumference of the branch. From this tunnel, a brood chamber is excavated in the longitudinal axis. A fine, white sawdust is produced from the entrance hole. The branch above the nest site gradually blackens and the leaves wither, and once this stage is reached the nest is abandoned and a fresh branch chosen.

Life cycle

Eggs are laid in the brood chamber and the larvae develop there. Food is provided by fungi that grow on the walls of the chamber, the so-called 'ambrosia' fungi. The fungal spores that initiate this growth are carried on the head of the female and deposited on the walls of the tunnel during nest preparation. The ambrosia fungus of *X. compactus* has been studied. In Côte d'Ivoire, Brader (1964) isolated a new genus of fungus from its galleries which he named *Ambrosiella xylebori* but, in the Seychelles, Brown (1954) found a mixture of *Cladosporium cladosporioides* (Fresen-)deVries and *Penicillium pallitans*, and reported that Muller had isolated *Monilia* and *Fusarium* spores from nests in Indonesia, so it is probable that several different fungi are implicated.

The length of the life cycle varies considerably, probably depending on the state of the branch. The findings of a number of authors have been gathered together by Entwistle (1972, Table 23.1). Preparation of the gallery takes 3–10 days, followed by a pre-oviposition of 1.4–9 days, incubation lasting 2–7 days, a larval period of 6–23 days and a pupal period of 3–9 days.

Natural enemies

The larvae of both the main species are parasitized by the Eulophid wasp, *Tetrastichus xylebororum* Domenichini, in Indonesia (Domenichini, 1960) and another *Tetrastichus* species similar to *T. xylebororum* is recorded from *X.*

compactus in India (Dhanam *et al.*, 1992). The Braconid, *Mesobraconoides psolopterus* (Wilkinson), attacks *X. compactus* in Sierra Leone (Wilkinson, 1931). Two predators of *X. compactus* are known from India, the Clerid beetle, *Callimerus* sp. (Sreedharan *et al.*, 1992), and the Eupelmid wasp, *Eupelmus* sp. (Vinodkumar *et al.*, 1986). Entwistle (1972) suggests that the reason why parasitism is not effective is because the female blocks the entrance to the nest with her body, preventing the wasp from reaching the larvae.

Control

Chemical control is difficult. In the past, reasonable control was achieved by the use of persistent insecticides such as dieldrin, which is no longer an option. In Florida, up to 83% control of *X. compactus* attacking flowering dogwood was obtained using chlorpyrifos (Mangold *et al.*, 1977). The alternative to chemical control is good hygeine: pruning off and burning infested branches, soil improvement, etc. In Madagascar, Frappa (1928) found small coffee plantations close to villages to be generally free of damage, which he attributed to a good manuring with offal and wood ash.

Red Branch Borer

Zeuzera coffeae Nietner [Lepidoptera: Cossidae]

Morphology

The adult is a fairly large moth with a wingspan of around 70–80 mm. The body of the male is dark brown and the wings clear except for faint black striae in the spaces between the veins in the forewing. The hindwing has marginal black spots. The veins and the leading edge of the forewing are orange-yellow. The female moth has a whitish body with four black spots on the thorax, and the black speckling on the forewing is more pronounced, with a larger black central spot (Fig. 3.7). The fully grown caterpillar is about 40 mm long, with a brown head and a brown shield on the thorax, the remainder of the body varying from pink to purplish or brick red.

Pest status and distribution

Distributed across South East Asia, from India and Sri Lanka to New Guinea and the Solomon Islands, and northward to China and Taiwan, the caterpillar of this moth is a pest of a range of woody plants including coffee, cocoa and tea. On coffee it has been recorded from India, Sri Lanka, Myanmar, Vietnam, Taiwan, Philippines, Malaysia, Indonesia (Java, Sumatra, West Irian) and Papua New Guinea. It is a borer of both branch and trunk and, being a fairly

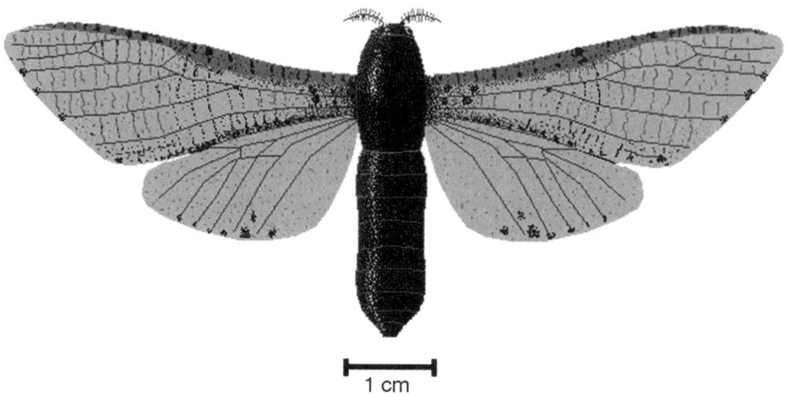

Fig. 3.7. Adult of *Zeuzera coffeae.*

large insect, can cause considerable damage to the tree. Nietner (1861), who described the species from Sri Lanka, reported that the pest enters the tree 15–30 cm from the ground and bores upwards, but it is more commonly a branch borer (CABI, 1973).

Life cycle

The reddish eggs are laid in strings in cracks of the bark of branches in large numbers. Kalshoven (1940) recorded numbers ranging from 348–966 eggs laid by four females in cages. The eggs hatch in about 10 d and the larvae at first feed in a group under a silk web. Later, they launch themselves on threads of silk and are widely dispersed by the wind. When they land on a suitable host they bore into it, often entering at a leaf node. The cylindrical tunnels of small larvae run down the centre of the branch, while those of older larvae are more irregular. The reddish frass is ejected through holes opened to the exterior at intervals.

The larval stage lasts 3–4 months, according to Beeson (1941) and, when ready to pupate, the larva cuts a chamber close to the exterior but which is still covered by a flap of bark. It pupates in this chamber with its head towards the exterior, this stage lasting 3–4 weeks. When ready to emerge as an adult, the pupa pushes its way past the bark flap so that half its body protrudes (see Fig. 3.8).

Natural enemies

Larvae are parasitized by several species of Braconid wasps, including *Glyptomorpha* sp. in Malaysia, *Amyosoma chinense* (Szepligeti) (no locality, CABI Compendium, 2003) and *Amyosoma leuzerae* Rohwer in India and

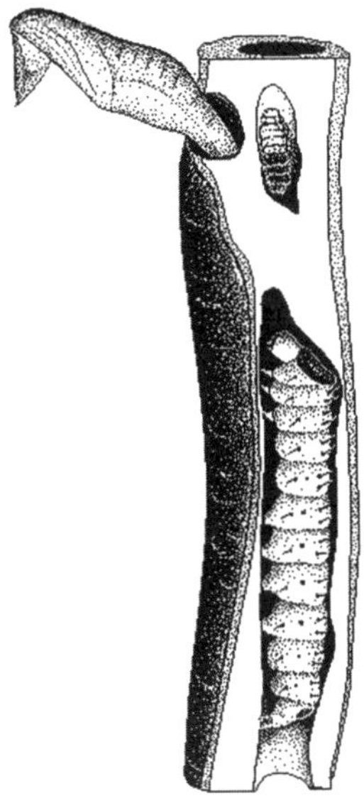

Fig. 3.8. *Zeuzera coffeae*: larva in burrow, cast larval skin and emtpy pupal case protruding from emergence hole.

Indonesia. Note that this species often appears under the name *zeuzerae*. *Bracon zeuzerae* Fahringer 1934 is now considered a subspecies of *A. leuzerae*, and the similarity of the names has caused some confusion in the literature.

Additionally, two species of Tachinid fly also attack the larvae in Indonesia, *Carcelia kockiana* Tours and *Isosturmia chatterjeeana* (Baranov). The larvae are also infected by the fungus *Beauveria bassiana*.

Control

In Malaysia, Khoo *et al.* (1991) advocate the pruning and destruction of smaller branches and the injection of an insecticide solution into the tunnels in larger branches or trunks. They recommend the use of 0.25% a.i. dieldrin or 2% a.i. chlorpyrifos. In China, up to 90% control has been achieved by the injection of 80% dichlorvos EC (Feng *et al.*, 2000) and, in India, Abraham and Skaria (1995) found that swabbing the main stem of allspice, *Pimenta dioica*, with 0.25% carbaryl was an effective prophylactive.

References

Abraham, C.C. and Skaria, B.P. (1995) New record of the red borer *Zeuzera coffeae* Nietner (Cossidae: Lepidoptera) as a pest of allspice, *Pimenta dioica* (Linn.) Merril. *Insect Environment* 1, 6.

Ahrens, L. and Vandenput, R. (1952) *La Lutte contre les Ennemis des Principales Cultures Pérennes de la Cuvette Central Congolaise*. Direction de l' Agriculture Ministère des Colonies Belgique, Brussels.

Anderson, T.J. (1919) Catalogue of the exhibit of entomology, Nairobi Agricultural Show 1919. *British East Africa Department of Agriculture*, 28 pp.

Aulmann, G. and La Baume, W. (1911) Die Schädlinge des Kaffees. *Die Fauna der deutschen Kolonien, V: Die Schädlinge der Kulturpflanzen*. Friedländer and Sohn, Berlin, 98 pp.

Báguena Corella, L. (1942) Observaciones sobre Longicornios perjudiciales a varios cultivos arbóreos y especialmente a los cacao y café en los Territorios Españoles del Golfo de Guinea. *Publicación, Direcccion de Agricultura de los Territorios Espanoles del Golfo de Guinea* 6, 39–91.

Beeson, C.F.C. (1919) The food plants of Indian forest insects Parts I–IV. *Indian Forester* 45, 49–56, 139–153, 312–323, 488–495.

Beeson, C.F.C. (1941) *The Ecology and Control of the Forest Insects of India and the Neighbouring Countries*. Vasant Press, Dehra Dun, India, 1007 pp.

Bhat, P.K. (1987) Current status and possibilities of integrated pest management in coffee with special reference to white stem-borer. *Journal of Coffee Research* 17, 126–128.

Brader, L. (1964) Étude de la relation entre le scolyte des rameaux du caféier, *Xyleborus compactus* et sa plante hôte. *Mededeelingen van de Landbouwhoogeschool te Wageningen* 64, 109 pp.

Bredo, H.J. (1939) Catalogue des pricipaux insectes et nematodes parasites des caféiers au Congo Belge. *Bulletin Agricole du Congo Belge* 30, 266–307.

Brown, E.S. (1954) *Xyleborus morstattii* (Coleoptera, Scolytidae), a shot-hole borer attacking avocado pear in the Seychelles. *Bulletin of Entomological Research* 40, 707–710.

Browne, F.G. (1968) *Pests and Diseases of Forest Plantation Trees*. Clarendon Press, Oxford, UK, 1330 pp.

CABI (1965) *Anthores leuconotus Pascoe*. *Distribution Maps of Pests* No. 196. CAB International, Wallingford, UK, 2 pp.

CABI (1971) *Xylosandrus morigerus* (Bldf.). *Distribution Maps of Pests* No. 292. CAB International, Wallingford, UK, 2 pp.

CABI (1973) *Zeuzera coffeae* Nietner. *Distribution Maps of Pests* No. 313. CAB International, Wallingford, UK, 2 pp.

CABI (1982) *Xylosandrus compactus* (Eichh.). *Distribution Maps of Pests* No. 244 (revised). CAB International, Wallingford, UK, 2 pp.

CABI (2003) *Crop Protection Compendium* 2003 edn, CABI, Wallingford, UK.

Crowe, T.J. (1959) The yellow-headed borer. *Kenya Coffee* 24, 13–20.

Crowe, T.J. (1962) The biology and control of *Dirphya nigrocornis* (Olivier), a pest of coffee in Kenya. *Journal of the Entomological Society of Southern Africa* 25, 304–312.

Dhanam, M., Raju, T. and Bhat, P.K. (1992) *Tetrastichus* sp. nr. *xylebororum* Domenichini (Hymenoptera: Eulophidae), a natural enemy of *Xylosandrus compactus* (Eichhoff). *Journal of Coffee Research* 22 (Suppl.), 127–128.

Domenichini, G. (1960) Descrizione di un nuovo *Tetrastichus* malese parassita di *Xyleborus* spp.. *Entomologische Berichten Amsterdam* 20, 256–258.

Duffy, E.A.J. (1957) *A Monograph of the Immature Stages of African Timber Beetles*. British Museum (Natural History), London, 338 pp.

Duffy, E.A.J. (1968) *A Monograph of the Immature Stages of Oriental Timber Beetles*. British Museum (Natural History), London, 434 pp.

Du Pasquier, R. (1930) Recherches sur les Hyménoptères parasites du 'Borer' (*Xylotrechus quadripes* Chev.). *Proceedings of the 4th Pacific Science Congress, Java, 1929*, pp. 519–524.

Duport, L. (1924) Rapport sur quelques-unes des recherches poursuivies à la Station. *Bulletin Périodique de la Station Entomologique Cho-Ganh* No. 19, 5 pp.

Entwistle, P.F. (1972) *Pests of Cocoa*. Longman, London, 779 pp.
Feng, R.Y., Guo, L.Z., Liang, E.Y. and Guan, H.Y. (2000) Bionomics of *Zeuzera coffeae* and its control. *Plant Protection* 26, 12–14.
Fonseca Ferrão, A.P.S. (1965) *Brocas do Tronco do Café*. Instituto de Investigação Agronómica de Chianga, Angola, 13 pp.
Forsyth, J. (1966) *Agricultural Insects of Ghana*. University of Ghana Press, Accra, 163 pp.
Frappa, C. (1928) Sur un ennemi du caféier a Madagascar (le bostriche du caféier). *Agronomie Coloniale* 133, 15–20.
Gowdey, C.C. (1914) The yellow-headed coffee borer (*Dirphya (Dirphya) princeps*, Jord.) *Bulletin of Entomological Research* 4, 279–281.
Hall, D.R., Cork, A., Phythian, S.J., Chittamuru, S., Jayarama, B.K., Venkatesha, M.G., Sreedharan, K., Vinod Kumar, P.K., Seetharama, H.G. and Naidu, R. (2006) Identification of components of male-produced pheromone of coffee white stem borer *Xylotrechus quadripes*. *Journal of Chemical Ecology* 32, 195–219.
Hancock, G.L.R. (1926) *Report of the Assistant Entomologist*. Report of the Department of Agriculture, Uganda 1925, pp. 25–28.
Harris, W.V. (1937) *Report of the Entomologist*. Report of the Department of Agriculture, Tanganyika Territory 1936, pp. 88–94.
Hill, D.S. (1975) *Agricultural Insect Pests of the Tropics and their Control*. Cambridge University Press, London, 516 pp.
Hillocks, R.J. (2000) Integrated crop management for smallholder farmers in Africa with special reference to Malawi. *Pest Management Science* 56, 963–968.
ICB (2002) White stem borer may hit coffee output. Indian Coffee Board web site. http://www.indiacoffee.org/news/5/61.html
Jover, H. (1954) Notes biologiques sur quelques Coléoptères xylophages de Basse-Côte d'Ivoire. Cerambycidae. *Revue de Pathologie Végetale et d'Entomologie Agricole de France* 32, 258–275.
Kalshoven, L.G.E. (1940) Observations on the red branchborer, *Zeuzera coffeae* Nietn. (Lep. Coss). *Entomologische Mededeelingen van Nederlandanasch-Indie*, 6, 50–56.
Kieffer, J.J. (1921) Sur divers Hymenopteres destructeurs des Cerambycides nuisibles au Caféier et au Bambou. *Bulletin Agricole de l' Insitute Scientifique* 3, 129–140.
Knight, C.D. (1939) Observations on the life history and control of white stem borer of coffee in Kenya. *East African Agricultural Journal* 5, 61–67.
Khoo, K.C., Ooi, P.A.C. and Ho, C.T. (1991) *Crop Pests and their Management in Malaysia*. Tropical Press, Kuala Lumpur, 242 pp.
Koningsberger, J.C. (1908) Tweede overzicht de Schadelijke en Nuttige Insecten. *Mededeelingen Uitgeven van het Departement van Landbouw in Nederlandsch-Indie* 6, 1–113.
Kung, P.C. (1977) Studies on two long-horned beetles infesting coffee trees in Kwangsi Autonomous Region. *Acta Entomologica Sinica* 20, 49–56.
Lan, C.C. and Wintgens, J.N. (2004) Major pests of coffee in the Asia-Pacific region. In: Wintgens, J.N. (ed.) *Coffee: Growing Processing and Sustainable Production*. Wiley-Verlach, Weinheim, Germany, 459–473.
Le Pelley, R.H. (1959) *Agricultural Insects of East Africa*. East African High Commission, Nairobi, 307 pp.
Le Pelley, R.H. (1968) *Pests of Coffee*. Longman, London, 415 pp.
Lepesme, P. (1953) Coléoptères Cerambycides (Longicornes) de Côte d'Ivoire. *Catalogue I.F.A.N.* 11, 103 pp.
Lepesme, P. and Paulian, R. (1943) Une Tachinaire parasite des larves du monohamme du caféier. *Revue de Zoologie et de Botanique Africaines* 37, 89–92.
Lepesme, P. and Villiers, A. (1941) Cerambycides recoltes par P. Lepesme sur caféier en A.E.F. et au Cameroun. *Bulletin de la Societe Entomologique de France* 46, 74–79.

Lepesme, M. and Villiers, A. (1944) *Les Longicornes du Caféier en Afrique Intertropicale.* Travailles de Section Technicale de Agriculture Tropicale Series 1, Paris, 27–70.

Le Poer Trench, A.D. and Anderson, T.J. (1930) A report on the campaign against *Stephanoderes* 1929. *Bulletin. Department of Agriculture, Kenya* 9, 19 pp.

Mallamaire, A. (1955) Les principaux insectes foreurs des tiges rameaux ou stipes des plantes cultiveés dans l'Ouest Africain. *Bulletin de la Protection Vegetaux* 2, 61–100.

Mangold, J.R., Wilkinson, R.C. and Short, D.E. (1977) Chlorpyrifos sprays for control of *Xylosandrus compactus* in flowering dogwood. *Journal of Economic Entomology* 70, 789–790.

Marchal, P. (1900) Sur quelques coléoptères nuisibles au caféier recuilles au Congo par M.F.Wisser. *Sixth Congress of International Agriculture*, Paris.

Mayne, R. and Donis, C. (1962) *Hôtes Entomologiques au Bois II. Distribution au Congo au Rwanda et au Burundi.* Publications de l'Institut national pour l'étude agronomique du Congo belge, 100, 514 pp.

Mertens, (1920) Les plantations de caféiers de Lula (Stanleyville). *Bulletin Agricole du Congo Belge* 9, 243–251.

Miwa, Y. (1936) Insect pests of coffee and coffee trees I. *Bulletin of the Government Research Institute, Formosa* 126, 33 pp.

Moffat, U.J. and Allan, W. (1934) A preliminary note on the white stem borer on coffee at Abercorn. *Bulletin of the Department of Agriculture, Northern Rhodesia* 1933, 39–41.

Morstatt, H. (1935) Kaffee-Schädlinge und Krantheiten Afrikas I. Kaffeebohrer (Stammbohrer). *Tropenpflanzer* 38, 413–431.

Munoz, R.H. (Undated typescript, Honduras) *Plagas insectiles del cafeto*, 32 pp.

Nietner, J. (1861) *Observations on the Enemies of the Coffee Tree in Ceylon*. Colombo, 31 pp.

Padi, B. and Adu-Acheampong, R. (2001) An integrated pest management strategy for the control of the coffee stem borer, *Bixadus sierricola* White (Coleoptera: Lamiidae). *Association Scientifique Internationale du Café, 19th International Scientific Colloquium on Coffee*, Trieste, Italy, 2001, 5 pp.

Reddy, K.B. and Bhat, P.K. (1987) Studies on the flight periods of coffee white stem-borer, *Xylotrechus quadripes* Chev. in Pulney and Shevaroy Hills. *Journal of Coffee Research* 17, 26–30.

Rhainds, M., Lan ChinChiew, Zhen MoLi and Gries, G. (2002) Incidence, symptoms and intensity of damage by three coffee stem borers (Coleoptera: Cerambycidae) in South Yunnan, China. *Journal of Economic Entomology* 95, 106–112.

Schoeman, P.S. (1995) Integrated control of arabica coffee pests (Part 2): an overview of measures to control the white coffee stem borer, *Monochamus leuconotus* (Pascoe) (Coleoptera: Cerambycidae). *Inligtingbulletion Institut vir Tropiese en Subtropiese Gewasse* 280, 11–20.

Schoeman, P.S. and Pasques, B. (1993) Chlordane stem treatments to combat white coffee stem borer. *Inligtingbulletion Institut vir Tropiese en Subtropiese Gewasse* 249, 17–19.

Schoeman, P.S. and Schoeman, M.H. (1997) Fungus could be used to control white stem borer. *Neltropika-Bulletin* 296, 36–42.

Schoeman, P.S., van Hamburg, H. and Pasques, B.P. (1998) The morphology and phenology of the white coffee stem borer, *Monochamus leuconotus* (Pascoe) (Coleoptera: Cerambycidae), a pest of Arabica coffee. *African Entomology* 6, 83–89.

Seetharama, H.G., Kumar, P.K.V. and Sreedharan, K. (2002) *Apenesia* sp.: a potential natural enemy for the biological control of coffee white stem borer. *Indian Coffee* 66, 24–25.

Sreedharan, K., Balakrishnan, M.M. and Bhat, P.K. (1992) *Callimerus* sp. (Coleoptera: Cleridae), a predator of the shot hole borer, *Xylosandrus compactus* (Eichh.). *Journal of Coffee Research* 22, 139–142.

Subramanyam, T.V. (1934) *The Coffee Stem Borer*. Entomological Series, Bulletin 11, Department of Agriculture, Mysore, India, 18 pp.

Tapley, R.G. (1960) The white coffee borer, *Anthores leuconotus* Pasc. and its control. *Bulletin of Entomological Research* 51, 279–300.

Tothill, J.D. (ed.) (1940) *Agriculture in Uganda*. Cambridge University Press, London, 551 pp.

Vayssiere, P. (1955) Les animaux parasites du caféier. In: Coste, R. (ed.) *Les Caféiers et les Cafés dans le Monde*. Larose, Paris, pp. 233–318.

Vega, F.E., Mercadier, G., Damon, A. and Kirk, A. (1999) Natural enemies of the coffee berry borer, *Hypothenemus hampei* (Ferrari) (Coleoptera: Scolytidae) in Togo and Côte d'Ivoire, and other insects associated with coffee beans. *African Entomology* 7, 243–248.

Vinodkumar, P.K., Balakrishnan, M.M. and Govindarajan, T.S. (1986) *Eupelmus* sp. (Hymenoptera Eupelmidae) – a new predator of *Xylosandrus compactus*. *Journal of Coffee Research* 16, 35–37.

Visitpanich, J. (1994a) The biology and survival rate of the coffee stem borer, *Xylotrechus quadripes* Chevrolat (Coleoptera, Cerambycidae) in Northern Thailand. *Japanese Journal of Entomology* 62, 731–745.

Visitpanich, J. (1994b) The parasitoid wasps of the coffee stem borer, *Xylotrechus quadripes* Chevrolat in northern Thailand. *Japanese Journal of Entomology* 62, 597–606.

Vrijdagh, J. (1930) Tableau systematique des insectes nuisibles aux plantes cultivées au Congo Belge. *Annales de Gembloux* 36, 425–434.

Wei JiaNing and Kuang RongPing (2002) Biological control of coffee stem borers, *Xylotrechus quadripes* and *Acalolepta cervinus*, by *Beauveria bassiana* preparation. *Entomologia Sinica* 9, 43–50.

Wilkinson, D.S. (1931) Four new species of Ichneumonoidea. *Bulletin of Entomological Research* 22, 393–397.

Wolcott, G.N. (1921) *El Calculo Taladrador del Tallo del Cafeto*. Porto Rico Insular Experimental Station Circular 48, 6 pp.

Wrigley, G. (1988) *Coffee*. Longman (Tropical agriculture series). Harlow, UK, 639 pp.

Zacher, F. (1921) Schädlinge der nutzpflanzen im West-Sudan. *Tropenpflanzer* 24, 97–108, 132–142.

4 Berry-feeding Insects

Introduction

Many of the insects that feed on green berries are also found on leaves or green shoots, and these are dealt with in Chapter 5. These include scale insects (Coccoidea) of various kinds, leaf-eating caterpillars (Lepidoptera) in several families, mites (Acari) and some Orthoptera, including grasshoppers and tree crickets. The insects dealt with in this chapter include the more important species that feed primarily on berries, although Antestia bugs can damage green shoots also.

Coffee Berry Borer

Hypothenemus hampei (Ferrari) [Coleoptera: Scolytidae]

Morphology

The female adult beetle is 1.5–2.0 mm long × 1 mm wide; the adult male is smaller, but this can vary somewhat from place to place, American specimens being somewhat smaller than Old World specimens. When first emerging, the beetle is brown, but as it matures over the next few days it becomes black, with a reddish tinge to the thorax. The pupae are a similar length to the adult and the larvae are a little smaller, having a white body and brown head (see Fig. 4.1).

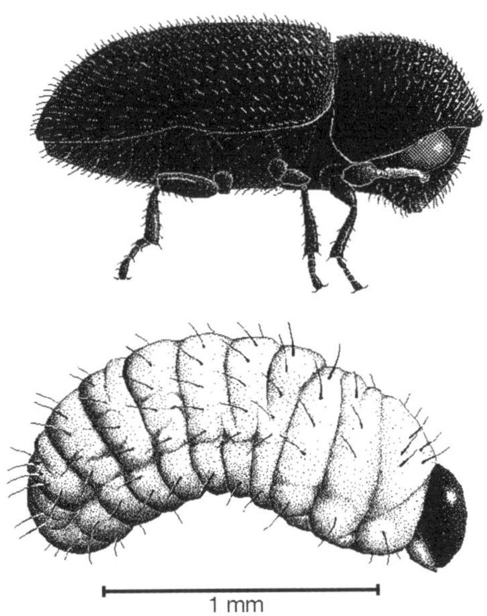

Fig. 4.1. Adult and larva of *Hypothenemus hampei.*

Pest status and distribution

A serious pest in many countries of low-altitude arabica and robusta coffee (CABI, 1981). Damage by the coffee berry borer is rarely severe at altitudes > 1370 m, and this species has not been found > 1680 m. The centre of origin of the species is in some doubt, but must have been somewhere in West or Central Africa. Morstatt (1941) considered it to have originated in the area around Lake Victoria, but Schedl (1961) disputes this. Bredo (1939) points out that at the time he was writing, the beetle had never been found in truly wild coffee in the Democratic Republic of Congo. The beetle was described by Ferrari (as *Cryphalus*) in 1867 from beetles found in stored produce, but the country of origin is not recorded.

The first field record of berry borer was from material collected by O.F. Cook in Liberia in 1897, and described as *Stephanoderes cooki* Hopkins (Hopkins, 1915), now considered to be a synonym of *H. hampei*. It was reported from Gabon in 1901, the Congo Republic and Chad in 1902–1904, Uganda in 1908, Angola in 1909 and the Democratic Republic of Congo in 1911 (Schedl, 1961). In 1914 in Tanzania, damage to coffee was restricted to robusta to the west of Lake Victoria (Morstatt, 1914), although he found it to occur in the Usambara Mountains but not on coffee. By 1929 it was affecting areas to the east of Lake Victoria, but did not reach the arabica-growing areas of Kilimanjaro until 1968. The first report of the insect in Kenya was in 1928 (Wilkinson, 1928).

It appeared at an early date in Indonesia, presumably as an import from

West Africa. Kalshoven (1950–1951) gives 1909 as the date of the first record from Java, but Schedl (1961) cites a report from there by Zimmermann in 1904. The first records from Sumatra are around 1919, from Malaysia 1928 (Corbett, 1933), from Sri Lanka 1935 (Hutson, 1936) and from the Philippines 1965 (Anon, 1965). After the Second World War it spread across the Pacific, being recorded from New Caledonia in 1948 (Cohic, 1958) and Tahiti in 1963 (Johnston, 1963). It was found in Irian Jaya in 1961 (Thomas, 1961), but so far has not reached Papua New Guinea. *Hypothenemus hampei* was not reported in India until 1990 but, 10 years later, some 36% of the coffee in the 'traditional' coffee areas was affected (Reddy and Rao, 1999).

The first report from the New World was in 1922, when a new Scolytid found boring into coffee twigs in the São Paulo region of Brazil was named *Xyleborus coffeicola* by De Campos Novaes (1922). This was shown by Da Costa Lima (1924) to be synonymous with *H. hampei*. It gradually spread to the other coffee-growing states of Brazil, and later to other South and Central American countries. It reached Peru in 1961 and had moved from there to Ecuador by 1981. It gradually spread north through Ecuador and reached Colombia in 1988 (Cárdenas and Posada, 2001) and Venezuela in 1995.

In Central America and the Caribbean, it was discovered first in Guatemala in 1971 (Hernandez Paz and Penagos Dardon, 1974), and from thence spread to Honduras in 1977, Mexico (Baker, 1984) and Jamaica (Reid, 1983) in 1978, El Salvador in 1981, Nicaragua in 1988, Cuba in 1994 (Vega *et al.*, 2002), the Dominican Republic in 1995 (Vega *et al.*, 2002) and Costa Rica in 2000. Extensive surveys reported by Vega *et al.* (2002) failed to find the pest in Puerto Rico, and the authors conclude that Le Pelley's inclusion of the island in the original distribution list was incorrect.

Damage

Hypothenemus hampei invades mostly older berries and cuts through the berry to penetrate the bean through a small hole at or near the apex of large green or ripe berries (see Fig 4.2). In Colombia, a high rate of survival and speedy development of the larvae required berries that were at least 120 days old (Baker, 1999). The larvae then consume one or both seeds (see Plate 4).

Damaged berries either fall or are rotted by secondary bacterial and fungal growth. Wet rot in the mesocarp of berries superficially damaged by *H. hampei* has been associated with the bacteria *Erwinia steartii* and *E. salicis* (Sponagel, 1994, cited in Damon, 2000). Attacks are more severe where coffee is grown under heavy shade and where pruning has been neglected.

Host range

Breeding takes place almost entirely in *Coffea* species, with *C. arabica* being the most attractive, followed by *C. canephora, C. dewevrei, C. dybowskii, C.*

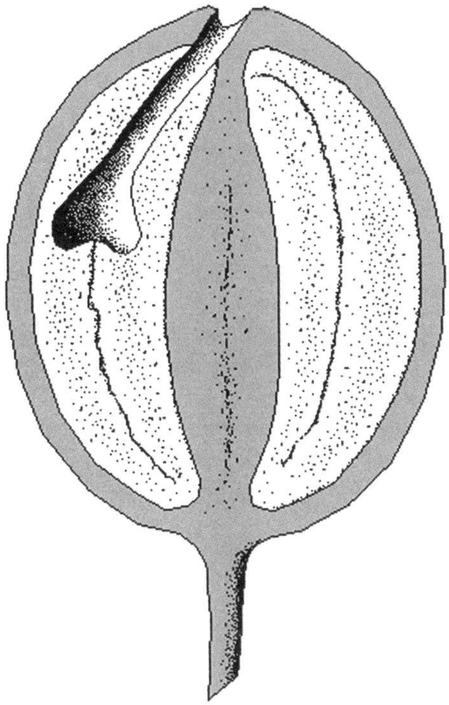

Fig. 4.2. Cross-section of coffee berry to show how the nest tunnel of *Hypothenemus hampei* is initiated from the apex of the fruit.

excelsa and finally *C. liberica* (Le Pelley, 1968). It should be noted that in a recent review of the genus *Coffea* by Davis *et al.* (see Chapter 1), the status of some of these species has changed, and *C. dewevrei*, *C. dybowskii* and *C. excelsa* are shown to be synonyms of *C. liberica* var. *dewevrei*. The fruits of wild coffee growing in dense forest are often heavily infested.

There is quite a long list of plants, other than coffee, from which *H. hampei* has been recorded (see Table 4.1), and it has been assumed by many authors (e.g. Le Pelley, 1968; Hill, 1975) that these are either exploratory attacks by the beetle on plants in which it cannot breed or that the beetle has been confused with other, similar, species of Scolytid. However, it seems likely that the beetle is able to breed in certain leguminous hosts.

Ghesquiere (1933) found all stages of the beetle in pods of *Dialium engleranum* in Democratic Republic of Congo, and Morallo-Rejesus and Baldos (1980) found eggs, larvae and pupae in *Leucaena leucocephala* and *Gliricidia sepium* in the Philippines, as well as in other non-leguminous hosts, whilst CABI (2003) reports that the beetle has been experimentally reared to adulthood on *Melicocca bijuga* in Colombia and on *Cajanus cajan* in Guatemala. It seems, therefore, that Leguminosae should not be overlooked as possible alternative hosts for the beetle.

Table 4.1. Host plants of *H. hampei* other than *Coffea* (from Schedl, 1961 except where indicated).

Family	Species	Reported locations
Apocynaceae		
	Pleiocarpa tubicina	Democratic Republic of Congo
Bignoniaceae		
	Spathodea campanulata	Democratic Republic of Congo
Dioscoriaceae		
	Dioscorea sp.	Philippines[a]
Guttiferae		
	Allanblackia floribunda	Democratic Republic of Congo
	Mammea africana	Democratic Republic of Congo
Leguminosae (*sensu lato*)		
	Caesalpinia pulcherrima	Democratic Republic of Congo
	Cajanus cajan	Guatemala
	Centrosema plumeri	Java
	Crotalaria sp.	Java
	Dialium engleranum	Democratic Republic of Congo
	Gliricidia sepium	Philippines[a]
	Leucaena leucocephala	Java, Philippines[a]
	Melicocca bijuga	Colombia[c]
	Oxystigma oxyphyllum	Democratic Republic of Congo
	Phaseolus lunatus	Uganda, Nigeria[b]
	Tephrosia candida	Sumatra
	Tephrosia sp.	Java
Malvaceae		
	Hibiscus sp.	Democratic Republic of Congo
Meliaceae		
	Trichilia gilgiana	Democratic Republic of Congo
Myristicaceae		
	Pycnanthus angolensis	Democratic Republic of Congo
Rosaceae		
	Eriobothrya japonica	New Caledonia
	Rubus sp.	Tanzania
Rubiaceae		
	Nauclea diderrichii	Democratic Republic of Congo
	Oxyanthus sp.	Uganda
	Psychotria (two spp.)	Philippines[a]
Sterculiaceae		
	Cola sp. near *lateritia*	Democratic Republic of Congo
	Theobroma cacao	Democratic Republic of Congo
Verbeneaceae		
	Vitex lanceolaria	Java

[a] Morallo-Rejesus and Baldos (1980).
[b] Ghesquiere (1933).
[c] CABI (2003).

Life cycle

The life cycle of *H. hampei* takes 28–35 d from egg-laying to the mature beetle, but the beetle often remains in the berry for 1–2 weeks before emerging. Having a preference for older berries, the life cycle may not be completed before harvest and there is therefore a danger of spreading the beetle within green beans (see Chapter 14). The female beetle bores through the berry into the bean and lays eggs in the tunnel. Over a period of 3–7 weeks, up to 60 eggs may be laid, which hatch in 5–9 d (Ticheler, 1961; Waterhouse and Norris, 1989). The larvae feed on the bean for 10–16 d. The females moult twice before pupation and the pupal stage lasts for 4–9 d.

The female becomes sexually mature soon after emergence, is fertilized by the male within the berry and is capable of laying eggs 3–4 d later. Males cannot fly and are capable of fertilizing up to 30 females during their lifespan (Bergamin cit. in Wrigley, 1988). Males live for 20–87 d and females up to 157 d (Barrera, 1994). Where coffee berries are present all year round, as in Uganda, *H. hampei* may produce eight or nine generations, completing its life cycle every 30 d (Hargreaves, H., 1935). In Colombia, however, only two or three generations are produced each year (Montoya and Cardenas-Murillo, 1994).

Rate of growth and reproduction is greatest at 29–33°C. The mating system appears to be inbreeding, with a male:female ratio of 1:10, and there are three instars (Vijayalakshmi *et al.*, 2002). Many species of Scolytidae rely on the presence of 'ambrosia' fungi within their burrows as a source of nutrients, and it has recently been shown (Morales-Ramos *et al.*, 2000) that *H. hampei* is dependent on *Fusarium solani* for successful breeding, and that the spores of this fungus accumulate in pits situated behind upward- and backward-pointing asperites on the prothorax of the female beetle, and so are carried by her to a fresh burrow.

Control

At altitudes where berry borer is a problem, the incidence of damage caused by the pest can be reduced by thinning of shade trees and pruning the coffee bushes to open the canopy. The crop should be picked at 2-week intervals during peak fruiting season, and all ripe or dried berries should be removed from the tree, cleared from the ground and destroyed.

Although there are several insecticides that are effective against berry borer, the beetle is protected to some extent from their effect by spending much of its life cycle within the berry. Spraying may do more harm than good, by destroying natural enemies. However, effective control has been achieved with endosulphan, and its use has been widely adopted in South America (Baker, 1999) and, more recently, its use has been evaluated in India (Rahman and Vijayalakshmi, 1999).

By the late 1980s, resistance to the insecticide had already begun to appear in New Caledonia following 10 years of use (Brun *et al.*, 1989; Brun and Suckling, 1992). Concerns over the human toxicity of endosulphan, and its misuse by poorly educated farmers in developing countries, have led to calls for it to be banned. The active ingredient was banned for use on coffee in Colombia in 1978, but it is only recently that steps have been taken to enforce the ban (Tovigan *et al.*, 2001). Chlorpyrifos is reported to be as effective as endosulphan in India (Balakrishnan *et al.*, 2001).

Hypothenemus hampei is indigenous to central Africa, where three important natural enemies have been known for more than 40 years. Two of these are Bethylid wasps: *Prorops nasuta* Waterston and *Cephalonomia stephanoderis* Betrem, and a Braconid: *Heterospilus coffeicola* Schmiedeknecht. *Prorops nasuta* acts as both parasite and predator, the adult feeding on eggs and young larvae while the larvae attack fully-grown larvae and pupae. *Prorops nasuta* has been introduced into Indonesia, Brazil and Sri Lanka and, although the number of infested berries was reduced, it was not sufficient to prevent serious losses from berry borer.

Prorops nasuta was introduced into Mexico in 1985 and Equador in 1987, where it was successfully reared and released (Murphy and Rangi, 1991), but it failed to establish in Mexico, due partly to the action of predatory ants and spiders (Infante *et al.*, 2003). *Cephalonomia stephanoderis* is the most important natural enemy in Côte d'Ivoire, where about half the berries attacked by *H. hampei* contained the parasite. The larvae parasitize the last larval stage of *H. hampei* and the adults feed on adult borers, resulting in considerable reduction in the borer population (Ticheler, 1961). *Cephalonomia stephanoderis* has been introduced into South America and it established well following its introduction into Mexico in 1985.

Rates of parasitism were high initially close to release sites, but fell to less than 10% after only a few months (Gutierrez *et al.*, 1998; Baker, 1999). *Heterospilus coffeicola* was discovered in Uganda and occurs elsewhere in Africa. The parasite larvae feed on the eggs of *H. hampei*, and sometimes on the young larvae. Several attempts to rear this parasite have failed.

Another wasp, *Phymastichus coffea* Lasalle, was reported first from Togo (LaSalle, 1990), but it is distributed throughout the coffee-growing areas of West Africa across to East Africa, where it has been recorded in Burundi and Kenya (Decazy, 1991). The larvae develop endoparsitically on *H. hampei* adults. *Phymastichus coffea* was introduced into Colombia in 1997 (Baker, 1999), where it established, and rates of parasitism as high as 67% have been recorded, indicating that this might be the most promising of the natural enemies for classical biological control of the borer.

In the late 1990s, *Cephalonomia hyalinipennis* Ashmead was discovered naturally attacking *H. hampei* in Mexico (Perez-Lachaud and Hardy, 1999), and preliminary investigations suggested that rates of attack were sufficiently high for *C. hyalinipennis* to be suitable for mass rearing and release. In Mexico, where both *C. stephanoderis* and *C. hylinipennis* have been released, they have been shown to compete with one another and it was concluded that *C. stehanoderis* is the more successful (Perez-Lachaud *et al.*, 2002). Damon and Valle (2002) estimated that, although kill rates with *C. stephanoderis* could be improved by releasing parasitized hosts rather than adults, 59 million parasitoids would be required per ha of coffee to obtain 65% kill. They concluded that this method of control on its own could never be economically viable.

In Colombia, Cenicafé has developed processes for the mass production of several coffee berry borer parasitoids: *Cephalonomia stephanoderis, Prorops nasuta* and *Phymastichus coffea*. The technology has been transferred to commercial laboratories and the commercial preparations are available to coffee farmers.

Wherever berry borer is found, the entomopathogenic fungus *Beauveria bassiana* is also present as a natural infection. Conidia of the fungus adhere to the cuticle of *H. hampei*, where they germinate and penetrate the cuticle, proliferating internally to destroy the insect. Commercial preparations of the fungus are widely available. Experiments conducted with *B. bassiana* in Colombia have shown mortality rates of berry borer to be over 70% but, in taking up to 30 d to achieve mortality after application of the fungus, damage to the bean was not always prevented. Using high-volume sprayers, it was concluded that the large spore numbers required for high mortality rates were uneconomic. The number of conidia required to reach 70% mortality could be decreased by using low-volume spinning disc spray technology, but this method would still require a higher-quality product than was commercially available (Baker, 1999).

Research continues in South America (e.g. Edgington *et al.*, 2000), India (e.g. Haraprasad *et al.*, 2001) and elsewhere to improve the quality of inoculum of *B. bassiana* and to improve delivery methods. Molecular markers have also been used in an attempt to identify those isolates with the greatest potential for control of berry borer (Velez-Arango *et al.*, 2001). Some success has also been reported with a *Fusarium* sp. (Diaz *et al.*, 2003).

Entomopathogenic nematodes have also been investigated for berry borer control: *Steinernema carpocapse* (Weisser), *Steinernema* spp. and *Heterorhabditis bacteriophora* (Poiner) have all showed high mortality rates on *H. hampei* larvae in laboratory tests (Castillo and Marban-Mendoza, 1996).

Salazar *et al.* (2003) emphasize the importance of postharvest control in Colombia, where farmers are encouraged to cover the harvested berries with plastic lids smeared with grease that prevent entry of, and trap, the borer.

Attempts to integrate cultural control with biological control using *B. bassiana* in Mexico showed that, as in Colombia, *B. bassiana* made little contribution to control due to the poor quality of the commercial product (Jarquin-Galvez *et al.*, 1999). It was concluded that in organic production systems, the most cost-effective measure was to ensure that all berries were removed from the tree and cleared from the ground after picking. In India, there are a number of cultural measures recommended to manage the disease. Spot-spraying with endosulphan has also been recommended where outbreaks occur, but only once, during April–May and, if necessary, once more in July–August. Application of endosulphan 35 EC was recommended at the rate of 340 ml/200 l of water, and applied only to the berries in the affected area (Reddy and Rao, 1999).

The use of traps containing alcohols as an IPM component in the management of borers was developed in Central America and has been adopted elsewhere (Dufour *et al.*, 2001). In India, following the introduction of IPM systems in Karnataka, a study of 476 estates conducted in Kodoga District found that 94% practised clean harvesting, 92% collected gleanings, 86% used picking mats, 30% had adopted the recently introduced 'broca' traps and only 16% resorted to spraying, which was discouraged by extension, except as a last resort (Nagarajaiah and Kumar, 2003).

Where cultural, chemical and biological control with *B. bassiana* have been integrated in Latin America, it has been found that the performance of the biological control agent is unreliable and that chemical control has to be used with cultural control to be fully effective. The main component giving the greatest economic benefit was cultural control, mainly gleaning of berries (Benavides *et al.*, 2002).

Berry Moth

Prophantis smaragdina (Butler), *P. octoguttalis* (Felder and Rogenhofer), *Thliptoceras longicornalis* (Mabile) [Lepidoptera: Pyralidae]

Morphology

The adult is a golden-brown moth with a wingspan of 1.3–2.0 cm. The larva is a pink/red caterpillar with dark markings and about 1.3 cm long when fully grown. The eggs are scale-like and usually found on green berries (Le Pelley, 1968; CRF, 1978) (see Fig. 4.3).

Pest status and distribution

There has been some confusion over the identity of coffee berry-boring Pyralids in the past, and they have been reported under a number of different names. *Prophantis smaragdina* is widespread across much of sub-Saharan Africa and the Indian Ocean islands, including Madagascar and parts of Asia.

On coffee, it has been recorded from Cameroon, Eritrea, Ethiopia, Kenya, Madagascar, Malawi, Mauritius, Nigeria, Réunion, São Tomé and Príncipe, Tanzania, Uganda, Yemen, Democratic Republic of Congo and Zimbabwe. *Prophantis octoguttalis* is found in Malaysia, Indonesia, the Philippines and Taiwan, and has been recorded from coffee in India. *Thliptoceras longicornalis* is known only from Madagascar and has been recorded on coffee in that country. A record of *P. octoguttalis* and its parasite, *Microbracon* sp., from the Democratic Republic of Congo (De Saeger, 1943) almost certainly refers to *P. smaragdina*.

Evidence of berry moth can usually be seen at low incidence in Tanzania and Malawi. However, severe attacks have been recorded at low altitude, with heavy loss of berries. Berry moth is said to have been the principal cause of the decline of coffee cultivation in Réunion (Chevalier, 1947), and is still a cause of

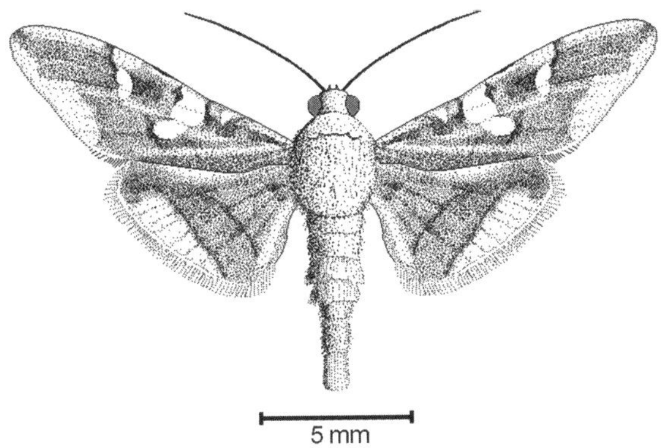

Fig. 4.3. *Prophantis smaragdina*, adult.

concern to the present revival of coffee on that island (Descroix, 2004). In Yemen, up to 60% infestation is reported (Ba-Angood and Al-Sunaidi, 2004). During the late 1970s, it was reported that *P. smaragdina* was becoming a more important coffee pest in Kenya (CRF, 1978). This suggests that regular spraying with insecticides may decrease control by natural enemies. Berry thinning caused by berry moth and similar pests may be regarded as beneficial when it results in thinning of overbearing branches.

Damage

The larvae bore into berries and destroy the seed. Sometimes whole berries may be consumed, and the larvae have been known to feed on the skin of older berries or flower buds and even on shoots. Silk webs around affected clusters of berries indicate the presence of this berry moth larva. Damage by berry moth has sometimes been confused with that caused by coffee berry disease, but the webbing on berry clusters is the distinguishing feature (see Plate 5).

Host range

Occurs mostly on *C. arabica*. The only recorded alternate hosts are the rubiaceous shrubs *Tricalysia* sp. (Le Pelley, 1959) – a genus which is widely distributed through Africa, and *Bertiera zaluzania*, endemic to Mauritius (Kaiser, 2005), a genus which is also widely distributed across Africa.

Life cycle

Eggs are laid usually singly, on green berries and hatch in about 6 days. The larvae bore into the berries to feed on the seed before moving on to feed on berries. As they move across the berry cluster they spin a web of silk, thus joining the cluster together. The larval period lasts about 14 d and the fully-grown larva passes through a resting stage before falling to the ground. Pupation occurs between two leaves stuck together and lasts for between six and 42 d depending on conditions (CRF, 1978). After the adult moth emerges, there is a pre-oviposition period of 3–4 d and the moth then lives for about 14 d.

Natural enemies

The larvae are heavily parasitized, but the natural enemies are not well studied due to the low pest status of berry moth. An egg parasite, *Trichogrammatoidea* sp., is reported from Sao Tome and Principe (Derron, 1977), and four species of Braconids, an Ichneumonid and two Tachinids attack the larvae (see Appendix A). Ndungi (1994) summarizes the situation in Kenya.

Control

Berry moth is probably kept in check by its natural enemies, and spraying is not usually required. If buds or young berries are being eaten in significant numbers soon after the main flowering period, it may be necessary to spray. Fenitrothion, fenthion, fenvalerate, chlorpyrifos and deltamethrine have been recommended in Kenya (Anon, 1992). To be effective, spraying must be carried out before most of the larvae have hatched and entered the berries.

Tapley and Materu (1961) experimented with a 1% spore suspension of *Bacillus thuringiensis* for control of *P. smaragdina* in Tanzania in small-scale tests and obtained fairly promising results both in the laboratory and in the field, but these results were never followed up. In Yemen an extract of *Ficus salicifolia* has been tested experimentally, apparently with some success (Ba-Angood and Sunaidi, 2004). There are no current recommendations for suitable insecticides, and local advice should thus be sought.

Berry Butterfly

Deudorix lorisona Hewitson [Lepidoptera: Lycaenidae]

Morphology

The male of this butterfly has a wingspan of around 25 mm. The forewings are dark brown, with two red spots near the middle of the wing, whilst the hind wings are red with a black border and short, hair-like tails. The wings of the female are typically brown, with a yellow spot in each wing, but the pattern can be variable. The larva is green–brown in colour, about 20 mm long and with bristly hairs.

Pest status and distribution

The berry butterfly is a minor pest of coffee. It occurs from West Africa through to East Africa and southwards into Southern Africa, and on coffee has been recorded from Guinea, Sierra Leone, Côte d'Ivoire, Ghana, Cameroon, Ethiopia, Kenya, Tanzania, Uganda and Yemen. In Kenya (Anon, 1967), it is said to be confined to arabica coffee, but in Sierra Leone, Hargreaves E. (1937) recorded an attack on robusta, excelsa (now = *C. liberica* var. *dewevrei*) and liberica coffee, with a preference for larger fruits. A secondary host is the rubiaceous shrub *Heinsia crinita* (Afzel.) G.Taylor.

Life cycle

The single egg is attached to the side of the green berry. Occasionally, two eggs are laid, but only one survives (Hargreaves, E., 1937). The hatching larva eats directly into the berry, where it completely consumes the contents, leaving an empty shell. The larval stage lasts around 21 d and during this time up to ten fruits may be eaten. Pupation takes place in a chamber hollowed out of dead wood (Anon, 1967) and lasts 10–11 d.

Natural enemies

A species of *Bracon* has been recorded as a natural enemy of the larva in Kenya (Le Pelley, 1959) and unspecified hymenopterous parasites in Uganda (Le Pelley, 1968) and Sierra Leone (Hargreaves, E., 1937).

Antestia Bug (variegated coffee bug)

Antestiopsis spp. Mainly *A. intricata* (Ghesquiere and Carayon) and sub-species of *A. orbitalis* (Westwood) [Hemiptera: Pentatomidae]

Morphology

The eggs are a dull white, about 1.2 × 1.0 mm and usually laid in groups of twelve. The adult bug has a flattened shield shape, is about 7–9 mm long (see Fig. 4.4) and dark brown in colour, with patches of orange, black and white. Colourings vary with race. Nymphs are similar to the adult in colour but are more circular in shape, lack wings and reach 3–4 mm in length (Greathead, 1966a; Wrigley, 1988).

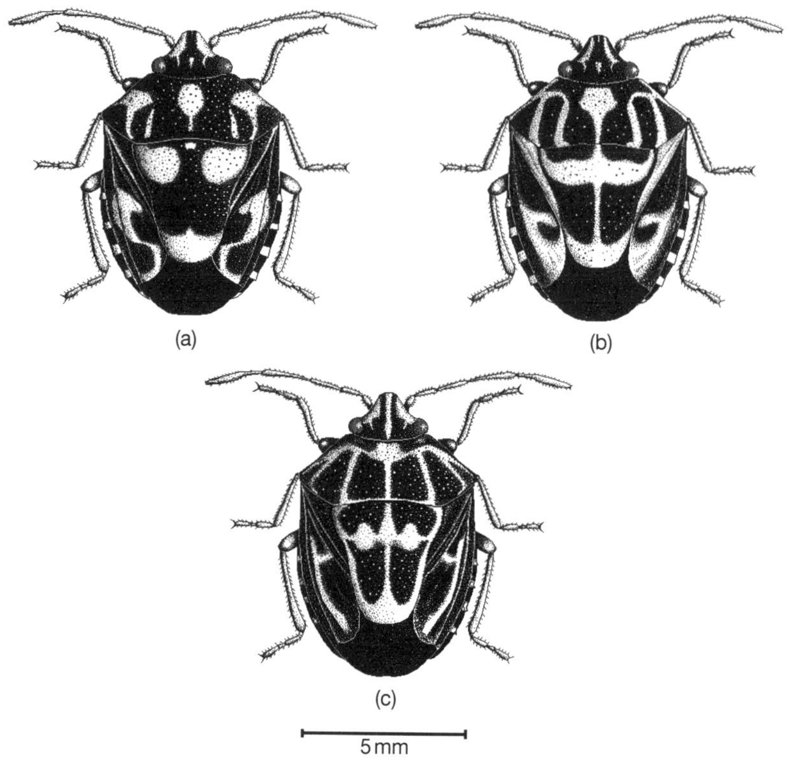

Fig. 4.4. Adults of the principal African *Antestiopsis* species: (a) *A. orbitalis bechuana*, (b) *A. orbitalis ghesquierei* and (c) *A. intricata.*

Pest status and distribution

Antestia bugs are a major pest of arabica coffee in all the main coffee-growing areas of Africa and minor pests in parts of Asia, but do not occur in the New World. The three main African species are: (i) *A. intricata* (Ghesquiere and Carayon), which has a generally western distribution (CABI, 1978a); (ii) the centrally located *A. orbitalis ghesquierei* Carayon; and (iii) *A. orbitalis bechuana* (Kirkcaldy) to the east (CABI, 1978b).

In addition, there are pockets of other species such as: (i) *A. facetoides* Greathead in the eastern parts of Ethiopia, Kenya and Tanzania; (ii) *A. clymeneis galtei* (Frappa) in Madagascar; and (iii) forms of *A. orbitalis orbitalis* in South Africa. Although *A. orbitalis* is distributed throughout South Africa, in many localities it does not occur on coffee (Greathead, 1966a, 1969) (see Fig. 4.5).

Apart from Africa, species of *Antestiopsis* and *Antestia* are found in India, Sri Lanka, Malaysia, Indonesia and Papua New Guinea. *Antestia partita* was considered to be a serious pest in Java at the beginning of the 20th century, but thereafter declined in importance (Kalshoven, 1950–51).

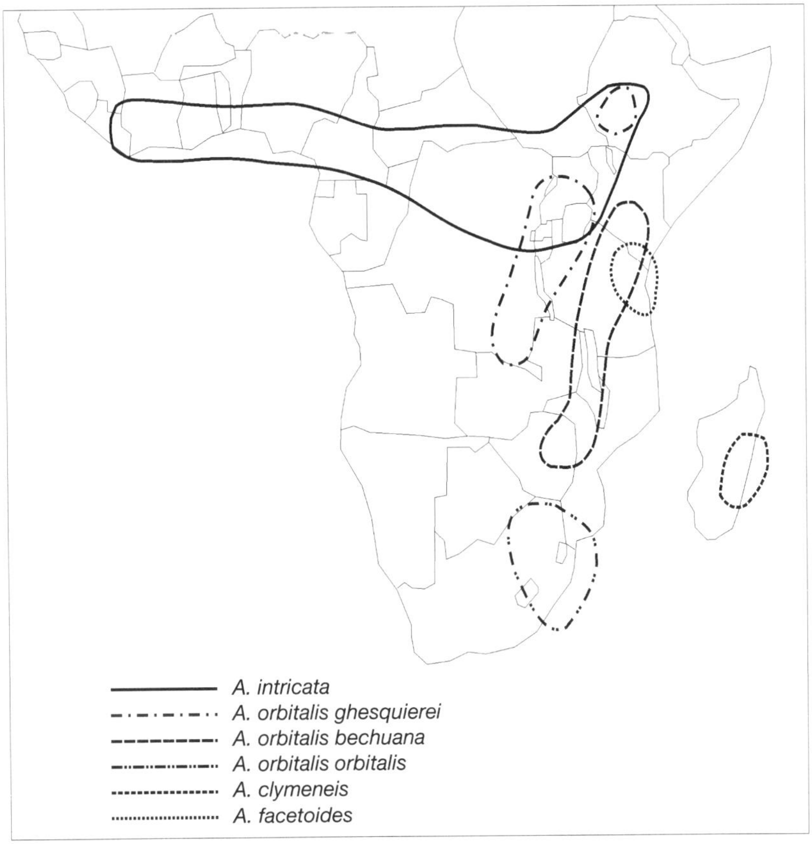

Fig. 4.5. Sketch map showing the approximate range of the main African species and subspecies of *Antestiopsis* (adapted from distribution maps in Greathead, 1966, 1969).

Records of *Antestiopsis* from coffee are as follows:

- *A. intricata*: Benin, Cameroon, Congo, Sierra Leone, Ethiopia, Gabon, Ghana, Côte d'Ivoire, Kenya, Nigeria, Sudan, Togo, Uganda, Democratic Republic of Congo
- *A. orbitalis bechuana*: Kenya, Malawi, Tanzania, Zambia, Zimbabwe
- *A. orbitalis ghesquierei*: Burundi, Ethiopia, Kenya, Rwanda, Tanzania, Uganda, Democratic Republic of Congo
- *A. orbitalis* (various forms): Mozambique, South Africa
- *A. crypta* Greathead: Democratic Republic of Congo
- *A. falsa* (Schouteden): Mozambique, South Africa
- *A. facetoides* Greathead: Ethiopia, Kenya, Tanzania, Zanzibar
- *A. clymeneis galtei* (Frappa): Madagascar
- *A. cruciata* (Fabricius): India, Malaysia, Sri Lanka
- *A. semiviridis* (Walker): Papua New Guinea
- *Antestia partita* (Walker): Indonesia Java and Sumatra.

Although encouraged by dense foliage on the coffee bush, Antestia prefers coffee grown without shade and at lower altitudes rather than the cooler, wetter conditions at higher altitude.

Damage

The adult bug prefers to feed on green berries and flower buds but, if necessary, it can also feed on green twigs. Feeding causes blackening of the flower buds and premature falling of berries. Large numbers of Antestia during the onset of the rains can greatly reduce the number of flowers that develop. The adult may also feed on mature berries, inserting its proboscis to suck out the sap. This causes little direct damage to the berry but may introduce two yeast-like fungi, *Nematospora coryli* and *N. gossypii*, and this fungal growth causes a sunken, discoloured patch on the berry surface.

Development of the fungus within the bean converts it to a soft, white paste, a condition known as 'posho beans' in Kenya, but the condition varies from complete rotting to slight black spotting. Such beans are of no value and the internal damage may not be seen until the berries are pulped. Antestia is very mobile and may probe several berries before feeding on one and, in this way, can quickly spread *Nematospora* to a large number of berries. After the crop has been harvested, Antestia bugs sometimes feed on the green shoots at the tips of the branches, causing shoot proliferation that gives a 'witches' broom' effect. If this is extensive, it can be difficult to restore normal branching, requiring several seasons of careful pruning (Le Pelley, 1968).

Host range

The preferred host is *Coffea arabica*, but both *A orbitalis ghesquierei* and *A. intricata* have been recorded from *Coffea canephora* and the latter from *C. liberica*, and doubtfully from excelsa (*C. liberica* var *dewevrei)*. Alternative host plants of *A. intricata* were identified by Taylor (1945) in Uganda, and a full list of the known host plants of all the African species is given by Greathead (1966a).

Life cycle

The duration of each stage of the life cycle varies considerably with temperature, being speeded up at high temperatures and slowed down at lower ones. Kirkpatrick (1937) found the development period of the egg to be 5–6 d at 24°C and 8–10 d at 19°C, while the total period of development from egg to adult varied from about 50 d at 24°C to 90 d at 19°C. The female does not begin to lay eggs until about 19 d after becoming an adult. Thereafter, it will lay about 160 eggs during its lifetime of about 3 months (Kirkpatrick, 1937; see Fig. 4.6).

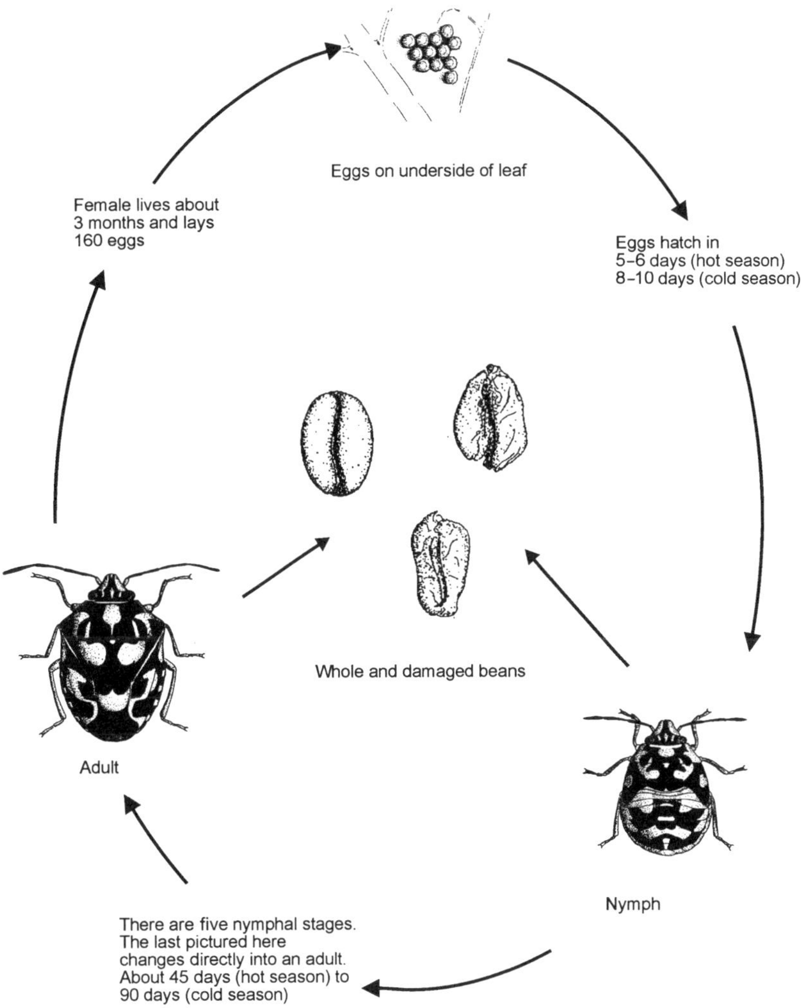

Fig. 4.6. Life cycle of *Antestiopsis orbitalis bechuana* in East Africa.

Natural enemies

Several species of hymenopterous parasites attack the eggs (see Appendix A for a full list) and, broadly speaking, these are common to the three main Antestia species (Greathead, 1996b). The Scelionids, *Telenomus seychellensis* Kieffer and *Gryon fulviventre* (Crawford), are the most important, (dealt with by Le Pelley, 1968 as *Asolcus seychellensis* and *Hadronotus antestiae*), causing between 63 and 90% mortality (Le Pelley, 1968).

A further set of parasites attack the adults and nymphs, including a Strepsipteran, *Corioxenos antestiae* Blair. This does not actually kill the adults, but renders them sterile. Some Reduviid predators of adults and nymphs of *A. intricata* have been recorded from West Africa (Carayon, 1954), but do not appear to have much effect as control agents.

Control

Insecticidal sprays should be avoided if possible in coffee gardens, but Antestia can cause losses at low population density. Two – or, in wetter areas, a single bug or nymph – per tree is the action threshold. With its long life and high fecundity, Antestia would be a much more serious pest if it were not for natural control by parasites.

However, natural control is sometimes insufficient to keep the pest population below the very low action threshold. If spraying becomes necessary, fenitrothion is the insecticide most commonly recommended. Pest levels may be kept below the action threshold, certainly in drier areas, by pruning the coffee bushes for an open canopy, as Antestia prefers dense foliage (Le Pelley, 1968).

Fruit Fly

Ceratitis capitata (Wiedemann), *Trirhithrum coffeae Bezzi*, *Anastrepha fraterculus* (Wiedemann) [Diptera: Tephritidae]

Morphology

Ceratitis capitata is a colourful fly, 5–6 mm long, with red eyes. The thorax is yellowish white with a number of black spots and patches, and with the abdomen predominantly yellow to orange brown. The wings are clear with brown patches.

The adult of *A. fraterculus* has a yellowish or orange brown thorax with paler longitudinal stripes laterally, and the transparent wings are patterned with tranverse brown stripes. *Trirhithrum coffeae* is dark brown in colour with a white face and irregular dark wing markings. The larvae of all species are typical legless white fly larvae, around 5 mm long when fully grown, with a cylindrical body tapering towards the head end and the extremity of the abdomen, where the spiracles are situated, flat. (see Fig. 4.7).

Pest status and distribution

A number of species of Tephritidae have been recorded from coffee in the genera *Anastrepha*, *Batrocera*, *Ceratitis* and *Trirhithrum*, but the most important are: (i) *Anastrepha fraterculus*, found throughout South and Central America; (ii) *Trirhithrum coffeae*, which is distributed across sub-Saharan Africa; and (iii)

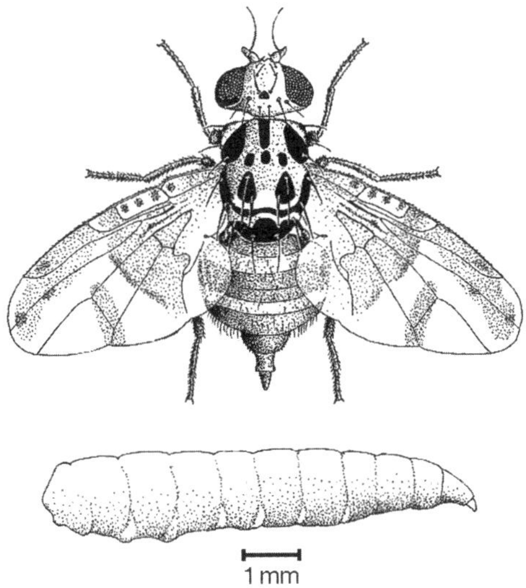

Fig. 4.7. Adult and larva of *Ceratitis capitata.*

Ceratitis capitata, the Mediterranean fruit fly, which is now distributed widely around the Mediterranean, Africa and Central and South America (CABI, 1984).

The latter was introduced into Hawaii around 1910 and Brazil about 1923, and has since spread through many South and Central American countries, reaching Costa Rica in 1955, El Salvador in 1975 and Mexico in 1977. Species of *Anastrepha* are difficult to distinguish and it is very likely that misidentifications have occurred in the past, so they have not been specified in the list below.

Records of the species found on coffee are as follows:

Anastrepha spp.

- Central America and Caribbean: Panama, Trinidad and Tobago, Cuba
- South America: Argentina, Brazil, Colombia, Venezuela

Ceratitis capitata

- Central America: Costa Rica, Guatemala, Mexico, Nicaragua
- South America: Argentina, Bolivia, Brazil, Colombia, Peru, Venezuela
- Atlantic Ocean: Canary Islands, Madeira, São Tomé and Príncipe
- West Africa: Cameroon, Togo
- Central Africa: Democratic Republic of Congo
- Eastern Africa: Ethiopia, Kenya, Tanzania, Uganda
- Southern Africa: South Africa, Zimbabwe
- Indian Ocean: Réunion, Yemen
- Pacific Ocean: Hawaii

Trirhithrum coffeae

- West Africa: Cameroon, Ghana, Nigeria, Sierra Leone, Togo
- Central Africa: Democratic Republic of Congo
- Eastern Africa: Ethiopia, Kenya, Tanzania, Uganda.

The larvae of fruit flies feed in the pulp of the ripe berries and do not directly damage the beans within. There is some controversy as to whether economic damage is caused to coffee by their feeding. In Brazil, Souza *et al.* (2005) considered that fruit flies cause premature dropping of fruit and decrease coffee quality, and in Colombia it was found that there was detriment to quality if attack began when the fruits were first ripening (Portilla *et al.*, 1995).

On the other hand, Abasa (1973) could not induce off-flavour in coffee liquor by exposing berries to *C. capitata* attack in Kenya, and premature fruit-fall as a result of this exposure was only 2.8%. Le Pelley (1968) quotes work by Stolp in the Democratic Republic of Congo in which he found off-flavour in Arabica coffee to be due to a bacterium that he thought was gaining entry to the cherry during oviposition by the fruit flies. There is no doubt, however, that the primary importance of coffee as a fruit fly host is because it provides a reservoir from which citrus and many other tropical and sub-tropical fruits are infested.

Life cycle

The eggs of *C. capitata* are laid within the pulp of the cherry and hatch in 2–4 d. The larvae feed on the pulp, leaving the epidermis and the beans intact for 6–11 d. When mature, they leave the cherries and drop to the ground, where they pupate under the soil. The pupal stage lasts a further 6–11 d.

Natural enemies

Because of the importance of *C. capitata* as a fruit pest in Hawaii, parasites have been sought for many years and, consequently, the parasites of African fruit flies have been well studied. *Diachasmimorpha tryoni* (Cameron) and *Dirhinus giffardii* Silvestri were introduced into Hawaii as early as 1913, and *Diachasmimorpha fullawayi* (Silvestri) in 1914. For a full list of parasites recorded from coffee, see Appendix A.

References

Abasa, R.O. (1973) Observations on the seasonal emergence of fruit flies on a Kenya coffee estate and studies of the pest status of *Ceratitis capitata* Wied. in coffee. *East African Agricultural and Forestry Journal* 39, 144–148.

Anon (1965) New country records in the region. *Quarterly Newsletter of the Plant Protection Commission for S.E. Asia* 8, 6–8.

Anon (1967) *Coffee Pests and their Control*. Coffee Research Foundation, Kenya, 90 pp.
Anon (1992) Control of Berry Moth, *Prophantis smaragdina* (Butler), in coffee. Coffee Research Foundation Technical Circular No 75. *Kenya Coffee* 57, 1427–1428.
Ba-Angood, S. and Al-Sunaidi, S.M.A. (2004) Assessment of damage caused by the coffee berry moth, *Prophantis smaragdina*, in different flowering and fruit setting dates in Yafe'a area, Republic of Yemen. *Yemeni Scientific Research Foundation, Science Conference Sana'a*, 11–13 October 2004.
Baker, P.S. (1984) Some aspects of the behaviour of the coffee borer in relation to its control in southern Mexico (Coleoptera, Scolytidae). *Folia Entomologia Mexicana* 61, 9–24.
Baker, P.S. (1999) The coffee berry borer in Colombia. *Final Report of the DFID–Cenicafe–CABI BioScience IPM for Coffee Project*. CABI BioScience, Ascot, UK, 144 pp.
Balakrishnan, M.M., Abjayan, K.J. and Sreedharan, K. (2001) Comparative efficacy of endosulphan 35 EC and chlorpyrifos 20 EC against coffee berry borer. *Journal of Coffee Research*, 29, 78–80.
Barrera, J.F. (1994) *Dynamique des populations du scolyte des fruits du caféier,* Hypothenemus hampei *et lutte biologique avec le parasitoide* Cephalonomia stephanoderis *au Chiapas, Mexique*. PhD Thesis, University Paul-Sabatier, Toulouse, France, 301 pp.
Benavides, M.P., Bustillo, P.A.E., Montoya, R.E.C., Cárdenas, M.R. and Megia, M.C.G. (2002) Participation of cultural, chemical and biological control in the management of coffee berry borer. *Revista Colombiana de Entomologia* 28, 161–165.
Bredo, H.J. (1939) Catalogue des principaux insectes et nematodes parasites des caféiers au Congo Belge. *Bulletin Agricole du Congo Belge* 30, 266–307.
Brun, L.O. and Suckling, D. (1992) Field selection for endosulphan resistance in coffee berry borer in New Caledonia. *Journal of Economic Entomology* 85, 325–334.
Brun, L.O., Marcillaud, C., Gausichon, V. and Suckling, D. (1989) Endosulphan resistance in *Hypothenemus hampei* in New Caledonia. *Journal of Economic Entomology* 82, 1311–1316.
CABI (1978a) *Antestiopsis intricata* (Ghesq. and Carayon). *Distribution Maps of Pests* No. 381, CAB International, Wallingford, UK, 2 pp.
CABI (1978b) *Antestiopsis orbitalis* (Westwood). *Distribution Maps of Pests* No. 382, CAB International, Wallingford, UK, 2 pp.
CABI (1981) *Hypothenemus hampei* (Ferrari). *Distribution Maps of Pests* No. 170 (revised), CAB International, Wallingford, UK, 2 pp.
CABI (1984) *Ceratitis capitata* (Westwood). *Distribution Maps of Pests* No. 1 (revised), CAB International, Wallingford, UK, 2 pp.
CABI (2003) *Crop Protection Compendium 2003 Edition*. CAB International, Wallingford, UK.
Carayon, J. (1954) Les *Antestiopsis* du Caféier en Afrique tropicale francaise. *Bulletin Scientifique. Ministere des Colonies de la France d'Outre Mer. Section Technique d'Agriculture Tropicale* 5, 363–373.
Cárdenas, M.R. and Posada, F.J. (2001) *Los Insectos y Otres Habitantes de Cafetales y Platanales*. Centro Nacional de Investigaciones de Café, Colombia, 250 pp.
Castillo, A. and Marban-Mendoza, N. (1996) Laboratory evaluation of steinernematid and heterorhabidid nematodes for biological control of the coffee berry borer. *Nematropica* 26, 101–109.
Chevalier, A. (1947) Les caféiers du globe fasc 3, systematique des caféiers et faux-caféiers maladies et insectes nuisibles. *Encyclopédie Biologique* 28, 356 pp.
Cohic, H. (1958) Le 'scolyte du grain de café' en Nouvelle-Calédonie. *Café, Cacao, Thé* 2, 10–14.
Corbett, G.H. (1933) Some preliminary observations on the coffee berry beetle borer *Stephanoderes (Cryphalus) hampei* Ferr. *Malayan Agricultural Journal* 21, 8–22.
CRF (1978) *An Atlas of Coffee Pests and Diseases*. Coffee Board of Kenya, Nairobi, 150 pp.

Da Costa Lima, A.M. (1924) Sobre a broca do café (*Stephanoderes coffeae*, Hag.). *Chacaras e Quintaes* 30, 413–416.

Damon, A. (2000) A review of the biology and control of the coffee berry borer, *Hypothenemus hampei* (Coleoptera:Scolytidae). *Bulletin of Entomological Research* 90, 453–465.

Damon, A. and Valle, J. (2002) Comparison of two release techniques for the use of *Cephalonomia stephanoderis* to control the coffee berry borer in Soconusco, southern Mexico. *Biological Control* 24, 117–127.

De Campos Novaes, J. (1922) Um broqueador do cafeeiro *Xyleborus coffeicola* n.sp. Fam. Ipidae. *Boletino de Agricultura Secretaria de Agricultura, Comercio e Obras Publicas, Sao Paulo* 23, 67–70.

Decazy, B. (1991) Investigaciones conducidas en Francia y en Africa sobre la broca del café. *Proceedings of the First Intercontinental Workshop on Coffee Berry Borer, Tapachula, Mexico*, pp. 47–49.

Derron, M. (1977) *Prophantis smaragdina* Butler and *Cryptophlebia colivora* Meyrick, two important pests on *Coffea arabica* L. on the island of São Tomé. *Mitteilungen der Schweizerischen Entomologischen Gesellschaft* 50, 149–151.

De Saeger, H. (1943) Quelques braconides nouveaux du genre Microbracon. *Revue de Zoologie et de Botanique Africaines* 36, 361–389.

Descroix, F. (2004) Une filière café gourmet à la Réunion: intérêt et viabilité. *Rapport Annuel CIRAD Réunion 2004*, 91–96.

Diaz, P.P., Gomez, Y.A., de Polania, I.Z. and Vargas, A. (2003) Evaluation of strains of *Fusarium* sp. for the management of the coffee berry borer, *Hypothenemus hampei*. *Revista Colombiana de Entomologia* 29, 71–76.

Dufour, B.P., Picasso, C. and Gonzales, M.O. (2001) Contribution au développement d'un piège pour capturer le scolyte du café *Hypothenemus hampei* Ferr. en El Salvador. *Proceedings of the 19th International Scientific Colloquium on Coffee*, Trieste, Italy, 14–18 May 2001. International Scientific Association of Coffee (ASIC), pp. 1–12.

Edgington, S., Segura, H., Rosa, W. de la and Williams, T. (2000) Photoprotection of *Beauveria bassiana*: testing simple formulations for control of coffee berry borer. *International Journal of Pest Management* 46, 169 –176.

Ghesquiere, J. (1933) Rôle des Ipides dans la destruction des végétaux au Congo Belge. *Annals de Gembloux* 39, 24–37.

Greathead, D.J. (1966a) A taxomomic study of the species of Antestiopsis associated with *Coffea arabica* in Africa. *Bulletin of Entomological Research* 56, 515–554.

Greathead, D.J. (1966b) The parasites of *Antestiopsis* spp. in East Africa and a discussion of the possibilities of biological control. *Technical Bulletin of the Commonwealth Institute of Biological Control* 7, 113–137.

Greathead, D.J. (1969) On the taxonomy of *Antestiopsis* spp. (Hem., Pentatomidae) of Madagascar, with notes on their biology. *Bulletin of Entomological Research* 59, 307–315.

Gutierrez, A.P., Villacorta, A., Cure, J.R. and Ellis, C.K. (1998) Tritrophic analysis of the coffee berry borer parasitoid system. *Annals of the Entomological Society of Brazil* 27, 357–385.

Haraprasad, N., Niranjana, S.R., Prakash, H.S., Shetty, H.S. and Wahab, S. (2001) *Beauveria bassiana*: a potential mycoparasite for the efficient control of coffee berry borer in India. *Biocontrol Science and Technology* 11, 251–260.

Hargreaves, E. (1937) Entomological work. *Report of the Department of Agriculture, Sierra Leone* 1937, 39–43.

Hargreaves, H. (1935) *Stephanoderes hampei* Ferr., coffee berry borer in Uganda. *East African Agriculture Journal* 1, 218–224.

Hernandez Paz, M. and Penagos Dardon, H. (1974) Evaluacion del sistina de aplicacion de bajo volumen en el control de la broca del fruto del café, *Hypothenemus hampei* (Ferrari). *Revista Cafetalera* 134, 15–21.

Hill, D.S. (1975) *Agricultural Insect Pests of the Tropics and their Control*. Cambridge University Press, London, 516 pp.

Hopkins, A.D. (1915) *Classification of the Cryphalinae with Descriptions of New Genera and Species*. United States Department of Agriculture Report 99, 75 pp.

Hutson, J.C. (1936) The coffee berry borer in Ceylon (*Stephanoderes hampei* Ferr.). *Tropical Agriculturist* 87, 378–383.

Infante, F., Mumford, J. and Garcia, B.A. (2003) Predation by native arthropods on the African parasitoid *Prorops nasuta* (Hymenoptera: Bethylidae) in coffee plantations in Mexico. *Florida Entomologist* 86, 86–88.

Jarquin-Galvez, R., Barrera-Gaytan, J.F., Nelson, K.C. and Martinez-Quezada, A. (1999) Metodos no quimicos contra la broca del café y su transferencia tecnologica en los altos de Chiapas, Mexico. *Agrociencia* 33, 431–438.

Johnston, A. (1963) *Stephanoderes hampei* in Tahiti. *Information Letter FAO Plant Protection Commission for S.E. Asia* No. 23, 4 pp.

Kaiser, C. (2005) Facultative or competitive interactions for pollination between an invasive and a native species: an experimental approach. *Abstracts of the Annual Meeting of the Swiss Zoological, Botanical and Mycological Societies*, 24–25 February 2005.

Kalshoven, L.G.E. (1950–1951) *De Plagen van de Cultuurgewassen in Indonesië*. Van Hoeve, Gravenhage (Netherlands)/Bandoeng (Indonesia), 1065 pp.

Kirkpatrick, T.W. (1937) Studies on the ecology of coffee plantations in East Africa. II. The autecology of *Antestia* spp. (Pentatoimidea) with a particular account of a Strepsipterous parasite. *Transactions of the Royal Entomological Society, London* 86, 247–343.

LaSalle, J. (1990) A new genus and species of *Tetrastichinae* (Hymenoptera: Eulophidae) parasitic on the coffee berry borer, *Hypothenemus hampei*. *Bulletin of Entomological Research* 80, 7–10.

Le Pelley, R.H. (1959) *Agricultural Insects of East Africa*. East African High Commission, Nairobi, 307 pp.

Le Pelley, R.H. (1968) *Pests of Coffee*. Longman, London, 415 pp.

Montoya, S.A. and Cárdenas-Murillo, R. (1994) Biologia de *Hypothenemus hampei* (Ferrari) en frutos de café de diferentes edades. *Cenicafé* 45, 5–13.

Morales-Ramos, J.A., Rojas, M.G., Sittertz-Bhaktar, H. and Saldana, G. (2000) Symbiotic relationship between *Hypothenemus hampei* (Coleoptera: Scolytidae) and *Fusarium solani* (Moniliales: Tuberculariaceae) *Annals of the Entomological Society of America* 93, 541–547.

Morallo-Rejesus, B. and Baldos, E. (1980) The biology of the coffee borer *Hypothenemus hampei* (Ferr.) (Scolytidae, Coleoptera). *Philippine Entomologist* 4, 303–316.

Morstatt, H.A. (1914) Kaffeekultur, kaffeeschädlinge und andere schädliche insekten im Bezirk Bukoba. *Der Pflanzer* 10, 133–141.

Morstatt, H.A. (1941) Über herkunft und verbreitung afrikanischer schädlinge. *Zeitschrift für Pflanzenkrankheiten. Pflanzenpathologie und Pflanzenschutz* 51, 209–217.

Murphy, S.T. and Rangi, D.K. (1991) The use of the African wasp, *Prorops nasuta* for the control of the coffee berry borer, *Hypothenemus hampei* in Mexico and Ecuador. *Insect Science and its Application* 12, 27–34.

Nagarajaiah, N.V. and Kumar, R. (2003) Position and management of coffee berry borer in South Kogoga. *Indian Coffee* 67, 10.

Ndungi, A.W. (1994) New natural enemies of the coffee berry moth, *Prophantis smaragdina* (Butler). *Kenya Coffee* 59, 1725–1726.

Perez-Lachaud, G. and Hardy, I.C.W. (1999) Reproductive biology of *Cephalonomia hyalinipennis* (Hymenoptera:Bethylidae), a native parasitoid of the coffee berry borer, *Hypothenemus hampei* in Chiapas, Mexico. *Biological Control* 14, 152–158.

Perez-Lachaud, G., Hardy, I.C.W., Lachaud, J.P. and Johnson, M.W. (2002). Insect gladiators: comparative interactions between three species of bethylid wasps attacking coffee berry borer. *Biological Control* 25, 231–238.

Portilla, M., González, G.G. and Nuñez-Bueno, L. (1995) Evaluación y descripción del dano ocasionado por *Ceratitis capitata* al café. *Revista Colombiana de Entomologia* 21, 15–24.
Rahman, P.A. and Vijayalakshmi, C.K. (1999) Evaluation of low-volume and high-volume application of endosulphan for the control of coffee berry borer in India. *Insects and Environment* 5, 120–121.
Reddy-Samba Murthy, A.G. and Rao, L.V.A. (1999) Incidence of coffee berry borer in non-conventional coffee area of Karnataka. *Indian Coffee* 63, 15–16.
Reid, J.E. (1983) Distribution of the coffee berry borer (*Hypothenemus hampei*) within Jamaica, following its discovery in 1978. *Tropical Pest Management* 29, 224–230.
Salazar, E.H., Aristizabal, A.L. and Mejia, C.G. (2003) Participatory research with coffee growers on post-harvest management of coffee berry borer. *Revista Colombiana de Entomologia* 29, 57–62.
Schedl, K.E. (1961) Scolytidae und Platypodidae Afrikas. Band I (Part). *Revista Entomologia de Mocambique* 4, 335–742.
Souza, S.A.S., Resende, A.L.S., Strikis, P.C., Costa, J.R., Ricci, M.S.F. and Aguiar-Menezes, E.L. (2005) Infestação natural de moscas frugivoras (Diptera: Tephritoidea) em café arábica, sob cultivo orgânico arborizado e a pleno sol, em Valença, RJ. *Neotropical Entomology* 34, 639–648.
Tapley, R.G. and Materu, M.E.A. (1961) Microbial control of lepidopteran coffee pests. *Kenya Coffee* 26, 213.
Taylor, T.C.H. (1945) Recent investigations of *Antestia* species in Uganda. *East African Agricultural Journal* 10, 223–233; 11, 47–55.
Thomas, R.T.S. (1961) Coffee pests in Netherlands New Guinea. *Bulletin of Economic Affairs Netherlands New Guinea (Agricultural Series)* 1961, 19–24.
Ticheler, J.H.G. (1961) An anlaytical study of the epidemiology of the coffee berry borer in Côte d'Ivoire. *Mededeelingen Landbouwhoogeschool, Wageningen* 61, 1–49.
Tovigan, S., Vodouhe, S.D. and Dinham, B. (2001) Cotton pesticides cause more deaths in Benin. *Pesticides News* 52, 12–14.
Vega, F.E., Franqui, R.A. and Benavides, P. (2002) The presence of the coffee berry borer in Puerto Rico: fact or fiction? *Journal of Insect Science, Tucson* 2, 1–3.
Velez-Arango, P.E., Estrada, V.M.N., Gonzales, G.M.T., Valderama, F.A.M. and Bustillo, P.A.E. (2001) Characterization of *Beauvaria bassiana* isolates for the control of coffee berry borer. *Manejo Integradas de Plantas* 62, 38–53.
Vijayalakshmi, C.K., Rahiman, P.A., Vijayan, V.A., Prakasan, C.B. and Sreedharan, K. (2002) Biology of the coffee berry borer, *Hypothenemus hampei* (Ferrari) under varying thermal conditions. *Journal of Entomological Research* 26, 67–70.
Waterhouse, D.F. and Norris, K.R. (1989) *Biological Control, Pacific Prospects*, Hypothenemus hampei (*Ferrari*). Australian Centre for Agricultural Research, Canberra, Suppl. 1, pp. 56–75.
Wilkinson, H. (1928) *The Coffee Berry Borer Beetle, Stephanoderes hampei* (Ferr.). Colony and Protectorate of Kenya, 10 pp.
Wrigley, G. (1988) *Coffee*. Longman (Tropical Agriculture Series), Harlow, UK, 639 pp.

5 Insects that Feed on Buds, Leaves, Green Shoots and Flowers

Introduction

A great many insects are to be found feeding on the leaves of coffee plants, including a number of serious pests. Additionally, many of these species may be found on green shoots and immature cherries.

A database of insects and mites recorded from coffee maintained by one of the authors contains records of 623 species attacking leaves, made up as follows:

Lepidoptera (Butterflies and moths)	218
Hemiptera (Coccids, aphids, leafhoppers etc.)	170
Coleoptera (Beetles)	161
Thysanoptera (Thrips)	26
Orthoptera (Grasshoppers)	16
Hymenoptera (Ants)	14
Acari (Mites)	14
Diptera (Flies)	4

Lepidopterous Caterpillars

Leaf miner

Leucoptera meyricki Ghesquiere, *L. caffeina* Washbourn, *L. coma* Ghesquiere, *Perileucoptera coffeella* (Guerin-Meneville) [Lepidoptera: Lyonetiidae]

Morphology

The adults are small, white moths with a wingspan of around 6–8 mm. The forewings have a brown fringe at their outer margin and dark spots near the hinder end, which are less pronounced in the case of *L. caffeina* (see Fig. 5.1). The larvae feed together within the palisade layer of the leaf, leaving the upper epidermis as a roof over the mine and producing a typical brown 'blotch' mine. They are around 5–6 mm long when fully grown, flattened and yellowish white in colour.

The damage caused is indirect, for although the mine itself reduces the photosynthetic efficiency of the leaf, the main economic loss is due to the leaf being shed prematurely. The white eggs, barely visible to the naked eye, are laid on the upper surface of the leaf. They are oval in shape, with a basal rim, steep sides and a convex dimple in the top. The yellowish white pupa is found beneath an X-shaped silk cocoon.

Pest status and distribution

Apart from a few species of minor importance, the principal leaf miners of economic importance are four closely related species belonging to the family Lyonetiidae. Because of their close superficial resemblance, taxonomy of these species has been confused in the past. The first species to be described from coffee was *Leucoptera coffeella* (Guerin-Meneville), which was transferred to the genus *Perileucoptera* by Silvestri in 1943 (Silvestri, 1943). This species is

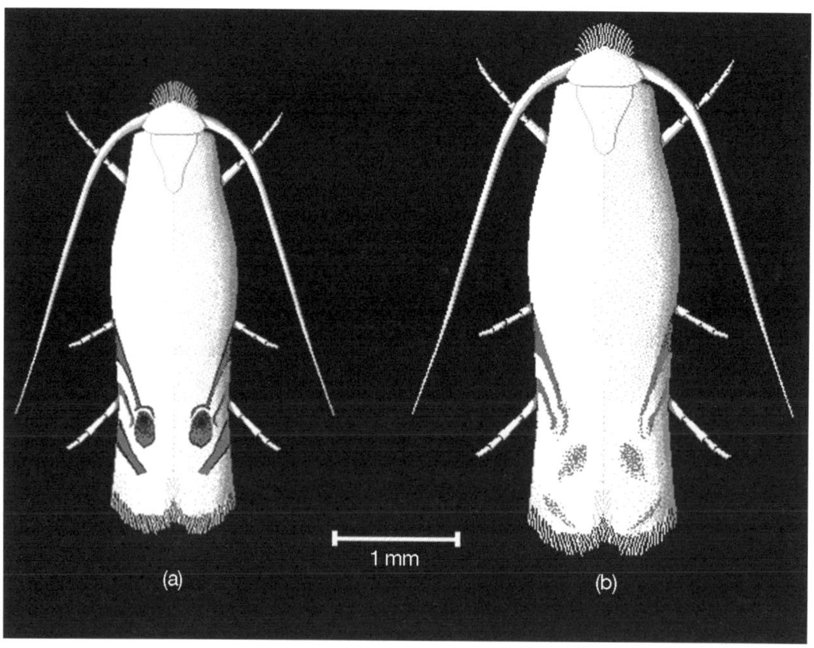

Fig. 5.1. Adults of (a) *Leucoptera meyricki*, (b) *L. caffeina*.

the coffee leaf miner of Central and South America and the Caribbean islands (CABI, 1973a).

Many authors do not recognize this change, including Bradley (1958), who distinguished the African *Leucoptera meyricki* Ghesquiere from it but considered the two species to be congeneric. The names *Perileucoptera coffeella* and *Leucoptera coffeella* have been freely used since.

In the Old World tropics, three species occur: (i) *Leucoptera meyricki* Ghesquiere, recorded from the Central African Republic, Ethiopia, Kenya, Tanzania, Uganda, Malawi and Zimbabwe (CABI, 1973b); (ii) *Leucoptera caffeina* Washbourn, from Angola, Democratic Republic of Congo, Ethiopia, Kenya, Tanzania, South Africa and Yemen; and (iii) *Leucoptera coma* Ghesquiere, from Tanzania, Uganda and Democratic Republic of Congo. The leaf miner recorded from the islands of Réunion and Madagascar is said to be *Perileucoptera coffeella*, but this may need confirmation.

Life cycle and damage

The life cycles of the various species are broadly similar (see Fig. 5.2). Females of *L. caffeina* lay their eggs in a closely grouped line along a vein, whilst those of *L. meyricki* and *P. coffeella* are scattered about the surface, but in loose groups, and are invariably on the upper leaf surface. *Leucoptera meyricki* prefers leaves of intermediate age for oviposition, the maximum number being found on leaves three to four nodes from the tip of the branch. The hatching larvae chew their way through the base of the egg and into the leaf tissue without appearing at the surface.

The duration of the egg and larval stages is dependent upon temperature. *Leucoptera meyricki* eggs hatch in 6 d at 28°C or 18 d at 16°C, whilst the larval life is passed in 11 d at 28°C or 24 d at 20°C, and the pupal stage, 5 d at 28°C or 8 d at 20°C (Notley, 1948, 1956; Bigger, 1967; Bigger and Tapley, 1969). There are four larval stages.

Tapley (1961), reported a natural mortality of eggs and first instar larvae of *L. meyricki* and *L. caffeina* for which he could find no explanation and which had previously been noticed by Notley (1956), and with *P. coffeella* by Jardine *et al.* (1941) in Puerto Rico. Bigger (1969a) found that the age of the leaf was important to the survival of larvae of *L. meyricki*, mortality being highest on the youngest leaves on a branch and decreasing with each successive node. Size of moths bred from these leaves also increased with distance from the tip of the branch, as did the fecundity of females. No causal factor for this antibiotic effect could be found.

An investigation of possible resistance to leaf miner attack in different coffee species and varieties proved inconclusive (Bigger, unpublished report). Once fully grown, the larva escapes through a slit bitten through the roof of the mine and proceeds to pupate under a silk shelter, either on the underside of a buckled leaf or in the leaf litter on the ground.

Moths of *L. meyricki* emerge early in the morning, but do not fly in search of a mate until the afternoon of the same day whilst egg-laying takes place about two nights later (Notley, 1948).

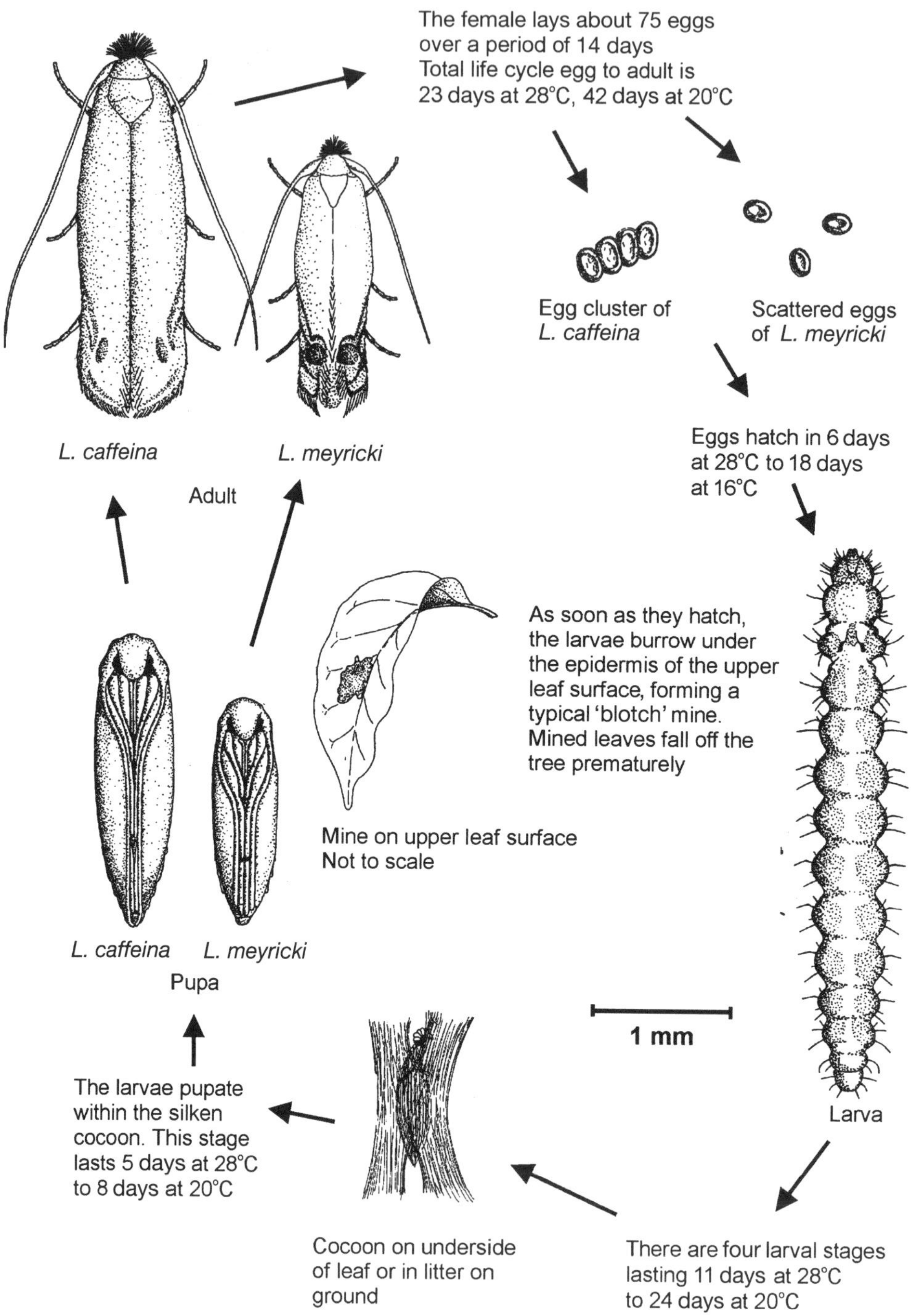

Fig. 5.2. The life cycle of coffee leaf miner (*Leucoptera* spp.) in East Africa.

An exhaustive study of leaf loss from arabica coffee caused by *L. meyricki* was made by Tapley (1965) in Tanzania. He found that there was an important relationship between the loss of leaf due to leaf miner attack and the loss of leaf due to rust (*Hemileia vastatrix*). By controlling leaf miners only, the average life of a leaf could be increased from 7 up to 8–9 months. By controlling rust as well as leaf miner, a leaf life of about 11–13 months could be obtained. He found that the extra life imparted to the leaf by the control of leaf miner did not increase yield, but the large increase which resulted from copper spraying could be further augmented by leaf miner control. The increase in leaf retention due to rust control also led to an increase in the incidence of leaf miner. His conclusion was that chemical control of leaf miner was uneconomic unless rust control was also employed.

Host range

Leucoptera meyricki and *L. caffeina* are pests chiefly of *C. arabica*, but *C. canephora* and other coffee species are also attacked. *Leucoptera meyricki* has been recorded from *Mussaenda arcuata* and *L. caffeina* from *Pavetta oliveriana*, *Pavetta ternifolia* and *Oxyanthus speciosus* (Ritchie, 1936). *Leocoptera coma* seems to be confined to *C. canephora*, whilst *P. coffeella* has been reported from a number of *Coffea* species, but no wild hosts are known for either.

Natural enemies

All species are attacked by a rich complex of larval and pupal hymenopterous parasites, many of which are themselves attacked by hyperparasites. These, generally speaking, keep the population size of the leaf miners well in check, but the complex dynamics of the interactions between the different species leads to unstable cycling in density, so that individuals of the leaf miner population tend to be in the same development stage at the same time, and thus distinct 'generation' peaks and troughs develop.

There is also an overall annual cycle in population density, with peak numbers occurring at the warmest and sunniest times of year and minimum numbers during wetter, cooler months. The rate of build-up of leaf miner outbreaks from these minimum levels is strongly dependent on the numbers of leaf miners and parasites surviving through the cool season. Where the population surviving the cool period is small, the subsequent build-up is rapid, and vice versa, due to the better survival of the parasite population (Bess, 1964; Bigger and Tapley, 1969; Bigger, 1973).

The very complicated relationships between parasite species gathered from studies of *L. meyricki* in East Africa by several workers are shown in Fig. 5.3 and summarized by Crowe and Greathead (1970).

There are no published records of predators of leaf miners from Africa, except for a report of spiders catching adult *L. coma* (Decelle, 1962), but from South America there are extensive records of predators of *P. coffeella*. These include mites (*Calvolia*, *Pyemotes*, *Tyrophagus*), lacewings (*Chrysopa*, *Chrysoperla*) and wasps (*Apoica*, *Brachygastra*, *Polistes*, *Polybia*, *Protonectarina*, *Protopolybia*). See Appendix A for a complete list.

Control

It has already been noted that leaf miner populations are, for the most part, kept in check by their natural enemies, but there are occasions where chemical control may be required. For many years, foliar sprays of organophosphorus insecticides were employed in East Africa for the control of leaf miner and Antestia, including fenitrothion and dicrotophos, but in the late 1980s insecticide resistance to these compounds was detected in northern Tanzania (Bardner and Mcharo, 1988). Similar resistance to disulfotan, ethion, methyl parathion and chloropyrifos by *P. coffeella* has been demonstrated in Brazil (Fragoso *et al.*, 2002).

An alternative to the use of foliar sprays is to use systemic insecticides applied to the soil. Disulfotan granules are registered for use against leaf miner

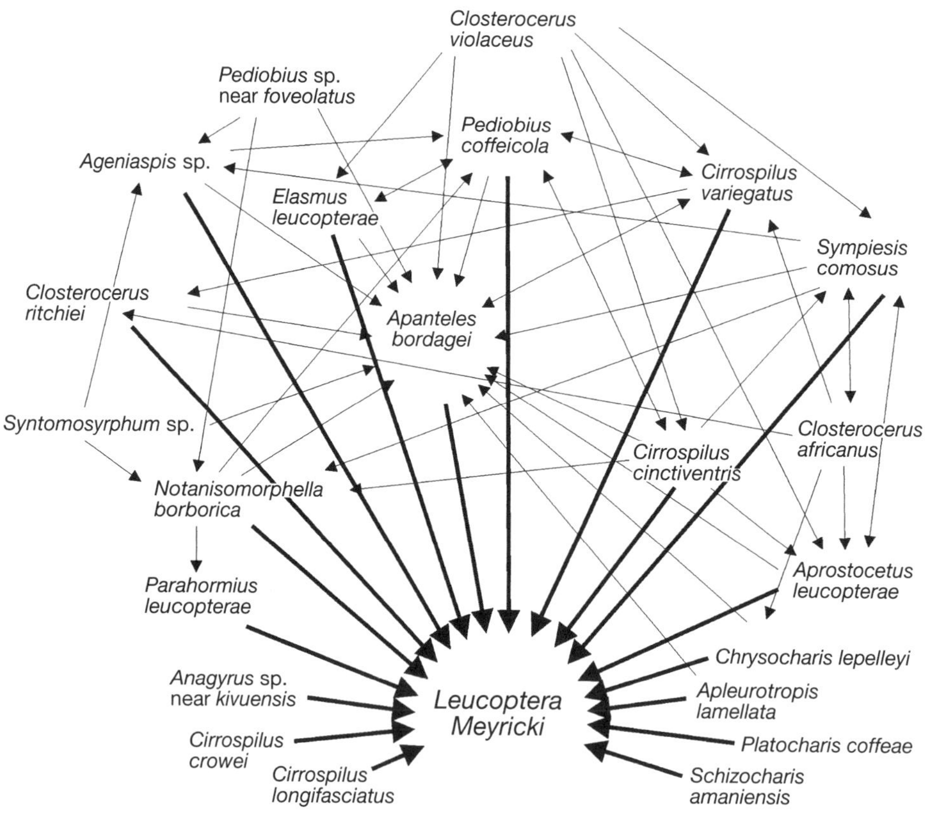

Fig. 5.3. The complex interactions between the hymenopterous parasites of *Leucoptera meyricki* in East Africa, showing primary parasites (thick lines) and hyperparasites (thin lines). Note that in some cases the parasitic species attack each other (e.g. *Apanteles bordagei* and *Cirrospilus variegatus*).

in Puerto Rico and in Kenya. Carbofuran granules have been recommended in Brazil and Kenya.

Guerreiro Filho *et al.* (1993) evaluated the resistance to *P. coffeella* of a range of 11 Madagascan *Coffea* species and *C. arabica* by rearing larvae in leaf discs maintained at a temperature of 30°C. Larval development on *C. arabica* was completed in 8.6 d, with a 91.7% survival rate. On *C. tetragona*, larval stadia lasted 21.3 d, whilst on *C. farafanganensis* and *C. resinosa* all larvae died.

A programme for the the genetic transformation of coffee for resistance to leaf miner by the introduction of the *Bacillus thuringiensis* gene was initiated in France by CIRAD in 1994 (Guerreiro Filho *et al.*, 1999). Field trials were initiated in French Guiana (Perthuis *et al.*, 2001), which unfortunately were destroyed by anti-GM protestors. It has been pointed out that the use of genetically modified coffee in the Old World might be unsafe due to the possibility of cross-pollination to wild *Coffea* species, which do not of course occur in the New World. Similar studies of Bt-modified coffee for control of *P. coffeella* are being carried out in Brazil.

Giant looper caterpillar

Ascotis selenaria reciprocaria Walker. [Lepidoptera: Geometridae]

Morphology

The adult is a white moth with brown markings and a wingspan of 4–5 cm. The larvae, which are up to 5 cm long when fully grown, have three pairs of legs near the head and two pairs of sucker feet near the rear end, and move in a characteristic looping fashion. At rest they extend their body between the stem and a leaf and can easily be mistaken for a dead twig (see Fig. 5.4).

Pest status and distribution

Ascotis selenaria is widely distributed through Europe, Africa and Asia and occurs in a number of subspecies. *A. selenaria reciprocaria* is distributed throughout much of sub-Saharan Africa and has been recorded on coffee from Ethiopia, Kenya, Tanzania, Uganda and Zimbabwe.

Wheatley (1964), showed that the build-up of *A. selenaria* as a pest of coffee in Kenya could be attributed to the widespread use of parathion for the control of other pests such as leaf miners, and a similar increase between 1963 and 1968 in Tanzania was demonstrated by Bigger (1969b).

Alternative host plants

The species is polyphagous, and the various subspecies have been recorded from a range of agricultural crops, as well as from many forest trees. Important

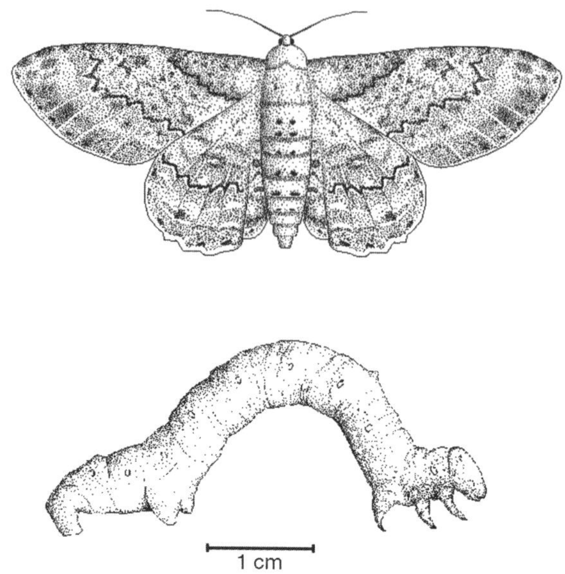

Fig. 5.4. Adult and larva of *Ascotis selenaria reciprocaria*.

horticultural host plants include cocoa, tea, avocado, citrus, apple, pear and mulberry, whilst field crops include cotton, asparagus, soya, groundnuts and lucerne.

Life cycle

Eggs are laid in bark crevices and hatch in 7–10 d. There are five larval stages lasting around 4–6 weeks, after which the larva leaves the tree and pupates in the soil, emerging as an adult about 10–14 d later (Wheatley, 1964).

Natural enemies

Larvae are attacked by several hymenopterous and dipterous parasites and are predated by at least two Hemiptera that suck the juices from larvae: *Macrorhaphis acuta* Dallas in Kenya (Wheatley, 1963) and Zimbabwe (Kutywayo, 1989) and *Rhynocoris segmentarius* (Germar), in Zimbabwe (Kutywayo, 1989). See Appendix A for a full list of parasites and predators.

Control

Bardner and Mathenge (1974) recommended the use of the organo-tin compounds, fentin hydroxide and fentin acetate, which have since remained effective (Mugo, 1992). More recently, trials of formulations of *Bacillus thuringiensis aizawai* and *B. thuringiensis kurstaki* have proved promising in Kenya (Mugo *et al.*, 1998). In Japan, a synthetic sex pheromone of *A. selenaria cretacea*, which is a serious pest of tea, has been synthesized and is proving promising for use in traps for male moths (Witjaksono *et al.*, 1999).

Tailed caterpillars

Epicampoptera andersoni (Tams), *E. ivoirensis* Watson, *E. marantica* (Tams), *E. strandi* (Bryk), *E. strandi glauca* (Hampson) [Lepidoptera: Drepanidae]

Morphology

The adults of all species are greyish brown moths with a wingspan of around 30–40 mm and are difficult to distinguish by external features alone. Griveaud (1967) presents illustrations and a key to adults of the three species found in Côte d'Ivoire. The fully grown caterpillar is about the same length and with a bulbous swelling on the thorax, and an appendage to the end of the abdomen forming a 'rat tail' about 10–15 mm long. The caterpillars are dark brown in the first instar, later turning green and finally red/brown or yellow/green. During the first larval instar, the tail is short and the thoracic hump not very pronounced (Fig. 5.5).

Pest status and distribution

Several species of *Epicampoptera* are found across tropical Africa: *E. andersoni* (Tams) has been recorded from Kenya, Uganda, Malawi and Democratic Republic of Congo; *E. ivoirensis* Watson from Côte d'Ivoire and Ghana;

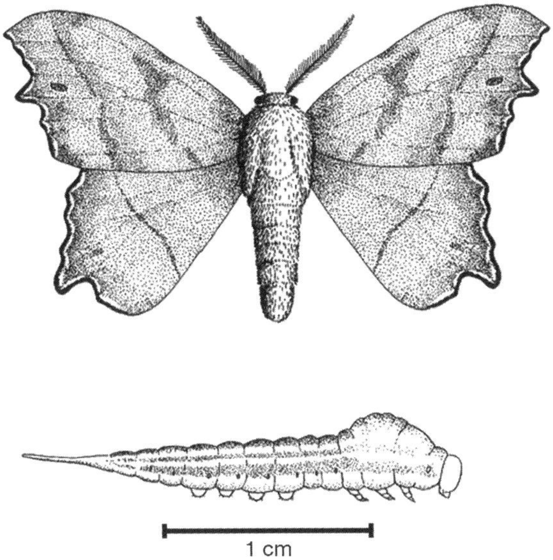

Fig. 5.5. Adult and larva of *Epicampoptera strandi glauca*.

E. marantica (Tams) from Tanzania, Uganda, Côte d'Ivoire, Cameroon and Democratic Republic of Congo; *E. pallida* Tams from Nigeria; *E. tamsi* Watson from Angola and *E. strandi* (Bryk) and its subspecies *E. strandi glauca* (Hampson) from Tanzania, Uganda, Democratic Republic of Congo, Cameroon, Nigeria, Ghana and Côte d'Ivoire. A record of *A. andersonii* from Côte d'Ivoire is probably a misidentification for *A. strandi glauca*.

Host range

Coffea is the only host plant known, and attempts by Griveaud (1967) to rear caterpillars of *Epicampoptera* spp. on wild species of Rubiaceae in the Côte d'Ivoire were unsuccessful. Both *C. arabica* and *C. canephora* are attacked. Defoliation of coffee can be severe, particularly where the coffee borders forest.

Life cycle

The brick-red eggs of *E. andersoni* are laid either in rows or scattered on the underside of the leaf and hatch in 8–9 d. Those of *E. strandi glauca* are arranged in an arching column of 6–45 eggs attached to the surface of the leaf by the lowest egg, and hatch in 2–3 d. They are pale at first but darken with age. The five larval stages last 18–20 d in the case of both *A. andersoni* and *A. strandi glauca*, and the pupal stage 9–11 d. On hatching, the larvae at first eat holes out of the leaf between the nerves but, as they grow bigger, more and more of the leaf is consumed until, by the last larval stadium the whole leaf may be eaten, including the veins. The caterpillar pupates in a cocoon comprising a rolled-up leaf.

Natural enemies

Most records of natural enemies are for *E. andersoni* in Kenya (Anderson, 1934; Evans, 1966). The larvae are parasitized by several species of wasps and Tachinid flies, and are predated by a number of Hemiptera that suck their juices. Egg parasites are recorded from *E. strandi glauca* in Nigeria (Okelana, 1985) and Côte d'Ivoire (Griveaud, 1967). See Appendix A for a full list.

Leaf-rolling caterpillars

Archips micaceana (Walker), *A. occidentalis* (Walsingham), *Homona coffearia* (Nietner), *H. fatalis* Meyrick, *H. trachyptera* Diakonoff, *Tortrix dinota* Meyrick. [Lepidoptera: Tortricidae]

The larvae of a number of species of moths in the family Tortricidae feed within shelters made by webbing together two leaves or rolling the edge of the leaf, and may also bore into green shoots and feed on green cherries and flower buds.

Morphology

Tortricid moths have a characteristic bell-shaped outline when at rest, and are patterned in patches and stripes of brown, grey and black. Females of *H. coffearia* (see Fig. 5.6), *A. occidentalis* and *T. dinota* have a wingspan of around 12–13 mm, with the males being smaller (8–10 mm). Eggs are flat and yellow or orange and are laid in overlapping masses on the upper leaf surface, hatching after about 10 d. Larvae of *H. coffearia* and *A. occidentalis* are pale in the first instar, but later become green with a black head, whilst those of *T. dinota* are reddish brown with darker stripes. Fully grown larvae of all species are about 25 mm long.

Pest status and distribution

Archips micaceana (Walker) is an Asian species recorded on coffee from India, Vietnam, Thailand and Indonesia, whilst the African *A. occidentalis* (Walsingham) is recorded from Kenya, Uganda, Zimbabwe, South Africa, Democratic Republic of Congo and the islands of São Tomé and Príncipe.

Homona is another Asian genus: *H. coffearia* (Nietner) is an important pest of tea, recorded on coffee from India, Sri Lanka, Java and Papua New Guinea (CABI, 1974b), whilst *H. fatalis* Meyrick is found in Malaysia and *H. trachyptera* Diakonoff in West Irian. *Tortrix dinota* Meyrick is an African species and has been recorded from Ethiopia, Kenya, Uganda, Zimbabwe and South Africa.

Life cycle and damage

The larvae feed within a shelter made by webbing two young leaves together. Once the leaves have been consumed, the larva prepares a new shelter. The five larval stages last 4–6 weeks. Pupation takes place within the last shelter, the moth emerging in around 10 d. The biology of the African species have been studied by Evans *et al.* (1968).

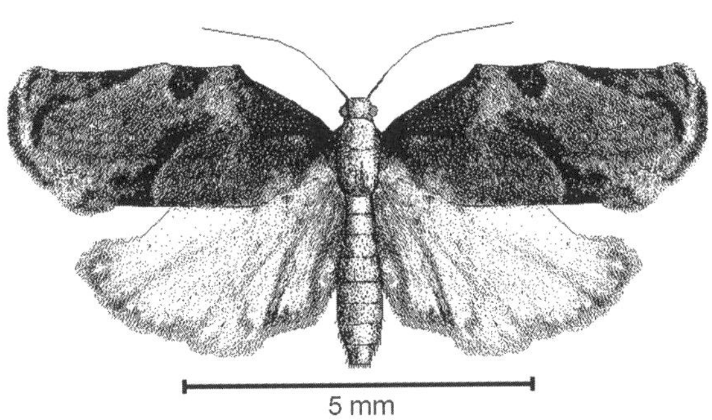

Fig. 5.6. Adult of *Homona coffearia*.

Natural enemies

Homona coffearia is generally now well controlled by its natural enemies. In Sri Lanka, where *H. coffearia* used to be a serious pest of tea, the parasite *Macrocentrus homonae* Nixon was introduced from Java during the 1930s, and has since greatly reduced the importance of the pest. In Papua New Guinea, this insect is adequately controlled by two Ichneumonid parasites, *Camptotypus sellatus* Kriechbaumer and *Theronia simillima* Turner. See Appendix A for a full list.

Control

Natural enemies should normally keep leaf-rolling caterpillars in check. If chemical control is needed, cypermethrin has been found to be effective.

Coffee leaf folder

Lamprosema crocodora (Meyrick) [Lepidoptera: Pyralidae]

Morphology

The adult is a small moth with golden-yellow wings striped with zigzag brown lines and black spots. Males have a wingspan of around 20 mm and females around 25 mm (see Fig. 5.7). The larva is around 20 mm long when fully grown, grey/green or yellowish in colour, with four dorsal black spots on each segment.

Pest status and distribution

The insect became a serious defoliator of *Coffea canephora* in the Democratic Republic of Congo following the use of persistent chemicals in the 1940s

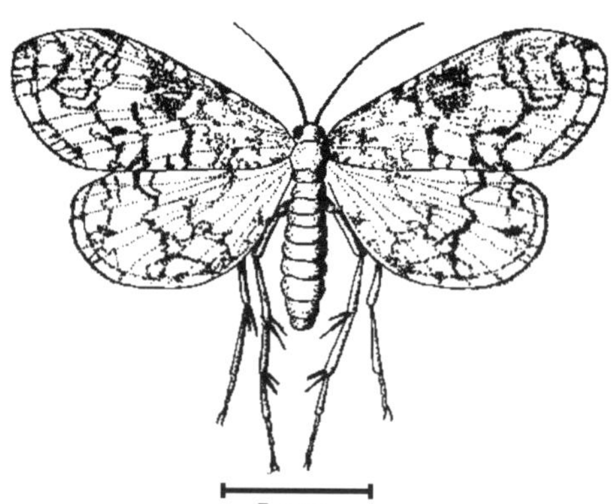

Fig. 5.7. Adult of *Lamprosema crocodora*.

(Schmitz, 1949). It is of less importance in Cameroon and Côte d'Ivoire because it prefers shaded conditions, and in those countries the crop is grown largely without shade (Bruneau de Mire and Muller, 1965). It has been recorded as a minor pest of coffee in Uganda (Le Pelley, 1959) and, more recently, in Yemen (Ba-Angood and Al-Sunaidi, 2004).

Life cycle

The eggs are laid in masses on the underside of the leaf, this stage lasting 11 d. After hatching, the larvae at first feed in a group within a shelter made by webbing together two leaves, but after the third instar they disperse and make individual shelters by folding part of a leaf. There are five larval stages lasting 2–5, 2–6, 4–7, 6–12 and 12–20 d, respectively. Pupation takes place among dead leaves at the base of the tree and lasts 21–30 d (Schmitz, 1949).

Natural enemies

Schmitz (1949) studied the natural enemies in the Democratic Republic of Congo in some detail. An egg parasite, *Trichogrammatoidea lutea* Girault, destroyed around 25% of eggs. Young larvae were parasitized by *Hypomicrogaster vacillatrix* (Wilkinson) and older ones by *Apanteles congoensis* De Saeger. The latter is, in turn, parasitized by three species of hyperparasite (see Appendix A). An attempt to move *L. crocodora* larvae parasitized by *A. congoensis* from heavily infested areas to more lightly infested coffee was partially successful in reducing the population.

Stinging caterpillars

Parasa lepida (Cramer), *Latoia vivida* (Walker) [Lepidoptera: Limacodidae]

Morphology

The adult *P. lepida* has a wingspan of around 40 mm. The basal half of the forewing is jade green and the distal half chocolate brown, and with a brown patch on the leading edge of the wing. The hind wings are pale brown. The head and thorax are green, with a brown central stripe (see Fig. 5.8).

By the 4th instar, the larvae are pale green with a double line of longitudinal blue spots dorsally and paler blue interrupted stripes laterally. There are conical projections protruding from the body that bear long, severely stinging hairs. When fully grown, the larvae are 20–30 mm long and the lateral protrusions are less pronounced, whilst the dorsal spots have coalesced into a single broad blue band (see Fig. 5.8). Pupation takes place within a tough, hemispherical, brown cocoon attached to the bark of the trunk, or sometimes 2–3 cm below the soil.

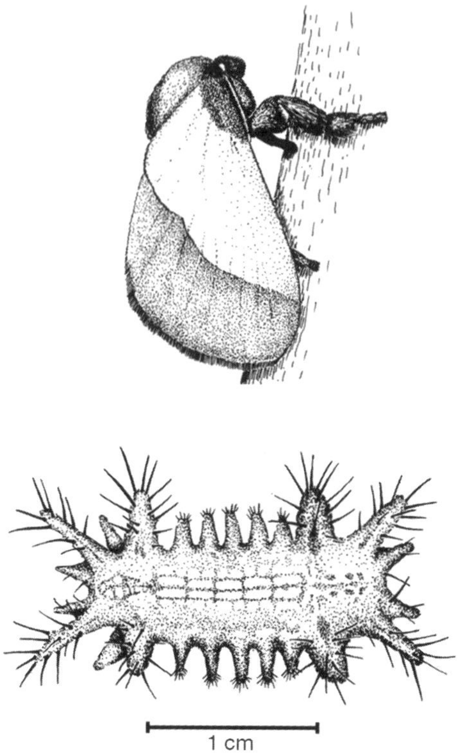

Fig. 5.8. Adult and larva of *Parasa lepida*.

Pest status and distribution

Parasa lepida is an Asian species, and has been recorded on coffee from India, Sri Lanka, Malaysia, Vietnam and Indonesia (CABI, 1986). *Latoia vivida* is African, recorded on coffee from Côte d'Ivoire, Ghana, Ethiopia, Kenya, Uganda, Tanzania, Malawi and Zimbabwe. At times these species can cause defoliation of coffee, but are more of a nuisance because of the severe sting that the caterpillar can inflict. Both are polyphagous. *Parasa lepida* is an important pest of coconut and oil palm, and is also found on other crops such as cocoa, tea and mango; *L. vivida* is found on cocoa, groundnut, cotton and castor, in addition to coffee.

Life cycle

The greenish, scale-like eggs are laid on the underside of the leaf, overlapping each other in a cluster. Those of *P. lepida* hatch in as little as 2 d on mango in India (Kapoor *et al.*, 1985), but elsewhere may take 5–7 d, whilst those of *L. vivida* hatch in 10 days on coffee in Kenya (Anon, 1967). *Parasa lepida* goes through six to eight larval instars lasting 30–48 d and a pupal stage of 16–97 d in China (Wu and Huang, 1983). The larval stage of *L. vivida* lasts a similar length of time, but the larva may go into a resting phase in the cocoon before pupating, so this stage can last for up to 134 d in Kenya (Anon, 1967) and 206–227 d in Malawi (Le Pelley, 1968).

Natural enemies

The larvae are attacked by a range of hymenopterous and dipterous parasites (See Appendix A). *Apanteles parasae* Rohwer is the most important hymenopterous parasite of *P. lepida.*

Two sucking bugs, the Pentatomid, *Macrorhaphis acuta* Dallas and the Reduviid, *Rhynocoris segmentarius* (Germar), are predators of *L. vivida* in Zimbabwe (Kutywayo, 1989).

Coffee shoot borer

Eucosma nereidopa Meyrick [Lepidoptera: Tortricidae]

Morphology

The adult moth is dark grey and about 8 mm long. The eggs are flat, circular, white and are laid singly or in small groups. The caterpillar is dark brown and about 12 mm when fully grown (Hill, 1975).

Pest status, distribution and damage

This is generally a minor pest, but it is sporadically serious in East Africa on high-altitude, shaded *C. arabica.* It has also been recorded on coffee from India. The caterpillar bores into the shoot tip a few cm behind the leaves and the tip rapidly withers. The larvae tunnel extensively in the sucker and one larva may attack several green shoot tips. The pest may also attack green berries. Damage is usually slight because pest numbers are low, but occasionally populations become high and damage can be extensive (Le Pelley, 1968).

Life cycle and host range

During daylight they rest on the trunks of shade trees near ground level. Eggs are laid on coffee trees, in grooves near the tip of a sucker or on the side of a berry, and they hatch about 14 d later. However, pupation takes place in a chamber made by the mature caterpillar in rough bark close to ground level. Pupation lasts about 36 d (Le Pelley, 1968).

Control

If large numbers of moths are observed, they should be killed before the eggs are laid using a suitable insecticide such as fenitrothion. This pest is normally seen only in shaded coffee, and may be sprayed while the insects rest by spraying the trunks of shade trees up to a height of 3 m. The intervention threshold is either: (i) more than five moths per shade tree before suckers have been thinned; or (ii) one moth per shade tree after thinning.

Coffee hawkmoth

Cephonodes hylas (Linnaeus), *C. picus* (Cramer) [Lepidoptera: Sphingidae]

Morphology

The adult *C. hylas* has a wingspan of around 60 mm, with transparent wings. The thorax and the first abdominal segment is grass-green above, whilst the remainder of the abdomen is yellow, divided from the green segment by a broad, dark red band. The larva, which grows to a length of 70 mm, is pale green with a bluish dorsal line and two lateral whitish lines punctuated by red spots. The end of the abdomen bears a vertical yellowish 'tail' covered in fine black spots (see Fig. 5.9).

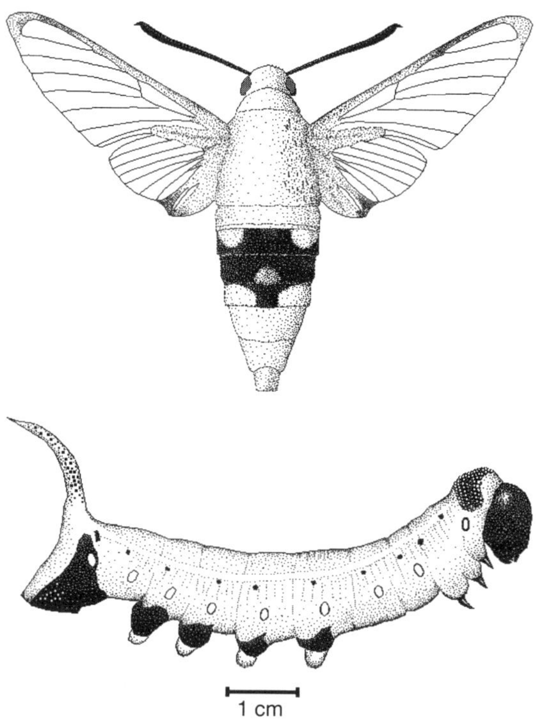

Fig. 5.9. Adult and larva of *Cephonodes hylas*.

Pest status and distribution

Cephonodes hylas has a wide distribution across Africa and South Asia, having been recorded on coffee from Côte d'Ivoire, Ghana, Benin, Nigeria, Cameroon, Democratic Republic of Congo, Ethiopia, Uganda, Tanzania, Madagascar, Malaysia and Vietnam (CABI, 1985), whilst *C. picus* is found in India, Malaysia and Indonesia.

Damage

Although the larvae are capable of causing considerable damage to coffee plantations, the seriousness of outbreaks varies considerably from place to place: from severe in Malaysia (Corbett and Yusope, 1932) to negligible in East Africa. Khoo *et al.* (1991) remark that the damage caused by this insect is one of the reasons for the failure of coffee cultivation to prosper in peninsular Malaysia.

Life cycle

Eggs are laid on the upper leaf surface and hatch in 4 d. The larval stage lasts around 3 weeks, after which pupation takes place in the soil at the base of the tree.

Natural enemies

Eggs of *C. hylas* are parasitized by *Ooencyrtus papilionis* Ashmead in Malaysia. Larval parasites of the same species are all Tachinid flies: *Actia heterochaeta* Bezzi in Uganda, *Blepharipa zebina* (Walker) in India and Malaysia and *Exorista bombycis* (Louis) in Malaysia. In addition, the Hemiptera, *Chrysocoris javanus* Westwood and *Sycanus leucomesus* Walker, have been recorded as predators in Democratic Republic of Congo and Malaysia, respectively, whilst the weaver ant, *Oecophylla smaragdina* (Fabricius), is a predator of larvae in Malaysia (see Appendix A).

Control

Khoo *et al.* (1991) recommend hand picking of the conspicuous larvae as a control measure and spraying with 0.1% a.i. trichlorfon where necessary.

Some Lepidoptera of lesser importance

Apart from the Lepidoptera already dealt with, coffee is host to a large number of leaf-eating caterpillars, most of which are of minor importance but may cause more serious defoliation on occasion. The geographical distribution of species in some of the larger genera is shown in Table 5.1.

Table 5.1. Records of some of the less important defoliating caterpillars.

Family/genus	Recorded from/by
Geometridae	
Epigynopteryx ansorgei (Warren)	Kenya (Le Pelley, 1932a)
E. stictigramma Hampson	Kenya (Leeuwangh, 1965)
E. tabitha Warren	Kenya (Le Pelley, 1959)
Hyposidra picaria Walker	Malaysia (Yunus and Ho, 1980)
H. talaca (Walker)	Malaysia (Corbett, 1935), Indonesia Java (Koningsberger, 1908), Papua New Guinea (Szent-Ivany and Stevens, 1966), Vietnam (Duport, 1912–1913)
H. infixaria Walker	Indonesia Java (Koningsberger, 1908), Turkmenistan (Zhang, 1994), Vietnam (Duport, 1912–1913)
Oxydia hispata Cramer	Colombia (Cárdenas and Posada, 2001)
O. noctuitaria Walker	Colombia (Cárdenas and Posada, 2001)
O. obrundata Guenee	Colombia (Cárdenas and Posada, 2001)
O. trychiata Guenee	Colombia (Cárdenas and Posada, 2001)
O. vesulia (Cramer)	Colombia (Cárdenas and Posada, 2001), Puerto Rico (Cotte, 1989)
Paragonia lanuginosa Schaus	Colombia (Cárdenas and Posada, 2001)
P. procidaria Herrich-Schaffer	Colombia (Cárdenas and Posada, 2001)
Limacodidae	
Cheromettia apicata Moore	India (Beeson, 1941), Indonesia Java (Van Hall, 1921)
C. ferruginea Moore	Sri Lanka (Sevastopulo, 1940)
C. lohor (Moore)	Indonesia Java (Koningsberger, 1908)
C. sumatrensis Heylaerts	Malaysia (Yunus and Ho, 1980)
Darna sordida Snellen	No locality (Cock *et al.*, 1987)
D. trima Moore	Indonesia Java (Koningsberger, 1908), Malaysia (Yunus and Ho, 1980)
Hindothosea cervina (Moore)	Vietnam (Vayssiére, 1955)
Niphadolepis alianta Karsch	Kenya (Coffee Research Foundation, 1961), Malawi (Smee, 1939), Tanzania (Ritchie, 1935), Zimbabwe (Weaving, 1972)
Thosea bipartita Hering	Thailand (Cock *et al.*, 1987)
T. lutea Heylaerts	Indonesia Sumatra (Kalshoven, 1950–1951)
T. mediostrigata Hering	Malaysia (Yunus and Ho, 1980)
T. sinensis (Walker)	Indonesia Java (Koningsberger, 1908), Vietnam (Duport, 1915)
T. unifascia (Walker)	Thailand (Anon, 1965)
T. vetusta Walker	Malaysia (Yunus and Ho, 1980)
Lymantridae	
Euproctis dewitzi (Grunberg)	Cameroon (De Fluiter, 1960), Uganda (Le Pelley, 1959)
E. flava (Fabricius)	Sri Lanka (Jepson, 1935)
E. fraterna Moore	India (Vayssiére, 1955), Sri Lanka (Vayssiére, 1955), Vietnam (Duport, 1915)

E. howra (Moore)	India (Sekhar, 1960)
E. taiwana (Shiraki)	Taiwan (Moriyama, 1941)
E. sp.(A)	Papua New Guinea (Szent-Ivany and Stevens, 1966)
Olene inclusa (Walker)	Indonesia Java (Koningsberger, 1908), Malaysia (Yunus and Ho, 1980)
O. mendosa (Hubner)	India (Beeson, 1941), Indonesia Java (Dammerman, 1929), Malaysia (Yunus and Ho, 1980), Uganda (Le Pelley, 1959), Vietnam (Duport, 1912–1913)
Orgyia araea Collenette	Malaysia (Yunus and Ho, 1980)
O. hopkinsi Collenette	Uganda (Le Pelley, 1959)
O. leucostigma plagiata (Walker)	Cuba (Bruner *et al.*, 1975)
O. osseata Walker	Malaysia (Yunus and Ho, 1980)
O. postica (Walker)	Indonesia Java (Koningsberger, 1908), Malaysia (Yunus and Ho, 1980), Papua New Guinea (Szent-Ivany, 1956), Sri Lanka (Nietner, 1861), Vietnam (Duport, 1912–1913)
Sphrageidus producta (Walker)	Uganda (Le Pelley, 1959), Tanzania (Le Pelley, 1959)
S. virguncula (Walker)	Indonesia Java (Kalshoven, 1950–1951), Sri Lanka (Nietner, 1861), Vietnam (Duport, 1912–1913)
S. xanthorrhoea (Kollar)	Indonesia Java (Koningsberger, 1908)
Psychidae	
Acanthopsyche emiliae (Heylaerts)	Kenya (Le Pelley, 1959), Zimbabwe (Kutywayo, 1989)
A. sierricola (White)	Nigeria (Entwistle, 1963)
Eumeta crameri (Westwood)	India (Kalshoven, 1950–1951), Indonesia Java (Kalshoven, 1950–1951), West Irian (Simon-Thomas, 1962), Sri Lanka (Kalshoven, 1950–1951)
E. minuscula Butler	S. Asia (Robinson *et al.*, 1994)
E. rougeoti Bourgogne	Nigeria (Entwistle, 1963)
E. variegata (Snellen)	China (Cheo, 1936), Indonesia Java (Kalshoven, 1950–1951), Malaysia (Yunus and Ho, 1980), Papua New Guinea (Bradley, 1987), Vietnam (Duport, 1912–1913)
Oiketicus geyeri Berg	Brazil (Dos Santos, 1988)
O. kirbyi (Guilding)	Brazil (Arce *et al.*, 1987), Cuba (Bruner, 1929), Honduras (Munoz, *c.*2000)
Pteroma pendula Joannis	Malaysia (Yunus and Ho, 1980), Papua New Guinea (Szent-Ivany and Stevens, 1966)
P. plagiophleps Hampson	India (De Fluiter, 1960), Indonesia (De Fluiter, 1960), Sri Lanka (De Fluiter, 1960)
P. sp.	Papua New Guinea (Szent-Ivany and Stevens, 1966)

Mention has already been made of the stinging caterpillars, *P. lepida* and *L. vivida*, in the family Limacodidae. Limacodids are fairly commonly found on coffee and over 60 species have been recorded, including several species of *Thosea*, *Cheromettia* and *Darna* from South East Asia and *Niphadolepis alianta* Karsch, the so-called 'jelly' caterpillar from Eastern Africa. They often cause problems to pickers because of their painful, stinging hairs.

Around 40 species of looper caterpillars (Geometridae) have been recorded from coffee and the giant looper, *Ascotis selenaria reciprocaria*, has already been dealt with. Old World genera include *Epigynopteryx* and *Hyposidra*, whilst *Oxydia* and *Paragonia* are found in the New World.

Caterpillars in the family Psychidae are known as 'bagworms', because the larvae live within a travelling case of silk to which fragments of leaf or stick have been attached, and around 30 species feed on coffee. Each species constructs a characteristic bag. Old World genera attacking coffee include *Acanthopsyche*, *Eumeta* and *Pteroma*, whilst *Oiketicus* species are found in the New World.

Tussock moths (Lymantriidae) have caterpillars with characteristic hair tufts and hair pencils, some of which also feed on berries and bark. Most of the species attacking coffee have an Old World distribution. Females of *Orgyia* are wingless.

Leaf-feeding Coleoptera

On emerging from their pupal stage, young adult beetles need to feed in order to become sexually mature, and often choose young leaves or green shoots as their source of food. Different families of beetles have characteristic patterns of damage.

Curculionidae

Adults of Curculionidae (weevils) characteristically make rounded notches on the margin of the leaf, whilst others ring bark young shoots. Around 70 species have been recorded as feeding on coffee leaves, but most do no significant damage.

Lachnopus coffeae Marshall in Puerto Rico and *L. buchanani* Marshall in Cuba can cause significant damage (Wolcott, 1922), whilst in Colombia, Cárdenas and Posada (2001) report leaf damage by species of *Compsus*, *Macrostylus* and *Epicaerus* feeding on seedling trees newly planted in the field. In addition to leaf damage, *Macrostylus* adults damage the growing point of the stem, causing bifurcation. The female *Compsus* lays its eggs in a folded leaf, from which the emerging larvae drop to the ground to feed in the soil on roots.

Lachnopus coffeae (see Fig. 6.5b) has a similar lifestyle to that of *Diaprepes abbreviatus* (see Fig. 6.5d) – dealt with under root pests in Chapter 6. In Papua New Guinea, species of *Apirocalus* cause leaf damage, including *A. canus* Thompson, *A. cornutus tenuiscapus* Thompson, *A. fallax* Thompson and *A. subcostatus* Thompson (Thompson, 1977).

Scarabaeidae

Larvae of Scarabaeidae are root feeders, the adults feeding on leaves but, unlike Curculionidae, they eat ragged holes in the body of the leaf, tending to leave the veins intact. Around 45 species feed on coffee as adults, mostly in the subfamilies Melolonthinae and Rutelinae. Species of *Adoretus*, *Anomala* and *Popillia* feed on coffee in Africa and Asia but are not generally considered serious pests. In Colombia, *Ancistrosoma rufipes* Latreille and *Platycoelia valida* Burmeister skeletonize leaves, feed on flowers and green cherries and cause defoliation under shade (Cárdenas and Posada, 2001).

Chrysomelidae

The adults of this family characteristically eat round to oval holes through the leaf lamella, which do not overlap. Around 40 species have been recorded, mostly in the subfamilies Eumolpinae and Galerucinae. Species of *Homophoeta*, *Cerotoma*, *Diabrotica* and *Colaspis* have been recorded as damaging newly planted coffee in Colombia, particularly in land adjoining fallow (Cárdenas and Posada, 2001).

Three species of *Dactylispa* that make furrows in the upper leaf surface have been recorded from East Africa, and *D. hirsuta* Gestro was very damaging to coffee in Tanzania during the 1920s, being described by Ritchie (1929, 1932) as the worst pest of bearing coffee at that time, but has not proved a problem since. In addition, *Dactylispa* sp. near *litigiosa* Peringuey and *Dactylispa* sp. near *tenuicornis* Chapuis are recorded by Le Pelley (1959) from Tanzania and Kenya, respectively.

Leaf-damaging Hymenoptera

Leaf-cutting ants

Atta spp., *Acromyrmex* spp. [Hymenoptera: Formicidae]

In South and Central America and the Caribbean, ants in the genera *Atta* and *Acromyrmex* cause considerable damage to a wide range of crops by removing sections of leaf. These are carried down into their underground nests and used as a substrate for the culture of fungi that the ants use as food. A good review of the biology of these ants is given by Entwistle (1972).

The species recorded as damaging coffee are:

- *Acromyrmex asperus* (F. Smith), Brazil (see Fig. 5.11a)
- *A. niger* (F. Smith), Brazil
- *A. octospinosus* (Reich), Brazil, Trinidad (see Fig. 5.11b)
- *A. subterraneus* (Forel), Brazil (see Fig. 5.11c)
- *Acromyrmex* sp., Peru

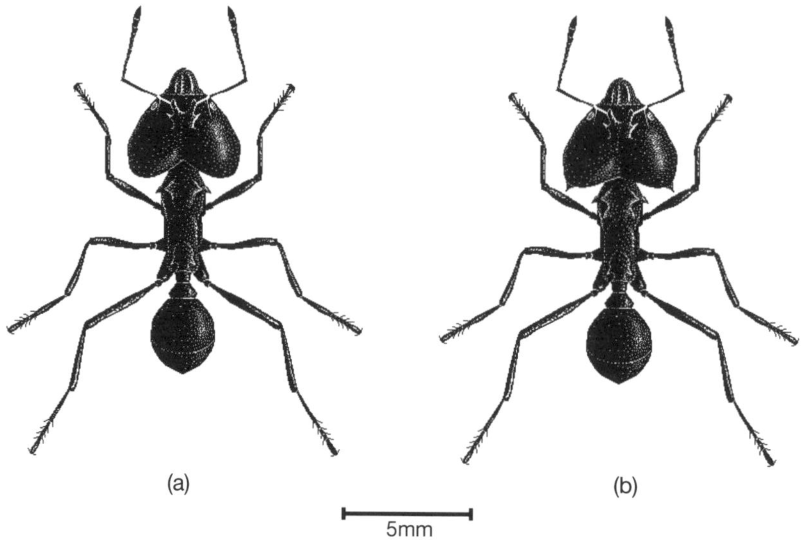

Fig. 5.10. Soldiers of (a) *Atta laevigata*, (b) *A. sexdens*.

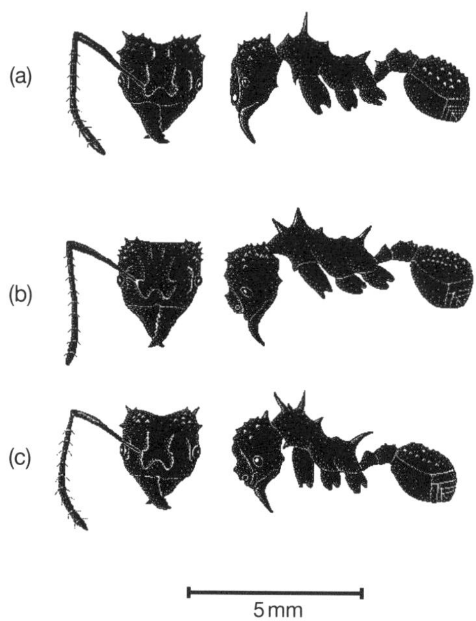

Fig. 5.11. Soldiers of (a) *Acromyrmex asperus*, (b) *A. octospinosus*, (c) *A. subterraneus*.

- *Atta cephalotes* (Linnaeus), Mexico, Honduras, Costa Rica, Trinidad, Surinam, Colombia (CABI, 1982)
- *A. insularis* Guerin-Meneville, Cuba
- *A. laevigata* (F. Smith), Brazil (see Fig. 5.10a)
- *A. mexicana* (F. Smith), Mexico
- *A. sexdens* (Linnaeus), Costa Rica, Surinam, Brazil, Peru (see Fig. 5.10b).

Because of the very large numbers of individuals in an ant colony and the fact that the workers may forage up to 200 m or more from the nest, damage to trees in the vicinty of a nest can be considerable, and defoliation is achieved in a very short time. Direct protection of the trees from attack is impractical, and control is usually attempted by destruction of the nest either by chemicals applied directly to the nest entrance or by baiting. Mirex is employed as a bait for the control of leaf-cutting ants in Brazil.

Leaf-cutting bees

Megachile lachesis F. Smith, *M. frontalis* (Fabricius), *M. tuberculata* F. Smith [Hymenoptera: Apidae]

Other Hymenoptera that damage leaves are leaf-cutting bees, which remove semicircular sections from the edge of the leaf to line the tunnels in which their larvae are raised. *Megachile lachesis* F. Smith and *M. frontalis* (Fabricius) are found in Papua New Guinea, and *M. tuberculata* F. Smith in Indonesia. None is a serious pest and some are valuable pollinators (see Appendix B).

Leaf-feeding Orthoptera

Long-horned grasshopper

Idiarthron subquadratum De Saussure and Pictet [Orthoptera: Tettigoniidae]

Morphology

The female is a green grasshopper with thin, whip-like antennae, much longer than the body, and a blade-like ovipositor projecting from the rear end of the abdomen. The overall length of the adult is around 50 mm (see Fig. 5.12).

Pest status and distribution

Long-horned grasshoppers in the family Tettigometridae have been recorded on coffee in many parts of the world, but *I. subquadratum* is the most important. It is found in Central and South America and has been recorded on coffee in Mexico, Costa Rica, El Salvador, Guatemala, Honduras and Colombia. The related *I. atrispinum* (Stal) is present in Nicaragua. Normally the insect is a minor pest, but serious outbreaks have been reported from El Salvador, Guatemala and Mexico. In Chiapas, Mexico, the pest was thought to be responsible for losses in crop of at least 50% during the period 1990–2000 (Zuniga *et al.*, 2002).

Nymphs and adults hide in tree holes during the day, coming out at night to feed on leaves, buds and green cherries. The outer part of the cherry is

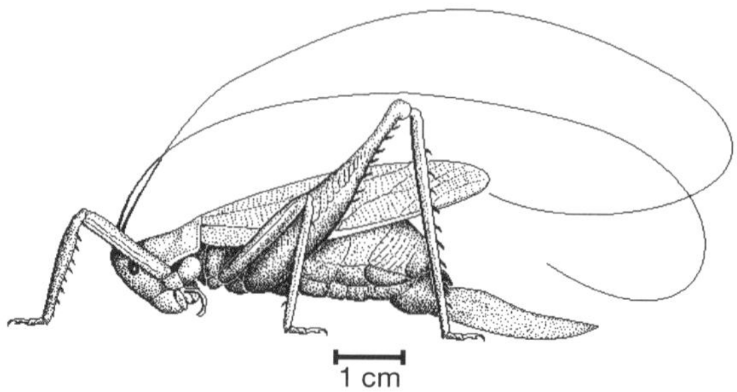

Fig. 5.12. Adult female of *Idiarthron subquadratum*.

eaten away, exposing the beans and causing the cherry to fall or, because of the damage, rendering the bean unusable.

Life cycle

The insect has one generation per year. Eggs are laid 5 cm deep in the ground from November to December, where they remain dormant until the onset of the rains in the following May to June. The eggs hatch about 2 weeks after the commencement of the rains. There are six nymphal stages. The life cycle from egg to adult lasts 78 d at 28°C (Barrera *et al.*, 2003).

Natural enemies

A Mymarid parasite, *Cleruchus* sp., attacks the insect in El Salvador (Berry and Abrego, 1953).

Control

Malathion is commonly used to control the insect, but it has been observed that the use of this insecticide has led to an upsurge in coffee leaf miner, *Perileucoptera coffeella* (Barrera *et al.*, 2003). Farmers in Mexico use a trap made from a bamboo internode to capture *I. subquadratum*.

Variegated grasshoppers

Zonoceros variegatus (Linnaeus), *Z. elegans* (Thunberg) [Orthoptera: Pyrgomorphidae]

Morphology

Both species are typical grasshoppers around 30–50 mm long and with distinctive coloration. *Zonoceros elegans* is slightly larger than *Z. variegatus* and is yellowish green with darker green/blue patches, whilst *Z. variegatus* is patterned in yellow and black, with a conspicuous yellow band around the hind legs (see Fig. 5.13).

Pest status and distribution

Variegated grasshoppers are generalist feeders and attack a wide range of crops across tropical Africa. *Zonoceros variegatus* is the more northerly species, ranging from West Africa to East Africa (CABI, 1974a), whilst *Z. elegans* is found in Southern Africa and Madagascar, ranging north into East Africa.

Zonoceros variegatus has been recorded on coffee from Guinea, Sierra Leone, Côte d'Ivoire, Ghana, Togo, Dahomey, Cameroon, Congo, Democratic Republic of Congo, Kenya, Tanzania, Uganda and Madagascar, and *Z. elegans* from Democratic Republic of Congo, Kenya and Tanzania. A record of *Z. variegatus* on coffee in Madagascar is probably a misidentification for *Z. elegans*.

When the eggs hatch at the beginning of the rains, the nymphs at first feed on weeds and nearby crops and do not move onto coffee until they are older or as adults. They then devour leaves, and sometimes green cherries and, being fairly large insects, can do considerable damage.

Life cycle

Zonoceros elegans has one generation per year, whilst *Z. variegatus* has one in drier areas and two where rain is more plentiful. Eggs are laid in the soil at the end of one rainy season and hatch at the beginning of the next. *Zonoceros elegans* has five nymphal stages whilst *Z. variegatus* has six, and in both cases the nymphal stage lasts around 100 d.

Control

A pesticide based on the fungus, *Metarhizium anisopliae var acridium*, and marketed as 'Green Muscle', has proved to be useful in the control of *Z. variegatus* (Muller *et al.*, 2002).

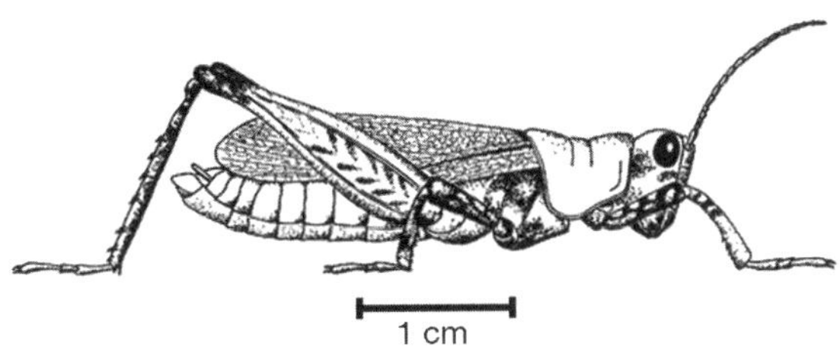

Fig. 5.13. Adult of *Zonocerus variegatus*.

Leaf- and Flowerbud-feeding Hemiptera

Lace bug

Dulinius unicolor (Signoret), *Habrochila ghesquierei* Schouteden, *H. placida* Horvath [Hemiptera: Tingidae]

Morphology

In all three species, the forewings of the adult present a reticulated pattern and there is a bulbous outgrowth, also reticulated, projecting forward over the head and leaf-like reticulated side lobes beside the thorax and head. The colour is yellowish brown. *Dulinius unicolor* is around 3 mm long and *H. ghesquierei* around 4 mm (see Fig. 5.14). The nymphs have knobbed projections bordering the sides of the abdomen and dorsally on the head, thorax and abdomen.

Pest status and distribution

Dulinius unicolor is confined to Madagascar, and the two *Habrochila* species are found from the Democratic Republic of Congo, through Ruanda and Burundi, to Uganda, Kenya and Tanzania. There is some doubt as to the identity of the lace bugs in East Africa, and Le Pelley (1968) considers that the species in Kenya may be neither *H. ghesquierei* nor *H. placida.*

Feeding takes place mainly on the underside of the leaf, leading to yellowing and even complete defoliation, but buds and young fruits are also

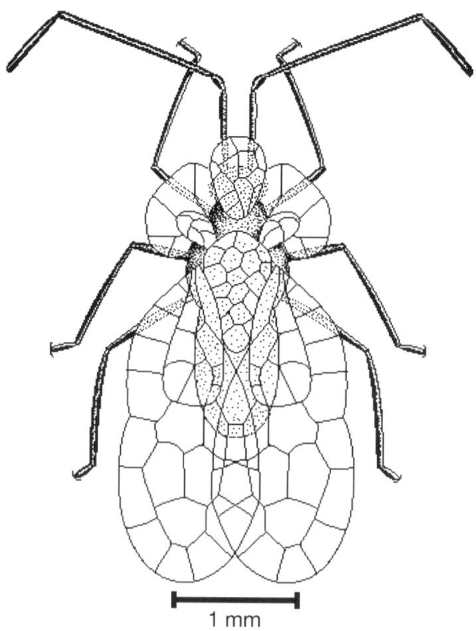

Fig. 5.14. Adult of *Habrochila ghesquierei.*

damaged. Most damage is caused by the nymphs, which feed in groups, whereas the adults are found only singly and do not stay on the plant for long. Attacks reach their maximum during the rainy season.

Life cycle

Eggs are embedded in the surface of the leaf or in green shoots and hatch in 13–16 d in the case of *H. placida* in Democratic Republic of Congo (Foucart, 1954) and 22–32 d in the case of *Habrochila* sp. in Kenya (Anon, 1967). *Habrochila placida* goes through four nymphal stages lasting 4–6, 3–6, 3–6 and 2–7 d, respectively, and the Kenyan species through five stages lasting a total of 16–36 d.

Natural enemies

No parasites have been recorded, but the insects are predated by the Mirid, *Stethoconus*. *Stethoconus frappai* Carayon attacks *D. unicolor* in Madagascar, whilst *S. distanti* (Schouteden) attacks *H. ghesquierei* in Ruanda and Democratic Republic of Congo, and *Stethoconus* sp. is recorded as attacking *Habrochila* in Kenya and Uganda.

Control

Natural control by predators is normally sufficient to keep lace bugs under control, but fenitrothion should prove effective if necessary.

Capsid bug

Ruspoliella coffeae (China), *Volumnus obscurus* Poppius [Hemiptera: Miridae]

Morphology

Both species are greenish brown Mirid bugs, but *R. coffeae* is about 6 mm long and smooth, whilst *V. obscurus* is 6–7 mm long and hairy. Nymphs are pale green (see Fig. 5.15).

Pest status and distribution

These bugs feed on anthers in the flower buds, causing the flower to abort. Damage can sometimes be serious, although it is thought that at lower altitudes attack may be beneficial by preventing the crop from overbearing (Le Pelley, 1968). Low populations of the bugs can cause significant damage, and in Central Africa a count of three to five individuals per tree was considered sufficient to make control measures necessary (Vayssiére, 1955).

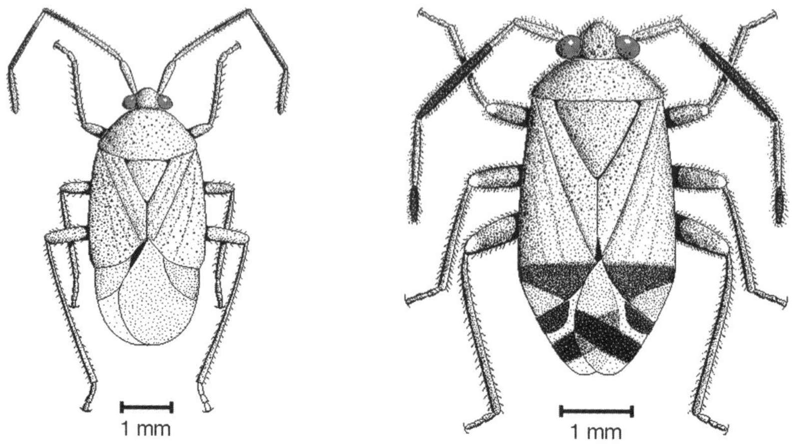

(a) (b)

Fig. 5.15. Adults of (a) *Ruspoliella coffeae*, (b) *Volumnus obscurus*.

Ruspoliella coffeae is important in Ethiopia, Kenya, Uganda and Tanzania, but has also been recorded from Cameroon, Democratic Republic of Congo and Madagascar, whilst *V. obscurus* is more important in West and Central Africa (Gabon, Cameroon, Democratic Republic of Congo) but has, in addition, been recorded from Uganda and Kenya.

Life cycle

Eggs are completely inserted into flower buds so they cannot be seen. There are five nymphal stages, lasting about 2 weeks.

Natural enemies

In Kenya, *R. coffeae* is predated by a Lygaeid bug, *Geocoris ruficeps* (Germar), and by an unidentified internal parasite (Le Pelley, 1932b).

Scale Insects, Mealy Bugs and Aphids

'Homoptera'[1] (Hemiptera: Sternorrhyncha) are small, mostly sedentary, insects which feed through tubular mouthparts which are inserted into the plant tissues and used to extract plant sap. To obtain their nutrient, a relatively large volume of liquid has to be passed through the digestive tract. Essential amino acids and some sugars are extracted, but many sugars are excreted and the so-called 'honeydew' provides food for other insects – particularly ants – which guard their source of food from predators (Adomako, 1972).

[1] The Hemiptera used to be divided into two suborders, the Homoptera and the Heteroptera but, following the work of Gullan (1999), the former is no longer recognized as a valid name.

Removal of sap by large numbers of insects is debilitating to the tree and may lead to serious chlorosis, die-back and, in extreme cases, death. Feeding is a two-way process, and whilst liquid is extracted, saliva is pumped into the plant to aid digestion. In some cases, this saliva is toxic to the plant and there is also the danger of transmitting diseases such as viruses. Scale insects belonging to several families such as Coccidae, Diaspididae, etc. live beneath a hard or soft scale. Mealy bugs in the family Pseudococcidae, on the other hand, have no scale but a covering of waxy filaments like cotton wool.

Soft green scales

Coccus alpinus De Lotto, *C. celatus* De Lotto, *C. viridis* (Green), *Pulvinaria psidii* (Maskell) [Hemiptera: Coccidae]

Morphology

Both adult females and nymphs live beneath a pale green, nearly flat, oval scale and are around 2–3 mm long when mature, usually clustered along the main vein on the underside of the leaf or on green shoots (see Plate 6). Females usually reproduce parthenogenetically. It had been thought that the male of *C. viridis* did not exist but, in 1974, males were observed for the first time in Cuba (Köhler, 1976). The greenish female of *P. psidii* extrudes a white, waxy ovisac that often lifts the hind end of the scale into the air and makes this species very conspicuous.

Pest status and distribution

Both *C. viridis* (see Fig. 5.16a) and *P. psidii* (see Fig. 5.16b) have a very wide distribution throughout the tropics and are found on a number of host plants, of which citrus and coffee are probably the worst effected (CAB International, 1955, 1972). There are doubtless other countries than those in the list below where they exist on coffee, but have not been reported.

Records of soft scales on coffee (C.a., *Coccus alpinus*; C.c., *Coccus celatus*; C.v., *Coccus viridis*; P.p., *Pulvinaria psidii*):

- Caribbean: Cuba (C.v., P.p.), Dominican Republic (C.v., P.p.), Guadeloupe (C.v.), Haiti (C.v.), Jamaica (C.v., P.p.), Martinique (C.v.), Puerto Rico (C.v., P.p.), Trinidad and Tobago (C.v.)
- Central America: Costa Rica (C.v.), Guatemala (C.v.), Honduras (C.v.), Panama (C.v.)
- South America: Bolivia (C.v.), Brazil (C.v.), Colombia (C.v.), Guyana (C.v.), Peru (C.v.), Surinam (C.v.), Venezuela (C.v.)
- Atlantic Ocean: Cape Verde Is. (C.v.), São Tomé and Príncipe (C.v.)
- West Africa: Ghana (C.v.), Guinea (C.v.), Côte d'Ivoire (C.v.)
- Central Africa: Democratic Republic of Congo (C.v., C.a., P.p.)
- Eastern Africa: Eritrea (C.v.), Ethiopia (C.v., C.a.), Kenya (C.v., C.a., C.c.,

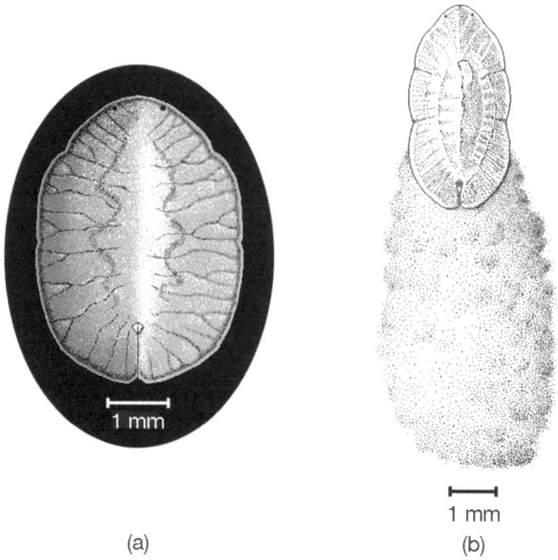

Fig. 5.16. Adult females of (a) *Coccus viridis*, (b) *Pulvinaria psidii* with ovisac.

P.p.), Tanzania (C.v., C.a., C.c., P.p.), Sudan (C.c.), Uganda (C.v., C.a., C.c., P.p.), Zanzibar (C.v.)
- Southern Africa: Malawi (C.a.), Zimbabwe (C.a.)
- Indian Ocean: Madagascar (C.v.), Mauritius (C.v.), Réunion (C.v.), Seychelles (C.v.), Yemen (C.v.)
- Asia: Cambodia (C.v.), China (P.p.), India (C.v., P.p.), Java (C.v., P.p.), Myanmar (C.v.), Philippines (C.v.), Sri Lanka (C.v.), Taiwan (C.v.), W. Irian (C.v., C.c.), W. Malaysia (C.v., C.c.), Vietnam (C.v.)
- Pacific Ocean: Cook Is. (C.v., P.p.), Federated States of Micronesia (C.v., P.p.), Fiji (C.v., P.p.), Hawaii (C.v., P.p.), New Caledonia (C.v., P.p.), Papua New Guinea (C.v., C.c., P.p.), Tonga (C.v.), Vanuatu (C.v.), Western Samoa (C.v.).

Coccus alpinus is restricted to Eastern and Central Africa and is recorded on coffee from Ethiopia, Kenya, Tanzania, Uganda, Malawi, Zimbabwe and Democratic Republic of Congo. It is very close to *C. viridis* in form, but is a highland species found above about 1200 m, whilst *C. viridis* tends to be found below this altitude.

Coccus celatus is also very similar to *C. viridis* in form, and may be difficult to distinguish from it in the field. It has a wider distribution on coffee than *C. alpinus* and is found in Africa in Sudan, Kenya, Uganda and Tanzania, but also in Malaysia, Papua New Guinea and West Irian.

Life cycle and damage

Laboratory rearing of *C. viridis* in Brazil at a temperature of 25°C showed the length of the life cycle from hatching of the egg to first egg production by the adult to be 47–51 d. The adult female *C. viridis* lived for a relatively long time, making the total life expectancy through all stages 116–134 d (Silva and Parra,

1982). The length of life cycle between generations of *P. psidii* is 43 d at 27°C (El-Minshawy and Moursi, 1976).

Severe outbreaks of green scales have been recorded in many countries: Le Pelley (1968) cites records of heavy damage by *C. viridis* in Sri Lanka, Java, India, Réunion, Cuba, Jamaica, Surinam and Brazil. Serious damage by green scale is mostly confined to nursery plants and young trees, but this is not always the case and it is a problem in field plantations in Papua New Guinea and West Irian, where *C. celatus* coexists with *C. viridis*.

Infestation results in general debilitation of the tree, with some leaf chlorosis and even death in extreme cases. Drought and poor soil nutrition exacerbates the decline. Trees infested with green scale can often be distinguished by their blackened, sticky leaves, where sooty moulds have grown on the 'honeydew' excreted by the large colonies of scales, and this must reduce the photosynthetic efficiency of the leaf.

Natural enemies

All three *Coccos* species are attacked by a rich assembly of Hymenopterous parasites, many of which are in turn attacked by hyperparasites (see Appendix A for full lists). In addition, there are numerous predators that feed on the scales. Most of the recorded predators are species of ladybirds (Coleoptera: Coccinellidae), of which both the adults and larvae are predaceous, but other predators exist, such as larvae of moths in the genera *Eublemma*, *Cryptoblabes* and *Coccidiphaga*, and larvae of Neuroptera. The fungus, *Septobasidium bogoriense* Pat, is recorded as attacking *C. viridis* in Indonesia.

Control

Colonies are commonly attended by ants, a number of species of which feed on the honeydew and may protect the scales against predators and parasites. In Venezuela for instance, it was observed that *Crematogaster* and *Camponotus* ants drove off Coccinellid beetles that were predating the scales and that numbers of *C. viridis* declined when these ants were excluded from the colonies. (Hanks and Sadof, 1990). If ants can be excluded from the scale colonies, the parasites and predators can soon effect adequate control. Since many of the attendant species are ground nesting, some sort of trunk barrier can be an effective deterrent.

Control of scales by the exclusion of attendant ants is not always so straightforward, however. On citrus trees in Trinidad, killing the ant *Azteca* sp. using a toxic bait led to an increase in numbers of the predaceous ladybird, *Azya* sp. and a decline in the population of *C. viridis*. But this had the unfortunate effect that the leaf-cutting ant *Atta cephalotes*, which had been kept off the trees by *Azteca*, soon defoliated them (Jutsum *et al.*, 1981).

Fungi, of which *Verticillium lecanii* is probably the most widespread, are also very important as control agents in conditions of high humidity, but of little significance during dry weather. Chemical control should be used as little as possible, in order not to upset the balance between the scales and their copious natural enemies.

Dark scales

Saissetia coffeae (Walker), *Saissetia oleae* (Olivier), *Parasaissetia nigra* (Nietner) [Hemiptera: Coccidae]

Morphology

Unlike the soft green scales that are relatively flat, the adult female dark scales are more or less dome-shaped and dark in colour. All are oval in shape. The adult female *S. coffeae* (see Fig. 5.17b) has a shiny, dark brown to black, domed scale which has given rise to its common name of 'helmet scale' or 'hemispherical scale'. The surface is covered by small cells.

Parasaissetia nigra (see Fig. 5.17c) is similar, but with a less domed scale. The colour varies from black, through brown to yellow. *Saissetia oleae* (see Fig. 5.17a) is also flatter, with two transverse ridges joining a median ridge in the form of an 'H'. Immature stages of *S. coffeae* also have this mark, but it disappears in the adult. All produce copious honeydew on which black sooty moulds grow, coating the leaves and hindering photosynthesis.

The nymphal stages of all species are pale and flattened. The scales are found on leaves, shoots and green berries, often in large numbers. The life cycle of *S. coffeae* (reared in the laboratory) was completed in 64 d at 25°C to 131 d at 17.5°C (Kozhechkin, 1984).

Pest status and distribution

All three species have a wide geographical and host range, being pests of many economic crops, including citrus and tea (CABI, 1973c, d).

Records of hard scales on coffee (S.c., *Saissetia coffeae*; S.o., *Saissetia oleae*; P.n., *Parasaissetia nigra*):

- Central America: Mexico (S.c.), Costa Rica (S.c., S.o.), El Salvador (S.c., P.n.), Guatemala (S.c., P.n.), Honduras (S.c.)

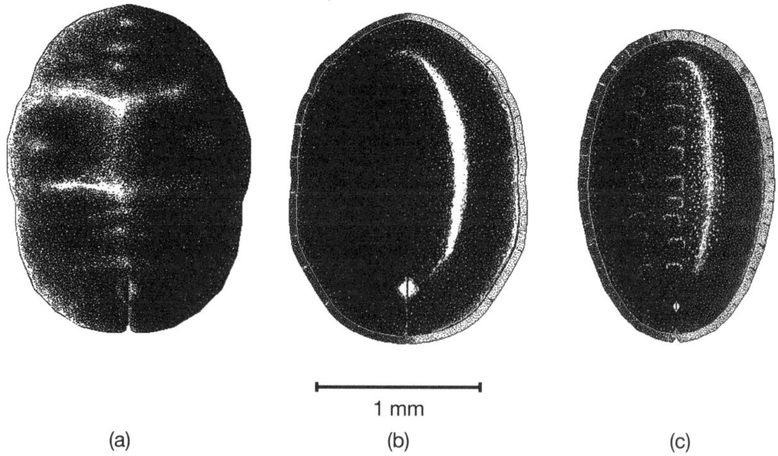

Fig. 5.17. Adult females of (a) *Saissetia oleae*, (b) *S. coffeae*, (c) *Parasaissetia nigra*.

- Caribbean: Puerto Rico (S.c., S.o., P.n.), Cuba (S.c., S.o.), Dominican Republic (S.c., S.o.), Jamaica (S.c., S.o., P.n.)
- South America: Surinam (S.c.), Peru (S.c., S.o.), Colombia (S.c.), Brazil (S.c., S.o.), Guyana (S.c., S.o.)
- Atlantic Ocean: St Helena (S.c.)
- West Africa: Côte d'Ivoire (S.c.), Nigeria (S.c.), Cameroon (S.c.)
- Central Africa: Congo (S.c.), Democratic Republic of Congo (S.c., P.n.), Equatorial Guinea (S.c.)
- Eastern Africa: Ethiopia (S.c.), Uganda (S.c., P.n.), Tanzania (S.c.), Kenya (S.c., P.n.)
- Southern Africa: Zimbabwe (S.c.)
- Indian Ocean: Madagascar (S.c.), Seychelles (S.c.), Reunion (S.c., P.n.), Mauritius (P.n.)
- Asia: China (S.c., S.o., P.n.), Sri Lanka (S.c., P.n.), Vietnam (S.c.), Malaysia (S.c., P.n.), India (S.c., P.n.), Indonesia (S.c., P.n.)
- Pacific Ocean: New Caledonia (S.c., P.n.), Papua New Guinea (S.c., P.n.), W. Irian (S.c.), Wallis Is. (S.c.), Western Samoa (S.c.), Federated States of Micronesia (S.c.), Fiji (S.c.), Tonga (P.n.), French Polynesia (P.n.), Vanuatu (P.n.), Cook Is. (P.p.), Hawaii (P.p.).

Natural enemies

Like the green scales, these species also are attacked by a range of natural enemies and solicited by ant species for their honeydew. The fungus, *Aschersonia sp.*, is recorded as attacking *S. coffeae* on tea in India (see Appendix A for full list).

Star scales

Asterolecanium coffeae Newstead, *A. hancocki* Laing, *A. pustulans princeps* Castel Branco [Hemiptera: Asterolecaniidae].

Morphology

The nymphal stages are yellowish in colour, whilst the adult female scale is about 1.5 mm long and pear-shaped in outline, reddish brown and covered in hair-like filaments that are particularly prevalent around the edge of the scale. For this reason, these scales are also known as 'fringed scales' (see Plate 7).

Pest status and distribution

Asterolecanium coffeae is found in East and Central Africa, mostly at lower altitudes. It can be a serious pest, and Crowe (1962a) considers it to be one of the few coffee pests that can kill the tree if not controlled. The scales are mostly found in crevices in the bark of the stem, although a few may be found on green wood in small pits. Trees coated with roadside dust are often more heavily infested than clean trees, which is thought to be due to suppression of natural enemies (Crowe, 1962a). Apart from *C. arabica*, the only known alternative host plants are *Jacaranda mimosifolia* and loquat, *Eriobotryia japonica*.

Asterolecanium pustulans princeps is known only from the islands of São Tomé and Príncipe, but the species as a whole is very widely distributed through the tropics and subtropics (CABI, 1984) and found on a range of cultivated and wild woody plants, including cocoa and tea. *Asterolecanium hancocki* has been recorded on coffee from Uganda (Laing, 1929) and an undetermined *Asterolecanium* species from the Cook Islands (Williams and Watson, 1990).

Life cycle

James (1932) found the life cycle of *A. coffeae* to last 60–70 d from hatching to oviposition, with the two nymphal stages lasting 7–12 and 18–25 d, respectively. Habib (1943) raised *A. pustulans* on potted *Nerium oleander* in Egypt. Eggs hatched in 18–29 d and the two nymphal stages lasted 2–6 and 10–17 d, respectively, whilst the adult females lived 73–87 d and oviposited only in the last 11–16 d of life.

Natural enemies

Both *A. coffeae* and *A. pustulans princeps* are attacked by Encyrtid parasites in the genus *Metaphycus* (see Appendix A), which appear to exert adequate control under normal circumstances but, unfortunately, they themselves are attacked by two Aphelinid hyperparasites, *Marietta buscki* (Howard) and *M. leopardina* Motschulsky.

Heavy outbreaks of *A. pustulans coffeae* on cocoa and coffee on Sao Tome in 1956 were attributed to attempts to control *Aspidiotus destructor* with insecticides that had destroyed the parasite *Encarsia citrina* – which had previously kept *A. pustulans* in check (Castel Branco, 1971). Several Coccinellid beetles predate the scales.

Control

Crowe (1962a) recommends painting infested bark liberally with 5% tar oil in water so that the mixture penetrates into cracks in the bark, but avoiding green bark or leaves, which can be scorched, and to prune infested trees to remove any colonies on green wood. In Sao Tome, stumping of coffee trees was recommended by Castel Branco (1971).

Mealy bugs

Planococcus citri (Risso), *P. kenyae* (Le Pelley), *P. lilacinus* (Cockerell), *P. minor* (Maskell), *Ferrisia virgata* (Cockerell) [Hemiptera: Pseudococcidae]

Morphology

Mealy bugs are small, wingless sucking insects covered by a dense covering of wooly wax, and usually live in quite dense colonies. Adult females of the *Planococcus* species are very similar in appearance in life (see Fig. 5.18a, c, d), being oval in shape, with a number of short, lateral, wax filaments and a light covering of powdery wax dorsally. *Ferrisia virgata* (see Fig. 5.18b) is more distinctive, having long tail filaments and sparse, hair-like wax dorsal filaments and several dark dorsal patches. The body length of the adult female is around 5–6 mm. Winged males do exist, but they are inconspicuous insects and, as they do not feed, are generally overlooked.

Pest status and distribution

Taxonomy of the *Planococcus* species has been confused in the past. The common coffee mealy bug in Kenya was thought for many years to be *P. citri*, but was later identified as *P. lilacinus*; however, in 1935, Le Pelley recognized it as a distinct species and renamed it *P. kenyae*. The root-living forms of *P. citri* in East Africa have also been shown to be distinct species, being named *P. fungicola* and *P. radicis*, respectively (see Chapter 6). In the Pacific region, the

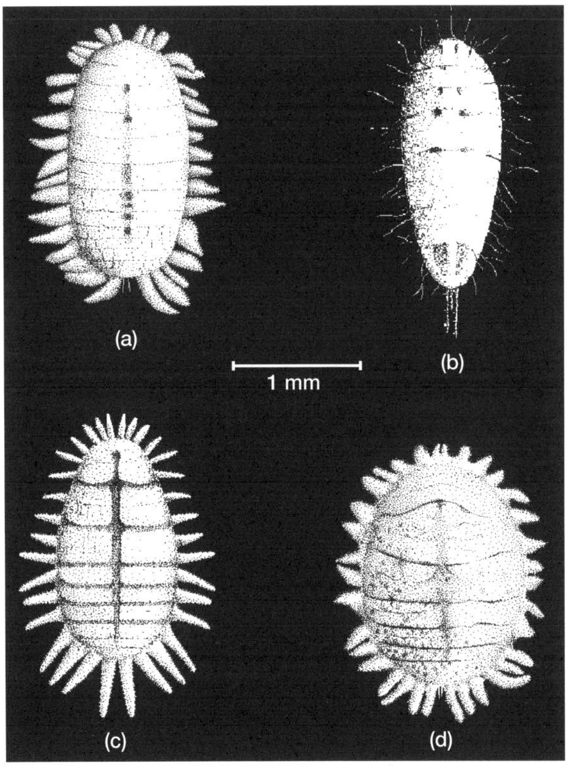

Fig. 5.18. Adult females of (a) *Planococcus citri*, (b) *Ferrisia virgata*, (c) *P. kenyae*, (d) *P. lilacinus*.

coffee mealy bug – once also thought to be *P. citri* – was recognized as distinct, being identified at first as *P. pacificus* and, later, as *P. minor*. See CABI Pest Distribution Maps for the general distribution of *P. citri* (CABI, 1969), *P. kenyae* (CABI, 1978), *P. lilacinus* (CABI, 1959) and *F. virgata* (CABI, 1966).

Records of mealy bugs on coffee are from (P.c., *Planococcus citri*; P.k., *Planococcus kenyae*; P.l., *Planococcus lilacinus*; P.m., *Planococcus minor*; F.v., Ferrisia virgata):

- Central America: Costa Rica (P.c., P.m.), El Salvador (F.v.), Guatemala (P.c., P.m., F.v.), Honduras (P.c.)
- Caribbean: Cuba (P.c.), Puerto Rico (P.c.)
- South America: Brazil (P.c., P.m.), Surinam (P.c.), Colombia (P.c., F.v.), Peru (P.c.), Argentina (P.m.)
- Atlantic Ocean: São Tomé and Príncipe (P.c.)
- West Africa: Sierra Leone (F.v.), Ghana (P.c.), Cameroon (F.v.), Ghana (F.v.), Togo (P.c.)
- Central Africa: Democratic Republic of Congo (P.c., P.k., F.v.)
- Eastern Africa: Tanzania (P.c., P.k., F.v.), Sudan (F.v.), Uganda (P.c., P.k., F.v.), Kenya (P.c., P.k., F.v.)
- Southern Africa: Malawi (P.c.), Zimbabwe (P.c.), Angola (P.c.), South Africa (P.c.)
- Indian Ocean: Madagascar (P.c., F.v.), Réunion (P.l.)
- Asia: China (P.c., F.v.), India (P.c., P.l., P.m., F.v.), Indonesia (P.c., P.l., F.v.), Philippines (P.c., F.v.), Taiwan (P.c., P.l.), Vietnam (P.c., P.l.)
- Pacific Ocean: Solomon Is (F.v.), Federated States of Micronesia (P.m., F.v.), Fiji (P.m., F.v.), New Caledonia (F.v.), Papua New Guinea (P.m., F.v.), Tonga (P.m.), Vanuatu (P.m.), W. Samoa (P.m.).

The species dealt with here feed on foliage, green shoots, flower and green cherry clusters – often in dense colonies and tended by ant species to a greater or lesser extent. Subterranean mealy bugs are dealt with in Chapter 6. The effect of large numbers of insects drawing off nutrients leads to debilitation of the plant and die-back of branches in extreme cases.

Life cycle

Eggs are laid in a woolly ovisac, which protrudes from the hind end of the body. Eggs of *F. virgata* hatch in 1–2 d, other species rather longer – 2–3 d in the case of *P. kenyae* and *P. minor*, and up to 10 d in the case of *P. citri*, depending on temperature. Females go through three nymphal stages, which average 7, 5.5 and 5 d for *P. citri* (Panis, 1969). The nymphal stages of the female *P. minor* at 26.4°C last a total of 19 d and the male, 23 d (Martinez and Suris, 1998).

Under laboratory conditions in East Africa, the first two female nymphal stages of *P. kenyae* were passed in 6–10 and 10–14 d, respectively, female development being completed in 36 d and the male in 33 d (Kirkpatrick, 1927). In Iraq, female nymphal development of *F. virgata* lasted from 43 d at 30°C to 93 d at 17°C (Awadallah *et al.*, 1979).

Natural enemies

Full records of natural enemies recorded on coffee from these species are shown in Appendix A. As with scale insects, the parasites and predators of mealy bugs are extremely important in checking population growth. During the early 1930s, devastating outbreaks of *P. kenyae* used to occur on coffee in the highlands of Kenya. Following unsuccessful attempts to introduce parasites of *P. lilacinus* from Indonesia, a search was made for parasites of *P. kenyae* in Uganda throughout 1937. *Anagyrus kivuensis* Compere was found to be very successful in controlling *P. kenyae* in Kenya, reducing the pest to minor status (Le Pelley, 1968).

A great many ladybird beetle species (Coccinellidae) predate *P. kenyae*, as well as larvae of Syrphid flies and carnivorous lepidopterous caterpillars.

Control

Like many other Coccoidea, mealy bugs produce copious amounts of honeydew rich in sugars, which are sought by various ant species as an important component of their diet. Many mealy bugs have specially adapted anal rings that concentrate the honeydew in a droplet that can easily be ingested by the ant. Ants help to keep the colonies free of parasites, predators and fungi and, if the ants are excluded, the colonies quickly become covered with sooty moulds.

Kirkpatrick (1927) found that the rate of increase of *P. kenyae* colonies on potted plants was three times greater when ants were present than when they were excluded. In India and Indonesia, *Anoplolepis gracilipes* (F. Smith) tends colonies of *P. citri*, whilst species of *Lepisiota*, *Monomorium* and *Pseudolasius* play the same role in East Africa. These same genera tend *P. kenyae* in East Africa (see Appendix A). If these ground-living ants can be excluded from the tree by the use of deterrent bands around the stem, then mealy bug numbers will decline.

In Colombia, where infestations of *P. citri* tend to occur as a result of frequent use of insecticides or fungicides, it is recommended that heavily infested trees be identified and infected branches drenched with a 0.2–0.5% solution of a short-lived contact insecticide in agricultural oil, whilst leaving less heavily infested trees as a reservoir for natural enemies (Cárdenas and Posada, 2001).

Fried egg scale

Aspidiotus sp. [Hemiptera: Diaspididae]

Morphology

Aspidiotus belongs to a group of scales called armoured scales that have a hard, semi-transparent covering to the soft body of the insect. That of fried egg scale is flat, about 2 mm in diameter, circular and with a yellowish centre surrounded by white.

Pest status and distribution

Fried egg scale appeared in Kenya in 1977 and built up to damaging proportions over the following years, but has since declined in importance, apparently being kept in check by its natural enemies. The insect has not been identified to species, but Crowe (2004) suggests that it might be *A. ruandensis* Balachowsky. If this is so, Kenya is a new locality for this species, which was described from Rwanda on *Euphorbia tommingi* (Balachowsky, 1955) and has additionally been recorded from Guinea and Cameroon, but never from coffee.

It is found primarily on leaves, being found on both surfaces, but will spread to other parts of the plant when infestation levels are high. Leaves become yellow and fall prematurely, and yield is affected.

Natural enemies

Mugo *et al.* (1997) have recorded several natural enemies from Kenya. The scale is parasitized by the Aphelinids, *Encarsia fuscus* (Compere) and *Marietta leopardina* Motschulsky; the latter is almost certainly a hyperparasite. They also record three Coccinellid predators, *Chilocoris nigripes* Mader, *Exochomus flavipes* Thunberg and *Hyperaspis senegalensis* Mulsant. To these can be added *Cybocephalus* sp., a Nitidulid beetle (Crowe, 2004).

Control

The scale probably came into prominence because of an imbalance due to insecticide use on coffee, and has since been adequately controlled by its natural enemies.

Black thread scale

Ischnaspis longirostris (Signoret) [Hemiptera: Diaspididae]

Morphology

The adult scale is black, with a brownish anterior end. It is long and thin, up to 3 mm long by about 0.5 mm wide and tapering towards the head end (see Fig. 5.19e). Eggs are yellow.

Life cycle

The eggs are laid under the female scale. Once hatched, the nymphs crawl out from under the mother and develop their own scale. Females go through two nymphal stages, but remain under the same scale throughout; males go through four moults before becoming adult. The life cycle from egg to adult female lasts around 30 d.

Distribution and pest status

The scale is widely distributed but is not generally a serious pest on coffee. Records of black thread scale from coffee are from:

- Caribbean: Dominican Republic, Cuba, Puerto Rico, Jamaica
- Central America: Mexico, Guatemala
- South America: Surinam, Guyana, Brazil, Colombia
- Atlantic Ocean: Sao Tome and Principe
- Central Africa: Democratic Republic of Congo
- Eastern Africa: Ethiopia, Kenya, Uganda, Tanzania, Zanzibar
- Indian Ocean: Seychelles, Mauritius
- Asia: India, Malaysia, Indonesia
- Pacific Ocean: Papua New Guinea, Tonga, French Polynesia, Fiji, Federated States of Micronesia, Western Samoa, Vanuatu.

Natural enemies

Natural enemies recorded from coffee include: (i) the parasite *Aphytis chrysomphali* (Mercet) from Kenya, Tanzania and Uganda; and (ii) the Coccinellid predators *Chilocorus distigma* Klug and *C. wahlbergi* Mulsant from Tanzania and *C. nigritus* (Fabricius) from India. In addition, these scales are attacked by fungi. *Septobasidium bogoriense* Pat. is recorded from Java and *Aschersonia* sp. from Colombia.

Wax scales

Ceroplastes destructor (Newstead), *C. brevicauda* Hall [Hemiptera: Coccidae]

Morphology

The adult females live beneath a white, waxy scale; they are broadly oval in outline, about 6 mm long by 5 mm wide and domed, but irregular in shape. They are found on green shoots, whereas the first-stage nymphs, described below, cluster on the leaf midribs.

Pest status and distribution

Both species occur naturally through many parts of Africa, including Madagascar, but *C. destructor* has been introduced to a number of Pacific countries (CABI, 1960). On coffee, there are records of *C. brevicauda* from Angola, Ethiopia, Ghana, Kenya, Nigeria, Uganda and Zimbabwe, and of *C. destructor* from Angola, Malawi, Uganda, Papua New Guinea and Democratic Republic of Congo.

In Africa, both species are usually of minor importance because they are kept in check by their natural enemies, but *C. destructor* is a pest of citrus and, where it became established as an introduced pest in the Pacific (Australia, Papua New Guinea, Solomon Islands and New Zealand), it has become a serious pest.

Starting in the 1930s, unsuccessful attempts have been made to introduce parasites into Australia from Kenya and Uganda, but *Anicetus communis* Annecke and *A. nyasicus* (Compere) were successfully introduced from South Africa between 1968 and 1973 (Sands *et al.*, 1986). In Papua New Guinea, it became a serious pest of coffee as well as of citrus, but the introduction of *A. nyasicus* in 1982 has brought it under control (Williams, 1987).

Life cycle

The life cycle of *C. brevicauda* on coffee in Kenya has been described by Crowe (1962b). The eggs are laid beneath the female, where they hatch into nymphs. The first nymphal stage is active, and the nymphs move away from the mother to become established beside the main vein of a hardened leaf, usually on the upper surface. The first waxy covering is produced after 2 d in the form of a central dome, with a number of lateral waxy plates described by Crowe as the 'cameo brooch stage (see Fig. 5.19a).

This stage is followed by a 'star stage', in which the lateral wax plates coalesce with the central dome and expand into a number of lateral triangular protruberances (see Fig. 5.19b). After this stage, the nymphs move to a green shoot, where the wax scale becomes cone-shaped and more circular (see Fig. 5.19c). Here, they finally settle down and the scale progressively thickens and becomes more rounded (see Fig. 5.19d). The entire life cycle lasts around 6 months.

Natural enemies

In Africa, both species are attacked by a number of parasitic wasps and in Kenya, *C. brevicauda* is predated by a Coccinellid beetle, *Cheilomenes lunata* (Fabricius) and two carnivorous moth larvae, *Coccidiphaga scitula* (Rambur) and *Eublemma costimacula* (Saalmuller) (see Appendix A).

Black citrus aphid

Toxoptera aurantii (Boyer de Fonscolombe). [Hemiptera: Aphididae]

Some other species of aphids have been recorded from coffee, including *Aphis gossypii* Glover and *Myzus persicae* Sulzer in Colombia, but the most commonly found and widespread species is *Toxoptera aurantii*.

Morphology

The adults, which are oval and black and about 1.5–2.0 mm long, come in two forms: those with wings and those without (see Fig. 5.19f).

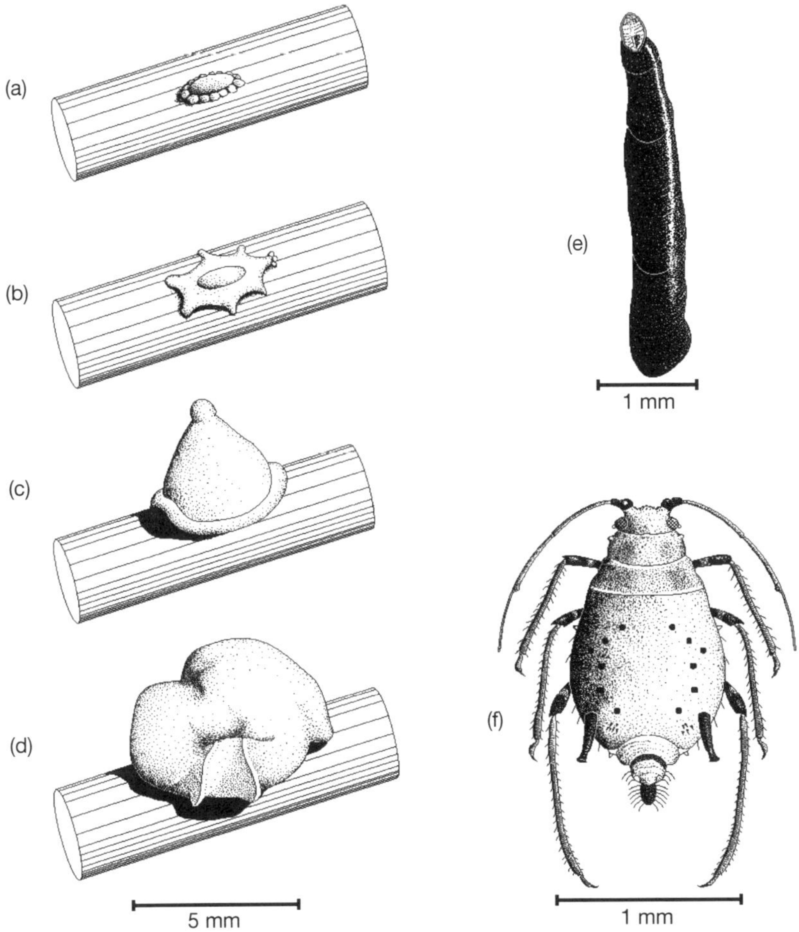

Fig. 5.19. *Ceroplastes brevicauda*: (a) 'cameo-brooch' stage, (b) star stage, (c) cone stage, (d) adult female. *Ischnaspis longirostris*: (e) adult female; *Toxoptera aurantii*: (f) apterous form.

Life cycle

Colonies are found on the underside of young leaves and on green shoots. The females reproduce almost entirely without the help of males, producing around 20 live offspring (there is no egg stage). Wingless females become adult within about 6 d and winged females in 7–8 d.

Pest status and distribution

Feeding results in some leaf distortion and malformation of shoots but, generally speaking, damage is minor on mature trees, but may be more significant on seedlings. The insect is said to be the vector of coffee ringspot virus of *Coffea liberica var. dewevrei* (*c. excelsa*) (CABI, 2003), which has also been shown to be transmitted by the mite, *Brevipalpus phoenicis*, in Brazil (see below).

The insect has a very wide geographical distribution (CABI, 1961a) and has been recorded from a large number of different host plants (at least 140 species according to CABI, 2003), including citrus, cocoa, mango and tea.

Records of black citrus aphid from coffee are from:

- Caribbean: Puerto Rico, Cuba, Guadeloupe, Martinique
- Central America: Mexico, Costa Rica, Honduras
- South America: Venezuela, Brazil, Colombia, Peru
- Atlantic Ocean: São Tomé and Príncipe
- West Africa: Côte d'Ivoire, Ghana
- Central Africa: Democratic Republic of Congo
- Eastern Africa: Ethiopia, Kenya, Tanzania, Uganda
- Southern Africa: Angola, Malawi
- Asia: India, Sri Lanka, Indonesia, Malaysia, Vietnam, Thailand
- Pacific Ocean: Hawaii, Papua New Guinea.

Natural enemies

The aphid is attacked by a few internal parasites (see Appendix A), but by a much larger assemblage of predators including Coccinellid beetles, Syrphid fly larvae and lacewings (Chrysopidae). Because of the short life cycle of the aphid, numbers can build up rapidly and predators struggle to keep ahead. The insects are commonly solicited by ants for their honeydew.

Control

Chemical control measures are not necessary on mature trees, but may sometimes be needed on nursery plants.

Thrips

Thrips are minute insects that have rasping mouthparts. Many species live on pollen and have been recorded from coffee flowers, including *Haplothrips tenuipennis* Bagnall (Sekhar and Sekhar, 1964), *Frankliniella parvula* Hood (Billes, 1941), *Thrips florum* Schmutz (Sekhar and Sekhar, 1964) and *Thrips parvispinus* Karny (Kalshoven, 1950–1951) but, because coffee flowering is such a short-lived phenomenon, they probably do little harm (see Appendix B). Around 30 species are leaf feeders and there are some predaceous species in the genus *Franklinothrips*.

Coffee thrips

Diarthrothrips coffeae Williams [Thysanoptera: Aeolothripidae]

Morphology

The adult is greyish brown, about 1.5–2.0 mm long and with fringed wings. Larvae are smaller and yellowish, similar in general appearance to the adult, but without wings.

Pest status and distribution

The species is confined to Eastern Africa and has been recorded on coffee from Ethiopia, Kenya, Tanzania, Uganda and Malawi. It seems to become troublesome as a pest only during or after prolonged spells of dry weather (Notley, 1936). Feeding on the epidermis of the underside of the leaf results in silvery patches appearing on the upper surface, leading eventually to defoliation, abortion of flower buds and crop loss.

Life cycle

The kidney-shaped eggs are laid in a slit cut by the female in the leaf. There are two larval stages passed on the leaf. At the end of the second stage, the larvae drop off the tree onto the ground and go through quiescent pre-pupal and pupal phases.

Natural enemies

A predaceous species of thrips, *Flanklinothrips megalops* (Trybom) (Aelothripidae), has been recorded attacking the insect in Kenya (Le Pelley, 1959).

Control

The ideal form of control is to try to avoid conditions which favour build-up of the pest, by ensuring that the trees do not suffer stress during dry conditions, i.e by mulching to improve humus content of the soil and the use of cover crops, etc. Fenitrothion is effective for chemical control should this become necessary.

Black tea thrips

Heliothrips haemorhoidalis Bouche [Thysanoptera: Thripidae]

Morphology

The adult is uniformly dark brown and about 1.5–2.0 mm in length. Mature larvae are yellow, with the end of the abdomen brown.

Pest status and distribution

On coffee, this species is distributed through South America and Asia into the Pacific, but not in Africa (CABI, 1961b). It has been recorded on coffee from Trinidad and Tobago, Surinam, Brazil, Colombia, China, Taiwan, India, Indonesia, Malaysia, Hawaii and Tonga. It is not usually a serious pest on coffee, and Le Pelley (1968) suggests that because it is often found in small numbers on coffee when this is growing close to heavily infested cocoa, coffee may not be a suitable host plant.

Life cycle

The eggs are laid singly underneath the leaf and hatch in around 14–15 d. There are two larval stages, lasting 9–11 and 10–16 d, followed by pre-pupal and pupal resting stages lasting 3–4 and 4–6 d, respectively. Unlike *D. coffeae*, all these stages are passed on the leaf. Feeding takes place on the underside of the leaf, leading to silvering of the top surface and leaf fall. The excrement produced by adults and nymphs persists as a black peppering on the undersurface. In Colombia, *Coffea canephora* is said to be the most badly affected species (Cárdenas and Posada, 2001).

Natural enemies

Megaphragma mymaripenne Timberlake (Trichogrammatidae) is recorded as an egg parasite in Hawaii (Pemberton, 1931). On cocoa, a Eulophid, *Goetheana shakespearei* Girault, is a parasite of this species (Callan, 1943) and two species of thrips are predators. *Franklinothrips tenuicornis* Hood (Aeolothripidae) predates it in Trinidad (Callan, 1943) and *Franklinothrips vespiformis* Crawford in Trinidad and Mexico (Johansen, 1976).

Red-banded thrips

Selenothrips rubrocinctus (Giard) [Thysanoptera: Thripidae]

Morphology

This is one of the more easily recognizable thrips species, because the pale yellowish nymphs have a conspicuous red band across the first two abdominal segments. Adults are brown and around 1.5–2.0 mm long and also have the red abdominal band (see Fig. 5.20).

Pest status and distribution

The red-banded thrips is better known as a major pest of cocoa, but it has been recorded from coffee in a number of countries, including Ghana, Indonesia,

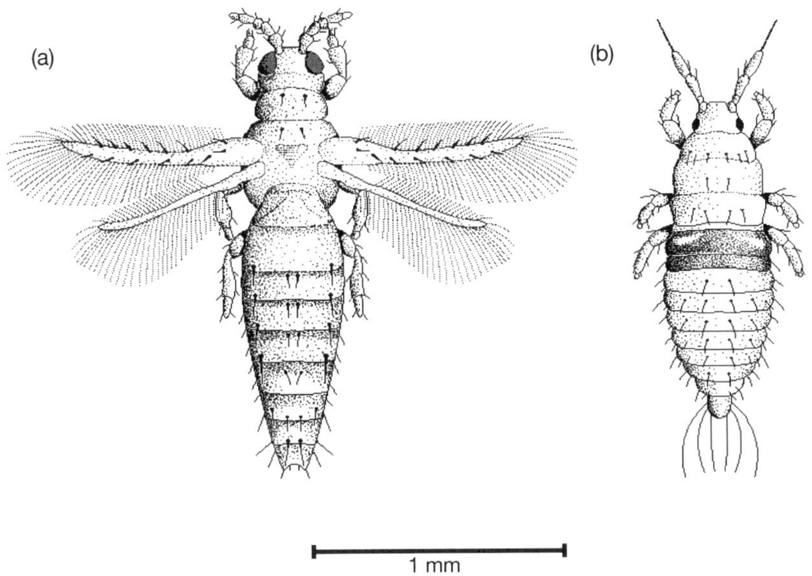

Fig. 5.20. Adult, (a) and nymph, (b) of *Selenothrips rubrocinctus*.

Trinidad and Tobago, Surinam, Brazil and Colombia, although it does not seem to be particularly serious (CABI, 1961c).

Feeding takes place on the underside of the leaf, leading to silvering of the top surface and leaf fall. The excrement produced by adults and nymphs persists as a black peppering on the under surface. In Colombia, *Coffea canephora* is said to be the most badly affected species (Cárdenas and Posada, 2001).

Life cycle

All stages of the insect are found on the leaf. The eggs are deposited singly and hatch in around 10–12 d. There are two larval stages, each lasting about 10 d, followed by pre-pupal and pupal resting stages of about 1 and 2–3 d, respectively, before the adult emerges.

Natural enemies

Eggs are parasitized by the Eulophid, *Goetheana shakespearei* Girault, in Ghana, Indonesia, Puerto Rico and Trinidad and Tobago, and the predatory thrip, *Franklinothrips vespiformis* (Crawford), has been recorded attacking it in Brazil (Vayssiére, 1955).

Tea thrips

Scirtothrips bispinosus (Bagnall) [Thysanoptera]

Pest status and distribution

The insect is a serious pest of tea and coffee in Southern India. It feeds on tender leaves, causing leaf distortion.

Life cycle

Eggs are inserted into the leaf and hatch in 7–11 d. There follow two larval feeding stages that last about 10 d, and then pre-pupal and pupal stages of around 7 d. Some individuals pass the latter stages on the leaf and some complete them on the ground.

Natural enemies

A predaceous thrip, *Franklinothrips* sp., and two mite species, *Typhlodromus* sp. and *Leptus* sp., attack this species in India.

Control

On tea, the use of a spore suspension of *Verticillium lecanii* at 1.5 kg/ha has been recommended for control of *S. bispinosus* and various chemicals are used, including chloropyrifos, quinalphos, fenthion, diazinon, dimethoate and endosulfan.

Mites

Also found in association with coffee are the mites belonging to the Class Acari (Acarina) of the Arachnide. Most are minute, many are phytophagous but others are predaceous, feeding on other mites and some insects.

Citrus flat mite, false spider mite

Brevipalpus californicus (Banks), *B. obovatus* Donnadieu, *B. phoenicis* (Geijskes) [Acari: Tenuipalpidae]

Pest status and distribution

Feeding by these mites causes silvering of leaves that later turn brown. Normally, the physical damage is not serious, but *B. phoenicis* is the vector in Brazil (Chagas *et al.*, 2003) and Costa Rica (Rodrigues *et al.*, 2002) of coffee ringspot virus, which causes leaf fall and off-flavour in coffee.

On coffee, *B. phoenicis* has been recorded from Mexico, Costa Rica, Brazil, Kenya, Tanzania and India (CABI, 1970), *B. obovatus* from Brazil and India (CABI, 1961d) and *B. californicus* from Brazil (CABI, 1975). The three species are very similar in appearance and have been confused with each other in the past. Welbourn *et al.* (2003) list characters that can be used to separate the species (see Fig. 5.21).

Life cycle

The life cycles of the three species are similar. All stages of *B. phoenicis*, including the egg, are scarlet red. Eggs are around 0.09 mm long and laid on the underside of the leaf or in bark crevices and hatch in 9–22 d. The larva is six-legged, 0.15 mm long and, after 3–10 d and a short resting period, changes into a protonymph. After a further 2–8 d, the protonymph, which is about 0.19 mm long, changes into a deutonymph, again after a short rest. This stage, which lasts 2–8 d with a further resting phase, then turns into an adult. The adult is around 0.28 mm in length, the female being slightly large than the male.

Natural enemies

In Brazil, *B. phoenicis* is attacked by several predatory mites including *Amblyseius herbicolus* (Chant) (Reis, 2002), *Euseius alatus* De Leon and *Iphiseiodes zuluagai* Denmark and Muma (Reis *et al.*, 2000).

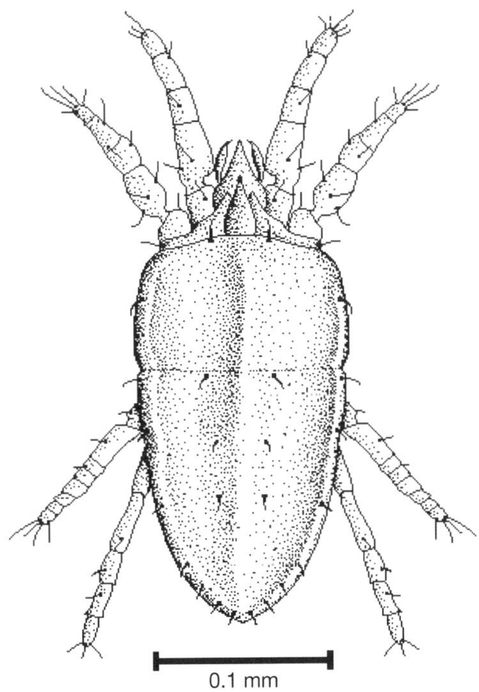

Fig. 5.21. Adult of *Brevipalpus phoenicis*.

References

Adomako, D. (1972) Studies on mealybug (*Planococcoides njalensis* (Laing)) nutrition: a comparative analysis for the free carbohydrate and nitrogenous compounds in cocoa bark and mealybug honeydew. *Bulletin of Entomological Research* 61, 523–531.

Anderson, T.J. (1934) *Entomology Section*. Report, Department of Agriculture, Kenya, 1933, pp. 137–145.

Anon (1965) *A Host List of Insects of Thailand*. Department of Agriculture of the Royal Thai Government and US Operations Mission to Thailand, 149 pp.

Anon (1967) *Coffee Pests and their Control*. Coffee Research Foundation, Kenya, 90 pp.

Arce, J.J.C., Peres Filho, O. and Berti Filho, E. (1987) Biologia do bicho do cesto *Oiketicus kirbyi* (Landes-Guilding 1827) (Lepidoptera Psychidae) em folhas de Eucalyptus spp. *Anais da Escola Superior de Agricultura 'Luiz de Queiroz', Universidade de São Paulo* 44, 341–357.

Awadallah, K.T., Ammar, E.D., Tawfik, M.F.S. and Rashid, A. (1979) Life history of the white mealy-bug *Ferrisia virgata* (Ckll.) (Homoptera, Pseudococcidae). *Deutsche Entomologische Zeitschrift* 26, 101–110.

Ba-Angood, S.A. and Al-Sunaidi, S.M. (2004) The biology of the coffee leaf miner, *Leucoptera caffeina* Wash. (Gracillariidae: Lepidoptera) and the effect of irrigation on its development on coffee leaves in Rubat and Hilam-Yahr areas at Yafe'a – Republic of Yemen. *University of Aden Journal of Natural and applied Sciences* 8, 195–203.

Balachowsky, A.S. (1955). Contributions à l'étude de la faune entomologique du Ruanda Urundi (mission P. Basilewsky 1935). Pt I. Homoptera Coccoidea. Sur un *Aspidiotus* Bouché (Diaspididae-Aspidiotini) nouveau du Ruanda. *Annales du Musée Royal du Congo Belge. Tervuren* 36, 390–393.

Bardner, R. and Mathenge, W.M. (1974) Organo-tin compounds against caterpillars on coffee leaves. *Kenya Coffee* 39, 257–259.

Bardner, R. and Mcharo, E.Y. (1988) Confirmation of resistance of the coffee leafminer *Leucoptera meyricki* Ghesquiere (Lepidoptera, Lyonetiidae) to organophosphate insecticide sprays in Tanzania. *Tropical Pest Management* 34, 52–54.

Barrera, J.F., Muñoz, J., Zúñiga, A., Morales, B., Ventura, S. and Campos, S. (2003) Aspectos bioecológicos del chacuatete *Idiarthron subquadratum* (Orthoptera: Tettigoniidae) en cafetales de Chiapas, México. *12th Latin American Coffee-growing Symposium, IHCAFE-PROMECAFE*, Honduras, 2003.

Beeson, C.F.C. (1941) *The Ecology and Control of the Forest Insects of India and the Neighbouring Countries*. Vasant Press, Dehra Dun, India, 1007 pp.

Berry, P.A. and Abrego, L. (1953) Insects and diseases affecting some crops in El Salvador. *F.A.O. Plant Protection Bulletin* 1, 151–153.

Bess, H.A. (1964) Populations of the leaf miner *Leucoptera meyricki* Ghesq. and its parasites in sprayed and unsprayed coffee in Kenya. *Bulletin of Entomological Research* 55, 59–82.

Bigger, M. (1967) Pest profiles. II – Leaf miners. *Tanganyika Coffee News* 7, 9–16.

Bigger, M. (1969a) Partial resistance of arabica coffee to the coffee leaf miner, *Leucoptera meyricki* Ghesq. (Lepidoptera, Lyonetiidae). *East African Agricultural and Forestry Journal* 34, 441–445.

Bigger, M. (1969b) Giant looper *Ascotis selenaria reciprocaria* Walk. in Tanzania. *East African Agricultural and Forestry Journal* 35, 49–51.

Bigger, M. (1973) An investigation by fourier analysis into the interaction between coffee leaf miners and their larval parasites. *Journal of Animal Ecology* 42, 417–434.

Bigger, M. and Tapley, R.G. (1969) Prediction of outbreaks of coffee leaf miners on Kilimanjaro. *Bulletin of Entomological Research* 58, 601–617.

Billes, D.J. (1941) Pollination of *Theobroma cacao* L. in Trinidad, B.W.I. *Tropical Agriculture* 18, 151–156.

Bradley, J.D. (1958) Taxonomic notes on *Leucoptera meyricki* Ghesquiere and *Leucoptera coffeella* (Guerin Meneville). *Bulletin of Entomological Research* 49, 417–419.
Bradley, J.D. (1987) *Card Catalogue of Identified Reared Material Received by Bradley for Identification 1955 to 1987*. Manuscript, British Museum of Natural History, London.
Bruneau de Mire, P. and Muller, R.A. (1965) Brève revue des problèmes phytosanitaire posés au Cameroun en matière de recherche sur le caféier. *Institut Francais du Café, du Cacao et autres Plantes Stimulantes*, July 1965, 11 pp. (typescript).
CABI (1955) *Pulvinaria psidii* Maskell. *Distribution Maps of Pests* No. 59. CAB International, Wallingford, UK, 2 pp.
CABI (1959) *Planococcus lilacinus* (Cockerell). *Distribution Maps of Pests* No. 101. CAB International, Wallingford, UK, 2 pp.
CABI (1960) *Ceroplastes destructor* Newstead. *Distribution Maps of Pests* No. 117. CAB International, Wallingford, UK, 2 pp.
CABI (1961a) *Toxoptera aurantii* (Boyer de Fonscolombe). *Distribution Maps of Pests* No. 131. CAB International, Wallingford, UK, 2 pp.
CABI (1961b) *Heliothrips haemorrhoidalis* Bouche. *Distribution Maps of Pests* No. 135. CAB International, Wallingford, UK, 2 pp.
CABI (1961c) *Selenothrips rubrocinctus* (Giard). *Distribution Maps of Pests* No. 136. CAB International, Wallingford, UK, 2 pp.
CABI (1961d) *Brevipalpus obovatus* Donnadieu. *Distribution Maps of Pests* No. 128. CAB International, Wallingford, UK, 2 pp.
CABI (1966) *Ferrisia virgata* (Cockerell). *Distribution Maps of Pests* No. 219. CAB International, Wallingford, UK, 2 pp.
CABI (1969) *Planococcus citri* (Risso). *Distribution Maps of Pests* No. 43 (revised). CAB International, Wallingford, UK, 2 pp.
CABI (1970) *Brevipalpus phoenicis* (Geijskes). *Distribution Maps of Pests* No. 106 (revised). CAB International, Wallingford, UK, 2 pp.
CABI (1972) *Coccus viridis* (Green). *Distribution Maps of Pests* No. 305. CAB International, Wallingford, UK, 2 pp.
CABI (1973a) *Leucoptera coffeella* (Guerin-Meneville). *Distribution Maps of Pests* No. 315. CAB International, Wallingford, UK, 2 pp.
CABI (1973b) *Leucoptera meyricki* Ghesquiere. *Distribution Maps of Pests* No. 316. CAB International, Wallingford, UK, 2 pp.
CABI (1973c) *Saissetia coffeae* (Walker). *Distribution Maps of Pests* No. 318. CAB International, Wallingford, UK, 2 pp.
CABI (1973d) *Saissetia oleae* (Olivier). *Distribution Maps of Pests* No. 24 (revised). CAB International, Wallingford, UK, 2 pp.
CABI (1974a) *Zonocerus variegatus* (Linnaeus). *Distribution Maps of Pests* No. 322. CAB International, Wallingford, UK, 2 pp.
CABI (1974b) *Homona coffearia* Nietner. *Distribution Maps of Pests* No. 330. CAB International, Wallingford, UK, 2 pp.
CABI (1975) *Brevipalpus californicus* (Geijskes). *Distribution Maps of Pests* No. 107 (revised). CAB International, Wallingford, UK, 2 pp.
CABI (1978) *Planococcus kenyae* (Le Pelley). *Distribution Maps of Pests* No. 384. CAB International, Wallingford, UK, 2 pp.
CABI (1982) *Atta cephalotes* (Linnaeus). *Distribution Maps of Pests* No. 444. CAB International, Wallingford, UK, 2 pp.
CABI (1984) *Asterolecanium pustulans* (Cockerell). *Distribution Maps of Pests* No. 460. CAB International, Wallingford, UK, 2 pp.
CABI (1985) *Cephonodes hylas* (Linnaeus). *Distribution Maps of Pests* No. 471. CAB International, Wallingford, UK, 2 pp.

CABI (1986) *Parasa lepida* (Cramer). *Distribution Maps of Pests* No. 363 (revised). CAB International, Wallingford, UK, 2 pp.

CABI (2003) *Crop Protection Compendium 2003 Edition*. CAB International, Wallingford, UK.

Callan, E.M. (1943) Natural enemies of the cacao thrips. *Bulletin of Entomological Research* 34, 313–321.

Cárdenas, M.R. and Posada, F.J. (2001) *Los Insectos y Otres Habitantes de Cafetales y Platanales*. Centro Nacional de Investigaciones de Café, Colombia, 250 pp.

Castel Branco, A.J.F. (1971) Cochonilhas de S.Tomé e Príncipe *Asterolecanium pustulans* (Cockerell) e *A. pustulans princeps* nov. var. (Hemiptera Homoptera). *Eos* 46, 27–36.

Chagas, C.M., Kitajima, E.W. and Rodrigues, J.C.V. (2003) Coffee ringspot virus vectored by *Brevipalpus phoenicis* (Acari: Tenuipalpidae) in coffee. *Experimental and Applied Acarology* 30, 203–213.

Cheo, M.T. (1936) A preliminary list of the insects and arachnids injurious to economic plants in China. *Peking Natural History Bulletin* 10, 5–37, 93–114, 167–182, 291–308; 11, 119–127, 281–286, 417–432.

Cock, M.J.W., Godfray, H.C.J. and Holloway, J.D. (1987) *Slug and Nettle Caterpillars. The Biology, Taxonomy and Control of the Limacodidae of Economic Importance on Palms in South East Asia*. CAB International, Wallingford, UK, 270 pp.

Coffee Research Foundation (1961) *An Atlas of Coffee Pests and Diseases*, Coffee Board of Kenya.

Corbett, G.H. (1935) Report of the Division of Entomology, 1934. *General Series. Department of Agriculture, Straits Settlements and Federated Malay States* 21, 43–56.

Corbett, G.H. and Yusope, M. (1932) The coffee clear wing hawk moth (*Cephanodes hylas* L.) *Malayan Agricultural Journal* 20, 508–517.

Cotte, O. (1989) *Oxydia vesulia* (Cramer) (Lepidoptera Geometridae): a geometrid caterpillar of coffee in Puerto Rico. *Journal of Agriculture of the University of Puerto Rico* 73, 159–160.

Crowe, T.J. (1962a) The star scale. *Kenya Coffee* 27, 9–11.

Crowe, T.J. (1962b) The white waxy scale. *Kenya Coffee* 27, 93–95.

Crowe, T.J. (2004) Coffee pests in Africa. In: Wintgens, J.N. (ed.) *Coffee: Growing, Processing, Sustainable Production. A Guidebook for Growers, Processors, Traders and Researchers*. Wiley, Chichester, UK, pp. 421–458.

Crowe, T.J. and Greathead, D.J. (1970) Parasites of *Leucoptera spp.* on coffee in East Africa: an annotated list. *East African Agricultural and Forestry Journal* 35, 364–371.

Dammerman, K.W. (1929) *Agricultural Zoology of the Malay Archipelago*. De Bussy, Amsterdam, 470 pp.

Decelle, J. (1962) La chenille des feuilles du caféier Robusta, *Leucoptera coma* Ghesq. *Parasitica* 18, 177–198.

De Fluiter, H.J. (1960) Krankheiten und schädlinge. In: Von Bernegg, S. (ed.) *Tropische und Subtropische Weltwirtschafts-pflanzen III. Tiel Band 2*. Kaffe, Stuttgart, Germany, pp. 96–102.

Dos Santos, B.B. (1988) Artropodos associados as plantas cultivadas no estado do Parana. II. Cafeeiro e mandioca. *Revista do Setor de Ciencias Agrarias* 10, 27–28.

Duport, M. (1912–1913) Notes sur quelques maladies et ennemis des plantes cultiveés en Extreme Orient. *Bulletin Économique de l'Indochine*, pp. 99, 102, 105, 147.

Duport, L. (1915) Rapport a M. le President de la Chambre d'Agriculture du Tonkin at du Nord-Annam sur les travaux effectues en 1914 a la Station entomologique de Cho-ganh. *Supplement de Bulletin 102 Chambre Agric. Tonkin, N. Annam*, 7 pp.

El-Minshawy, A.M. and Moursi, K. (1976) Biological studies on some soft scale-insects (Hom., Coccidae) attacking guava trees in Egypt. *Zeitschrift fur Angewandte Entomologie* 81, 363–371.

Entwistle, P.F. (1963) Observations on the biology of four species of Psychidae (Lepidoptera) on *Theobroma cacao* L. in Western Nigeria. *Proceedings of the Royal Entomological Society of London (A)* 38, 145–152.
Entwistle, P.F. (1972) *Pests of Cocoa*. Longman (Tropical Science Series), London, 779 pp.
Evans, D.E. (1966) *Entomology*. Report. Coffee Research Foundation, Kenya, 1965–1966.
Evans, D.E., Andrade, O. and Mathenge, W.M. (1968) The biology and control of *Archips occidentalis* (Wals.) and *Tortrix dinota* Meyr. (Lepidoptera Tortricidae) on coffee in Kenya. *Journal of the Entomological Society of Southern Africa* 31, 133–140.
Foucart, G. (1954) Un nouvel ennemi du caféier d'Arabie au Kivu *Habrochila placida* (note préliminaire). *Bulletin INEAC* 3, 51–62.
Fragoso, D.B., Guedes, R.N.C., Picanco, M.C. and Zambolim, L. (2002) Insecticide use and organophosphate resistance in the coffee leaf miner *Leucoptera coffeella* (Lepidoptera: Lyonetiidae). *Bulletin of Entomological Research* 92, 203–212.
Griveaud, P. (1967) Lépidoptères defoliateurs des rubiacées en Côte d'Ivoire. *Cahiers ORSTOM Série Biologie* 3, 3–41.
Guerreiro Filho, O., Eskes, A.B., Decazy, B. and Charrier, A. (1993) Résistance de *Coffea* spp. à la mineuse des feuilles du caféier, *Perileucoptera coffeella* (Guerin-Meneville, 1842) (Lepidoptera, Lyonetiidae) *15th Colloque Scientifique International sur le Café*, Montpellier, France, 6–11 June 1993, pp. 783–785.
Guerreiro Filho, O., Silvarolla, M.B. and Eskes, A.B. (1999) Expression and mode of inheritance of resistance in coffee to leaf miner, *Perileucoptera coffeella. Euphytica* 105, 7–15.
Gullan, P.J. (1999) Why the taxon Homoptera does not exist. *Entomologica* 33, 101–104.
Habib, A. (1943) The biology and bionomics of *Asterolecanium pustulans* Ckll. (Hemiptera, Coccidae). *Bulletin de la Société Fouad Ier d'Entomologie* 27, 87–111.
Hanks, L.M. and Sadof, C.S. (1990) The effect of ants on nymphal survival of *Coccus viridis* (Homoptera: Coccidae). *Biotropica* 22, 210–213.
Hill, D.S. (1975) *Agricultural Insect Pests of the Tropics and their Control*. Cambridge University Press, London, 516 pp.
James, H.C. (1932) The Control of *Asterolecanium*. *Bulletin, Department of Agriculture, Kenya* 23, 4 pp.
Jardine, J.T., Frome, F.D. and Knight, H.L. (1941) *Puerto Rico. Report on the Agricultural Experimental Stations*, 1940, 116–117.
Jepson, F.P. (1935) Report of the Entomology Division. *Administrative Report of the Director of Agriculture, Ceylon* 1934, D132–D147.
Johansen, R.M. (1976) Algunos aspectos sobre la conducta mimetica de *Franklinothrips vespiformis* (Crawford) (Insecta: Thysanoptera). *Anales del Instituto de Biologia, Universidad Nacional Autonoma de Mexico, Serie Zoologia* 47, 25–50.
Jutsum, A.R., Cherrett, J.M. and Fisher, M. (1981) Interactions between the fauna of citrus trees in Trinidad and the ants *Atta cephalotes* and *Azteca* sp. *Journal of applied Ecology* 18, 187–195.
Kalshoven, L.G.E. (1950–1951) *De Plagen van de Cultuurgewassen in Indonesië*. Van Hoeve, Gravenhage, Netherlands/Bandoeng, Indonesia, 1065 pp.
Kapoor, K.N., Deobhakta, S.R. and Dhamdhere, S.V. (1985). Bionomics of the slug caterpillar, *Latoia lepida* (Cramer) (Lepidoptera: Limacodidae) on mango. *Journal of Entomological Research* 9, 235–236.
Khoo, K.C., Ooi, P.A.C. and Ho, C.T. (1991) *Crop Pests and their Management in Malaysia*. Tropical Press, Kuala Lumpur, 242 pp.
Kirkpatrick, T.W. (1927) The common coffee mealy bug (*Pseudococcus lilacinus Ckll.*) in Kenya Colony. *Bulletin. Department of Agriculture, Kenya* 18, 110 pp.
Köhler, G. (1976) Beitrag zur kenntnis des Männchens der Grünen Kaffeechildlaus, *Coccus viridis* (Green) (Hemiptera: Coccinea, Coccidae). *Beiträge zur Entomologie* 26, 471–477.

Koningsberger, J.C. (1908) Tweede overzicht de schadelijke en nuttige insecten. *Mededeelingen Uitgeven van het Departement van Landbouw in Nederlandsch-Indie* 6, 1–113.

Kozhechkin, O.A. (1984) Biological principles for mass-rearing of the hemisperical scale. *Zashchita Rastenii* 1984 No. 1, 42 pp.

Kutywayo, D. (1989) An annotated list of parasitoids and predators of coffee insect pests in Zimbabwe. *Zimbabwe Journal of Agricultural Research* 27, 11–20.

Laing, F. (1929) Descriptions of new, and some notes on old, species of Coccidae. *Annals and Magazine of Natural History* 4, 465–501.

Leeuwangh, J. (1965) The biology of *Epigynopteryx stictigramma* (Hampson), a pest of coffee in Kenya. *Journal of the Entomological Society of Southern Africa* 28, 21–31.

Le Pelley, R.H. (1932a) On the pest status of certain coffee-feeding insects, with records of some insects newly recorded from coffee in Kenya. *Journal of the East Africa and Uganda Natural History Society* 41, 67–77.

Le Pelley, R.H. (1932b) *Lygus*, a pest in coffee in Kenya Colony. *Bulletin of Entomological Research* 23, 85–99.

Le Pelley, R.H. (1959) *Agricultural Insects of East Africa*. East African High Commission, Nairobi, 307 pp.

Le Pelley, R.H. (1968) *Pests of Coffee*. Longmans, London, 590 pp.

Martinez, M. de los A. and Suris, M. (1998) Biologia de *Planococcus minor* Maskell (Homoptera: Pseudococcidae) en condiciones de laboratorio. *Revista de Proteccion Vegetal* 13, 199–201.

Moriyama, T. (1941) Insects injurious to coffee berries. *Formosan Agricultural Review* 37, 14–20.

Mugo, H.M. (1992) Possibility of using natural enemies to control Giant looper, *Ascotis selenaria reciprocaria* Walk. in Kenya. *Kenya Coffee* 57, 1453–1455.

Mugo, H.M., Omondi, G.O. and Ndugo, S.M. (1997) Biological control of coffee insect pests in Kenya. *Dix-septième Colloque Scientifique International sur le Café*, Nairobi, Kenya, 20–25 July 1997, pp. 646–652.

Mugo, H.M., Dooso, B.S. and Ndoiru, S.N. (1998) Laboratory and field performance of *Bacillus thuringiensis* (Berliner) formulations against giant looper; *Ascotis selenaria reciprocaria* (Lepidoptera, Geometridae) in Kenya. *Kenya Coffee* 63, 2763–2768.

Muller, D., De Groote, H., Gbongboui, C. and Langewald, J. (2002) Participatory development of the biological control strategy of the variegated grasshopper in the humid tropics in West Africa. *Crop Protection* 21, 265–275.

Munoz, R.H. (*c*.2000) *Plagas Insectiles del Cafeto* (undated typescript), Nicaragua, 32 pp.

Nietner, J. (1861) *Observations on the Enemies of the Coffee Tree in Ceylon*. Colombo, 31 pp.

Notley, F.B. (1936) Coffee thrips. A contribution to our knowledge of the cause of outbreaks. *East African Agricultural Journal* 1, 283–292.

Notley, F.B. (1948) The *Leucoptera* leaf miners of coffee on Kilimanjaro I. *Leucoptera coffeella* Guer. *Bulletin of Entomological Research* 39, 399–416.

Notley, F.B. (1956), The *Leucoptera* leaf miners of coffee on Kilimanjaro II. *Leucoptera caffeina* Wshbn. *Bulletin of Entomological Research* 46, 899–912.

Okelana, F.A. (1985) Oviposition pattern, hatching and parasitism of eggs of *Epicampoptera strandi* Byrk. ssp *glauca* Hmps. (Lepidoptera: Drepanidae) on coffee in Nigeria. *Café-Cacao-Thé* 29, 273–276.

Panis, A. (1969) Observations faunistiques et biologiques sur quelques Pseudococcidae (Homoptera, Coccoidea) vivant dans le Midi de la France. *Annales de Zooogie et Ecologie Animale* 1, 211–244.

Pemberton, C.E. (1931) An egg parasite of Thrips in Hawaii. *Proceedings of the Hawaiian Entomological Society* 7, 481–482.

Perthuis, B., Philippe, R., Pradon, J.L., Dufour, M. and Leroy, T. (2001) Premières observations sur la résistance au champ de plantes de *Coffea canephora* génétiquement modifieés contre la mineuse des feuilles *Perileucoptera coffeella* Guérin-Méneville. *Association Scientifique Internationale du Café, 19th International Colloquium on Coffee*, Trieste, Italy, 2001.

Reis, P.R. (2002) *Brevipalpus phoenicis*, ácaro vetor da mancha-anular em cafeeiro: bioecologia, dano e controle. I *Simposio de Pesquisa dos Cafés do Brasil, Palace Cassino, Pocos de Caldas, MG*, Brazil, 26–29 September 2000, pp. 257–280.

Reis, P.R., Teodoro, A.V. and Neto, P.M. (2000) Predatory activity of phytoseiid mites on the developmental stages of coffee ringspot mite (Acari: Phytoseiidae: Tenuipalpidae). *Neotropical Entomology* 29, 547–553.

Ritchie, A.H. (1929) *Control of Plant Pests and Diseases*. Report. Department of Agriculture, Tanganyika Territory 1927–1928, pp. 34–40.

Ritchie, A.H. (1932) *Report of the Entomologist*. Report. Department of Agriculture, Tanganyika Territory 1931, pp. 83–86.

Ritchie, A.H. (1935) *Report of the Entomologist*. Report. Department of Agriculture, Tanganyika Territory 1934, pp. 73–83.

Ritchie, A.H. (1936) *Report of the Entomologist*. Report. Department of Agriculture, Tanganyika Territory 1935, pp. 95–103.

Robinson, G.S., Tuck, K.R. and Shaffer, M. (1994) *A Field Guide to the Smaller Moths of South East Asia*. Malaysian Nature Society, Kuala Lumpur, 309 pp.

Rodrigues, J.C.V., Rodriguez, C.M., Moreira, C.M., Villalobos, W., Rivera, C. and Childers, C.C. (2002) Occurrence of coffee ringspot virus, a *Brevipalpus* mitebourne virus in coffee in Costa Rica. *Plant Disease* 86, 564.

Sands, D.P.A., Lukins, R.G. and Snowball, G.J. (1986) Agents introduced into Australia for the biological control of *Gascardia destructor* (Newstead) (Hemiptera: Coccidae). *Journal of the Australian Entomological Society* 25, 51–59.

Schmitz, G. (1949) La pyrale du caféier robusta *Dichocrocis crocodora* Meyrick. Biologie et moyens de lutte. *Publications de l'Institut National Agronomique du Congo Belge* 41, 132 pp.

Sekhar, P.S. (1960) Life cycle of *Euproctis howra* a pest on coffee. *Current Science* 29, 230–231.

Sekhar, P.S. and Sekhar, S. (1964) Investigations on thrips occurring on arabica coffee. I. Bionomics. *Indian Coffee* 28, 173–179.

Sevastopulo, D.G. (1940) On the food plants of Indian Bombyces (Heterocera). *Journal of the Bombay Natural History Society* 41, 817–827.

Silva, C.G. and Parra, J.R.P. (1982) Biologia e danos de *Coccus viridis* (Green 1889) (Homoptera, Coccidae) em mudas de café (*Coffea* spp.). *Anais da Sociedade Entomologica do Brasil* 11, 181–195.

Silvestri, F. (1943) *Compendio di Entomologia Applicata (agraria-forestale-medica-vetinaria). Parte Speciale*, Vol. 2, Portici, pp. 1–512.

Simon-Thomas, R.T. (1962) *Checklist of Pests on Some crops in West Irian*. Bulletin of the Department of Economic Affairs, West Irian, Papua New Guinea.

Smee, C. (1939) Gelatine grub (*Niphadolepis alianta*). *Nyasaland Tea Association Quarterly Journal* 3, 5–7.

Szent-Ivany, J.J.H. (1956) New insect pest and host plant records in the Territory of Papua and New Guinea. *Papua and New Guinea Agricultural Journal* 11, 6 pp.

Szent-Ivany, J.J.H. and Stevens, R.M. (1966) Insects associated with *Coffea arabica* and some other crops in the Wau-Bulolo area of New Guinea. *Papua and New Guinea Agricultural Journal* 18, 101–119.

Tapley, R.G. (1961) Natural mortality of eggs and early instars of leaf miner. *Research Report 1961, Coffee Research Station, Lyamungu, Tanganyika*, pp. 48–49.

Tapley, R.G. (1965) Premature leaf fall in arabica coffee in Tanganyika. London University PhD Thesis.

Thompson, R.T. (1977) A revision of the New Guinea weevil genus *Apirocalus* Pascoe (Coleoptera Curculionidae). *Bulletin of the British Museum. Entomology Series* 36, 193–280.
Van Hall, C.J.J. (1921) Ziekten en plagen der cultuurgewassen in Nederlandsch-Indie in 1920. *Mededeelingen van het Instituut voor Plantenziekten, Buitenzorg* 46, 50 pp.
Vayssiére, P. (1955) Les animaux parasites du caféier. In: Coste, R. (ed.) *Les Caféiers et les Cafés dans le Monde*. Larose, Paris, pp. 233–318.
Weaving, A.J.S. (1972) A preliminary screening trial of insecticides for the control of the coffee leaf miner, *Leucoptera meyricki* Ghesq. (Lepidoptera Lyonetiidae) in Rhodesia. *Rhodesian Journal of Agricultural Research* 10, 79–89.
Welbourn, W.C., Ochoa, R., Kane, E.C. and Erbe, E.F. (2003) Morphological observations on *Brevipalpus phoenicis* (Acari: Tenuipalpidae), including comparisons with *B. californicus* and *B. obovatus*. *Experimental and Applied Acarology* 30, 107–133.
Wheatley, P.E. (1963) The giant coffee looper *Ascotis selenaria reciprocaria*. *East African Agricultural and Forestry Journal* 29, 143–146.
Wheatley, P.E. (1964) The giant looper. *Kenya Coffee* 29, 141–143.
Williams, D.J. (1987) Scale insects (Homoptera: Coccoidea) on coffee in Papua New Guinea. *Papua New Guinea Journal of Agriculture, Forestry and Fisheries* 34, 1–7.
Williams, D.J. and Watson, G.W. (1990) *The Scale Insects of the Tropical South Pacific Region. Part 3. The Soft Scales (Coccidae) and other Families*. CAB International Institute of Entomology, Wallingford, UK, 268 pp.
Witjaksono, Ohtani, K., Yamamoto, M., Miyamoto, T. and Ando, T. (1999) Responses of Japanese giant looper male moth to synthetic sex pheromone and related compounds. *Journal of Chemical Ecology* 25, 1633–1642.
Wolcott, G.N. (1922) The status of entomology in Porto Rico. *Journal of the Department of Agriculture, Porto Rico* 6, 11 pp.
Wu, J.F. and Huang, Z.H. (1983) A preliminary study of the eucleid *Latoia lepida* (Cramer). *Acta Entomologica Sinica* 26, 36–41.
Yunus, A. and Ho, T.H. (1980) *List of Economic Pests, Host Plants, Parasites and Predators in West Malaysia (1920–1978)*. Bulletin of the Malaysian Department of Agriculture No. 153, 538 pp.
Zhang, B.C. (1994) *Index of Economically Important Lepidoptera*. CAB International, Wallingford, UK, 599 pp.
Zuniga, J.A., Barrera, J.F., Valle, J. and Williams, T. (2002) Estimating populations of *Idiarthrom subquadratum* (Orthoptera: Tettigoniidae) using mark–recapture methods in coffee plantations in Chiapas, Mexico. *Environmental Entomology* 31, 515–522.

6 Root- and Collar-feeding Insects

Introduction

Root-feeding insects fall into six main groups, many of which also damage the collar region of the stem close to the ground, particularly with nursery plants:

- Sucking insects
- Larvae of Coleoptera (chafers, weevils, etc.)
- Larvae of Lepidoptera (cut worms)
- Crickets and mole crickets
- Termites
- Larvae of Diptera.

Additionally, the larvae of some wood-boring insects, which live within tunnels in the lower part of the stem, extend their activities into the principal roots. These include the cerambycid borers, *Bixadus sierricola*, *Monochamus leuconotus*, *Neonitocris princeps* and *Xylotrechus quadripes*, and the lepidopterous borer, *Zeuzera coffeae*. These are dealt with in Chapter 3.

Sucking Insects

Insects of the order Hemiptera feed by imbibing plant juices through their tubular mouthparts. Species in several families feed on roots and can cause debilitation and even death of the tree by removing nutrients from the phloem sap.

Cicadas, Hemiptera: Cicadidae

Status and distribution

Insects belonging to this family are important pests of coffee in Central and South America, particularly Brazil. The more important American species are contained in the genera *Quesada*, *Carineta* and *Fidicinoides* (see Table 6.1). The nymphal stage of the insect may feed on roots for several years before emerging as an adult. The adults feed exclusively on aerial parts of the plant but are not normally a problem, although adults of *Ueana lifuana* Montrouzier have been reported as causing damage in New Caledonia by ovipositing within small branches, which caused the twigs to fall (Risbec, 1936).

Morphology

The most studied species is *Quesada gigas*. The adult is a large insect, being around 70 mm long including the wings. It is greenish grey in colour, with clear wings (see Fig. 6.1). The nymphs are white, becoming somewhat darker in the later stages and with a foreleg strongly developed for digging. The average length of the first instar nymph is 2.2 mm, the second 4.2 mm, the third 8.2 mm and the fourth 14.9 mm. By the fifth instar the sexes can be

Table 6.1. The distribution of species of Cicadidae recorded from roots of coffee.

Species	Recorded from/by
Baeturia maddisoni Duffels	Fiji (Duffels, 1988)
Carineta fasciculata (Germar)	Brazil (Da Fonseca and Araujo, 1932)
C. matura (Distant)	Brazil (Martinelli and Zucchi, 1989b)
C. spoliata (Walker)	Brazil (Martinelli and Zucchi, 1989b)
C. sp.	Honduras (Munoz, *c.*2000)
Dorisiana drewseni (Stal)	Brazil (Da Fonseca and Araujo, 1932)
D. viridis (Olivier)	Brazil (Martinelli and Zucchi, 1989a)
D. sp.	Honduras (Munoz, *c.*2000)
Dundubia rufivena Walker	Malaysia (Vayssière, 1955)
Fidicina mannifera (Fabricius)	Brazil (Da Fonseca, 1934)
Fidicinoides pronoe (Walker)	Brazil (Martinelli and Zucchi, 1984), Peru (Escalante, 1974)
F. pullata (Berg)	Brazil (Hempel, 1913)
F. sp.	Honduras (Munoz, *c.*2000)
Fingeriana dubia Cavichioli	Brazil (Cavichioli, 2003)
Quesada gigas (Olivier)	Brazil (Da Fonseca and Araujo, 1939), Peru (Escalante, 1974)
Q. sodalis (Walker)	Brazil (Da Fonseca, 1934)
Q. sp.	Honduras (Munoz, *c.*2000)
Ueana lifuana Montrouzier	New Caledonia (Risbec, 1936)
Yanga guttulata Signoret	Madagascar (Breniere and Syfrig, 1965)
Zammara sp.	Puerto Rico (Wolcott, 1924)

distinguished, and these average 27.7 mm for the male and 26.9 for the female (Maccagnan and Martinelli, 2004).

Life cycle and damage

The adults are active in September and October in Brazil (Martinelli and Zucchi, 1987) The female makes an incision in the bark, into which the eggs are laid. Within a few days they hatch and the young nymphs drop to the ground and burrow down to the feeder roots, where they insert their mouthparts and suck the sap. They are generally found in the superficial layers of the soil, but may penetrate as much as 60 cm (Wille, 1952).

Feeding leads to a general debility of the tree and, with a severe attack, can result in leaf and fruit fall and even death. Some of the damage is because the cavities which the nymphs form in the soil become filled with excreta, encouraging disease organisms which, in turn, infect the roots through the feeding wounds (Da Fonseca and Araujo, 1939). There are five nymphal

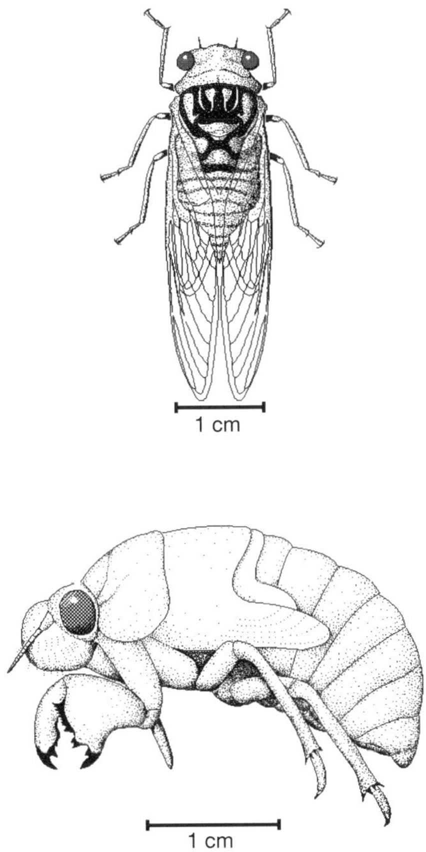

Fig. 6.1. Adult and nymph of *Quesada gigas*.

stages. The total life cycle is thought to last 3–4 years, but nymphs of all stadia can be found under the same plant (Maccagnan and Martinelli, 2004).

Scale insects, Hemiptera: Coccidae and Ortheziidae

Some scale insects, more often found on aerial parts of the plant, have also been recorded from roots. These include *Coccus brasiliensis* Da Fonseca in Brazil (Da Fonseca, 1958) and *Saissetia coffeae* (Walker), which Le Pelley (1968) claims has been found on roots but does not mention where. The Ortheziids, *Mixorthezia fodiens* (Giard) in Guadeloupe (Vayssière, 1955) and *Mixorthezia reynei* (Laing) in Surinam (Laing, 1925) and Trinidad (Morrison, 1952), were both described from specimens obtained from coffee roots. None of these species appears to be of importance as a root pest.

Mealy bugs, Hemiptera: Pseudococcidae

Status and distribution

Mealy bugs of the family Pseudococcidae are found on the roots of coffee and can be serious pests in many parts of the tropics. They are particularly prevalent in Central and South America and the Caribbean islands (see Table 6.2).

The insects live in association with a fungus, *Diacanthodes novoguineensis* (*Polyporus coffeae*) in East Africa and *Septobasidium* sp. in South America, which forms a rubbery sheath of mycelial hyphae covering both the roots and the mealy bug colonies. The combination of feeding by the mealy bugs and the presence of a fungal barrier between the roots and the soil weakens the tree and may kill it if under water stress.

Mealy bugs obtain their food from the phloem sap of the plant and excrete large quantities of undigested amino acids and sugars (Adomako, 1972). This exudate is an important constituent in the diet of several ant species that attend the ant colonies and remove droplets of the 'honeydew'. This service helps to keep the mealy bugs free of fungal contaminants. On aerial parts of the tree, the presence of ants also helps to deter attack by natural enemies. Underground colonies are not much troubled by natural enemies, but whether these are prevented from reaching the colonies by the protection of the ants or merely because of inaccessibility is not clear.

For many years, the mealy bug species on coffee roots in East Africa was thought to be a form of *Planococcus citri* (Risso) that is common on aerial parts of the tree, but it has been shown (Watson and Cox, 1990) that the subterranean root mealy bug in Kenya, Tanzania, Uganda, Zimbabwe and Democratic Republic of Congo is a separate species, *Planococcus fungicola* Watson & Cox. The same authors described another new root-feeding species, *Planococcus radicum* Watson & Cox, from Tanzania and Nigeria.

Planococcus citri has been recorded as a root-feeding mealy bug in some

Table 6.2. The distribution of Pseudococcidae recorded from roots of coffee.

Species	Recorded from/by
Benedictycoccina ornata (Hambleton)	Trinidad and Tobago (Williams and Granara de Willink, 1992)
Capitisetella migrans (Green)	Surinam (Bünzli, 1935)
Cataenococcus podagrosus (Green)	Surinam (Green, 1933)
Coccidella globoculus (Hambleton)	Trinidad and Tobago (Hambleton, 1946)
Dysmicoccus brevipes (Cockerell)	Brazil (Williams and Granara de Willink, 1992), Colombia (Cárdenas and Posada, 2001), Costa Rica (Zamora, 1998), El Salvador (Berry and Abrego, 1953), Guatemala (Williams and Granara de Willink, 1992)
D. debregeasiae (Green)	India (Williams, 2004)
D. grassi (Leonardi)	Colombia (Williams and Granara de Willink, 1992), Costa Rica (Williams and Granara de Willink, 1992)
D. probrevipes (Morrison)	Guatemala (Hamilton, 1967)
D. radicis (Green)	Brazil (Da Costa Lima, 1936), Surinam (Bünzli, 1935)
D. subterreus Williams	India (Williams, 2004)
D. texensis (Tinsley)	Brazil (Santa-Cecilia *et al.*, 2002)
Formicococcus greeni (Vayssière)	Madagascar (Frappa, 1934)
F. robustus (Ezzat & McConnell)	China (Lan and Wintgens, 2004)
Geococcus coffeae Green	Brazil (Da Costa Lima, 1936), Costa Rica (Williams and Granara de Willink, 1992), El Salvador (Hambleton, 1946), Ghana (Strickland, 1947, Guatemala (Hambleton, 1946), Honduras (Munoz, *c.* 2000), Surinam (Hambleton, 1946)
Neochavesia caldasiae (Balachowsky)	Colombia (Balachowsky, 1957)
N. eversi (Beardsley)	Colombia (Beardsley, 1970)
Paraputo leveri (Green)	Papua New Guinea (Williams and Watson, 1988)
P. sp.	Honduras (Munoz, c2000)
Planococcus angkorensis (Takahashi)	India (Williams, 2004)
P. citri (Risso)	Costa Rica (Williams and Granara de Willink, 1992), India (Anstead 1919), Philippines (Teodoro and Gomez, 1926), Uganda (De Lotto, 1964), Democratic Republic of Congo (De Lotto, 1964)
P. fungicola Watson & Cox	Kenya (Watson and Cox 1990), Tanzania (Watson and Cox 1990), Uganda (Watson and Cox 1990), Democratic Republic of Congo (Watson and Cox 1990), Zimbabwe (Watson and Cox 1990)
P. lilacinus (Cockerell)	India (Sekhar, 1964), Java (Begemann, 1929), Taiwan (Takahashi, 1934), Vietnam (Kalshoven, 1950–1951)
P. radicum Watson & Cox	Nigeria (Watson and Cox 1990), Tanzania (Watson and Cox 1990)
Pseudococcus cryptus Hempel	Brazil (Pickel, 1928), Colombia (Cárdenas and Posada, 2001), Honduras (Munoz, *c.* 2000)
P. pseudocitriculus Betrem	Java (Betrem, 1937)
P. sp.	Colombia (Roba, 1938)

Table 6.2. *Continued*

Species	Recorded from/by
Pseudorhizoecus proximus Green	Colombia (Williams and Granara de Willink, 1992), Ecuador (Williams and Granara de Willink, 1992), Guatemala (Williams and Granara de Willink, 1992), Surinam (Bünzli, 1935)
Puto antioquensis (Murillo)	Colombia (Roba, 1938), Honduras (Munoz, *c.*2000)
P. mexicanus (Cockerell)	El Salvador (De Fluiter, 1960), Guatemala (Williams and Granara de Willink, 1992)
P. sp.	Costa Rica (Zamora, 1998)
Rhizoecus americanus (Hambleton)	Colombia (Williams and Granara de Willink, 1992), Ecuador (Hambleton, 1946)
R. arabicus Hambleton	Colombia (Hamon, 1982), Costa Rica (Hamon, 1982)
R. caladii Green	Surinam (Bünzli, 1935)
R. coffeae Laing	Brazil (Da Costa Lima, 1936), Colombia (Roba 1938), Costa Rica (Williams and Granara de Willink, 1992), Surinam (Van Dinther, 1960), Venezuela (Fernandez, 1957)
R. divaricatus Hambleton	Nicaragua (Hambleton, 1978)
R. eloti Giard	Guadeloupe (Williams and Granara de Willink, 1992)
R. falcifer Kunckel d'Herculais	Surinam (Green, 1933)
R. nemoralis (Hambleton)	El Salvador (Guerrero Berrios and Hanania, 1979), Honduras (Munoz, *c.*2000)
Ripersiella andensis (Hambleton)	Colombia (Hambleton, 1946)
R. kondonis (Kuwana)	Guatemala (Williams and Granara de Willink, 1992)

Asian countries, including, India (Anstead, 1919), the Philippines (Teodoro and Gomez, 1926) and Vietnam (Du Pasquier, 1932; Hanson, 1963) but Williams (2004), in his recent review of Oriental mealy bugs, did not cite any material from roots in India or the Philippines, and remarks that some reports of *P. citri* may be misidentifications of *P. minor* (Maskell).

In Vietnam, it is found on the roots of both *C. arabica* and *C. liberica* (Du Pasquier, 1932). Although the aerial form of *P. citri* is reported as being common on coffee in Indonesia, the species has not been recorded there on coffee roots (Kalshoven, 1950–1951), and Williams (2004) is of the opinion that it is a misidentification of *P. minor*. *Planococcus minor* appears to be primarily an aerial species, and this may account for the observations in Indonesia.

However, the species was described (as *Dactylopius calceolariae* var. *minor*) by Maskell from a specimen found on the roots of onion grass in Mauritius (Williams, 2004), so it is evident that it can be subterranean. However, *Pseudococcus pseudocitriculus* Betrem has been recorded from the roots of *C. canephora* in Java (Betrem, 1937). From India, Williams (2004) has described a new species, *Planococcus subterreus* Williams, from the roots of *C. canephora*, and also records *Dysmicoccus debregeasiae* (Green) from roots of *C. arabica* and *Planococcus angkorensis* (Takahashi) from roots of *C. canephora*.

Two important genera of mealy bugs found on coffee roots in South and Central America are *Dysmicoccus* and *Rhizoecus*. It should be noted that several of the species that were previously in *Rhizoecus* have recently been moved to other genera, namely *Benedictycoccina ornata* (Hambleton), *Capitisetella migrans* (Green), *Ripersiella andensis* (Hambleton) and *R. kondonis* (Kuwana). *Dysmicoccus texensis* (Tinsley) occurs on coffee roots in Brazil (Santa-Cecilia *et al.*, 2002). *Dysmicoccus brevipes* (Cockerell) is well known as a pest of pineapple and has a widespread tropical distribution (see Table 6.2) (CABI, 1972). It is found mainly on the collar region of the trunk and on the principal roots.

By contrast, *Pseudococcus cryptus* Hempel lives further from the trunk within root nodules formed by the fungus *Septobasidium* sp. (Cárdenas and Posada, 2001). Another important species is *Geococcus coffeae* Green, widely distributed throughout Central and South America on roots of coffee and citrus, among other hosts (CABI, 1971a), and also recorded from Ghana (Strickland, 1947), where it attacks the roots of both coffee and cocoa.

Morphology

The winged males of mealy bugs are rarely seen and, as they do not feed, are not of economic importance. The adult females and the immature stages are wingless, soft-bodied, oval-shaped in outline in most species and domed (see Figs 5.18a, d, 6.2). They excrete wax through pores in the integument, which forms a powdery covering to the body that in many species develops into lateral filaments.

Females of *Neochavesia caldasiae* are rather more distinct, having a swollen cephalothorax, but with a narrower and longer abdomen ending in a pair of rounded lobes (see Fig. 6.3). The mouthparts are developed into tubes, often much longer than the body of the insect; these can be coiled when not in use and can easily penetrate plant tissues to obtain liquid nutrients, typically from the phloem vessels.

Life cycle

In some species of mealy bugs, such as *Planococcus citri*, the eggs are laid in a woolly ovisac that remains attached to the rear end of the female. Others, such as *Dysmicoccus brevipes* and *Planococcus lilacinus*, are oviviviparous and the eggs hatch immediately upon laying. *Planococcus citri* eggs hatch after 3–5 d.

Upon hatching, the female nymphs of most mealy bugs go through three stages before changing directly into an adult, but *Neochavesia* females have a pupal stage. Males have two nymphal stages followed by a prepupal and pupal stage. The female immature stages of *D. brevipes* on pineapple in Malaysia last

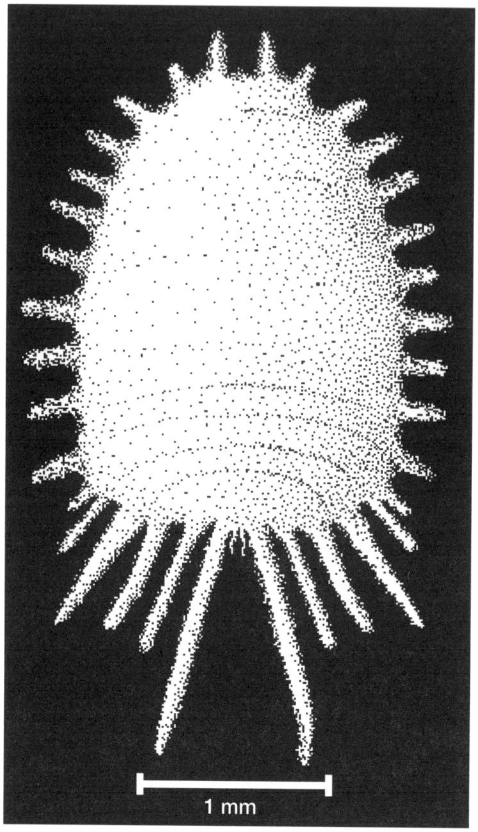

Fig. 6.2. Adult female of *Dysmicoccus brevipes*.

10, 6.7 and 7.9 d, respectively, and the male 9.9, 5.8, 2.5 and 3.7 d, totalling around 24 d (Lim, 1973).

The length of the nymphal stages of *P. citri* on cocoa in West Africa is around 32–38 d (Entwistle, 1972), the complete life cycle therefore lasting 37–43 d. On coffee in Indonesia, the whole life cycle lasts 28–49 d (Betrem, 1936). The life cycle of *P. lilacinus* in Malaysia lasts 37–50 d (Chong *et al.*, 1991).

Ant attendance

The South American mealy bugs are attended by various species of ants of the genus *Acropyga*. *Acropyga parimaribensis* Borgmeier attends root mealy bugs in Colombia and Surinam, *A. pickeli* Borgmeier in Brazil, Colombia and Surinam, *A. robae* Donisthorpe in Colombia and *A. rutgersi* Bünzli in Surinam (see Table 6.3). *Acropyga* species have a very close relationship with the mealy bugs that they tend, and the queens carry a mealy bug in their jaws during their nuptial flight. This relationship has existed for a very long time, and alate queen *Acropyga* and *Azteca* ants have been found preserved in amber of the

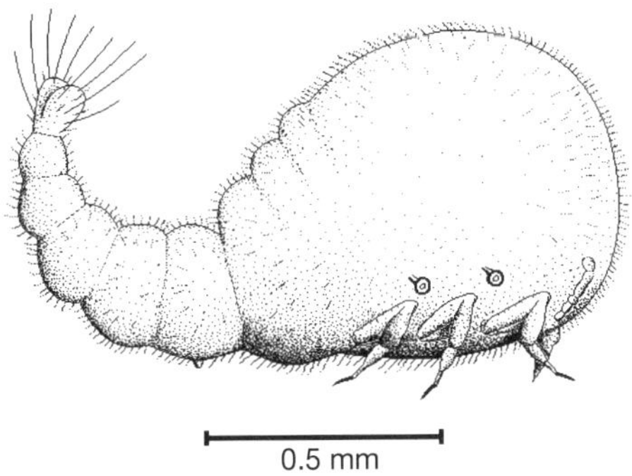

Fig. 6.3. Adult female of *Neochavesia caldasiae*.

Miocene era in the Dominican Republic holding mealy bugs in their jaws (Johnson *et al.*, 2001). Honeydew appears to be the principal food for the ants and it is said that the workers very rarely leave the underground tunnels where the mealy bugs are tended (Johnson *et al.*, 2001).

Solenopsis sp. attends *Dysmicoccus grassii*, *D. brevipes* and *P. cryptus* in Colombia (Cárdenas and Posada, 2001). In East Africa, root mealy bugs are attended by *Pheidole megacephala* (Fabricius), *Pseudolasius gowdeyi* (Wheeler), *Solenopsis puctaticeps* Mayr and *Monomorium pharoensis* (Linnaeus).

Control

Because the mealy bugs are found in inaccessible sites and are to a large extent protected by the presence of attendant ant species, their numbers are not kept in check by natural enemies so, in order to achieve control, recourse must be made to chemicals. In Costa Rica, turbufos (Counter), diazinon, phorate (Thimet) and chlorpyrifos (Lorsban) have been recommended (Zamora, 1998), and diazinon in Colombia (Cárdenas and Posada, 2001). In India, isofenphos (Oftanol) gave control of *Planococcus lilacinus*, whilst a mixture of this chemical with the fungicides triadimefon (Bayleton) or oxycarboxin (Plantvax) gave control of both the mealy bugs and the fungus (Chacko and Sreedharan, 1981).

In Brazil, control of *Pseudococcus cryptus* was obtained using granules of 10% aldicarb, but reinfestation occurred within about 1 month after using disulfotan, phorate or methomyl (Aphidan) (Cavalcante, 1975). More recently (Andalo *et al.*, 2004), the possible use of *Beauvaria bassiana* as a control agent for *Dysmicoccus texensis* has being explored. In Kenya, the recommendation is that infested trees should be uprooted and burnt, and the soil treated with a drench of carbofuran (Furadan) or aldicarb (Temik) (Anon, 1991).

Warning: it must be emphasized that aldicarb, turbufos and disulfotan fall into the WHO category Class 1A (highly hazardous), and isofenfos, methomyl and carbofuran into Class IB. Therefore, these substances may not be registered for use in all countries and, because of their toxicity, are likely to be more restricted in their use in the future. Diazinon and chlorpyrifos are classified as moderately hazardous (Class II).

Table 6.3. Ant species associated with mealy bugs on roots of coffee.

Species of ant	Recorded on mealy bugs from/by
Acropyga fuhrmanni (Forel)	*Neochavesia caldasiae*, Colombia (Roba, 1936)
Acropyga paramaribensis Borgmeier	*Geococcus coffeae*, Surinam (Bünzli, 1935); *Pseudorhizoecus proximus*, Surinam (Bünzli, 1935); *Rhizoecus caladii*, Surinam (Bünzli, 1935); *Rhizoecus coffea*, Colombia (Roba, 1936); Surinam (Bünzli, 1935); *Rhizoecus falcifer*, Surinam (Bünzli, 1935)
Acropyga pickeli Borgmeier	*Rhizoecus coffeae*, Brazil (Pickel, 1928); Surinam (Laing, 1925); *Pseudococcus cryptus*, Brazil (Hambleton ,1935); *Ripersiella andensis*, Colombia (Hambleton, 1946)
Acropyga robae Donisthorpe	*Rhizoecus coffeae*, Colombia (Roba, 1938); *Neochavesia caldasiae*, Colombia (Balachowsky, 1957)
Acropyga rutgersi Bünzli	*Capitisetella migrans*, Surinam (Bünzli, 1935); *Pseudorhizoecus proximus*, Surinam (Bünzli, 1935); *Dysmicoccus radicis*, Surinam (Bünzli, 1935)
Acropyga sp.	*Puto* sp., Costa Rica (Zamora, 1998)
Lepisiota capensis Mayr	*Planococcus fungicola* (as *P. citri*), Kenya (Anderson, 1927)
Monomorium pharaonis (Linnaeus)	*Planococcus fungicola* (as *P. citri*), Kenya (Le Pelley, 1959)
Pheidole megacephala (Fabricius)	*Planococcus fungicola*, Kenya (Baum, 1968)
Pheidole sp.	*Rhizoecus coffeae*, Brazil (Pickel, 1928); *Puto* sp., Costa Rica (Zamora, 1998)
Pseudolasius gowdeyi (Wheeler)	*Planococcus fungicola* (as *P. citri*), Uganda (Hargreaves, 1924)
Solenopsis geminata (Fabricius)	*Puto* sp., Costa Rica (Zamora, 1998)
Solenopsis punctaticeps Mayr	*Planococcus fungicola* (as *P. citri*), Kenya (James, 1933)
Solenopsis sp.	*Dysmicoccus brevipes*, Colombia (Cárdenas and Posada, 2001); *Dysmicoccus grassi*, Colombia (Cárdenas and Posada, 2001); *Pseudococcus cryptus*, Colombia (Cárdenas and Posada, 2001)

Larvae of Coleoptera

Three families of beetles contain important root-feeding species, particularly on nursery plants: Tenebrionidae, Curculionidae and Scarabaediae.

Tenebrionidae

Distribution and status

Known as 'dusty brown beetles', the adults of the genus *Gonocephalum* feed on the bark around the collar of the plant whilst the larvae – which are known as 'false wireworms' – feed on the roots. The species are widely distributed throughout Central, Eastern and Southern Africa, India, Indonesia, the Philippines, Australia and parts of Europe, and attack a variety of of crops such as maize, groundnuts, tobacco and sugar cane. They are not found in the Americas.

Species recorded from coffee include *G. aequatoriale* Blanchard and *G. depressum* (Fabricius) from Indonesia, *G. bilineatum* Walker from India, *G. biroi* Kaszab from Papua New Guinea and *G. simplex* (Fabricius) from Ethiopia, Kenya, Tanzania, Uganda, Malawi, Zimbabwe and Democratic Republic of Congo. The latter species has a number of subspecies, many described by Ferrer (2000), who reviews the genus in Africa and Europe.

Morphology and life cycle

Adult beetles of *G. simplex* and *G. depressum* are about 9 mm long, dull brown and rather flattened. The larvae are elongated, about 11–14 mm long, shiny, rather hard and tan-coloured (see Fig. 6.4). The adults hide under leaf litter during the day and come out to feed after dark. The eggs are laid loosely in the soil. Under laboratory conditions in Madagascar under a constant temperature of 28°C and 75% relative humidity, *G. simplex* eggs hatched in an average of 4.4 d, followed by a larval stage lasting 110.4 d during which the insects passed through ten instars and a pupal stage of 7.3 d but, in the field, the larvae were observed to pass through 11–13 instars (Brénière, 1960). As many as 300 larvae/m^2 were recorded by Senstius (1915) in a coffee nursery in Java.

The adults of *G. simplex*, as well as feeding on bark in the collar region, will also climb the tree and eat the bark of primaries and secondaries – and even ring bark them (Anderson, 1924).

Control

No parasites appear to have been recorded for *G. simplex* but, in Kenya, the Carabid beetles, *Carabomorphus brachycerus* Gerstaecker and *Scarites madagascariensis* Dejean, are predators (Le Pelley, 1968). The fungi, *Metarhizium anisopliae* and *Beauveria bassiana*, attack *Gonocephalum* larvae in Australia. Baits have proved effective in controlling the adult beetles. In Australia, a bait consisting of 100 ml 50% chlorpyrifos and 125 ml sunflower oil in 2.5 kg cracked sorghum or wheat is used (Murray and Wicks, 1984).

Curculionidae

The larvae of some weevils feed on roots, the adults feeding on aerial parts of the plant. In the West Indies, *Diaprepes famelicus* Olivier (see Fig. 6.5 d) is an

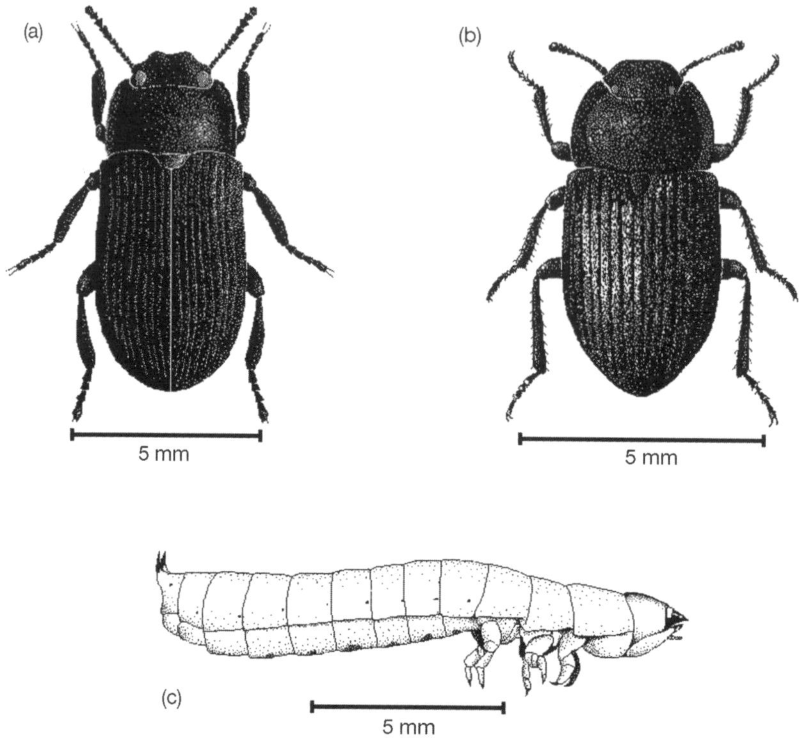

Fig. 6.4. Darkling beetles: (a) adult of *Gonocephalum depressum*; (b) adult of *G. simplex*; (c) larva of *G. depressum*.

important pest of sugar cane, and in Guadeloupe and Martinique it has caused problems with young coffee trees planted in land cleared of sugar cane (Le Pelley, 1968).

The larvae of *Lachnopus buchanani* Marshall in Cuba, and *L. coffeae* Marshall (see Fig. 6.5b) in Puerto Rico have been recorded as attacking coffee roots (Le Pelley, 1968), as have *Pantomorus cervinus* (Bohemann) in Brazil, Hawaii, Costa Rica and Nicaragua and *Ellemenistes laesicollis* Fabricius in South Africa (Milne, 1973).

Scarabaeidae

Larvae of the Dynastid beetle, *Heterolygus meles* (Billberg) (see Fig. 6.5a, c) have been recorded attacking coffee roots in Cameroon (Vayssière and Galland, 1948). Other members of this family belong to the subfamily Melolonthinae and are known as 'cockchafers'. In all cases it is the larvae that feed on roots and they are particularly damaging to nursery plants. On emerging from the ground when mature, the adults feed on leaves.

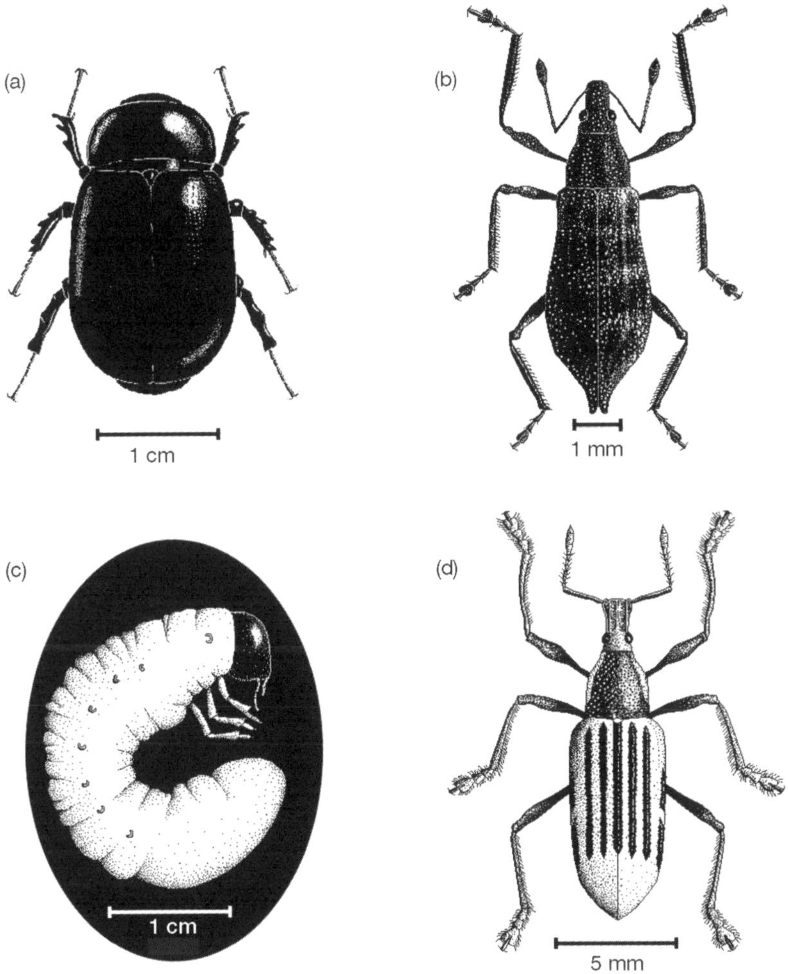

Fig. 6.5. (a) Adult of *Heterolygus meles*; (b) adult of *Lachnopus coffeae*; (c) larva of *H. meles*; (d) adult of *Diaprepes famelicus* Olivier.

Status and distribution

There has been some confusion over the identity of species in the genera *Holotrichia*, *Lachnosterna* and *Phyllophaga*, which are very similar. *Holotrichia farinosa* Nonfried and *H. sculpticollis Blanchard* have been recorded from India (Patil and Veeresh, 1988) and *H. mindanaona* Brenske from the Pacific island of Guam (Oakley, 1945).

Phyllophaga constricta (Burmeister) *P. leucophthalma* (Wiedemann) and *P. severini* (Brenske) are pests in Indonesia (De Fluiter, 1960), *P. menetriesi* (Blanchard) in Costa Rica, El Salvador and Colombia, *P. vicina* (Moser) in Costa Rica (Abarca and Quesada, 1997), *P. plaei* (Blanchard) in the Dominican

Republic (Krug and De Poerck, 1968) and *P. hondura* (Saylor) in Honduras (Munoz, *c.*2000).

In addition, unspecified *Phyllophaga* and *Lachnosterna* are recorded from Sri Lanka (Nietner, 1861), India (D'Souza *et al.*, 1970), Indonesia (De Fluiter, 1936), Puerto Rico (Van Zwaluwenburg, 1917), El Salvador (Guerrero Berrios, 1980), Haiti (Aitken-Soux, 1985), Colombia (Cárdenas and Posada, 2001) and Costa Rica (Zamora, 1998). *Leucopholis pinguis* Burmeister is a pest in India (Beeson, 1921) and Sri Lanka (Green, 1916) and *L. rorida* (Fabricius) in Indonesia (Dammerman, 1929).

Phyllophaga larvae are quite large insects, and can do considerable damage, killing even mature trees. E.E. Green, in a revision of Nietner's (1861) pamphlet, reported on an estate in Sri Lanka where the pest had been at work for some years: ' … the grubs making their way from one end of the estate to the other, devouring everything before them, at a rate of about eight or nine acres per annum.'

Morphology and life history

Adult beetles are brown and around 20–25 mm long and are mostly active at night. The larvae are white and C-shaped, with well-developed legs (see Fig. 6.6). The emergence of adults is seasonal, usually towards the beginning of the rainy season, when large numbers appear and feed on foliage.

In Indonesia, *P. severini* females lay around 15 eggs at a depth of 10–20 cm. These hatch in 10–12 d. The larval stage lasts around 4 months, followed by a pupal stage of around 13 d. The adults do not emerge immediately, but remain deep in the soil at depths of up to 1 m until the end of the dry season, when they move closer to the surface and emerge when conditions are favourable (Kalshoven, 1950–1951).

The life history of *P. menetriesi* in Costa Rica has been described by King and Saunders (1984), who provide good illustrations of larvae, pupae and adults. The white eggs are laid in the soil singly or in small groups 2–10 cm deep and hatch in 10–12 d. The larvae reach a size of 35–40 mm when fully grown and go through three larval instars in 8–9 months. At the end of the third instar, an earthen cell is prepared and the larva remains quiescent for some time, before pupating in January or February. Adults emerge shortly after the start of the first rains, in late April and May.

Control

Records of parasites are sparse. In Puerto Rico, two Tachinid flies, *Cryptomeigenia aurifacies* Walton and *Eutrixoides jonesi* Walton, were found to parasitize adults (Van Zwaluwenburg, 1917), and in Mexico the nematodes *Steinernema carpocapsea* and *Heterorhabditis bacteriophora*, as well as suspensions of the fungus *Metarhizium anisopliae*, have shown promise in the control of *Cyclocephala comata* Bates, an important pest of maize (Ponce *et al.*,

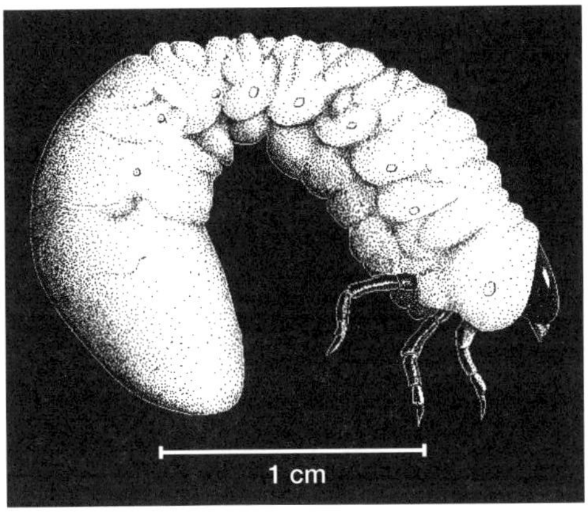

Fig. 6.6. Larva of *Phyllophaga* sp.

2004). Fipronil is used for chemical control of white grubs on maize in Brazil (Ceccon *et al.*, 2004). Fironil falls into the WHO moderately hazardous Class II category.

Larvae of Lepidoptera

The larvae of several soil-living species of moths in the family Noctuidae – known as 'cutworms' – are notorious pests of a very wide range of agricultural and horticultural crops. The larvae emerge from the soil during the night and tend to cut off the plants at ground level. As pests of coffee they are important in nursery beds, where they damage very young plants.

Species recorded from coffee include *Agrotis biconica* Kollar from Kenya, *A. ipsilon* (Hufnagel) from Indonesia, Malaysia, Papua New Guinea and Côte d'Ivoire (CABI, 1969), *A. segetum* (Denis and Schiffermuller) from China, Indonesia, India, Sri Lanka, Kenya, Malawi and Democratic Republic of Congo, *Agrotis* sp. from Costa Rica and Colombia, *Feltia* sp. from Costa Rica, *Spodoptera* sp. from Costa Rica and *Tycomarptes inferior* (Guenee) from Kenya and Democratic Republic of Congo.

Crickets and Mole Crickets

Crickets belonging to the family Gryllidae of the Orthoptera and mole crickets belonging to the family Gryllotalpidae are relatively large insects which live in burrows in the soil, coming out at the surface only at night. Both the adults and the young stages feed on roots and may also damage the collar region of the plant at or close to ground level, especially with seedling plants.

Control

Crickets and mole crickets are generalist feeders, attacking a wide range of cultivated plants and, with coffee, they are harmful mostly in nurseries. It has been pointed out by Crowe (2004) that the modern practice of raising seedlings in polythene pots has largely overcome the problem of these pests. Should chemical control become necessary, chlorpyrifos granules have proved successful in controlling mole crickets in Senegal (and also cutworms) (Collingwood *et al.*, 1980), and bran-based baits incorporating carbaryl or bifenthrin are used in Florida. All these chemicals fall into the WHO moderately hazardous (Class II) category.

Some success has been achieved in the control of *Scapteriscus* species in Florida by the use of the parasitic nematode, *Steinernema scapterisci* Nguyen & Smart, and the Tachinid, *Ormia depleta* (Wiedemann), both of which were introduced to the USA from South America (Parkman *et al.*, 1996).

Gryllotalpidae

Two species of mole crickets are recorded as damaging coffee: (i) *Gryllotalpa africana* Palisot de Beauvois (see Fig. 6.7) in the Old World (China, Indonesia, Papua New Guinea, Kenya, Uganda, Côte d'Ivoire, Democratic Republic of Congo) (CABI, 1971b); and (ii) *Scapteriscus didactylus* (Latreille) in the New World (Colombia, Surinam, Puerto Rico). A report of a *Gryllotalpa* sp. from Colombia (Cárdenas and Posada, 2001) probably refers to the latter species. Mole crickets are primarily pests of nursery plants. They live in undergound burrows and feed on roots. Adult *G. africana* are soft brown insects around 35–40 mm long, with prominent, finger-like processes on the forelimbs used for digging through soil. A good account of the biology of the species is given by De Graaf *et al.* (2004).

Gryllidae

Once again, the family is split into Old World and New World species. *Gryllus assimilis* (Fabricius) is a New World species recorded on coffee from Colombia, Costa Rica, Haiti and Jamaica, whilst *G. bimaculatus* De Geer has been recorded from Kenya and Uganda.

Brachytrupes membranaceus (Drury) is largely African, with records from Kenya, Tanzania, Uganda, Zambia, Madagascar, Democratic Republic of Congo and Côte d'Ivoire. *Brachytrupes portentosus* (Lichtenstein) is Asian in distribution, with records from China, Vietnam, Malaysia, Indonesia and Papua New Guinea.

Other records include *Anurogryllus abortivus* (Saussure) from Cuba, *Teleogryllus mitratus* (Burmeister) from Indonesia and Vietnam, *Gymnogryllus commodus* (Walker) from Papua New Guinea and *Ceuthophilus* sp. from Honduras. They have the same habits as mole crickets, living in underground burrows from which they emerge at night to feed.

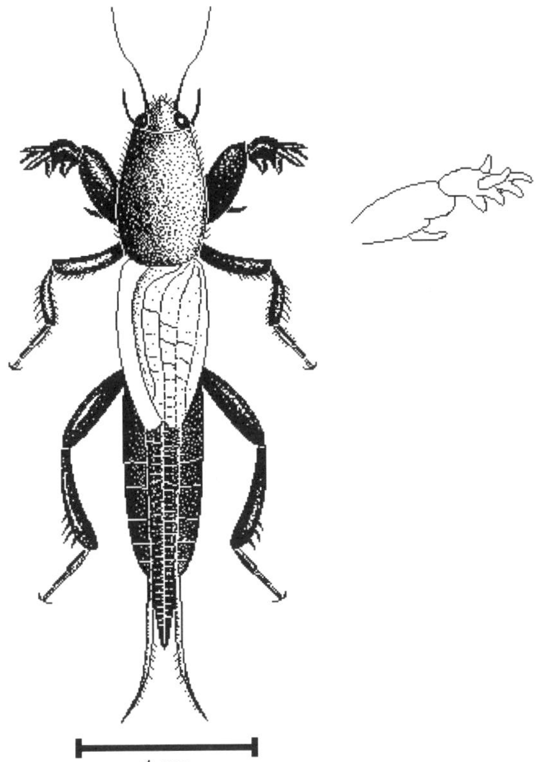

Fig. 6.7. Adult of *Gryllotalpa africana*.

Termites

Termites are a pantropical problem with woody plants, but there are surprisingly few records of damage to coffee trees in which the species of termite is identified: a list is given in Table 6.4. Damage is normally to the bark in the collar region, and sometimes to the main roots. Vayssière (1955) remarks that, in West Africa, infestation by termites often occurs where the tree has already been damaged by stem borers such as *Bixadus sierricola*.

Larvae of Diptera

The Stratiomyiid fly, *Chiromyza vittata* Wiedemann, is a serious problem in the State of Minas Gerais in Brazil, first noticed in 1986 (D'Antonio, 1991). The larvae occur in large numbers around the feeder roots, damaging these and causing a general debility of the plant and allowing the entry of pathogenic fungi such as *Fusarium*. A survey carried out during 2001 on 24 plantations showed the fly to be present in 83% of them. Two wasps in the family

Table 6.4. Termite species recorded as pests of coffee.

Species	Recorded from/by
Ancistrotermes latinotus (Holmgren)	Democratic Republic of Congo (Harris, 1969)
Comatermes perfectus (Hagen)	Colombia (Weidner, 1980)
Coptotermes curvignathus Holmgren	Indonesia (De Fluiter, 1960), Malaysia (Kalshoven, 1950–1951)
C. sjostedti Holmgren	Guinea (Mallamaire, 1937), Côte d'Ivoire (Mallamaire, 1937)
C. testaceus (Linnaeus)	Surinam (Reyne, 1920)
Glyptotermes dilatatus (Bugnion & Popoff)	Sri Lanka (Jepson, 1929)
G. pubescens Snyder	Puerto Rico (Wolcott, 1951)
Macrotermes bellicosus (Smeathman)	Uganda (Le Pelley, 1959), Yemen (Ba-Angood and Al-Sunaidi, 2004)
M. gilvus (Hagen)	Indonesia (De Fluiter, 1960), Malaysia (Yunus and Ho, 1980)
Microcerotermes edentatus Wasmann	Côte d'Ivoire (Mallamaire, 1937)
Microtermes insperatus Kremner	Indonesia Java (De Fluiter, 1960), Malaysia (De Fluiter, 1960)
M. pallidus (Haviland)	Malaysia (Corbett, 1935)
M. yemenensis Wood	Yemen (Ba-Angood and Al-Sunaidi, 2004)
Nasutitermes rippertii (Rambur)	Brazil (Caminha Filho, 1926)
Odontotermes badius (Haviland)	Kenya (Le Pelley, 1959), Tanzania (Morstatt, 1913)
Procornitermes triacifer (Silvestri)	Brazil (Harris, 1969)

Monomachidae have been identified as larval and pupal parasites: *Monomachus eurycephalus* Schletterer and *M. fuscator* (Perty) (Musetti and Johnson, 2004).

References

Abarca, G. and Quesada, M. (1997) Especies del complejo de jobotos (*Phyllophaga* spp, *Anomala* spp y *Cyclocephala* spp) associadas a cultivos en la Valle Central y Pacifico seco de Costa Rica. *Agronomia Mesoamericana* 8, 44–53.

Adomako, D. (1972) Studies on mealy bug (*Planococcoides njalensis* (Laing)) nutrition: a comparative analysis for the free carbohydrate and nitrogenous compounds in cocoa bark and mealy bug honeydew. *Bulletin of Entomological Research* 61, 523–531.

Aitken-Soux, P. (1985) Quelques maladies et pestes courantes du caféier. *Feuille d'Extension* 59, 5 pp.

Andalo, V., Moino, A., Santa-Cecilia, L.V.C. and Souza, G.C. (2004) Compatibilidade de *Beauveria bassiana* com agrotoxicos visando o controle da cohonilha-da-raiz-do-cafeeiro *Dysmicoccus texensis* Tinsley (Hemiptera: Pseudococcidae). *Neotropical Entomology* 33, 463–467.

Anderson, T.J. (1924) Annual report of the Division of Entomology. Kenya, *Annual Report of the Department of Agriculture* 1922, 105–110.

Anderson, T.J. (1927) Report of the entomologist. *Kenya, Annual Report of the Department of Agriculture 1926*, 135–147.

Anon. (1991) The root mealy bug, *Planococcus fungicola*, damage and control. *Kenya Coffee* 56, 1025.

Anstead, R.D. (1919) The coffee planting industry in Southern India. *Agricultural Journal of India* 14, 578–585.

Ba-Angood, S.A. and Al-Sunaidi, S.M. (2004) The biology of the coffee leaf miner, *Leucoptera caffeina* Wash. (Gracillariidae: Lepidoptera) and the effect of irrigation on its development on coffee leaves in Rubat and Hilam-Yahr areas at Yafe'a – Republic of Yemen. *University of Aden Journal of Natural and Applied Sciences* 8, 195–203.

Balachowsky, A.S. (1957) Sur un nouveau genre aberrant de cochenille radicole myrmecophile nuisible au caféier en Colombie. *Revue de Pathologie Végetale et d'Entomologie Agricole de France* 36, 157–164.

Baum, H. (1968) The root mealy bug complex. *Kenya Coffee* 33, 4 pp.

Beardsley, J.W. (1970) Three new species of *Chavesia* Balachowsky from tropical America (Homoptera Coccoidea). *Proceedings of the Hawaiian Entomological Society* 20, 509–520.

Beeson, C.F.C. (1921) The food plants of Indian forest insects. *Indian Forester* 47, 247–252.

Begemann, H. (1929) Over schildluizen van de koffie. *Mededeelingen van het Proefstation, Malang* 71, 11 pp.

Berry, P.A. and Abrego, L. (1953) Insects and diseases affecting some crops in El Salvador. *F.A.O. Plant Protection Bulletin* 1, 151–153.

Betrem, J.G. (1936) Gegevens omtrent de biologie van de dompolanluis en de lamtoroluis. *Archief voor de Koffiecultuur in Nederlandsch-Indische* 10, 43–84.

Betrem, J.G. (1937) De morphologie en systematiek van enkele van de vooraamste witte-luizensoorten van Java. *Archief voor de Koffiecultuur in Nederlandsch-Indische* 11, 1–118.

Brénière, J. (1960) Note sur *Gonocephalum simplex* Fabr. coléoptère ténébrionide nuisible au tabac de Madagascar. *Agronomie Tropicale* 15, 414–433.

Brénière, J. and Syfrig, J. (1965) Un nouvel ennemi de la canne a sucre a Nossi-Be, *Yanga guttulata* Signoret. *Congres de la protection des cultures tropicales. Compte Rendu des Travaux, 23–27 Mars* 1965, 407–410.

Bünzli, G.H. (1935) Untersuchungen über coccidophile ameisen aus den kaffiefeldern von Surinam. *Mitteilungen der Schweizerischen Entomologischen Gesellschaft* 16, 453–593.

CABI (1969) *Agrotis ipsilon* (Hufnangel). *Distribution Maps of Pests* No. 261. CAB International, Wallingford, UK, 3 pp.

CABI (1971a) *Geococcus coffeae* Green. *Distribution Maps of Pests* No. 285. CAB International, Wallingford, UK, 2 pp.

CABI (1971b) *Gryllotalpa africana* Palisot de Beauvois. *Distribution Maps of Pests* No. 293. CAB International, Wallingford, UK, 2 pp.

CABI (1972) *Dysmicoccus brevipes* (Cockerell). *Distribution Maps of Pests* No. 50 (revised). CAB International, Wallingford, UK, 2 pp.

Caminha Filho, A. (1926) Relatorio do mez de septembro de 1925 do auxiliar de inspector agricola de Defensa do Café na zona da Matta. *Boletim do Ministerio da Agricultura, Industria e Comercio* 15, 122–127.

Cárdenas, M.R. and Posada, F.J. (2001) *Los Insectos y Otres Habitantes de Cafetales y Platanales*. Centro Nacional de Investigaciones de Cafe, Colombia, 250 pp.

Cavalcante, R.D. (1975) Control da 'cochonilha da raiz' do cafeeiro – *Planococcus cryptus* Hempel, com insecticidas sistemicos. *Fitossanidade* 1, 68–69.

Cavichioli, R.R. (2003) *Fingeriana dubia gen. nov. e sp. nov.* de Cicadellini (Hemiptera, Auchenorryncha, Cicadellidae) do sudesta e sul do Brasil. *Revista Brasiliera de Zoologia* 20, 247–249.

Ceccon, G., Raga, A., Duarte, A.P. and Siloto, R.C. (2004) Efeito de inseticidas na semeadura sobre pragas iniciais e produtividade de milho safrinha em platio direto. *Bragantia* 63, 227–237.

Chacko, M.J. and Sreedharan, K. (1981) Control of *Planococcus lilacinus* and *Diacanthodes* sp. associated with coffee roots. *Journal of Coffee Research* 11, 76–80.
Chong, K.K., Ooi, P.A.C. and Tuck, H.C. (1991) *Crop Pests and their Management in Malaysia.* Tropical Press Sdn. Bhd., Malaysia, 242 pp.
Collingwood, E.F., Bourdouxhe, L. and Diouf, M. (1980) *Protection des Vegetaux. Rapport des Essais Pesticides 1976–1979. Premier Parte: Insecticides.* Centre pour le Développement de l'Horticulture de Cambrene, Dakar, Senegal, 93 pp.
Corbett, G.H. (1935) Report of the Division of Entomology, 1934, General Series. *Department of Agriculture, Straits Settlements and Federated Malay States* 21, 43–56.
Crowe, T.J. (2004) Coffee pests in Africa. In: Wintgens, J.N. (ed.) *Coffee: Growing, Processing, Sustainable Production. A Guidebook for Growers, Processors, Traders and Researchers.* Wiley, Chichester, UK, 1022 pp.
Da Costa Lima, A. (1936) *Terceiro Catalogo dos Insectos que Viviem nas Plantas do Brasil.* Ministério de Agricultura, Rio de Janeiro, 460 pp.
Da Fonseca, J.P. (1934) Relação das principais pragas observadas nos anos de 1931, 1932 e 1933 nas plantas de maior cultivo no estado de São Paulo. *Archivos do Instituto Biologico* 5, 263–289.
Da Fonseca, J.P. (1958) Três novas espécies de coccideos do Brasil, sôbre cafeeiro. *Archivos do Instituto Biológico* 24, 123–135.
Da Fonseca, J.P. and Araujo, R.L. (1932) Principales pragas do café no Estado de São Paulo. *Instituto Biológico de Defensa Agricola e Animal*, 87 pp.
Da Fonseca, J.P. and Araujo, R.L. (1939) Informações sôbre a praga das cigarras em S. Paulo e sôbre as possibilidades de seu combate. *Biológico* 5, 285–291.
Dammerman, K.W. (1929) *Agricultural Zoology of the Malay Archipelago.* De Bussy, Amsterdam, 470 pp.
D'Antonio, A.M. (1991) A descorberta da mosca das raizes. *Correio Agricola* 2, 8–9.
De Fluiter, H.J. (1936) Voorloopige mededeeling i.z. het onderzoek over een engerlingplaag in de Java-koffie. *Bergcultures* 10, 8 pp.
De Fluiter, H.J. (1960) In: Von Bernegg, S. (ed.) *Tropische und Subtropische Weltwirtschafts-Pflanzen III: Tiel Band 2 Kaffe*, Stuttgart, Germany, pp. 96–102.
De Graaf, J., Schoeman, A.S. and Brandenburg, R.L. (2004) Seasonal development of *Gryllotalpa africana* (Orthoptera: Gryllotalpidae) on turfgrass in South Africa. *Florida Entomologist* 87, 130–135.
De Lotto, G. (1964) Observations on African mealy bugs. *Bulletin of the British Museum of Natural History, Entomology* 14, 341–397.
D'Souza, G.I., Shamanna, H.V. and Venkataramaiah, G.H. (1970) Control of cockchafers in coffee fields with systemic insecticides. *Indian Coffee* 34, 268–269
Duffels, J.P. (1988) The cicadas of the Fiji, Samoa and Tonga islands, their taxonomy and biogeography (Homoptera, Cicadoidea). *Entomonograph* 10, 108 pp.
Du Pasquier, R. (1932) Principales maladies parasitaires du théier et du caféier en Extrême-Orient *Bulletin Économique de l'Indochine* 35, 689B–720B.
Entwistle, P.F. (1972) *Pests of Cocoa.* Longman (Tropical Science Series), London, 779 pp.
Escalante, G.J.A. (1974) Apuntes sobre insectos del Alto Urubamba, Cusco. *Revista Peruana de Entomologia* 17, 120–121.
Fernandez, A. (1957) Las 'escamas palomillas' del cafeto en Venezuela y avances logrados en su control quimico. *Agronomie Tropicale* 7, 7–17.
Ferrer, J. (2000) Révision des èspeces Africaines et Europèennes appartenant au genre *Gonocephalum* Solier (Coleoptera, Tenbrionidae, Opatrini). 3ème partie. *Atti del Museo Civico di Storia Naturale di Trieste* 48, 69–153.
Frappa, C. (1934) Les insectes nuisibles au caféier à Madagascar. *Bulletin Économique de Madagascar* 88, 296–305.

Green, E.E. (1916) On some animal pests of the Hevea rubber tree. *Transactions of the 3rd International Congress of Tropical Agriculture, London*, pp. 608–636.
Green, E.E. (1933) Notes on some Coccidae from Surinam Dutch Guiana with descriptions of new species. *Stylops* 2, 49–58.
Guerrero Berrios, M. (1980) Evaluacion de dosis de insecticideas en el combate de orugas (*Phyllophaga* spp.) *Boletin Tecnico Instituto Salvadoreno de Investigaciones del Café* 5, 12 pp.
Guerrero Berrios, M. and Hanania, C.A. (1979) Evaluacion de plaguicidas contra el piojo blanco de la cabellera *Geococcus coffeae* Green y *Rhizoecus nemoralis* Hambleton. *Boletin Tecnico Instituto Salvadoreno de Investigaciones del Café* 2, 16 pp.
Hambleton, E.J. (1935) Notas sobre Pseudococcinae de importancia economica no Brasil. *Archivos do Instituto Biológico* 6, 105–120.
Hambleton, E.J. (1946) Studies in hypogenic mealy bugs. *Revista de Entomologia* 17, 1–77.
Hambleton, E.J. (1978) Three new neotropical *Rhizoecus* (Homoptera, Pseudococcidae). *Proceedings of the Entomological Society of Washington* 80, 156–163.
Hamilton, D.W. (1967) Injurious and beneficial insects in coffee plantations of Costa Rica and Guatemala 1964. *Journal of Economic Entomology* 60, 1409–1413.
Hamon, A.B. (1982) *Rhizoecus arabicus* Hambleton, a root mealy bug in Florida (Homoptera: Coccoidea:Pseudococcidae). *Entomological Circular, Division of Plant Industry, Florida Department of Agriculture and Consumer Services* No. 238, 2 pp.
Hanson, H.C. (1963) *Diseases and Pests of Economic Plants of Vietnam, Laos and Cambodia.* American Institute of Crop Ecology, Washington, DC, 155 pp.
Hargreaves, H. (1924) Report of the Government Entomologist. *Report of the Department of Agriculture, Uganda*, 1922, 29–32.
Harris, W.V. (1969) *Termites as Pests of Crops and Trees.* Commonwealth Institute of Entomology, London, 41 pp.
Hempel, A. (1913) As Cigarras do Caffeeiro. *O Fazendeiro* 6, 92–93.
James, H.C. (1933) Taxonomic notes on the coffee mealy bugs of Kenya. *Bulletin of Entomological Research* 24, 429–436.
Jepson, F.P. (1929) The control of 'Calotermes' in living plants. *Bulletin of the Department of Agriculture, Ceylon* 86, 11 pp.
Johnson, C., Agosti, D., Delabie, J.H., Dumpert, K., Williams, D.J., Von Tshirnhaus, M. and Maschwitz, U. (2001) *Acropyga* and *Azteca* ants (Hymenoptera: Formicidae) with scale insects (Sternorrhyncha: Coccoidea): 20 million years of intimate symbiosis. *American Museum Novitiates* 3335, 18 pp.
Kalshoven, L.G.E. (1950–1951) *De Plagen van de Culturgewassen in Indonesie.* Van Hoeve, S-Gravenhage, Netherlands, 1065 pp.
King, A.B.S. and Saunders, J.L. (1984) *The Invertebrate Pests of Annual Food Crops in Central America.* Overseas Development Administration, London, 166 pp.
Krug, C.A. and De Poerck, R.A. (1968) *World Coffee Survey.* FAO Agricultural Studies No. 76.
Laing, F. (1925) Descriptions of two species of Coccidae feeding on roots of coffee. *Bulletin of Entomological Research* 15, 383–384.
Lan, C.C. and Wintgens, J.N. (2004) Major pests of coffee in the Asia–Pacific region. In: Wintgens, J.N. (ed.) *Coffee: Growing Processing and Sustainable Production.* Wiley-Vch, Weinheim, Germany, pp. 459–473.
Le Pelley, R.H. (1959) *Agricultural Insects of East Africa.* East African High Commission, Nairobi, 307 pp.
Le Pelley, R.H. (1968) *Pests of Coffee.* Longmans London, 590 pp
Lim, W.H. (1973) Studies on the bisexual race of *Dysmicoccus brevipes* Ckll.: its bionomics and economic importance. *Malaysian Agricultural Journal* 49, 254–267.
Maccagnan, D.H.B. and Martinelli, N.M. (2004) Descrição das ninfas de *Quesada gigas* (Olivier) (Hemiptera: Cicadidae) associados ao cafeeiro. *Neotropical Entomology* 33, 439–446.

Mallamaire, A. (1937) Les principaux nématodes, myriapodes et insectes parasites des caféiers cultivés dans l'Ouest african français. *Annales Agricoles de l'Afrique Occidentale Français et Étrangère* 1, 1–45.

Martinelli, N.M. and Zucchi, R.A. (1984) Ocorrencia de *Fidicina pronoe* (Walker) (Homoptera Cicadidae) em cafeeiros. *Proceedings of the 11th Brasilian Entomology Congress, Londrina, Brasil*, 22–27 July 1984.

Martinelli, N.M. and Zucchi, R.A. (1987) Cigarras associadas ao cafeeiro. I. Genero *Quesada* Distant, 1905 (Homoptera: Cicadidae:Cicadinae). *Anais da Sociedade Entomologica do Brasil* 16, 51–60.

Martinelli, N.M. and Zucchi, R.A. (1989a) Cigarras associadas ao cafeeiro. III. Genero *Dorisiana* Metcalf 1952 (Homoptera, Cicadidae, Cicadinae). *Anais da Sociedade Entomologica do Brasil* 18 (Suppl.), 5–12.

Martinelli, N.M. and Zucchi, R.A. (1989b) Cigarras associadas ao cafeeiro. IV. Genero *Carineta* A&S 1843 (Homoptera, Tibicinidae, Tibicininae). *Anais da Sociedade Entomologica do Brasil* 18 (Suppl.), 13–22.

Milne, D.L. (1973) Insect pests of subtropical fruit crops. *Entomology Memoir, Department of Agriculture Technical Services, Republic of South Africa* 33, 11–19.

Morrison, H. (1952) Classification of the Ortheziidae. *U.S. Department of Agriculture Technical Bulletin* 1052, 80 pp.

Morstatt, H. (1913) Bemerkungen zur cultur und den krankheiten des Kaffee am Meru., *Der Pflanzer* 9, 63–77.

Munoz, R.H. (*c*.2000) *Plagas Insectiles del Cafeto*, Honduras, 32 pp.

Murray, D.A.H. and Wicks, R. (1984) Baiting for soil-dwelling insects. *Proceedings of the 4th Australian Applied Entomological Research Conference, Adelaide, 24–28 September 1984*, pp. 268–273.

Musetti, L. and Johnson, N.F. (2004) Revision of the New World species of the genus *Monomachus* Klug (Hymenoptera: Proctotrupoidea, Monomachidae). *Canadian Entomologist* 136, 501–552.

Nietner, J. (1861) *Observations on the Enemies of the Coffee Tree in Ceylon*. Colombo, 31 pp.

Oakley, R.G. (1945) Preliminary life-history studies in Guam of the Scarab beetle *Ancylonycha mindanaona* (Brenske). *Journal of the Washington Academy of Sciences* 35, 7–12.

Parkman, J.P., Frank, J.H., Walker, T.J. and Schuster, D.J. (1996) Classical biological control of *Scapteriscus* spp. (Orthoptera Gryllotalpidae) in Florida. *Environmental Entomology* 25, 1415–1420.

Patil, S.P. and Veeresh, G.K. (1988) Description of the third stage larva of *Holotrichia farinosa* Nonfried (Melolonthinae: Scarabaeidae:Coleoptera). *Bulletin of Entomology, New Delhi* 29, 61–64.

Pickel, B. (1928) Alguns parasitos radicicolas da cafeeiro em Pernambuco. *Chacaras y Quintaes* 37, 369–370.

Ponce, P.P., Ramirez, J.L.M., Amaya, J.C., Arevalo, J.C.S., Luna, F.A.P. and Rodriguez, E.G. (2004) Susceptibility of *Cyclocephala comata* Bates (Coleoptera: Scarabaeidae) to different biopesticides. *Resistant Pest Management Newsletter* 13, 20–22.

Reyne, A. (1920) Verslag van den Entomoloog., *Verslagen Departement van den Landbouw, Suriname 1919*, 20–24.

Risbec, J. (1936) Les parasites du caféier en Nouvelle Calédonie. *Agronomie Coloniale* 226, 105–123.

Roba, R.P. (1936) La hormiga de Amaga (*Acropyga fuhrmanni*). *Revista Cafetera de Colombia* 6, 2023–2034.

Roba, R.P. (1938) Notes sur quelques insectes s'attaquant au caféier en Colombie. *Bulletin et Annales de la Societé Royale Entomologique de Belgique* 78, 235–239.

Santa-Cecília, L.V.C., Reis, P.R. and Souza, J.C. (2002) Sobre a nomenclatura das espécies de cochonilhas-farinhentas do cafeeiro nos Estados de Minas Gerais e Espírito Santo. *Neotropical Entomology* 31, 333–334.

Sekhar, P.S. (1964) Pests of coffee. In: Pant, N.C. (ed.) *Entomology in India 1938–1963. Indian Journal of Entomology, Silver Jubilee edn*, pp. 99–109.

Senstius, M.W. (1915) Over eene recente insectenplaag op Koffiezaadbedden. *Medeedelingen op het Proefstation, Malang* 7, 5–19.

Strickland, A.H. (1947) Coccids attacking cacao (*Theobroma cacao*) in West Africa with descriptions of five new species. *Bulletin of Entomological Research* 38, 497–523.

Takahashi, R. (1934) Observations on the Coccidae of Formosa. Part IV. *Report of the Government Research Institute, Formosa* No. 63, 1–38.

Teodoro, N.G. and Gomez, E.T. (1926) Coffee diseases and their control. *Philippines Agricultural Review* 19, 249–257.

Van Dinther, J.B.M. (1960) Insect pests of cultivated plants in Surinam. *Bulletin Landbouwproefstation, Surinam* 76, 159 pp.

Van Zwaluwenberg, R.H. (1917) Insects affecting coffee in Porto Rico. *Journal of Economic Entomology* 10, 513–517.

Vayssière, P. (1955) Les animaux parasites du caféier. In: Coste, R. (ed.) *Les Caféiers et les Cafés dans le Monde.* Larose, Paris, pp. 233–318.

Vayssière, P. and Galland, H. (1948) Sur trois insectes d'importance économique encore peu connues en Afrique francaise. *Proceedings of the 8th International Congress of Entomology*, Stockholm, 1948, pp. 699–701.

Watson, G.W. and Cox, J.M. (1990) Identity of the African coffee root mealy bug, with descriptions of two new species of *Planococcus* (Homoptera Pseudococcidae). *Bulletin of Entomological Research* 80, 99–105.

Weidner, H. (1980) Termiten aus Kolumbien nach Beobachtungen vo German O. Valenzuela und Fritz Schremmer. *Anzeiger fur Schädlingskunde Pflanzenschutz Umweltschutz* 53, 65–69.

Wille, J.E. (1952) *Entomologia Agricola del Peru. Manual para Entomólogos, Ingenieros, Agronómos, Agricultores y Estudientes de Algricultura.* Junta de Sanidad Vegetal, Ministerio de Agricultura, Lima, Peru, 544 pp.

Williams, D.J. (2004) *Mealy Bugs of Southern Asia.* Natural History Museum, London and Southdene Sdn. Bhd., Kuala Lumpur, 896 pp.

Williams, D.J. and Granara de Willink, M.C.G. (1992) *Mealy Bugs of Central and South America.* CAB International, Wallingford, UK, 635 pp.

Williams, D.J. and Watson, G.W. (1988) *The Scale Insects of the Tropical South Pacific Region. Part 2 The Mealy Bugs (Pseudococcidae).* CAB International Institute of Entomology, Wallingford, UK, 260 pp.

Wolcott, G.N. (1924) *Insectae Portoricensis.* A preliminary annotated check-list of the insects of Porto Rico with descriptions of some new species. *Journal of the Department of Agriculture Puerto Rico* 7, 1–313.

Wolcott, G.N. (1951) The insects of Puerto Rico, *Journal of Agriculture of the University of Puerto Rico* 32, 1–975.

Yunus, A. and Ho, T.H. (1980) List of economic pests, host plants, parasites and predators in West Malaysia (1920–1978). *Bulletin of the Malaysian Department of Agriculture* No. 153, 538 pp.

Zamora, Q.L. (1998) *Manual de Recomendaciones para el Cultivo del Café.* Instituto del Café de Costa Rica, 195 pp.

III Diseases and their Management

Although large numbers of fungi and bacterial species have been recorded from coffee, few of these can be regarded as pathogens. There are some 380 records of different taxa from coffee in the CABI herbarium database (excluding hymenomycetes and bacteria), and undoubtedly many more occur. CABI (2005) lists 39 fungi, six bacteria and two viruses associated in some way with coffee diseases. Holliday (1980) lists 25 fungal pathogens, and most general coffee texts give accounts of fewer than 20 diseases on coffee.

This is in contrast to the large number of arthropods recorded from coffee, a large proportion of which may feed on the plant and cause damage. However, although there are relatively few serious pathogens of coffee, those that do attack coffee are intricately linked to the plant through their habitat and life cycles. They may damage plants as outright killers – such as wilts – or as insidious, yield-reducing parasites such as rust that are often difficult to control.

Most of the significant diseases of coffee are caused by co-evolved pathogens that have developed as specialized coffee pathogens with the evolution of the genus *Coffea* in its centre of diversity in Africa. So-called 'new encounter' pathogens that are not co-evolved are less host-specific, but they can cause significant damage in some areas: South American leaf spot (*Mycena citricolor*) and many nematodes are examples.

A significant feature of plant diseases – and particularly those affecting coffee – is the major effect that environmental conditions have on their development, both directly and indirectly through effects on host physiology. This explains the major effect that cultural conditions have on the severity of coffee disease and often enables the more serious effects of diseases to be alleviated by appropriate agronomic practices.

Climatic change will undoubtedly influence the incidence and severity of coffee diseases because of the close linkage with the environment. Even a small increase in temperature is likely to result in an increase in rust severity on coffee grown at high elevations. Reduced rainfall will place coffee grown at lower altitudes under stress, leading to higher levels of diseases such as *Fusarium* bark disease, while wetter conditions are likely to exacerbate diseases of berries and shoots, such as coffee berry disease and web blights.

Coffea arabica is a naturally in-breeding species and, as a result, the genetic base of the arabica coffees grown around the world is narrow. In addition, coffee is often grown under conditions far removed from those to which it is adapted in its centre of diversity. This may explain why diseases have had such an impact on the development and distribution of the crop. Rust caused the demise of the coffee industry in Sri Lanka (then Ceylon) in the late 19th century. The disease threatened coffee cultivation in Latin America when it first spread there during the 1960s. Coffee berry disease appeared in Africa later than rust, but soon became the more destructive disease on arabica coffee grown at high altitude.

Fortunately, this disease has not been recorded in Latin America, and some selections from Catimor populations have proved to have some resistance to coffee berry disease. Other diseases have also threatened coffee growing, primarily in Africa. *Fusarium* bark disease temporarily halted development of the arabica coffee industry in Malawi during the 1950s, and *Fusarium* wilt almost achieved the same for robusta coffee in West Africa around the same time. Around 20% of the robusta coffee trees in Uganda have been destroyed since the resurgence of *Fusarium* wilt in the region in the mid-1990s.

Management of coffee diseases – particularly rust and berry disease – with copper-based and other fungicides – has become very costly on arabica in some parts of the world. This has encouraged the switch to Catimor-type cultivars that are often grown without shade trees, with consequent implications for biodiversity in coffee-based agro-ecosystems. Although fungicides have been widely used for management of the major diseases, the production and use of new cultivars with durable resistance to many diseases is taking over as the main method for disease management. Crop husbandry is also important, because environmental factors play a crucial role in the development of diseases.

There are relatively few texts devoted to coffee diseases, and those listed below provide some information.

References

CABI (2005) *Crop Protection Compendium 2005 Edition*. CAB International, Wallingford, UK.

CRF (1964–1977) *An Atlas of Coffee Pests and Diseases*. Coffee Board of Kenya, Nairobi.

Gil, L.F., Castro Caicedo, B.L. and Gomez, P. (2004) *Enfermedades del Cafeto en Colombia*. Cenicafe, Bogota.

Holliday, P. (1980) *Fungus Diseases of Tropical Crops*. Cambridge University Press, Cambridge, UK.

Kranz, J., Schmutterer, H. and Koch, W. (1977) *Diseases, Pests and Weeds in Tropical Crops*. Verlag Paul Parey, Berlin.

Muller, R.A., Berry, D., Avelino, J. and Bieysse, D. (2004) *Coffee Diseases*. In: Wintgens, J.N. (ed.) *Coffee: Growing, Processing and Sustainable Production*. Wiley-Verlach, Weinheim, Germany, pp. 491–545.

Waller, J.M. (1987) Coffee diseases: current status and recent developments. *Review of Tropical Plant Pathology* 4, 1–33.

7 Foliage and Shoot Diseases

Rust

Pathogen: *Hemileia vastatrix* Berkeley & Broome [Basidiomycota: Uredinales].

Coffee leaf rust, CLR or orange rust (Spanish: roya del cafeto; French: rouille orangée, rouille de la feuille du caféier) is a 'classic' among plant diseases. It was one of the first whose aetiology was fully elucidated (Ward, 1882), after it caused havoc in Ceylon (now Sri Lanka) and from where it was first recorded and characterized in 1869.

The fungus is a co-evolved pathogen of *Coffea* spp. in Africa and affects both wild and cultivated *Coffea* species, but causes most damage to *C. arabica*. It spread rapidly throughout Africa and Asia as the coffee industries of these countries developed, and eventually reached the New World in 1971. It now occurs in almost all coffee-producing countries (CMI, 1989). There is an extensive bibliography on the disease. Wrigley (1988) provides a summary of the spread of coffee rust. Reviews of the disease have been produced by Rayner (1972), Holliday (1980), Waller (1982), Fulton (1984), Wrigley (1988) and Kushalappa and Eskes (1989a).

The history and spread of coffee rust

Coffee rust is thought to have developed on wild arabica coffee in its centre of diversity in Ethiopia (Wellman, 1970). A detailed account of how the disease might have reached Sri Lanka has been produced by Wrigley (1988), who provides evidence that it had been in the southern Asia region for some time. From specimens sent from Sri Lanka, the fungus was described and named by Berkley and Broom in the November 1869 edition of the *Gardeners Chronicle*.

When the disease first appeared on coffee estates in Sri Lanka and South India, it caused enormous damage to productivity and resulted in the collapse or conversion of many estates, as reduced yields made coffee growing uneconomic. The coffee industry in Sri Lanka was virtually finished by 1890, although coffee is still grown there in some areas. Many of the estates were replanted with tea, which is supposed to have reinforced the tea-drinking habit in the UK.

As the disease subsequently spread throughout the warm, moist coffee-growing areas in South East Asia, it continued to cause problems for the coffee industry (for full accounts see Large, 1940 and Carefoot and Sprott, 1969). However, coffee grown at the higher altitudes of these countries (e.g. western Ghats in India) largely escaped the ravages of the disease (Muthiah, 1993), and the situation eased as control measures based on shade management and use of fungicides were instigated. The disease was recorded from Java in 1876 and had spread as far as Fiji by 1879. Papua New Guinea remained free from the disease until 1986, but some earlier small, isolated outbreaks had been quickly identified and successfully eradicated (Shaw, 1977).

There was little commercial coffee grown in Africa before the early 20th century, but the first confirmed report of coffee rust appears to be from Natal in 1878. D'Oliveira (1971) indicates that the earliest observed occurrence was on material collected in the Lake Victoria area in 1861. By 1920, the disease occurred in most of the countries in central and eastern Africa, where arabica coffee cultivation was rapidly expanding, but it did not reach much of West Africa until the 1950s.

In 1970, coffee rust eventually reached Brazil and spread rapidly through the coffee-growing areas at a rate of around 500 km/year, so that by 1975 the disease had reached all coffee areas. Bowden *et al.* (1971) suggested that uredospores of the fungus could have been carried by upper air currents from Africa to Brazil, but it is more likely that the disease was carried across the Atlantic by human agency. There was an isolated occurrence in the Carazo area of Nicaragua in 1976 and this was contained, but not eradicated, despite a determined programme (Waller, 1979). By 1981 the disease had spread to most coffee-growing areas in Central and South America, reaching Costa Rica and Colombia in 1983, and Jamaica and Cuba in 1986 (Waller, 1972; Wrigley, 1988).

Coffee rust is now endemic in all major coffee-producing countries, although absent from coffee-growing areas in Australia, Hawaii and some other Pacific islands. Figure 7.1a and b show the progressive spread of coffee rust from 1970 to 1990.

The disease requires control wherever arabica coffee is grown under warm, humid conditions. It is most severe on coffee grown at lower altitudes (below about 1500 m), where warmer temperatures permit greater infection during wet periods and a shorter latent period (Rayner, 1961a). However, it is of little significance at higher, cooler altitudes (> 1700 m in equatorial areas) such as occur in the East African highlands, in Andean countries of South America and elsewhere.

The spread of coffee rust has not had the devastating effect on the industry that was predicted from earlier accounts, and the disease has generally failed to live up to the legend that was built around it by some earlier authors.

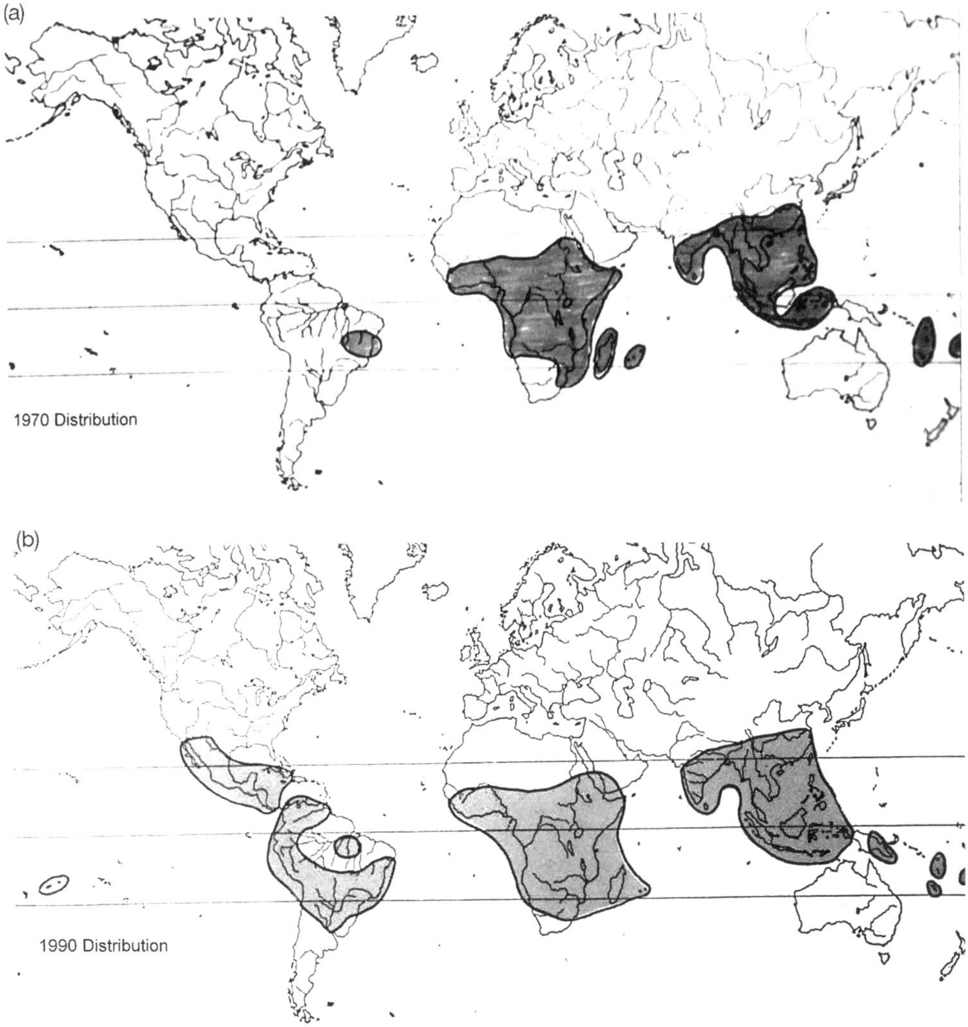

Fig. 7.1a, b. Maps showing the spread of coffee rust between 1970 and 1990.

Nevertheless, the impact of the disease on many facets of man's activities has been appreciable, and McCook (2006) presents convincing evidence of how the impact, both real and potential, of coffee rust has helped to shape the coffee environment and economy of the industry over time.

Economic impact

The economic impact of coffee rust occurs not only through reduction of both quantity and quality of yield, but also through the need to undertake expensive control measures on susceptible cultivars. Because of the difficulty of accurately partitioning and measuring losses caused by coffee rust from those caused by other pests and disease, agronomic factors and their interactions, there are relatively few records of accurately quantified yield losses caused by rust.

Most published data come from the early experiences in Ceylon. Following the first report of rust in Ceylon in 1869, coffee production was reduced by 75% during the next 10 years. Between 1871 and 1878, yields declined from 550 to 240 kg/ha and the area under coffee had decreased from 68,787 to 14,170 ha by the 1890s as uneconomic plantations were abandoned. Amongst other knock-on effects, the Oriental Bank collapsed (Large, 1940).

Monaco (1977) estimated a 30% reduction in coffee yields in Brazil if no control measures were undertaken. Costs of control using a complete fungicide control schedule were estimated at US$67/ha or US$74m for the whole coffee area of Brazil under threat, representing some 9% of the value of coffee exports (Wallis, quoted by Schieber, 1972b).

Costs of chemical control vary between countries but, relative to coffee prices, the costs of inputs in the form of fungicides, sprayers and labour have increased relentlessly. These were estimated to have risen in yield equivalent 'farm gate' costs per hectare from approximately one 60 kg bag of coffee to four bags between 1976 and 1982 (Waller, 1982) as coffee prices fell and input costs rose.

In Brazil (Schieber, 1972b; Eskes, 1989a) and India (Narasimhaswamy, 1961), control costs were estimated to be about 10% of production costs. In Kenya, Nutman and Roberts (1970) estimated that the annual cost of protecting 18,700 ha of coffee at most risk from rust was US$810,000 for a yield benefit worth US$2.9 million.

Estimates of yield loss vary widely, from 15–80%, but worldwide losses are thought to average around 15% (Kushalappa and Eskes, 1989a). Eskes (1989a) estimated overall global costs of the disease at between US$1 and 3 billion per year. However, these estimates will vary according to the market price of coffee. To these costs must be added the long-term research and development costs borne by the industry, national governments and international aid agencies in producing and deploying rust resistant varieties and, although difficult to calculate, these must be very large.

Symptoms

Yellow to orange powdery spots appear on the underside of leaves, with corresponding chlorotic patches on the upper side (see Plates 8a, b). Initially, these are only 2–3 mm in diameter, but steadily expand and can eventually reach a diameter of several centimetres. Young lesions may appear as small chlorotic spots before sporulation occurs. The centres of older lesions become necrotic and the sporulating zone is restricted to the outermost zone. On older leaves, several lesions may merge to produce irregular diseased areas covering much of the leaf.

However, diseased leaves are usually shed before this stage and a major effect of rust is to cause defoliation (see Plate 9). Hyperparasitic fungi such as *Darluca* and *Verticillium* spp. grow over the older sporulating parts of lesions, especially under humid conditions, producing a pale mycelial growth (see

Plate 10). Lesions may also occur on cotyledons and, very occasionally, on young green stems and berries. Grey or powdery rust of *Coffea* spp., caused by *H. coffeicola*, is a related disease of minor significance and restricted to hot, humid areas of central and western Africa. It characteristically produces scattered uredia over the entire leaf surface, in contrast to the discrete, sub-circular blotches produced by *H. vastatrix*; uredospores of *H. coffeicola* also have fewer and larger spines.

Pathogen morphology

Rust lesions consist of many individual uredia that grow through the stomata on the underside of leaves. These are produced from substomatal aggregations of thickened fungal hyphae, from which finer feeding hyphae extend into the leaf tissue. Uredospores are produced in spherical clusters, on the ends of sporogenous cells that emerge through the stomata. The developing uredospores are packed together somewhat like segments of an orange (see Fig. 7.2). This gives them a characteristic, rather reniform, shape, with a curved surface covered in short spines and a smooth, flattened surface – hence the generic name *Hemileia*, meaning 'half smooth' (see Fig. 7.3.).

Urediniospores are 18–28 × 28–36 μm, yellow, with an attachment scar on their smooth surface. Pedicellate, subspherical or napiform, smooth-walled teliospores of about 20–25 μm diameter may be produced within the uredia. These occasionally germinate *in situ* to produce a promycelium, from which four basiospores are formed (Coutinho *et al.*, 1995). However, the role of the basidiospores is not known, as they do not infect coffee and no alternate host for coffee rust has been found. See also Laundon and Waterston (1964a).

Epidemiology

A number of factors influence disease incidence and severity: (i) susceptibility of the variety grown; (ii) races of the pathogen present; (iii) age and vigour of the crop; (iv) presence or absence of shade; (v) bearing level in relation to photosynthetic capacity; and (vi) climatic conditions. On a susceptible variety, severe infection depends on early establishment and spread of the disease within the canopy, which requires a good source of rust spores carried over from the previous season, and subsequently on spread between trees. Under optimal climatic conditions for epidemic development, the disease progresses more rapidly on old or poorly managed trees.

Heavy bearing predisposes the trees to rust unless they are well fertilized, and trees grown under shade tend to be less affected by rust. There is a complex interaction between depth of shade, bearing level and nutrient status of the individual tree. Due to differing bearing levels and foliation, rust can be severe on one tree and slight – or even absent – from a neighbouring tree with similar depth of shade. Kushalappa (1989a) provides a detailed analysis of the processes involved in the epidemiology of the disease.

Fig. 7.2. Scanning EM photo of uredium of *Hemileia vastatrix* protruding through stoma of coffee leaf.

Fig. 7.3. Scanning EM photo of spores of *Hemileia vastatrix* showing characteristic morphology.

Infection

Uredospores of *H. vastatrix* require liquid water for germination and an optimum temperature in the range 21–25°C, with a maximum of 28°C and minimum of 15°C. Club-shaped hyaline appressoria are produced at the end of germ tubes (see Fig. 7.4) and are usually produced within 10 h of spore germination. De Jong *et al.* (1987) found that the temperature requirements for appressorial formation were lower than those for spore germination and that infection could take place after 6 h from spore germination. Appressoria are produced mostly above stomatal openings that occur on the undersides of leaves, and through which infection takes place.

Rainy weather is generally necessary for the production of the under-leaf wetness required for infection to take place. Germination is inhibited by strong light, but not by the low light levels occurring on the undersides of leaves in the coffee canopy during daylight. Nevertheless, in East Africa, Rayner (1961a) found that most infection occurs at night, providing ambient temperatures are high enough, because this is when wetness is of sufficient duration. After spores have been dispersed, viability falls rapidly; few remain viable under field conditions after a week or so. Young, incompletely expanded leaves are resistant to infection; leaves are most susceptible when just fully expanded at the second or third fully expanded node, while old leaves are resistant.

Leaves on heavily bearing trees or branches are more susceptible than leaves on lower-yielding trees, and removal of immature berries has been shown to decrease disease severity (Monaco, 1977). Rust severity tends to be less under shaded coffee and this may be related to yield levels, as fully exposed coffee produces higher yields and is more susceptible. Susceptibility has also been related to the mineral content of leaves (Carvalho *et al.*, 1996)

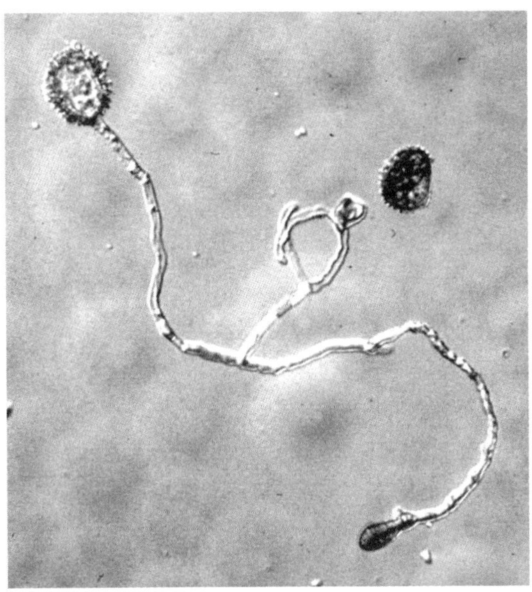

Fig. 7.4. Germinating uredospore of *Hemileia vastatrix* with appressorium.

and to starch content (Zambolin *et al.*, 1992). Stomatal frequency is also much higher on exposed leaves (Alvim, 1958), and this could permit greater infection frequency than on shaded leaves. However, difference in stomatal frequency between cultivars has not been related to difference in susceptibility to rust.

The interaction between cropping levels and disease severity may lead to a cyclical alternation between seasons with heavy yields and associated high disease intensity leading to extensive defoliation, followed by seasons with low yield and low disease levels when the plant recovers to produce more vegetative growth. This enables rust levels to rise again, as more leaves are available for infection and a heavier crop produced in the following year when rust again becomes more severe.

The latent period of the disease is influenced both by intrinsic host factors and environmental conditions. It is shortest (1–2 weeks) on susceptible cultivars carrying a heavy crop and when conditions are most conducive to rapid host growth, and is longest (up to several months) on older leaves of more resistant cultivars under cool, dry conditions (Rayner, 1961a; Eskes, 1982).

First symptoms appear 1–3 weeks after infection, and sporulation usually occurs from about 2 weeks up to 2 months after infection. Rayner (1961a) correlated the mean latent period with mean daily maximum and minimum temperatures, and used this to predict the potential severity of coffee rust epidemics under different climatic conditions. A similar relationship was found in Brazil (Moraes *et al.*, 1976), and the length of the latent period has a major bearing on the progress of coffee rust epidemics. A single lesion or pustule produces four to six crops over a 3–5 month period, releasing 300,000–400,000 spores (Nutman and Roberts, 1963).

Dispersal

The mechanism of spore dispersal was the subject of much debate in the 1960s. Based on inferences from laboratory and field observations, Nutman *et al.* (1960) claimed that spores were dispersed primarily by rain-splash, whereas others (Burdekin, 1960; Rayner, 1961b) produced evidence to indicate that spores were primarily wind dispersed. Earlier, both Ward (1882) and Mayne (1930) had caught airborne uredospores on sticky slides placed some distance from diseased coffee. Subsequent work by Becker *et al.* (1975) and Martinez *et al.* (1977), using spore-trapping devices, confirmed that spores were primarily wind dispersed, and subsequent work showed that aerial spore loads were greatest when wind speeds were highest.

However, uredospores of *H. vastatrix* tend to occur in clumps and are produced under rather sheltered condition on the undersides of leaves in coffee canopies, and it seems likely that rain may assist in spore release through the 'tap and puff' mechanism described by Hirst and Steadman (1963). Spores can also be disseminated by insects (Crowe, 1963), and dispersal by human agency also plays a part. Evidence from the Nicaraguan campaign showed that rust foci occurring after the coffee harvest were most frequent close to paths, suggesting the role of human agency in spore dispersal.

Becker and Kranz (1977) assessed the relative importance of the various methods of spore dispersal and concluded that wind was the most significant factor overall. The spread of coffee rust to Brazil and within Latin America – and the role which spore dispersal mechanisms play – has been discussed by Bowden *et al.* (1971), Waller (1979) and Schieber and Zentmeyer (1984).

Disease progress

Disease severity depends on the rate of increase of disease both within and between plants. Epidemic development of coffee rust is rapid, and a typically polycyclic or logarithmic rate of increase characterizes the early stage of the epidemic under favourable conditions of temperature and wetness. Mayne (1930) established that seasonal epidemics of coffee rust in India commenced at the beginning of the monsoon season, with an initial extensive phase, and then increased logarithmically until the end of the rains, when shedding of diseased leaves reduced disease incidence. He likened the disease increase to that of compound interest – at least 30 years before this idea became popular with plant pathologists.

Subsequently, Bock (1962) elucidated the main factors influencing disease development as rainfall, residual inoculum remaining at the beginning of the rainy season and the degree of foliar density on the tree. Rapid disease progress generally starts soon after commencement of rainy seasons; maximum disease levels (and defoliation) occur after the rains have finished (see Fig. 7.5). The loss of diseased leaves at the end of the wet season and the lack of new infections during the dry season reduce the amount of inoculum available at the beginning of the next season, and can delay disease build-up at the start of the next rainy season.

Inter-seasonal survival of rust occurs through discrete lesions remaining on leaves and through prolonged latency of infection on older leaves under dry conditions. In areas where there is no clear dry season to halt epidemic development, disease increase is related to cropping and foliation patterns, with progress being fastest soon after periods of leaf growth and crop development. Details of the epidemic process in Brazil have been reviewed by Kushalappa (1989a, b). He integrated a series of factors affecting disease progress, including inoculum levels (proportion of rusted leaves), cropping levels (affecting susceptibility) and climatic parameters (affecting infection and latent period), to produce a predictive system used to refine chemical applications.

Temperature has a major influence on disease progress, with the cooler conditions pertaining above about 1500 m in equatorial areas, or during winter months in seasonal climates, inhibiting infection during wet periods and extending the latent period. Climate change that may result in increased temperatures in montane areas is thus likely to result in an increase in rust severity at higher altitudes. The formulae of Rayner (1961a) and Moraes *et al.* (1976) can be used to predict the potential severity of coffee rust epidemics under changing climatic conditions.

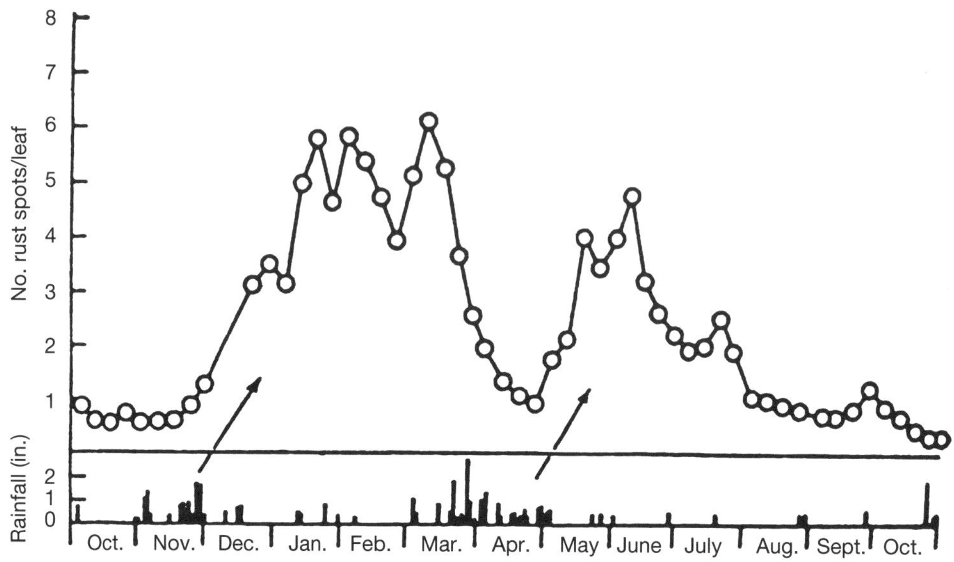

Fig. 7.5. Epidemic development of rust in Kenya (east rift areas) showing relationship between rainfall and disease increase (after Bock, 1962).

The epidemiology of coffee rust on *Coffea canephora* was studied in Uganda (Hakiza, 1997), where its behaviour on a series of eight robusta clones was compared to that on a rust-susceptible arabica cultivar (SL14). The disease was slower to progress than on the arabica cultivar, with a lower infection frequency, longer latent period and slower sporulation rate, even on clones which appeared to be more susceptible in the field. Although the genetic basis of this resistance was not established, there does appear to be a level of field or horizontal resistance present in robusta coffee that was not apparent in comparison with the arabica cultivar.

Effect on yield

The major effect of coffee rust is to cause premature shedding of leaves; this reduces the photosynthetic capacity of the plant and restricts the growth of new stems on which the next season's crop is borne. Disease severity in one year therefore directly affects the cropping potential in the following year, and the disease has an insidious, debilitating effect on the plant over successive seasons. The disease can render coffee cultivation uneconomic wherever it reaches epidemic proportions. Severe disease can also affect the crop of the current season, as defoliation causes carbohydrate starvation of heavily bearing trees. This leads to premature ripening of berries that produce poor-quality, 'light' coffee beans.

Because developing berries are a strong physiological sink for nutrients, a condition known as overbearing dieback may occur when shoots and roots dieback as nutrients are preferentially translocated to berries. This may eventually kill the tree. The tendency of the disease to be more severe on heavily bearing trees can lead to a biennial bearing cycle, where seasons of vegetative growth alternate with seasons of heavy cropping and severe rust.

However, the relationship between disease severity and yield loss is complex, not only because of the inter-seasonal effects but also because of other interacting factors affecting tree vigour (Kushalappa, 1989a). Attempts have been made to develop regression coefficients between disease severity and yield loss, but these are have produced variable results (Chalfoun, 1981). Generally, low levels of rust have no discernible effect and, as a rule of thumb, 5–10% incidence 3 months after flowering has been taken as a threshold level (Waller, 1982; Sierra *et al.*, 1995).

Host–pathogen interactions and pathogen specialization

Hemileia vastatrix occurs in a number of physiologic races (see Table 7.1) that are pathogenic on a range of different *Coffea* spp.: *C. arabica* and *C. racemosa* are attacked by many races, whereas fewer affect *C. canephora* and *C. liberica*. A significant feature of the history of coffee rust is the progressive failure of attempts to control coffee rust by the use of resistant varieties as new virulent races of the pathogen became predominant.

Table 7.1. Current status of knowledge of physiological races and their associated virulence factors (from Varzea and Marques, 2005).

Physiological race	Virulence factors (genotype)	Physiological race	Virulence factors (genotype)
I	v2,5	XXIII	v1,2,4,5
II	v5	XXIV	v2,4,5
III	v1,5	XXV	v2,5,(6)?
IV	v?	XXVI	v4,5,(6)
VI	v?	XXVII	v1,4,(6)
VII	v3,5	XXVIII	v2,4,(5,6)
VIII	v2,3,5	XXIX	v5,(6,7,8,9)
X	v1,4,5	XXX	v5,(8)
XI	v?	XXXI	v5,(6,9)
XII	v1,2,3,5	XXXII	v(6)?
XIII	v5,?	XXXIII	v5,(7) or v5,(7,9)
XIV	v2,3,4,5	XXXIV	v2,5,(7) or v2,5,(7,9)
XV	v4,5	XXXV	v2,4,5,(7,9)
XVI	v1,2,3,4,5	XXXVI	v2,4,5,(8)
XVII	v1,2,5	XXXVII	v2,5,(6,7,9)
XVIII	v?	XXXVIII	v1,2,4,5,(8)
XIX	v1, 4,?	XXXIX	v2,4,5,(6,7,8,9)
XX	v?	XL	v1,2,4,(6)
XXI	v?	XLI	v2,5, (8)
XXII	v5, (6)	XLII	v2,5,(7,8) or v2,5,(7,8,9)

? = Unknown virulence factor(s).
( ) = Resistance factors from HdT derivatives.

The existence of different pathogenic races of *H. vastatrix* was first detected in India by Mayne (1932) using the cultivars Coorg and Kent's (a high-yielding and previously rust-resistant selection found by Mr Kent in his plantation near Mysore) as differentials to identify four races in India. Pathogenic variation on different coffee cultivars was also found in East Africa at around the same time, and there seemed to be no reliable resistance to the disease. Because of the threat posed to the New World by coffee rust, the Centro Internacional de Ferrugems do Cafeeiro (CIFC – International Coffee Rust Centre) was established at Oeiras in Portugal in 1955, with a remit to study the host–pathogen interactions in more detail.

Different races of coffee rust are detected by their reactions on a series of differential coffee selections. This forms the basis of the gene-for-gene relationship in the *Hemileia vastatrix*/*Coffea arabica* pathosystem in which virulence genes (v) in the pathogen interact with resistance (S_H) genes in the host, either singly or in combination. Four major resistance genes were initially detected in *C. arabica* collections from their interactions with rust collections and, including all combinations, gave 16 possible S_H/v interactions. This earlier work, forming the basis of understanding the variability of the rust pathogen, has been reviewed by Rodrigues *et al.* (1975). Other races able to infect diploid coffee species, such as *C. canephora* and interspecific hybrids, indicated the presence of further S_H genes with corresponding v genes in pathogen races (Eskes, 1989a).

Races are designated by a Roman numeral, with race II being the most widespread and having the $v5$ virulence factor. As more resistant cultivars have been deployed over large areas in an effort to control the disease, selection pressure on the pathogen has resulted in the development of further pathogenic races and, currently, some 40 races are known to exist (Varzea and Marques, 2005) (see Table 7.1). Only a few of these races occur widely, and it is race II that predominates on *C. arabica* (see Table 7.2).

Current evidence indicates that the interaction between host resistance and pathogen virulence operates through a protein recognition system mediated by resistance genes in the host that are able to detect elicitor proteins produced by

Table 7.2. Frequency of isolation of the first eight of 30 races of *H. vastatrix* (from Rodrigues *et al.*, 1975)

Race No.	Countries (*n*, 33)	Isolation frequency (*n*, 788)[a]	(%)
I	15	117	15
II	30	442	56
III	10	70	9
IV	3	9	1
V	2	3	0.4
VI	3	26	3
VII	2	2	0.3
VIII	2	4	0.5

[a] Total number of isolates tested from which 30 races were identified.

avirulence (*avr*) genes in the pathogen. This recognition signal then initiates a resistance reaction. Virulence in the pathogen is due to the lack of production of these proteins, or of the avirulence gene producing them, enabling it to escape detection by the host (Van der Vossen, 2005).

The relevance of this is that the more *avr* genes the pathogen must lose to overcome more S_H factors in the host, the fewer elicitor proteins it produces. This is associated with a lack of vigour or fitness in the pathogen (such as reduced growth and sporulation). This explains why coffee rust races with many virulence factors are comparatively rare, are not as pathogenically aggressive as simpler races and soon lose virulence unless maintained on the relevant host plant genotype. Histological details of host-resistant responses during the infection process have been studied by Silva *et al.* (2002).

The mechanism for production of new races is not known, even though these clearly have appeared (and continue to do so) in Asia and Africa, and arose within a few years of the fungus reaching Brazil. Gouveia *et al.* (2005) found that the variability of *H. vastatrix* based on RAPD markers was greatest in Africa and Asia and least in South America. Overall genetic diversity in the pathogen is low and genetic differentiation among populations suggests that asexual reproduction is likely to play an important role in the biology of the fungus. If new races arise by the deletion or suppression of an *avr* gene producing an elicitor protein, this could occur during somatic cell replication without a requirement for sexual recombination. However, Rajendran (1967) reported meiosis occurring in germinating 'uredinioid teliospores', and this would enable the recombination of the genes determining races to arise.

Control

Chemical control

Application of fungicides is still the most widely used method of control, and has been reviewed by Muthappa *et al.* (1989). The widespread use of chemical sprays to control coffee rust began in the 1930s in India and East Africa, and depended on high-volume applications of Bordeaux mixture. After the 1940s, copper oxides and copper oxychloride replaced Bordeaux mixture, and copper fungicides still remain the most effective and economical fungicides for the control of coffee rust (Waller, 1982; Kushalappa and Eskes, 1989a).

These materials are also relatively safe, no resistance has evolved in the pathogen and they adhere well to leaf surfaces. They are also effective against a range of other fungal pathogens and are used at doses of about 0.3–0.7% active ingredient (a.i.) at conventional application volumes.

Numerous fungicides have been tested for rust control since the 1950s, mainly in India and East Africa, then later in Brazil and other Latin American countries after rust had spread there in the 1970s (see Table 7.3 for the main fungicides currently in use). Organic protectants such as dithiocarbamates give some control, but are less effective than copper compounds when disease pressure is high, but-tin based fungicides are effective (Waller, 1982).

Table 7.3. Some fungicides commonly recommended for rust control.

Common name	Formulation	Application rate (for conventionsal knap sack sprayer)
Contact fungicides		
Cuprous oxide	50% Cu WP	3.8[a]
	75% Cu WP	2.4[a]
Cupric hydroxide	50% Cu WP	3.8[a]
Copper oxychloride	50% Cu WP	3.8[a]
Fentin hydroxide	47.5% WP	2.75[a]
Systemic fungicides		
Triadimefon	50% WP	2.0[a]
Triadimenol	1% GR	20–30[b]
Propiconazole	25% SC	0.75[c]
Hexaconazole	5% EC	2.2[c]
Tebuconazole	25% EC	1–2[c]
Flutriafol	1% GR	19[b]
Fosethyl	80% WW	2–3[a]
Azoxystrobin	25% SC	1–2[c]
Organic and copper mixture		
Chlorothalonil/Copper	25/30% WP	7.7[a]

[a] kg/ha.
[b] g per tree.
[c] l/ha.
WP, wettable powder.
SC, suspension concentrate.
EC, emulsifiable concentrate.
WW, water-dispersible granules.
GR, granules.

The advent of systemic compounds in the 1970s provided the opportunity for both protective and curative action. Pyracarbolid was one of the first systemics to be widely used for rust control, especially in East Africa, while oxycarboxin was promoted in India. These have now been superseded by the more effective and safer triazole compounds such as triadimefon, propiconazole and tebuconazole.

Strobilurins such as azoxystrobin are also effective. These are generally used at much lower doses, of about 0.1–0.2% a.i.. These chemicals have a relatively low toxicity, being usually classified as WHO group III substances (see Chapter 15). Certain quaternary ammonium compounds are also effective (Guzzo *et al.*, 1999), and chemicals that activate natural host defensive mechanisms are likely to become more widely used for rust control.

Alternation of systemic compounds with copper fungicides is often recommended, especially where disease pressure is high. Growers also need to consider the concurrent control of other leaf diseases and, in Africa, the control of coffee berry disease (*Colletotrichum kahawae*) and the interactions that might occur (Vine *et al.*, 1973). Fungicide application has a 'tonic' effect on coffee, enhancing yields even in the absence of disease (Rayner, 1957). This is thought to occur through a reduction in leaf microflora that would otherwise induce leaf abscission (Van der Vossen and Browning, 1978; Griffiths, 1981). This effect also exacerbates coffee berry disease (see Chapter 8).

Tank mixtures with copper compounds and organics such as chlorothalonil are often recommended where the need is to control both rust and berry disease (e.g. Anon., 1998). At higher altitudes in equatorial areas (e.g. above 1700 m in Kenya), rust is seldom severe enough to warrant chemical control, because of cool temperatures.

Research on spray application has focused on obtaining effective rust control while reducing application rates. While application methods on large estates can be mechanized and access to water may not be a problem, much of the world's coffee is grown by smallholders for whom fetching and carrying large volumes of water is a major constraint. The trend over the years, therefore, has been to develop spraying methods that require reduced volumes of water.

Fungicide spray application should protect the vulnerable part of the crop (lower surface of the leaf), and spraying techniques should deposit the chemical in this region, although redistribution in rainwater from deposits on the upper surfaces of leaves can be significant (Rayner, 1962). Knapsack mist-blowers provide better penetration and overage in the coffee canopy and use less volume of liquid (50–100 l/ha) than the older mechanized hydraulic machines (> 1000 l/ha).

Ultra-low-volume (ULV) spraying techniques have also been tested, and some of these are effective using minimal amounts of chemical and a much reduced labour input, which is an important factor in coffee grown in mountainous areas. In Brazil, multi-row sprayers applying fungicide over the tops of coffee trees gave adequate control at volumes as low as 50 l/ha. Work on ULV spraying in Brazil cited by Waller (1982) also produced good results with knapsack mist-blowers adapted for ULV application, hand-held ULVA sprayers or tractor-drawn ULV applicators that required only 1.5–3.0 kg copper oxychloride in 8–10 l/ha of a 20:80 oil:water emulsion.

The performance of several air-assisted, hand-held ULV sprayers was tested in Colombia (Aston *et al.*, 1991). A back-mounted, air-assisted spinning disc sprayer was developed to give control at volumes as low as 20 l/ha using dose rates of about 1% a.i. (Waller *et al.*, 1994).

However, many smallholders in Africa still use the conventional knapsack sprayers that require up to 0.5 l per tree to give adequate cover (depending on size and foliation), equivalent to 500–700 l /ha with application rates of 2.5–7.5 kg/ha, depending on the fungicide and formulation used (0.5–2.5 l/ha for systemic compounds). However, Wallis and Firman (1962) showed that lever-operated hydraulic knapsack sprayers can give effective coverage at low volumes of about 100 l/ha, provided they are carefully operated.

With the increasing costs of fungicides relative to prices paid to producers for their crop, spraying for rust control is unlikely to be economic for small-scale growers, unless the coffee plantation is well managed and a berry yield of at least 400–500 kg/ha is normally obtained.

Granular formulations of systemic compounds, such as triadimenol, that can be applied to the soil are also available. These are taken up by the root system and translocated to the leaves, and may offer a way of reducing the frequency and cost of spraying in areas prone to severe rust. They have been used with some success in Latin America. A 1% granular formulation used at 20–30 g/plant gave good control of rust and controlled *Cercospora coffeicola* and *Corticium salmonicolor* as well (Villarraga, 1987).

Studies in Kenya, with a 1% granular formulation of flutriafol, showed that a single application, at 19 g product per tree along the drip-line on irrigated coffee, controlled rust better than five applications of cuprous oxide (Masaba and King'ori, 1995). The protective effect lasted for 12 months and, although the yield boost derived from the tonic effect of copper was absent, it was suggested that ground-applied systemics could be cost-effective in eliminating the need for spray machinery. Although ground applications would have little effect on coffee surface microflora, they do have non-target effects in the soil.

Application schedules usually involve monthly or 6-weekly applications throughout the rainy season, but can be modified to fit local conditions and related to disease epidemiology, equipment, type of fungicide and other factors (Kushalappa, 1989b). In East Africa, where rainfall is bimodal, two sprays are required in each of the rainfall periods, with perhaps a fifth spray if the rains persist well into April. In India, two applications of Bordeaux mixture in the pre- and post-monsoon periods are generally effective in shaded coffee. In some Latin American countries with more continuous rainfall patterns, up to seven sprays may be required for heavy-bearing coffee (Kushalappa and Eskes, 1989b).

The relationship between disease severity and yield loss is a complex one, and it can be difficult to measure the yield benefits from spraying for rust control (Vine *et al.*, 1973). These complications also make it difficult to design a simple forecasting system to obtain the most economic use of fungicides. This has been attempted in Minas Gerais, Brazil, where the simplest of the forecasting systems required the farmer to estimate percentage leaf cover or leaf area diseased and to decide whether the expected yield would be high or low. These factors then determined if a spray was required during the cropping period of August–March (see Table 7.4) (Kushalappa and Eskes, 1989b).

It has been shown that the percentage of infected leaves correlates closely with the proportion of leaf covered by rust pustules, so that the incidence of leaf infection – which are the simpler data to collect – can be used as an estimate of disease severity (Silva-Acuna *et al.*, 1999). A level of 5% of leaves infected is generally regarded as an appropriate threshold for spraying to commence (e.g. Sierra *et al.*, 1995). In Colombia, it was determined that yield would not be affected in the same year or following year, provided the percentage of infected leaves did not exceed 13% in March–June and 5% in July–October (Villarraga and Baeza, 1987).

Table 7.4. A simple forecast system to time fungicide applications for rust control in Minas Gerais, Brazil (modified from Kushalappa and Eskes, 1989b).

	Infected leaves (%) and yield level							
Month	5–9		10–14		15–19		> 29	
	Low	High	Low	High	Low	High	Low	High
S	–	–	–	–	–	–	–	–
O, A	–	–	–	–	–	X	X	X
N, M	–	–	–	X	X	X	X	X
D, F	–	X	X	X	X	X	X	X
J	X	X	X	X	X	X	X	X

S, September; O, October; A, April; N, November; M, March; D, December; F, February; J, January; X, months when spray application required.

Resistant cultivars

Attempts to use plant resistance to control coffee rust have continually been thwarted by the development of new virulent races of the pathogen. In Indonesia, *C. liberica* remained free of the disease for some years in the late 19th century but eventually succumbed, and in India the resistance in first Corg and then Kent's eventually became ineffective. Other selections in Africa that initially appeared resistant also succumbed to the disease after a few years. Nevertheless, some selections such as K7 and Rume Sudan continue to present a degree of partial resistance to the disease that offers sufficient protection when disease pressure is not too high.

Reactions to rust infection vary from minute flecking – indicative of a hypersensitivity reaction or tumefaction of the leaf cells, both representing a resistant reaction – to vigorous sporulating rust lesions indicative of full susceptibility. Slowly developing rust pustules with poor sporulation and tissue necrosis are indicative of an intermediate reaction, often characteristic of partial resistance. The main effect of this type of resistance is to delay the build-up of the disease, so that it may not reach a damaging level during the season. This operates by reducing the epidemiological components of the disease such as lower infection frequency, longer incubation period and lower sporulation.

Major gene race-specific resistance (vertical resistance *sensu* Van der Plank, 1963) conferred by S_H genes occurs, together with varying degrees of basic, multigenic race-, non-specific resistance (horizontal resistance of Vanderplank). When major gene resistance is breached by corresponding pathogen races, any remaining basic resistance is usually partial, giving an intermediate reaction. Resistance that is incomplete is often evident in hybrid populations (Eskes and Da Costa, 1983), but some apparently race-specific resistance may also be incomplete (Eskes *et al*., 1990).

Coffee varieties and species have been categorized into a series of physiologic groups according to their reaction to different coffee rust races and the particular S_H genes they contain. Those in group A have resistance to all known races of the pathogen, whereas those in group F are susceptible to all races and have no S_H genes. Most of the traditional cultivars such as Caturra, Mondo Novo, SLs and Bourbon belong to group E, with the S_H5 gene; the Kent's derivatives belong to group D and have S_H2 and S_H5 genes. Many of the varieties collected from Ethiopia have the S_H1 or S_H4 gene. Derivatives from interspecific hybrids have also added more S_H genes.

In India, a natural hybrid between arabica and liberica was crossed with Kent's to produce the S795 selection that contained the S_H3 gene from liberica, in addition to S_H2 and S_H5 from Kent's. Resistance to rust remained effective for more than a decade after the cultivar was released for widespread planting in the 1950s, but eventually race VIII of the pathogen appeared that was able to overcome the S_H genes of S795. The selection still remains the most widely grown cultivar in India, and some residual resistance is still apparent.

A major advance in breeding for resistance came when selections of the Timor hybrid (HdT) – a natural *C. arabica/canephora* cross – was found to have group A resistance and was crossed with Caturra in Brazil to produce the Catimor hybrids that also had group A resistance. A similar hybrid was produced in Colombia, and HdT was also crossed with Villa Sarchi to produce the Sarchimor hybrids. Other *C. arabica/canephora* hybrids have also produced useful resistance to the disease, including Devamachy in India and Icatu in Brazil. (Carvalho *et al.*, 1989; Eskes, 1989a). The incomplete resistance exhibited by the Icatu population has been characterized by Eskes and Da Costa (1983), and its inheritance studied by Eskes *et al.* (1990).

The Catimor and similar *C. arabica*/HdT hybrids have been used in breeding programmes to produce rust-resistant cultivars such as Cauvery in India, the multiline 'Colombia' in Colombia and the F1 hybrid Ruiru II in Kenya (King'ori and Masaba, 1994). However, new races of the pathogen have emerged that can breach this resistance, especially in India (Prakash *et al.*, 2005), and further work showed that there were another four S_H genes derived from *C. canephora* in these cultivars that could be counteracted by corresponding virulence factors (loss of *avr* genes) in the pathogen. Hence 40 physiologic coffee resistance groups have now been characterized at CIFC (Varzea and Marques, 2005).

Pathogen races able to overcome this resistance are mostly uncommon and not aggressive, so that their potential for damage under field conditions may not be great. This may also be due to the remaining levels of basic, non-race-specific resistance that these hybrids possess and which Eskes and Da Costa (1983) demonstrated as occurring in *C. arabica/canephora* hybrids in Brazil. Pelaez and Gil (2000) also found that advanced progenies of the Caturra/HdT hybrids in Colombia showed characteristics typical of partial non-race-specific resistance.

Nevertheless, some lines from these hybrids still posses grade A resistance, although it seems likely that new races of coffee rust will continue to emerge, and the durability of resistance to coffee rust still awaits clarification. Various aspects of the current situation regarding the search for durablility of resistance to rust have been reviewed by Zambolim *et al.* (2005).

There are strategies that can be used for the enhancement of resistance durability. Breeding programmes that aim to combine as many S_H genes as possible into one cultivar (gene pyramiding) may provide such a considerable barrier to the pathogen that complex races able to overcome this resistance will lose so much fitness as to be epidemiologically unable to sustain a damaging epidemic (Ram, 2001).

The multiline approach adopted in Colombia (Castillo, 1989), where a range of different resistance genotypes are deployed, can also reduce the epidemiological potential of the disease. There is also the level of basic, non-race-specific resistance, already apparent in the *C. arabica/canephora* hybrids (Eskes, 1989a, 2005) and in *C. arabica* selections such as Rume Sudan and Tefarikella. But this appears to be polygenic, and thus difficult to manipulate in breeding programmes.

Hemileia vastatrix also occurs on robusta coffee (*C. canephora*), but is generally less severe – although this varies. The Congalese group (including the Robusta and Kouillou varieties) has a high level of resistance conferred by S_H genes and/or polygenes, but the Guiniensis group is more susceptible (Eskes, 2005). In Uganda, certain clones of *C. canephora* originally selected for good field performance appeared to become susceptible, with moderate defoliation leading to some yield reduction (Hakiza, 1997). Although there was evidence of loss of resistance in some of these clones during the 1980s, lesion types suggest that there is a reasonable degree of basic, non-race-specific resistance.

Cultural and biological control

As referred to earlier, susceptibility to rust is significantly affected by cultural conditions. Factors such as exposure to full sun, heavy bearing and other forms of physiologic stress apparently reduce the efficiency of the basic, non-race-specific resistance that is present to some extent in all cultivars. Therefore, some shade with adequate pruning and soil fertility to prevent over-bearing are significant elements in managing the disease, and form important elements in any integrated control package. Various theoretical aspects of how the environment and production systems interact to influence rust severity and management have been discussed by Avelino *et al.* (2004).

Several hyperparasitic fungi, such as *Darluca filum* (Biv.) Castagne, several species of *Verticillium*, *Cladosporium hemileiae* Steyeart and *Paranectria hemileiae* Hansf., have been recorded from rust lesions. These hyperparasites are more abundant on rust lesions on coffee grown in shaded, humid conditions and may contribute to the lower rate of rust development observed in these situations. *Verticillium lecanii* Zimm. is particularly common, and its use as a biocontrol agent has been reviewed by Eskes (1989b); Alarcón and Carrion (1994) report its use in the field in Mexico.

There may be more scope for work in this area, perhaps integrating the use of *V. lecanii* with copper fungicides if copper-resistant strains of *V. lecanii* could be developed (Eskes *et al.*, 1991; Gonzalez *et al.*, 1995). Dagmar *et al.* (1989) reported that rust could be controlled with sprays containing *Bacillus thuringiensis*. A commercial product containing the bacterium was effective at 5–20 mg/ml, giving 90% protection that was systemic and lasted for 5 weeks.

Certain *Pseudomonas* spp. may also induce resistance (Porras *et al.*, 1999), and there is some evidence that endophytic fungi may induce systemic resistance in similar crops (Arnold *et al.*, 2003). A range of insects and mite species also feed on rust spores in older lesions. However, it is not clear whether any of these naturally occurring agents could be used in an enhanced biological control programme.

Economic considerations

Before the advent of effective resistance in the form of Catimor hybrids and similar derivatives, much effort was put into improving the economic effectiveness of chemical control through more effective fungicides, improved timing of applications and more efficient spray application techniques.

Problems are particularly acute for smallholder growers who can ill-afford investment in spraying machinery, chemicals and labour before receiving payment (often delayed through complex marketing channels) for their crop, in marginal enterprises where potential yields are low and in mountainous areas where labour costs of spraying are high. Lack of rural credit facilities and falling or static prices on the international coffee markets exacerbate the problem. In many areas, substantial savings can be made in the costs of fungicide application through improved application technology, such as the use of low-volume controlled droplet applications mentioned above.

Integrated management of coffee rust using a combination of shade, some chemical control and cultivars showing some resistance may be effective, but both the chemical applications and the reduced yielding potential of shaded coffee impose an economic cost. The deployment of effective rust-resistant cultivars clearly provides the most economic solution (assuming continued durability of resistance) as, apart from initial replanting and lower yields from young plants, they require no economic input for rust control from the farmer.

Grey Rust

Pathogen: *Hemileia coffeicola* Maublanc & Roger [Basidiomycota Uredinales]; synonym: *Uredo coffeicola* Maublanc & Roger [anamorph].

This leaf disease, also known as powdery rust of coffee or rouille farineuse (Fr), is confined to the wetter parts of West and Central Africa. It was first recorded in Cameroon in 1929 (Coste, 1992). The disease is of local importance in some West African countries but is currently not a major problem. It has been reported from Cameroon, Central African Republic, Democratic Republic of the Congo, Nigeria, Uganda, Côte d'Ivoire, Togo and Sao Tome and Principe (CMI, 1985), and is a potential threat to coffee production in other warm, humid, coffee-producing areas.

Symptoms

Hemileia coffeicola produces pale yellow to orange uredia that are spread over the underside of the leaf to give a dusty coating. This is in contrast to *H.*

vastatrix infestation in which uredia occur in rounded blotches; the symptoms of *H. coffeicola* are less conspicuous and defoliation is less evident. Diseased leaves eventually become chlorotic, then desiccated and are shed. Often, the first sign of the disease might be the appearance of chlorotic spots on the leaves; on some varieties, sporulation might be sparse. The disease is usually most obvious on the older leaves towards the bases of branches inside the canopy. Most cultivars of *C. arabica* are susceptible, as are *C. canephora*, *C. liberica* and several wild species, although some resistance has been reported (Rodrigues, 1957; Tarjot and Lotode, 1979).

Pathogen morphology

Uredia are produced through the leaf stomata from swollen 'feeder' hyphae in the substomatal cavity. The mycelium is sparse, with a few large feeder hyphae that swell to 30 μm in diameter. The tips of these cells bear numerous pedicels from which the spores are produced in spherical clusters, giving the spores a characteristic reniform shape with spines on the outer surface.

Uredospores are 34–40 × 20–28 μm, with a hyaline wall and the spines on the convex surface, but are more sparse and slightly larger than those of *H. vastatrix*. Subspherical, smooth, teliospores, 20–26 μm in diameter, may also be produced, but basidiospores do not infect coffee and the rust has no known alternative host (Laundon and Waterston, 1964b).

Hemileia coffeicola is distinguished from *H. vastatrix* by: (i) the uredia being scattered over the leaf surface rather than being confined to defined spots; (ii) having fewer swollen feeder hyphae; and (iii) the uredospores bearing larger but fewer spines. Saccas (1972) provides a review.

Epidemiology

There is little published information on its biology, although basic features of the pathogen and disease have been reviewed by Saccas (1972). There is no known alternative host and the disease is maintained through infection by uredospores: although teliospores are produced, they appear to be functionless. The disease occurs from sea level to above 1500 m and appears less constrained by altitude than does *H. vastatrix*. Epidemic development begins at the start of the rainy season and the disease is encouraged by high humidity.

Infection occurs through stomata on the underside of coffee leaves. Mycelium becomes systemic in the leaf tissues, with subsequent production of uredia from stomata over the whole underside of the leaf. Uredospores are presumed to be wind dispersed and germinate to infect the host through stomata on the lower surface of the leaf; general behavioural characteristics are broadly similar to *H. vastatrix*, but *H. coffeicola* is less aggressive and more restricted in its distribution. Water balance and phosphate levels of leaf tissues were shown to be correlated with resistance to the disease, and infection influenced phosphate and calcium metabolism in leaves (Massaux *et al.*, 1978).

Isolates from Côte d'Ivoire are considered to be different from those in Central Africa, and isolates from different wild hosts also differed in their pathogenicity to a range of coffee cultivars (Lourd and Huguenin, 1982). In Cameroon, the disease is more prevalent on robusta coffee (Tarjot and Lotode, 1979). Coste (1992) considers that the importance of the disease in limiting yield is often underestimated.

Control

Copper fungicides provide effective control (Saccas, 1972), and *H. coffeicola* may be controlled along with the more virulent *H. vastatrix* by the use of the same fungicidal sprays. However, the disease is not usually a sufficiently important disease to justify control. Some species and varieties have some resistance to the disease (Tarjot and Lotode, 1979).

American Leaf Spot

Pathogen: *Mycena citricolor* (Berkeley and Curtis) Saccardo [Basidiomycete Tricholomatacea.] Synonyms: *Agaricus citricolor* Berkeley and Curtis; *Omphalia flavida* Maublanc & Rangel; *Stilbum flavidum* Cooke [anamorph].

This is a common and conspicuous leaf spot of coffee under cool, humid conditions in the Caribbean and South and Central America, where it is commonly known as Ojo de gallo (cock's eye). *Mycena citricola* has been recorded on a wide range of hosts: Carvajal (1939) and Sequiera (1959) provide extensive lists and, according to Wellman (1972), over 550 plant species can be infected. It is common on woody perennials in cool, humid equatorial forests, several of which are shade trees in coffee plantations. Because of its potential to cause significant damage to coffee under certain conditions, *M. citricolor* is a potential threat to all arabica coffee-producing areas where it does not occur (Africa, Asia and Oceania) (CMI, 1996).

Economic impact

Mycena citricolor can cause significant defoliation in some localities. The result of this is reduced growth of coffee, producing less bearing wood to carry the following season's crop. Severe defoliation may also cause berry shedding and reduced quality of the current season's crop. It may also infect green berries. In humid, shaded coffee plantations in some Central American countries, losses of up to 20 or 30% have been estimated (Wellman, 1972), but Bonilla (1982) estimated losses of up to 73% in areas of El Salvador.

Symptoms

On coffee, sub-circular spots – initially brown, becoming pale brown to straw-coloured – are produced mainly on leaves (see Plate 11) and are often numerous and conspicuous. The spots have a distinct margin and are 6–13 mm in diameter but with no halo. The fungal mycelium tends to be luminous, giving the spots a bright appearance in the shade. Mature spots become lighter and develop minute, yellow, pin-like gemmifers 1–4 mm long (anamorph *Stilbum* state) consisting of a thin pedicel surmounted by a spherical gemma about 0.4 mm in diameter. Gemmifers are mostly produced on the upper surface of the spots.

The centres of older leaf spots may disintegrate, giving a 'shot hole' appearance. Similar spots may be produced on stalks and berries (see Plate 12). The main effect is to cause leaf fall, with consequent reduction in growth and yield of the coffee tree. Symptoms on other hosts are broadly similar (Wellman, 1972).

Pathogen morphology

The sporophore of the perfect state (which occurs rarely on coffee) consists of a minute agaric with a thin, membranous, campanulate pileus. It is 1.5–2.5 mm in diameter, radially striate with a few, distinct, waxy gills and is sulphur yellow in colour. The stipe is 1.0–1.5 cm × 0.25 mm, straight with no swollen base. Basidiospores are hyaline, ovoid 4–5 × 2.5μm. The gemmifer of the anamorph consists of a stalk (pedicel) and a head (gemma). The yellow stalk is some 2.0 × 0.1 mm, cylindrical and often curved at maturity. The gemma is a tough, pseudoparenchymatous oblate sphere averaging 0.36 mm in diameter and covered with protruding filaments.

Further detail may be found in Wellman (1972). Coffee leaf spot caused by *Cercsopora coffeicola* bears some resemblance, but the leaf spots are smaller, reddish brown in colour although paler in the centre. The parasitic alga *Cephaleuros virescens*, which may also occur on leaves, also produces stalked gemmae but lesions are quite different.

Epidemiology

The gemmae are the infective propagules, the perfect state rarely occurs on disease lesions and basidiospores are not thought to be a significant factor in disease epidemiology. However, the perfect state can be readily produced in culture by growing with *Penicillium oxalicum* and some other *Penicillium* spp. (Salas and Hancock, 1972), and probably occurs saprophytically in forest conditions.

The disease is distributed as localized foci in coffee plantations and spreads slowly, indicative of the relative large, infective propagules that are distributed by wind and rain (Wellman, 1950). Disease foci in coffee frequently occur beneath shade trees that can act as wild hosts and sources of inoculum. There is no evidence for long-distance dispersal, and *M. citricolor* is an example of an airborne pathogen that only spreads very slowly. Dispersed gemmae germinate under wet conditions on leaf surfaces, with the production of numerous infective hyphae.

In Costa Rica the life cycle is about 8 days (Bonilla, 1982). The disease is most prevalent in heavily shaded, high-rainfall locations and there is some evidence that injury to leaves facilitates infection (Tewari *et al.*, 1986). The pathogen produces calcium oxalate that apparently has a major role in pathogenesis. (Rao and Tewari, 1989). Athough the disease may affect coffee berries, there is no evidence for seed-borne transmission.

Control

Cultural control measures aimed at reducing inoculum sources such as badly diseased trees and infected shade trees have had some success (Wellman, 1950). As the disease spreads slowly, sanitation methods should have a significant impact but may be difficult to execute effectively in smallholder coffee. Measures to reduce shade and improved aeration by judicious pruning also help to reduce infection.

Application of fungicides continues to be used in areas where the disease is problematic. Lead arsenate sprays have been shown to have an eradicant effect (Echandi and Segall, 1958), and became commonly used in Central America before being superseded by safer, protective compounds such as Bordeaux mixture and other copper fungicides. Triazole systemic fungicides are now most effective (Vargas *et al.*, 1990), and application of calcium hydroxide has been shown to suppress symptoms due to the neutralization of oxalic acid produced by the pathogen (Rao and Tewari, 1988).

Cyproconozole prepared as an alkaline spray was most effective in Costa Rica (Mora, 2000). There have been several investigations of the use of biological control agents: (i) Vargas (1984) reduced the disease by applying coffee grindings inoculated with *Trichoderma harzianum*; (ii) antagonistic bacteria isolated from diseased coffee were used by Calvo and Vargas (1989) and Mora *et al.* (1989); and (iii) a *Bacillus* sp. was used by Quesada and Jimenez (1996). Mora-Bolanos *et al.* (2000) found that the antibiotic validomycin was particularly efficient and prevented berry infection.

Brown Eye Spot

Pathogen: *Cercospora coffeicola* Berk. & Cook (syn. *C. coffea* Zimm.; *C. herrerana* Farn.) [fungus imperfectus].

Brown eye spot is a very common leaf spot of coffee of worldwide distribution: also known as Cercospora leaf spot, mancha de hierro (Spanish) and La maladie de yeux brun, cercosporosis (French). The pathogen also causes berry blotch and red blister on robusta berries.

Symptoms

The leaf spots begin as a small, chlorotic spot that expands to up to 1.5 cm in diameter, becoming brown to reddish brown, often with an ash-grey centre, roughly circular but may appear angular, being delimited by the main veins (see Plate 13). The spot may be surrounded by a yellow halo, and sporulating areas in the grey centre of the spot appear olive green in colour. Brown eye spot lesions are more pronounced on the upper lamina. The disease rarely causes losses in mature arabica coffee, but can reduce the vigour of seedlings in the nursery and young plants and may cause defoliation, although infection of berries (berry blotch or red blister on *C. canephora*) can be troublesome (see Chapter 8).

Pathogen morphology

Cercospora coffeicola produces conidia, mainly on the upper lamina. Conidiophores arise in fascicles of three to 30, and are sometimes branched, septate and 20–275 × 4–6 μm (see Fig. 7.6). Conidia are hyaline, elongated, straight or slightly curved with a conspicuous hilum, multiseptate and 40–150 × 2–4 μm (Mulder and Holiday, 1974). No physiological races have been reported. *Mycosphaerella coffeicola* (Cook) J.A. Stev. & Wellman has been recorded as the teleomorph.

Epidemiology

Sporulation occurs readily under humid conditions. Conidia may survive on the leaf surfaces for up to 2 months, but fallen leaves – in which the pathogen can remain viable for 9 months – can also be a source of infection at the end of the dry season. Infection takes place when conidia germinate and penetrate the stomata on the underside of the leaf. Optimum temperature for germination is 27°C in free water, when germ tubes form in 2 h, but germination occurs in the range 15–30°C. Symptoms begin to appear about 28 days after infection, and sporulation on the lesion occurs 3 weeks later under suitable conditions. Damage to berries predisposes them to infection, and sun scorch is often a major factor in this. The incubation period is also much shorter on berries.

Infection is more prevalent and disease development is more rapid in situations where the plant is under stress, such as on unshaded coffee and where nutrients are in short supply. Junior *et al.* (2003) found that the balance of potassium and calcium was critical, and Pozza *et al.* (2004) found that silicon was important. The disease is more prevalent at lower altitudes and where temperatures are higher. Spread within the canopy takes place by rain-splash, and between trees by wind-borne conidia. General accounts of the disease are given by Echandi (1959), Subramanian and Sridhar (1966) and Siddiqi (1970).

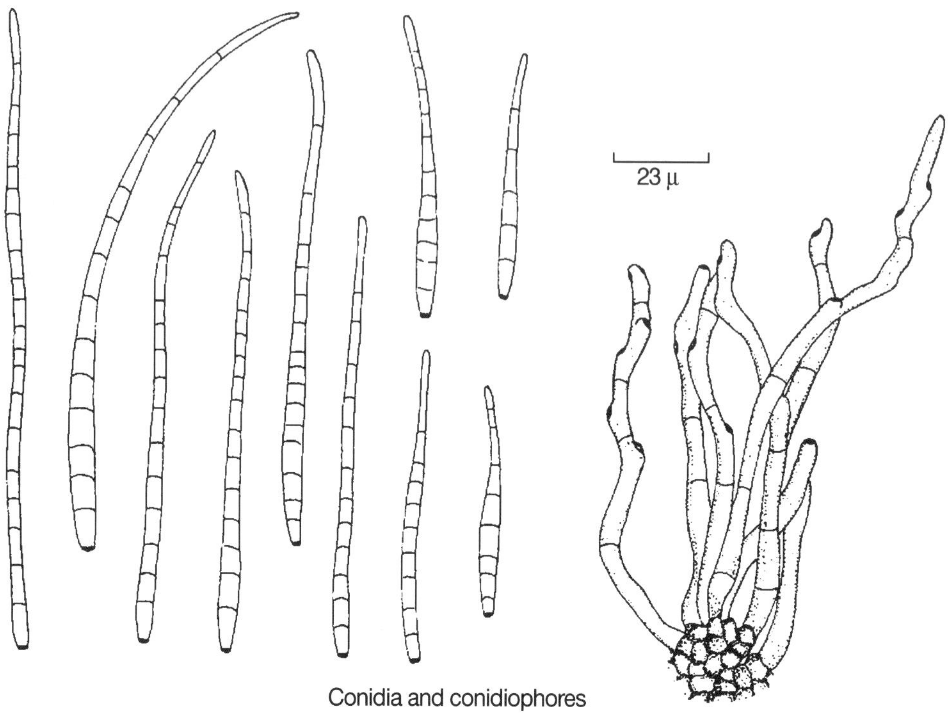

Fig. 7.6. *Cercospora coffeicola* (Mulder and Holliday, 1974).

Control

The disease rarely requires control in fully grown arabica, but in the nursery it is often necessary to control the disease on seedlings by fungicidal application (Siddiqi, 1970). The disease was reported as causing significant defoliation of *C. arabica* in the Indian State of Karnataka (Nataraj and Srinivasan, 1996).

Prevention of predisposition is a major factor in control and includes provision of shade, and adequate fertilization – especially of potassium and nitrogen – is needed; seedlings planted out without shade are particularly vulnerable to the disease. As the disease has a long incubation period, protective fungicides must be applied from the first leaf stage. Several fungicide trials have been undertaken for controlling the disease, especially in nurseries. Copper-based protectants are efficient. In Brazil, Pozza *et al.* (1997) found that chlorothalonil and mancozeb with mineral oil were most effective and, in India, captafol was recommended for eye spot control (Hanumantha *et al.*, 1994).

There is some variation in susceptibility to brown eye spot in *C. arabica*. Some resistant lines were identified in Costa Rica from genotypes with resistance to rust, and the variety Catuai was also among the cultivars least affected (Izquierda-Barroa *et al.*, 1993) but some lines of *C. arabica* × *canephora* hybrids can be particularly susceptible.

Bacterial Blight

Pathogen: *Pseudomonas syringae pv garcae* (Amaral, Teixeira & Pinheiro) Young, Dye & Wilkie [bacterium: Pseudomonadales].

Also known as Elgon or Solai dieback or BBC in East Africa. This bacterial disease causes a leaf and twig blight of coffee in parts of South America and East Africa. The disease was previously attributed to *Colletotrichum*. A related strain of the bacterium also causes halo or aureola leaf spot in Brazil. In Puerto Rica, *Pseudomonas cichorii* (Swingle) Stapp also causes a leaf spot (Sanchez *et al.*, 2003).

Symptoms

Lesions on the leaf begin as water-soaked spots that turn black and expand, so that the leaves become necrotic and curl inwards, but remain attached to the plant. When the terminal bud on the shoot tip becomes infected, the infection moves down the twig causing dieback (see Plate 14). Axial buds on young branches some distance from the tip are also vulnerable to infection. The lesion can girdle the branch, damaging the vascular tissue and causing the branch above the blackened node to wilt and die.

In severe cases, the plant can appear to have been badly scorched. The flowers and pinhead-sized berries can also be infected (Ramos and Shavida, 1976). The bacterium can infect all the young, aerial parts of the coffee tree. The halo spot reported from Brazil begins as a small, water-soaked lesion that develops into a dark, irregular, necrotic lesion surrounded by a chlorotic halo.

Pathogen characteristics

The causal bacterium is *Pseudomonas syringae* pv. *garcae*, first described from Brazil as *P. garcae* then renamed as a pathovar of *P. syringae*. The Brazilian halo spot pathogen differs in several respects from the Kenyan strain (Ramos, 1979). The bacterial blight pathogen can be distinguished from *P. syringae* and *P. garcae* by a combination of biochemical and pathogenicity tests (Kairu, 1997). *Pseudomonas syringae* has a wide host range but *P.s.* pv. *garcae* is pathogenic only on *C. arabica*.

Isolation of the bacteria on nutrient sucrose agar can be easily carried out from diseased shoot material. After incubation at 23°C for 48 h, colonies of the pathogen are opaque, ash-white, levan-forming, mucoid and with or without yellow fringe. A method for diagnosis of the bacterium is the appearance of yellow-green pigmentation of the colony under UV light. Artificial inoculation with a coffee isolate of the pathogen produces no symptoms on citrus (*Citrus sinensis*), peach (*Prunus persica*) or Lima bean (*Phaseolus lunatus*), though these crops are preferred hosts of *P. syringae* (Kairu, 1997).

Epidemiology

Although the disease does not affect more than 5% of the crop in Kenya, it can cause total crop loss in some areas, and severely affected trees sometimes have to be destroyed. This disease is confined to some highland areas of Kenya at altitudes above 1800 m, on Mount Elgon, Solai and parts of Nyeri District. It is most serious where coffee is exposed to cold night temperatures. The bacteria are spread by rain-splash and the disease develops on new growth produced during the rainy season.

The bacterium enters the plant through pruning wounds, storm damage or directly into the stomata. Firman (1963), who studied the problem before its bacterial aetiology had been resolved, found that hail was a major factor in severe attacks. Infection through young apical leaves or young axial buds leads to systemic invasion of the vascular tissue in young leaves and twigs, resulting in dieback. Infections on older leaves remain localized and do not become systemic. The pathogen persists as an epiphyte in the bark and, in common with other *P. syringae* pathogens, in inconspicuous stem cankers. Humans and farm implements can be responsible for considerable disease spread within the plantation (Ramos, 1979).

Control

Application of bactericidal chemicals such as copper or dithianon have given effective control in localities where the disease is severe. These can also be used for concurrent control of coffee berry disease (CBD). An upsurge of this disease in the 1970s was due to the replacement of copper fungicides by organic compounds for control of CBD, and captafol was shown to stimulate the disease (Kairu *et al.*, 1984). Copper sprays remain the most effective way of controlling the disease and control has been good with some low-copper bactericidal formulations (Kairu *et al.*, 1991). Kasugamycin is also recommended.

Spray application must start before the rains begin and continue through the wet season; this this may involve 12 or more applications. In Brazil, a range of antibiotic substances has shown effectiveness against aureola leaf spot (Paradela *et al.*, 2000). Geisha has some resistance to the disease (Firman, 1963).

Other Foliage and Shoot Diseases

Koleroga

Pathogen: *Corticium koleroga* (Cooke) Hoenel. (Syn. *Pellicularia koleroga* Cooke; *Corticium stevensii* Burt; *Ceratobasidium stevensii* (Burt) Talbot &

Venkat.) [Basidiomycete: Corticiaceae]. The disease is also known as black rot, thread blight, mal de hilachas (Spanish) and koleroga des agrumes (French).

This pathogen has a very wide host range, including many tropical tree crops and some herbaceous crops such as cucurbits and pigeon pea. *Coffea robusta* is more susceptible than *C. arabica*. The disease is rarely serious, but is found throughout the coffee areas of Latin America and the Caribbean. It is widespread in India and has been recorded in South-east Asia, some of the Pacific Islands and in parts of Africa: Côte d'Ivoire, Congo and Madagascar.

Symptoms consist of greyish white, thread-like fungal growth on the underside of the leaves and on berries. This coating later becomes brown, then the leaves yellow and die, becoming black (see Plate 15). Dead leaves often remain attached to the branch by the fungal web and the fungus survives the dry season as sclerotia on dead host material. Coleman *et al.* (1923) made the first detailed study of the disease, which can be troublesome in the monsoon period in India (Narasimhan, 1933).

The disease is favoured by cool, wet weather and high humidity, so is more common at higher altitudes. It is most prevalent under humid, heavily shaded, dense coffee, where surfaces remain wet for long periods; densely planted semi-dwarf cultivars are prone to attack and the disease has become more prevalent in Central America (Muller *et al.*, 2004). The removal and destruction of affected stems is important, and open planting, adequate pruning and lightening of heavy shade help to prevent the disease. In India, copper sprays (e.g. 1% Bordeaux mixture) have been recommended, with or without carbendazim (0.03% a.i.).

Similar thread blight diseases are caused by several species of *Marasmius* and *Crinipellis* (Holliday, 1980). Horse hair blight, so named because of the black fungal strands that spread across foliage, is commonly caused by *Marasmius crinisequi* F. Muell. ex Kalch.

Pink disease

Caused by *Erythricium salmonicolor* (Berk. & Broome) Burds. (Syn. *Corticium salmonicolor* Berk. & Br.; *Pellicularia salmonicolor* (Berk. & Broome) Dastur). [Basidiomycete: Corticiaceae]. Also known as enfermedad rosada; mal rosado (Spanish), maladie rose (French).

This minor disease of coffee has been recorded in almost every country where coffee is grown. Pink disease has a wide host range, having been recorded on over 500 woody plants. Among primary hosts are *Coffea*, *Hevea brasiliensis*, *Camellia sinensis*, *Citrus*, *Eucalyptus*, *Theobroma* and *Chinchona* (Mordue and Gibson, 1976; Holliday, 1980). The disease is more important on rubber, cocoa, citrus and *Eucalyptus* spp. – on which it can cause major damage – than on coffee.

The pathogen is related to that causing koleroga, and the disease has some resemblance to it. An encrustation of dirty white to pale pink surface mycelium on affected stems and branches is characteristic of the disease. This is associated with some swelling of the bark, and parts of the tree distal to the infected area dieback. On bearing branches, the disease may spread to the berries. Basidiospores are produced from the mycelium on diseased tissues and there is a conidial *Necator* stage also produced in the bark as emergent pustules. Spore release is favoured by prolonged rainfall, although basidiospores are subsequently more abundant after light rainfall (Sneider-Christians *et al*., 1986). Spores can directly penetrate stem tissues.

The disease is favoured by wet conditions and moderate temperatures (28°C). Copper sprays are effective against pink disease, and infected twigs should be removed and burned. As with koleroga, the disease is more severe on densely planted, heavily shaded coffee.

Anthracnose

Pathogen: *Glomerella cingulata* (Stonem.) Spauld & Shrenk [Ascomycete], *Colletotrichum gloeospoioides* Penz. [anamorph] (see Fig. 7.7) often referred to as *Colletotrichum coffeanum* Noack on coffee (Firman and Waller, 1977).

Leaf anthracnose is common, can cause severe defoliation and twig dieback (Saccas and Charpentier, 1969). Necrotic, brown lesions are usually produced on leaf margins (see Plate 16) and, as they spread and become older, they may develop concentric rings in which acervuli may be visible as black dots on the upper leaf surface. This is not normally serious, but has been known to be problematic in nurseries, causing heavy defoliation (Saccas and Charpentier, 1969). On young, green twigs lesions start as depressed, yellow, elongated spots that rapidly become dark brown and may be surrounded by a yellow halo. These may girdle the stem. As the lesion ages, the cortex splits and

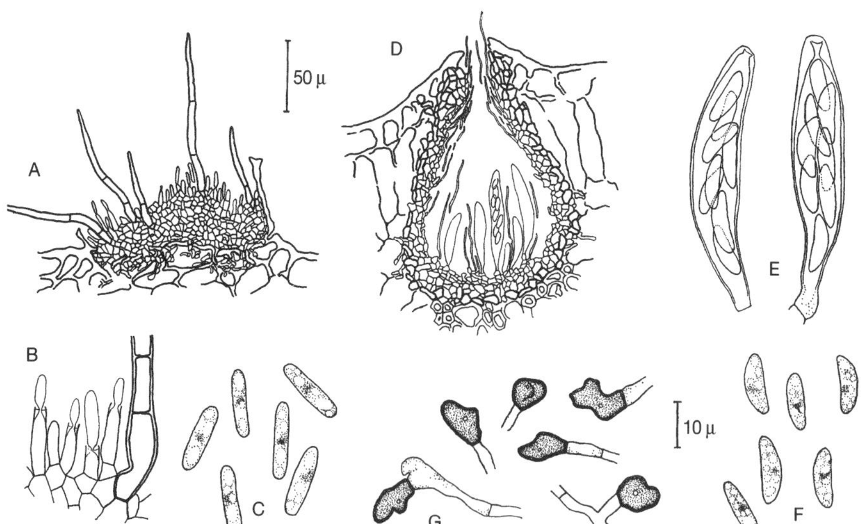

A, T.S. acervulus; B, conidiophores; C, conidia; D, T.S. perithecium; E, asci; F, ascospores; G, appressorium formation from hyphae.

Fig. 7.7. *Glomerella cingulata* (Mordue, 1971).

acervuli, which sporulate abundantly, form in the cracks. The affected trees appear to be suffering from dieback (Firman and Waller, 1977).

The disease is referred to as brown blight in India and can be controlled with foliar sprays of 0.5% Bordeaux mixture if required (CCRI, 1999). Anthracnose also affects fruit (see Chapter 8). *Colletotruchum gloeosporioides* is also a common secondary invader of damaged tissue, frequently colonizing old rust lesions, is a regular inhabitant of the bark of coffee stems and is probably endophytic in coffee tissues. Significant damage by anthracnose to vegetative parts of the tree is probably associated with other debilitating factors. In Latin America, *Colletotrichum* is associated with blister spot (see below).

Leaf blight and dieback

This disease symptom is caused by several fungi: *Ascochyta tarda* Stewart [Ascomycete] (Syn. *Phoma tarda* (Stewart) H.Verm.) (Stewart 1957) (see Fig. 7.8); *Phoma costarricensis* Echandi (Echandi, 1957) and *Ascochyta coffeae* P. Henn. (Schieber, 1972a). Also known as quema, derrite (Spanish).

Ascochyta tarda occurs in Africa and South-east Asia, whereas *P. costarricensis* is reported from Central America and India. Both are very similar, and unpublished studies at CAB International Bioscience suggested that they were synonymous. Symptoms begin as a necrotic spotting of young leaves that expands to form brown leaf lesions covering much of the lamina. Older lesions exhibit round, black pycnidia in the necrotic tissue, and the disease develops into a general blight of young shoots; affected shoots appearing brown and scorched (see Plate 17).

There is a superficial resemblance of leaf lesions to those of anthracnose (above), and damage can be confused with, and extend damage initially caused by, the shoot tip borer, *Eucosma nereidopa* Meyrick. The disease may occur in the higher-altitude coffee areas of Kenya (Firman, 1965) and elsewhere, where it can damage young stems and seedlings (Figueiredo *et al.*, 1976). Removal of diseased shoots is the primary control option (Vermeullen

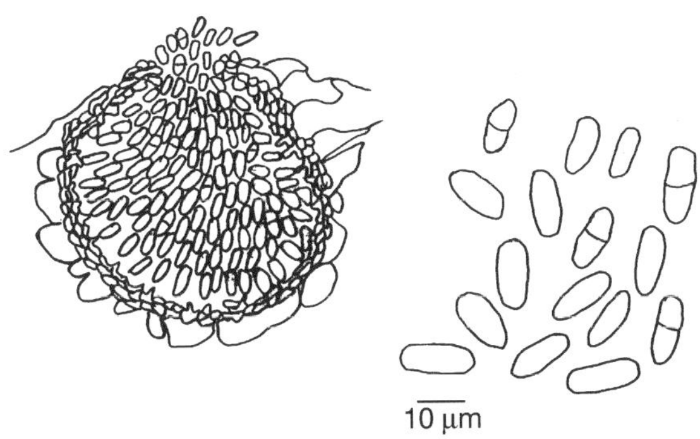

Fig. 7.8. *Ascochyta tarda.*

and Patwa, 1966). Other similar fungi have been recorded from coffee (Punithalingam, 1981).

Blister spot

Also known as greasy spot, or mancha mantecosa (Spanish), the cause is attributed to *Colletotrichum gloeosporioides* Penz. (see anthracnose) (Vargas and Gonzales, 1972) but, as this fungus is ubiquitous in coffee tissues, there are probably other factors influencing the aetiology of this disease. The disease was earlier thought to be of viral aetiology (Wellman, 1957) and has been confused with ring spot (see below).

Symptoms are chlorotic, circular spots often slightly raised and well defined (see Plate 18); these may spread and eventually kill the leaf. Sunken spots also occur on fruit and develop into light brown lesions. The condition is often most evident on seedlings, but may be restricted to certain plants. It has been reported from many Central and South American countries and from Papua New Guinea, where Shaw (1968) found no evidence for a viral aetiology.

Other leaf spot fungi

Myrothecium roridum Tode ex Fries [fungus imperfectus] causes a zonate leaf spot and stem necrosis of young coffee plants and seedlings, and was considered an important disease in Guatemala that could require control by copper fungicides (Schieber and Zentmeyer, 1968). The same fungus – reported as a synonym *Myrothecium advena* Sacc. – causes a target spot of coffee in India that occurs mostly on seedlings, but can affect mature plants (Nag Raj and George, 1958). The same authors (1962) recorded *Cephalosporium* (= *Acremonium*) *zonatum* (Sawada) W. Gams causing a leaf spot of coffee. *Guignardia coffea* Punith. and Lee Boon Siew (conidial state: *Phyllosticta coffea-libericae* Sawada) and *Phyllosticta bokensis* (P. Henn.) Van der Aa have also been associated with leaf lesions (Punithalingam, 1981). *Phyllosticta usteri* Speg. has also been isolated from diseased coffee shoots.

Algal leaf spot

Caused by *Cephaleuros virescens* Kunze, a parasitic alga, this disease can be prevalent on the leaves of coffee and a wide range of other tropical evergreen plants. Symptoms consist of light, green–yellow more or less circular, raised blotches on upper leaf surfaces. The disease is most common on older leaves; older lesions take on a reddish tinge. The alga propagates by gemmae and these can be seen as hair-like outgrowths on the lesion (see Plate 19). The alga does not penetrate into the leaf tissue and causes no significant damage.

Viral diseases

Coffee ring spot virus (CoRSV) or mancha annular (Spanish) was first described by Bitancourt (1938) from Brazil, and is apparently limited to Brazil, Costa Rica (Rodrigues *et al.*, 2002) and the Philippines, where seed transmission has been reported (Reyes, 1961). It is the only virus known to naturally infect coffee. Symptoms are chlorotic rings or curves on lower leaves and fruit, which may also show sunken lesions.

The virus is a member of the Rhabdovirus group and consists of rod or bullet-shaped particles containing single-stranded RNA (Boari *et al.*, 2004). It is spread by the mite, *Brevipalpus pheonicis* Geijskes (p. 136). The virus presumably spread from a native host in South America. It is of low incidence and causes no significant damage to coffee (Kitajima and Chagas, 2004).

References

Alarcón, R. and Carrión, G. (1994) The use of *Verticillium lecanii* in coffee plantations as biological control of coffee rust. *Fitopatología* 29, 82–85.

Alvim, P. (1958) Recent advances in our knowledge of coffee trees. 1 Physiology. In: Sachs, B. (ed.) *Advances in Crop Production Technology*. Coffee and Tea Industries 81, 17–25.

Anon (1998) Control of coffee berry disease and leaf rust in 1998. Coffee Research Foundation Technical Circular No. 70. *Kenya Coffee* 64, 2829–2836.

Arnold, E., Mejia, L.C., Kyllo, D., Rojas, E.I., Maynard, Z., Robbins, N. and Herre, E.A. (2003) Fungal endophytes limit pathogen damage in a tropical tree. *Proceedings of the USA National Academy of Sciences* 100, 15,649–15,654.

Aston, R.P., Sharp, D.G. and Clayton, J.S. (1991) The use of air-assisted hand-held ULV sprayers for control of coffee leaf rust in Colombia. British Crop Protection Council Monograph No. 46. *Proceedings of a Symposium Organized by the BCPC and the Association of Applied Biologists*, University College Swansea, *UK*. pp. 151–160.

Avelino, J., Willocquet, L. and Savary, S. (2004) Letter to the editor. Effects of crop management patterns on coffee rust epidemics. *Plant Pathology* 53, 542–547.

Becker, S. and Kranz, J. (1977) Comparative investigation on the spread of *Hemileia vastatrix* in Kenya. *Zeitschrift fur Pflanzenkrankheiten, Pflazenpathologie und Pflanzenschutz* 84, 526–539.

Becker, S., Mulinge, S.K. and Kranz, J. (1975) Evidence that uredospores of Hemileia vastatrix Berk. and Br. are wind-borne. *Phytopathologische Zeitschrift* 82, 359–360.

Bitancourt, A.A. (1938) A mancha anular, uma nova doença do caffeeiro. *O Biológico* 4, 404–405.

Boari, A.J., Freita-Astua, J., Ferreira, P.T.O., Nader, D.J., Nogueira, N.L., Rossi, M.L. and Kitajima, E.W. (2004) Prurification and serology of the coffee ringspot virus. *Summa Phytopathologica* 30, 453–458.

Bock, K.R. (1962) Seasonal periodicity of coffee leaf rust and factors affecting the severity of outbreaks in Kenya colony. *Transactions of the British Mycological Society* 45, 289–300.

Bonilla, G.J.C. (1982) Estudio del ojo de gallo causado por el hongo *Mycena citricolor. Third Latin American Symposium on Coffee Cultivation*, Tegucigalpa, Honduras, 9–10 December 1980, pp. 177–188.

Bowden, J., Gregory, P.H. and Johnson, C.G. (1971) Possible wind transport of coffee leaf rust across the Atlantic Ocean. *Nature* 229, 500–501.

Burdekin, D.A. (1960) Wind and water dispersal of coffee leaf rust in Tanganyika. *Kenya Coffee* 25, 212–213.

Calvo, S. and Vargas, E. (1989) Efecto de diferentes adherentes y formulaciones de una bacteria parasitica en el combate dei ojo de gallo *Mycena citricolor* (Berk & Curt) Sacc. en el Cafeto. *Turrialba* 39, 328–334.

Carefoot, G.L. and Sprott, E.R. (1969) *Famine in the Wind – Plant Disease and Human History*. Angus and Robertson, London.

Carvajal, B.F. (1939) 'Ojo de Gallo' (*Omphalia flavida*). *Revista del Instituto de Defensa de Café del Costa Rica* (Suppl. 7) (52), 40 pp.

Carvalho, A., Eskes, A.B., Castillo, Z.J., Streenivasan, M., Echeverri, J.H., Fernandez, C.E. and Fazouli, L.C. (1989) Breeding programmes around the world (Chapter 7). In: Kushalappa, A.C. and Eskes, A.B. (eds) *Coffee rust: Epidemiology, Resistance and Management*. CRC Press, Boca Raton, Florida, pp. 293–335.

Carvalho, V.L. de, Chalfoun, S.M., Castro, H.A. de and Carvalho, V.D. de (1996) Influencia da producao na incidencia da ferrugem do cafeeiro. *Pesquisa Agropecuaria Brasileira* 31, 401–405.

Castillo, Z.J. (1989) Breeding for rust resistance in Colombia. In: Kushalappa, A.C. and Eskes, A.B. (eds) *Coffee Rust: Epidemiology, Resistance and Management*. CRC Press, Boca Raton, Florida, pp. 307–316.

CCRI (1999) Diseases of coffee in India and their management practices. *Indian Coffee* 63, 14–16.

Chalfoun, S.M. (1981) Relacao de diferentes indices de infeccao de ferrugem (*Hemileia vastatrix* Berk. & Br.) sobre a producao de cafeeiros (*Coffea arabica* L.) em algumas localidades do Estado de Minas Gerais. *Fitopatologia Brasileira* 6, 137–142.

CMI (1985) *Distribution Maps of Plant Diseases, No. 470*. CAB International, Wallingford, UK.

CMI (1989) *Distribution Maps of Plant Diseases No. 5* (edn 8). CAB International, Wallingford, UK.

CMI (1996) *Distribution Maps of Plant Diseases No. 9* (edn 5). CAB International, Wallingford, UK.

Coleman, L.C., Venkata Rao, M.K. and Narasimhan, M.J. (1923) *Black Rot or Koleroga of Coffee in Mysore*. Department of Agriculture, Mysore State, Mycological Series Bulletin 5, 15 pp.

Coste, R. (1992) *Coffee, the Plant and the Product*. Macmillan, London.

Coutinho, T.A., Rijkenberg, F.H.J. and Asch, M.A.J. van (1995) Teliospores of *Hemileia vastatrix*. *Mycological Research* 99, 932–934.

Crowe, T.J. (1963) Possible insect vectors of uredospores of *Hemileia vastarix* in Kenya. *Transactions of British Mycological Society* 46, 24–26.

Dagmar, S.R., Teixeira, A.R.R. and Morales, W.B.C. (1989) *Bacillus thuringiensis* – a new perspective for an induced protection to coffee leaf rust. *Journal of Phytopathology* 126, 149–159.

De Jong, E.J., Eskes, A.B., Hoogstraten, J.G.J. and Zadoks, J.C. (1987) Temperature requirements for germination, germ tube growth and appressorial formation of urediospores of *Hemileia vastatrix*. *Netherlands Journal of Plant Pathology* 93, 61–71.

D'Oliveira, B. (1971) *Report of the Coffee Rusts Research Centre*, Oeiras, Portugal.

Echandi, E. (1957) La quema de los cafetos causada por *Phoma costarricensis* n. sp. *Revista Biologica Tropical* 5, 81–102.

Echandi, E. (1959) La chasparria de los cafetos causada por el hongo *Cercospora coffeicola* Berk. and Cook. *Turrialba* 9, 54–67.

Echandi, E. and Segall, R.H. (1958) The effectivenes of certain eradicant fungicides on inhibition of gemmae of *Mycena citricolor*. *Phytopathology* 48, 11–14.

Eskes, A.B. (1982) Incomplete resistance to coffee leaf rust. In: Lamberti, F., Waller, J.M. and Van der Graaff, N. (eds) *Durable Resistance in Crops*. Plenum, New York, pp. 291–315.

Eskes, A.B. (1989a) Resistance. In: Kushalappa, A.C. and Eskes, A.B (eds) *Coffee Rust: Epidemiology, Resistance and Management*. CRC Press, Boca Raton, Florida, pp. 171–291.

Eskes, A.B. (1989b) Natural enemies and biological control. In: Kushalappa, A.C. and Eskes, A.B. (eds) *Coffee Rust: Epidemiology, Resistance and Management*. CRC Press, Boca Raton, Florida, pp. 161–169.

Eskes, A.B. (2005) Phenotypic expression of resistance to coffee leaf rust and its possible relationship with durability. In: Zambolim, L., Zambolim, E.M. and Varzea, V.M.P. (eds) *Durable Resistance to Coffee Leaf Rust*. Univ. Fed. Vicosa, Minas Gerais, Brazil, pp. 305–331.

Eskes, A.B. and da Costa, W.M. (1983) Characterisation of incomplete resistance to *Hemileia vastatrix* in the Icatu coffee population. *Euphytica* 32, 649–657.

Eskes, A.B., Hoogstraten, J.G.J., Toma-Braghina, M. and Carvalho, A. (1990) Race specificity and inheritance of incomplete resistance to coffee leaf rust in some Icatu coffee progenies and derivatives of Hibrido de Timor. *Euphytica* 47, 11–19.

Eskes, A.B., Mendes, M.D.L. and Robbs, C.F. (1991) Laboratory and field studies on parasitism of *Hemileia vastatrix* with *Verticillium lecanii* and *V. leptobactrum*. *Café-Cacao-Thé* 35, 275–282.

Figueiredo, P., Mariotto, P.R. and Figueiredo, M.B. (1976) *Ascochyta coffeae* P. Henn. agente causal de uma nova doenca dos viveiros de cafe no Brasil. *Biologico* 42, 211–213.

Firman, I.D. (1963) Elgon die-back. *Kenya Coffee* 28, 9–14.

Firman, I.D. (1965) Some investigation on a disease of *Coffea arabica* caused by *Ascochyta tarda*. *Transaction of the British Mycological Society* 48, 161–165.

Firman, I.D. and Waller, J.M. (1977) *Coffee Berry Disease and other Colletotrichum Diseases of Coffee*. Phytopathological paper No. 20, CMI, Kew, UK, 53 pp.

Fulton, R.H. (1984) *Coffee Rust in the Americas. Symposium Book No. 2*. American Phytopathological Society, St Paul, Minnesota.

Gonzalez, E., Palenzuela, I. and Argudin, Z. (1995) The effect of Bayleton and copper oxychloride on *in vitro* growth of *Verticillium lecanii*. *Revista de Proteccion Vegetal* 10, 13–18.

Gouveia, M.M.C., Ribeiro, A., Varzea, V.M.P. and Rodrigues Junior, C.J. (2005) Genetic diversity in *Hemileia vastatrix* based on RAPD markers. *Mycologia* 97, 396–404.

Griffiths, E. (1981) Iatrogenic plant diseases. *Annual Review of Plant Pathology* 19, 69–82.

Guzzo, S.D., Harakava, R., Kida, K., Martins, E.M.F. and Roveratti, D.S. (1999) Protecao de cafeeiros contra *Hemileia vastatrix* por chloreto de bezalconio (composto de amonio quaternario). *Summa Phytopathologica* 25, 339–345.

Hakiza, G.J. (1997) Characterisation of the epidemiology of coffee leaf rust caused by *Hemileia vastatrix* on robusta coffee (*Coffea canephora*) in Uganda. PhD thesis, University of Reading, UK.

Hanumantha, B.T., Kamanna, B.C., Nimala, K., Govindarajan, T.S. and Kannan, N. (1994) Effect of chlorothalonil and captafol on brown eye spot disease. *Journal of Coffee Research* 24, 15–21.

Hirst, J.M. and Steadman, O.J. (1963) Dry liberation of fungal spores by rain drops. *Journal of General Microbiology* 33, 335–344.

Holliday, P. (1980) *Fungus Disease of Tropical Crops*. Cambridge University Press, Cambridge, UK.

Izquierda-Barroa, J.E., Cabrera-Ochoa, M., Pantoja-Fajardo, R. and Martinez-Flores, L. (1993) Severidad de la cercosporiosis en genotipos de cafetos portadores de genes de resistencia a *Hemileia vaststrix* B.B. *Centro Agricola* 20, 59–64.

Junior, D.G., Pozza, E.A., Pozza, A.A.A., Souza, P.E., Carvalho, J.G. and Balieiro, A.C. (2003) Incidencia e severidade da cercosporiose do cafeeiro em funcão do suprimento de potassio e calcio em solucão nutritiva. *Fitopatologia Brasileira*. 28, 286–291.

Kairu, G.M. (1997) Biochemical and pathogenic differences between Kenyan and Brazilian isolates of *Pseudomonas syringae pv. garcae*. *Plant Pathology* 46, 239–246.

Kairu, G.M., Nyangena, C.M.S., Javed, Z.U.R. and Crosse, J.E. (1984) Iatrogenic effects of captafol on bacterial blight of coffee. *Plant Pathology* 33, 131–132.

Kairu, G.M., Nyangena, C.M.S., Muthamia, J.E. (1991) The performance of copper-based bactericides in the control of bacterial blight of coffee and coffee berry disease in Kenya. *Tropical Pest Management* 37, 1–5.

King'ori, P.N. and Masaba, D.M. (1994) Current status of coffee rust (*Hemileia vastatrix*) in Kenya. *Kenya Coffee* 59, 1877–1887.

Kitajima, E.W. and Chagas, C.M. (2004) Viral disease in coffee. In: Wintgens, J.N. (ed.) *Coffee: Growing, Processing and Sustainable Production*. Wiley-Verlach, Weinheim, Germany, pp. 546–550.

Kushalappa, A.C. (1989a) Biology and epidemiology. In: Kushalappa, A.C. and Eskes, A.B. (eds) *Coffee Rust: Epidemiology, Resistance and Management*. CRC Press, Boca Raton, Florida, pp. 13–80.

Kushalappa, A.C. (1989b) Rust management: an epidemiological approach and chemical control. In: Kushalappa, A.C. and Eskes, A.B. (eds) *Coffee Rust: Epidemiology, Resistance and Management*. CRC Press, Boca Raton, Florida, pp. 81–139.

Kushalappa, A.C. and Eskes, A.B. (1989a) *Coffee Rust: Epidemiology, Resistance and Management*. CRC Press, Boca Raton, Florida.

Kushalappa, A.C. and Eskes, A.B. (1989b) Advances in coffee leaf rust research. *Annual Review of Phytopathology* 27, 503–531.

Large, E.C. (1940) Coffee rust in Ceylon. In: *Advance of the Fungi*. Jonathan Cape, London, pp. 196–207.

Laundon, G.F. and Waterston, J.M. (1964a) *Hemileia vastatrix*. CMI Descriptions of Pathogenic Fungi and Bacteria, No. 1. CAB International, Wallingford, UK.

Laundon, G.F. and Waterston, J.M. (1964b) *Hemileia coffeicola*. CMI Descriptions of Pathogenic Fungi and Bacteria, No. 2. CAB International, Wallingford, UK.

Lourd, M. and Huguenin, B. (1982) La rouille farineuse de caféiers, *Hemileia coffeicola*, en Côte d'Ivoire. Etude de sa repartition et de son pouvoir pathogene. *Garcia de Orta, Estudos Agronomicos* 9(1/2), 71–82.

Martinez, J.A., Palazzo, D.A. and Krazawa, M. (1977) Importancia do vento na liberacão e disseminacão dos esporas de *Hemileia vastatrix* Berk & Br. *Fitopatologia Brasileira* 2, 35–42.

Masaba, D.M. and King'ori, P.N. (1995) Flutriafol, a ground-applied fungicide for the control of coffee leaf rust in Kenya. *Kenya Coffee* 60, 2059–2064.

Massaux, F., Misse, C., Tchendji, C., Lacroix, M. and Tarjot, M. (1978) Contribution a l'étude des relations hôte–parasite entre la rouille farineuse (*Hemileia coffeicola*) et le caféier (*Coffea canephora var robusta*) au Cameroun. *Cafe Cacao Thé* 22, 37–56.

Mayne, W.W. (1930) Seasonal periodicity of coffee leaf disease (*Hemileia vastatrix* Berk. & Br.). *Bulletin of the Mysore Coffee Experiment Station*, No. 4.

Mayne, W.W. (1932) Physiologic specialisation of *Hemiliea vastatrix* Berk at Br. *Nature* 129, 510.

McCook, S.G. (2006) The Global Rust Belt: *Hemileia vastatrix* and the ecological integration of world coffee production since 1850. *Journal of Global History* 1, 177–195.

Monaco, L.C. (1977) Consequences of the introduction of coffee rust into Brazil. In: Day, P.R. (ed.) *The Genetic Basis of Epidemics in Agriculture*. Annals of the New York Academy of Sciences No. 287, 57–71.

Mora, A.O. (2000) La utilizacion de fuentes alcalinas para el control del ojo de *gallo* (*Mycena citricolor*) en mezcla con cyproconazole. *XIX Simposio Latino Americano de Caficultura – Memoria, Instituto del Café de Costa Rica, San Jose, Costa Rica*, pp. 349–354.

Mora, F., Ramirez, C., Vargas, E. and Rodriguez, T. (1989) Application of antagonistic bacteria for the control of American leaf spot disease of coffee caused by *Mycena citircolor*. *Turrialba* 39, 346–352.

Mora-Bolanos, J., Arturo-Solorzano, J. and Vargas, L.G. (2000) Combate preventivo del ojo de

gallo (*Mycena citricolor*) en el cultivo de cafe. In: *XIX Symposio Latino Americano de Cafeculturu – Memoria*, Instituto del Café de Costa Rica, San Jose, Costa Rica, pp. 443–441.
Moraes, S.A., de Sugimori, H.M., Ribeiro, I.J.A., Ortolani, A.A. and Pedro, M.J., Jr (1976) Periodo de incubacão de *Hemiliea vastatrix* Berk et Br. em tres regiões do estado de São Paulo. *Summa Phytopathologica* 2, 32–38.
Mordue, J.E.M. (1971) *Glamerella cingutata*. CMI Descriptions of Pathogenic Fungi and Bacteria No. 315, CAB International, Wallingford.
Mordue, J.E.M. and Gibson, I.A.S. (1976) *Corticium salmonicolor*. CMI Descriptions of Pathogenic Fungi and Bacteria No. 511. CAB International, Wallingford.
Mulder, J.L. and Holliday, P. (1974) *Cercospora coffeicola*. CMI Descriptions of Pathogenic Fungi and Bacteria No. 415. CAB International, Wallingford.
Muller, R.A., Berry, D., Avelino, J. and Bieysse, D. (2004) Coffee diseases. In: Wintgens, J.N. (ed.) *Coffee: Growing Processing and Sustainable Production*. Wiley-Verlach, Weinheim, Germany, pp. 491–545.
Muthappa, B.N., Kannan, K., Javed, J.B., Matiello, J.B. and Mansk, Z. (1989) Chemical control practices around the world. In: Kushalappa, A.C. and Eskes, A.B. (eds) *Coffee Rust: Epidemiology, Resistance and Management*. CRC Press, Boca Raton, Florida, pp. 141–159.
Muthiah, S. (1993) *A Planting Century: the First Hundred Years of the United Planters Association of Southern India*. Affiliated East-West Press, Madras, India.
Nag Raj, T.R. and George, K.V. (1958) 'Target spot' disease of coffee. I. Occurrence, symptoms and etiology. *Indian Phytopathology* 11, 153–158.
Nag Raj, T.R. and George, K.V. (1962) Occurrence of zonal leaf spot of coffee caused by *Cephalosporium zonatum*. *Current Science* 31, 104–105.
Narasimhan, M.J. (1933) Black rot of coffee in Mysore. *Phytopathology* 23, 875–886.
Narasimhaswamy, R.L. (1961). La herrumbre del café (*Hemileia*) in la India. *Café* (Turrialba) 3, 41–45.
Nataraj, T. and Srinivasan, C.S. (1996) Extent of defoliation by leaf rust and other foliar diseases on S795 arabica coffee. *Indian Coffee* 60, 9–10.
Nutman, F.J. and Roberts, F.M. (1963) Studies on the biology of *Hemiliea vastatrix*. *Transcations of the British Mycological Society* 46, 27–48.
Nutman, F.J. and Roberts, F.M. (1970) Coffee leaf rust. *PANS* 16, 607–624.
Nutman, F.J., Roberts, F.M. and Bock, K.R. (1960) Methods of uredospore dispersal of coffee leaf rust fungus, *Hemileia vastatrix*. *Transactions of the British Mycological Society* 43, 509–515.
Paradela, A.L., Paradela, F.O., Scachetti, A.P. and Oliveira, J.R.S. (2000) Eficiencia de alguns fungicidas-bactericidas no controle de mancha aureolada (*Pseudomonas syringae pv. garcae*) em cafe (*Coffea arabica* L.). *Ecossistema* 25, 21–23.
Pelaez, P.M.J. and Gil, V.L.F. (2000) Evaluacion en laboratorio y en el campo de la resistencia incompleta a *Hemileia vastatrix* Berk. & Br. en progenies de Caturra x Hybrido de Timor. *Fitopathologia Colombiana* 24, 61–72.
Porras, N.C., Leguizamon, C.J.E and Martinez, M.M. (1999) Induccion de resistencia por *Pseudomonas* spp. en plantulas de café contra la roya del cafeto *Hemileia vastarix* Berk. y Br. *Ascolfi-Informa* 25, 12–14.
Pozza, A.A.A., Zambolin, L., Pozza, E.A., Costa, H. and Vale, F.X.R. do-. (1997) Controle quimico da mancha de olho pardo (*Cercospora coffeicola*) do cafeeiro em condicoes de viveiro. *Fitopatologia Brasileira* 22, 543–545.
Pozza, A.A.A., Alves, E., Pozza, E.A., Carvalho, J.G. de, Montanari, M., Guimaraes, P.T.G. and Santos, D.M. (2004) Efeito do silicio no controle da cercosporiose em tres variedades de cafeeiro. *Fitopatologia Brasileira* 29, 185–188.
Prakash, N.S., Ganesh, D. and Bhat, S.S. (2005) Population dynamics of coffee leaf rust (*Hemiliea vastarix* Berk. & Br.) and recent advances in rust research in India. In: Zambolim, L., Zambolim, E.M. and Varzea, V.M.P. (eds) *Durable Resistance to Coffee Leaf Rust*. Univ. Fed. Vicosa, Minas Gerais, Brazil, pp. 411–442.

Punithalingam, E. (1981) Studies on Sphaeropsidales in culture III. *Mycological Papers* 149, 42 pp.
Quesada, C.A. and Jimenez, U.F. (1996) In vitro evaluation of a *Bacillus* sp. for the biological control of the coffee phytopathogen *Mycena citricolor*. *World Journal of Microbiology and Biotechnology* 12, 97–98.
Rajendran, R.B. (1967) A new type of nuclear life cycle in *Hemileia vastatrix*. *Mycologia* 59, 279–285.
Ram, A.S. (2001) Breeding for rust resistance in coffee – the gene pyramid model. *Journal of Plantation Crops* 29, 10–15.
Ramos, A.H. (1979) Bacterial blight of coffee: inoculum supply and avenues of infection. *Plant Disease* 63, 6–10.
Ramos, A.H. and Shavida, L.D. (1976) A die-back of coffee in Kenya. *Plant Disease* 60, 831–835.
Rao, D.V. and Tewari, J.P. (1988) Suppression of the symptoms of American leaf spot of coffee with calcium hydroxide. *Plant Disease* 72, 688–690.
Rao, D.V. and Tewari, J.P. (1989) Occurrence of magnesium oxalate crystals on lesions incited by *Mycena citricolor* on coffee. *Phytopathology* 79, 783–787.
Rayner, R.W. (1957) Tonic spraying of coffee. *Proceedings of the Nairobi Scientific and Philosophical Society* 9, 6–12.
Rayner, R.W. (1961a) Germination and penetration studies on coffee rust (*Hemileia vastatrix* Berk. & Br.). *Annals of Applied Biology* 49, 497–505.
Rayner, R.W. (1961b) Spore liberation and dispersal of coffee rust, *Hemileia vastatrix*. *Nature*, 191, 725.
Rayner, R.W. (1962) The control of coffee rust in Kenya by fungicides. *Annals of Applied Biology* 50, 245–261.
Rayner, R.W. (1972) *Micologia Historia y Biologia de la Roya del Cafeto*. Publicacion Miscelanea No. 94, IICA, Turrialba, Costa Rica.
Reyes, T.T. (1961) Seed transmission of coffee ringspot in Excelsa coffee (*Coffea excelsa*). *Plant Disease Reporter* 45, 185.
Rodrigues, C.J. (1957) Nota sobre a resistencia de algumas especies de coffea a *Hemileia coffeicola* Maub. & Rog. *Revista Cafe Portuguesa* 3, 48–71.
Rodrigues, C.J. Jr, Bettancourt, A.J. and Rijo, L. (1975) Races of the pathogen and resistance to coffee rust. *Annual Review of Phytopathology* 13, 49–70.
Rodrigues, J.C.V., Rodriguez, C.M., Moreira, L., Villalobos, W., Rivera, C. and Childers, C.C. (2002) Occurrence of coffee ringspot virus, a *Brevipalpus* miteborne virus in coffee in Costa Rica. *Plant Disease* 86, 564.
Saccas, A.M. (1972) La rouille 'farineuse' des caféiers due à *Hemileia coffeicola* Maub. & Rog. *Bulletin Institute Francais du Café et du Thé* 11, 5–68.
Saccas, A.M. and Charpentier, J. (1969) L'anthracnose des caféiers robusta et excelsa due à *Colletotrichum coffeanum* Noack en République Centrafricaine. *Café Cacao Thé* 13, 221–230.
Salas, T.A. and Hancock, J.G. (1972) Production of the perfect stage of *Mycena citricolor*. *Hilgardia* 41, 213–234.
Sanchez, L., Zapata, M.P., Rodriguez, R. del, Beaver, J.S. (2003) Distribucion y patogenicidad de *Pseudomonas cichorii* (Swingle) Stapp en el cultivo del cafe en Puerto Rico. *Journal of Agriculture of the University of Puerto Rico* 87, 123–135.
Schieber, E. (1972a) Leaf blight incited by *Ascochyta coffeae* of coffee in Guatemala. *Plant Disease Reporter* 56, 753–75.
Schieber, E. (1972b) Economic impact of coffee rust in Latin America. *Annual Review of Phytopathology* 10, 491–510.
Schieber, E. and Zentmeyer, G.A. (1968) Myrothecium stem necrosis and leaf spot – an important coffee disease in Guatemala. *Plant Disease Reporter* 52, 115–117.
Schieber, E. and Zentmyer, G.A. (1984) Coffee rust in the western hemisphere. *Plant Disease* 68, 89–93.

Sequiera, L. (1959) Host range of *Mycena citricolor* (Berk. & Curt.) Sacc. *Turrialba* 8, 136–147.
Shaw, D.E. (1968) Wellman's leaf and fruit spot of arabica coffee in New Guinea. *Papua and New Guinea Agricultural Journal* 19, 152–166.
Shaw, D.E. (1977) Report on continuing eradication of coffee rust in two areas in Papua New Guinea. *Papua New Guinea Agricultural Journal* 28, 27–32.
Siddiqi, M.A. (1970) Incidence, development and symptoms of *Cercospora* disease of coffee in Malawi. *Transactions of the British Mycological Society* 54, 415–421.
Sierra, C.A., Montoya, E.C. and Velez, R.C. (1995) Nivel de dano y umbral economico para la roya del cafeto (*Hemileia vastatrix* Berk. y Br.). *Fitopatologia Colombiana* 19, 43–48.
Silva, M.C., Nicole, M., Guerra-Guimaraes, L. and Rodrigues Junior, C.J. (2002) Hypersensitive cell death and post-haustorial defence responses arrest the orange rust (*Hemileia vastatrix*) growth in resistant coffee leaves. *Physiological and Molecular Plant Pathology* 60, 169–183.
Silva-Acuna, R., Maffia, L.A., Zambolin, L. and Berger, R.D. (1999) Incidence–severity relationships in the pathosystem *Coffea arabica–Hemileia vastatrix*. *Plant Disease* 83, 186–188.
Sneider-Christians, J., Fliege, T. and Sclossen, E. (1986) On the release, survival and importance of basidiospores of *Corticium salmonicolor*, the pathogen causing pink disease. *Zeitschrift fur Pflanzenkrankenheiten und Pflanzenschutz* 8, 200–216.
Stewart, R.B. (1957) Leaf blight and stem die-back of coffee caused by an undescribed species of *Ascochyta*. *Mycologia* 49, 430–433.
Subramanian, S. and Sridhar, T.S. (1966) Studies on brown eye spot of coffee I–IV. *Revista Patologia Vegetale* Serie IV 2, 127–149.
Tarjot, M. and Lotode, R. (1979) Contribution à l'étude des rouilles orangée et farineuse du caféier au Cameroun. *Café-Cacao-Thé* 23, 103–118.
Tewari, J.P., Rao, D.V. and Vargas, E. (1986) Estudio preliminar sobre el modo de penetracion de *Mycena citricolor* en la hoja de cafeto. *Agronomia Costarricense* 10, 199–202.
Van der Plank, J.E. (1963) *Plant diseases: Epidemics and Control*. Academic Press, New York.
Van der Vossen, H.A.M. (2005) State-of-the-art of developing durable resistance to biotrophic pathogens in crop plants, such as coffee leaf rust. In: Zambolim, L., Zambolim, E.M. and Varzea, V.M.P. (eds) *Durable Resistance to Coffee Leaf Rust*. Univ. Fed. Vicosa, Minas Gerais, Brazil, pp. 1–29.
Van der Vossen, H.A.M. and Browning, G. (1978) Prospects of selecting genotypes of *Coffea arabica* which do not require tonic sprays of fungicide for increased leaf retention and yield. *Journal of Horticultural Science* 53, 225–233.
Vargas, G.E. (1984) Interaccion de tratamiento biologico y quimico en el combate del ojo de gallo (*Mycena citricolor*) en el cafeto. *Agronomia Costarricense* 8, 91–97.
Vargas, G.E. and Gonzales, U.L.C. (1972) La mancha mantecosa del café causada por *Colletotrichum*. *Turrialba* 22, 129–135.
Vargas, G.E., Gonzales, L.M., Umana, R.G. and Vargas, V.L. (1990) Nuevas alternativas de combate quimico del ojo de gallo (*Mycena citricolor*). *Boletin de Promecafe* 47, 9–14.
Varzea, V.M.P. and Marques, D.V. (2005) Population variability of *Hemileia vastatrix vs.* coffee durable resistance. In: Zambolim, L., Zambolim, E.M. and Varzea, V.M.P. (eds) *Durable Resistance to Coffee Leaf Rust*. Univ. Fed. Vicosa, Minas Gerais, Brazil, pp. 53–74.
Vermeullen, H. and Patwa, N. (1966) The control of *Ascochta tarda* under field conditions in Kenya. *Kenya Coffee* 31, 409–411.
Villarraga, A.L.A. (1987) Dosis y frequencia de applicacion de fungicidas sistemicos granulados para el control de la roya del café (*Hemileia vastatrix*) en Colombia. *Ascolfi Informa* 13, 49–50.
Villarraga, A.L.A. and Baeza, A.C. (1987) Relaciones de infeccion y severidad con defoliacion y producion en *Coffea arabica* var. Caturra por accion de *Hemiliea vaststrix*. *Ascolfi Informa* 13, 50–57.
Vine, B.H., Vine, P.A. and Griffiths, E. (1973) Some problems of evaluating fungicides for use on coffee in Kenya. *Annals of Applied Biology* 75, 377–385.

Waller, J.M. (1972) Coffee rust in Latin America. *PANS* 18, 402–408.
Waller, J.M. (1979) The recent spread of coffee rust (*Hemileia vastatrix* Berk. & Br) and attempts to control it. In: Ebbels, D.L. and King, J.E. (eds) *Plant Health*. Blackwell, Oxford, UK, pp. 275–283.
Waller, J.M. (1982) Coffee rust – epidemiology and control. *Crop Protection* 1, 385–404.
Waller, J.M., Leguizamon, J., Gill, L.F., Aston, R.A., Cookman, G.P., Sharp, D.G., Ford, M.G. and Salt, D.W. (1994) Laboratory and field development of a CDA spraying system for control of coffee leaf rust (*Hemileia vastatrix*): an overview. In: Hewitt, H.G., Caseley, J., Copping, L.G., Grayson, B.T. and Tyson, D. (eds) *Comparing Glasshouse and Field Pesticide Performance II*. British Crop Protection Council, Farnham, UK, pp. 261–266.
Wallis, J.A.N. and Firman, I.D. (1962) Spraying Arabica coffee for the control of leaf rust. *East African Agriculture and Forestry Journal* 28, 89–104.
Ward, H.M. (1882) Researches on the life history of *Hemileia vastatrix*. *Journal of the Linnaean Society* 19, 299–335.
Wellman, F.L. (1950) Dissemination of *Omphalia flavida* leaf spot of coffee. *Turrialba* 1, 12–17.
Wellman, F.L. (1957) Blister spot of arabica coffee from virus in Costa Rica. *Turrialba* 7, 13–15.
Wellman, F.L. (1970) The rust *Hemileia vastatrix* now firmly established on coffee in Brazil. *Plant Disease Reporter* 54, 539.
Wellman, F.L. (1972) *Tropical American Plant Diseases*. Scarecrow Press, Metuchen, New Jersey, pp. 409–410, 606–627.
Wrigley, G. (1988) *Coffee*. Longman, Harlow, UK.
Zambolim, L., Silva-Acuna, R., Rena, A.B. and Chaves, G.M. (1992) Relacao de producao de graos aos teores foliares de amido e de acucares e seus efeitos subsequentes no desenvolvimento da ferrugem do cafeeiro. *Fitopatologia Brasileira* 17, 23–27.
Zambolim, L., Zambolim, E.M. and Varzea, V.M.P. (eds) (2005) *Durable Resistance to Coffee Leaf Rust*. Univ. Fed. Vicosa, Minas Gerais, Brazil.

8 Berry Disease

Coffee Berry Disease

Pathogen: *Colletotrichum kahawae* Waller & Bridge (Syn. *Colletotrichum coffeanum* Noack var. *'virulans'* Rayner (1952); *Colletotrichum coffeanum* Noack 'sensu Hindorf' (Hindorf, 1970).

Coffee berry disease, popularly known as CBD, also as anthracnose de baies du caféier (French), antracnosis del cafeto (Spanish) occurs only in Africa and only on *C. arabica*. CBD is regarded as one of the main threats to the coffee industries in Latin America and Asia, although good sources of resistance are currently available.

History and spread

From the earliest report of CBD from Western Kenya in 1922 (McDonald, 1926), the disease spread more slowly than rust, being recorded only in Rwanda, Congo, Uganda and Angola before 1950 and Cameroon in 1955. CBD appeared in northern Tanzania in 1964 (Tapley, 1964) and Ethiopia in 1971, not reaching Malawi and Zimbabwe until 1985 and Zambia in 1986. (Firman and Waller, 1977; Masaba and Waller, 1992) (Fig. 8.1). The disease is particularly severe at higher altitudes (above about 1600 m in Kenya).

The disease did not occur in Ethiopia, the centre of diversity of *C. arabica*, until 1971 indicating that it is not a co-evolved pathogen of the species. It most probably originated as part of the *Colletotrichum* complex of the natural mycobiota on the diploid progenitors of the crop such as *C. eugenioides* or *C. canephora*, both of which are native to areas close to where CBD was first encountered (Waller and Bridge, 2000). No alternative host of the fungus was found from flora associated with coffee in Kenya (Hindorf, 1974).

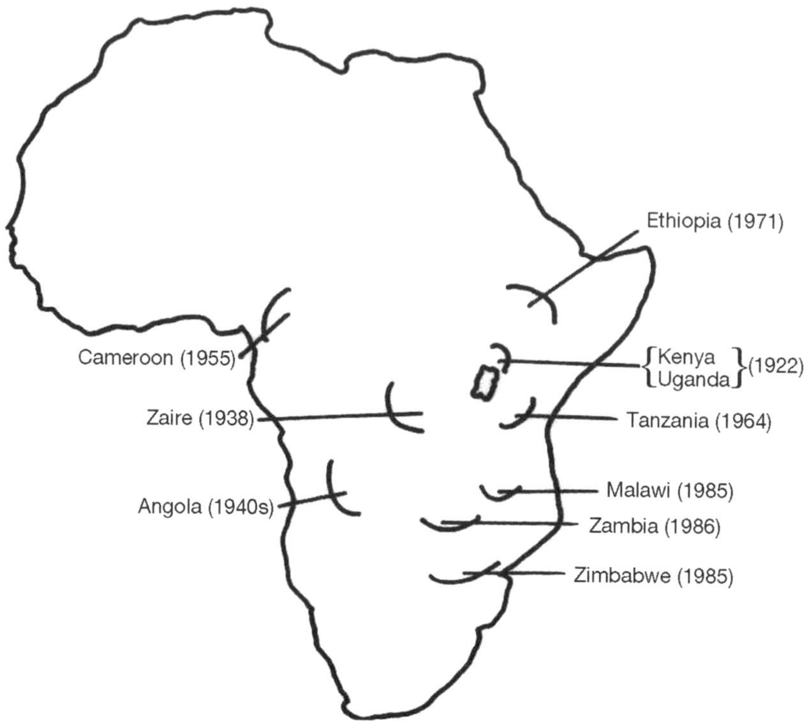

Fig. 8.1. Map of Africa showing the progressive spread of CBD.

In Kenya, CBD was rampant in the West Rift District during the 1930s, reaching the East Rift areas in 1951 and had spread to most of the coffee growing areas by 1964. Early attempts at control using fungicides gave inconsistent results and it was not until the late 1960s that the problem was resolved by a change in spraying strategy. It was suggested that change in rainfall patterns had made the disease worse after the late 1950s.

While erratic rainfall patterns can disrupt normal flowering patterns, it seemed more likely that a change from single-stem to multiple-stem pruning that allowed two overlapping crops each year was the main factor exacerbating the disease (Griffiths and Waller, 1971).

Economic impact

Because the disease directly destroys the berries, losses can be dramatic. McDonald (1926) reported that coffee berry disease caused losses of up to 75% and resulted in the abandonment of coffee plantations in several districts of western Kenya soon after it first appeared in the region. By 1964, losses in Kenya were estimated at 19%, equivalent to UK£6 million (Nutman, 1966, cited by Firman and Waller, 1977). A severe epidemic in central Kenya in 1967 caused the loss of entire crops, and overall losses were in excess of 30% (Griffiths, 1969).

In Cameroon, losses were estimated to be as high as 80% (Muller *et al.*, 2004). In areas where conditions favour the disease, control of CBD with fungicides can double or triple the yields (Griffiths *et al.*, 1971) but, as with rust control, yield increments from the tonic effect of fungicide sprays are difficult to separate from gains due to disease control. Cramer (1967) estimated the total annual loss to CBD in Africa at around 20%.

Fungicide regimes that effectively control CBD are costly, but in the late 1980s these could give cost:benefit ratios of as much as 1:5 in Kenya (Masaba and Waller, 1992) but, with increasing input, costs of labour and materials coupled with falling coffee prices, this no longer applies. Spraying in Kenya (mostly for CBD) now accounts for one-third of all production costs and, coupled with low prices reaching the farmer, is making coffee growing barely economic for smaller growers (see Chapter 2).

Symptoms

The characteristic symptom of CBD is a progressive anthracnose of young, expanding coffee berries, but the production of scab lesions and the shedding of young berries are also significant features (Bock, 1956). Active anthracnose lesions commence as small, water-soaked spots that rapidly become dark and sunken. These expand, causing a rot of the whole berry and, under humid conditions, pink spore masses become visible on the lesion's surface (see Plate 20).

The scab lesions are pale straw in colour and suberized (see Plate 21). These develop more slowly, fungal growth is restricted and sporulation is rare. Lesions may also occur on young berry stalks, causing them to be shed before lesions are evident on the berry itself. Although young infected berries are readily shed, those that are older may remain on the stalk, becoming black and mummified.

Immature and ripening berries and flowers are susceptible to infection by CBD. The unexpanded pinhead-sized berry is resistant, but becomes highly susceptible during the expansion phase, from 4–14 weeks after flowering (Mulinge, 1970). Fully expanded green berries are again resistant to infection but become susceptible as they begin to ripen. Scab lesions form if infection takes place as the berry is reaching the end of its susceptible phase, or if weather conditions become unsuitable for fungal growth.

Scab lesion formation is a resistant host response and is more common on cultivars possessing some resistance and on unsprayed coffee. They may completely heal, or remain dormant until the berry begins to ripen, when they may develop into active anthracnose lesions. The disease also affects ripening berries, causing a 'brown blight' phase as typical dark, sunken anthracnose lesions envelop the red berry. Anthracnose of ripe berries is also commonly caused by the weakly pathogenic *Colletotrichum gloeosporioides*. The mature seeds (beans) are not destroyed by this phase of the disease, so that it is of less significance; *C. kahawae* infects flowers under very wet conditions, causing brown lesions on petals.

Colletotrichum gloeosporioides can cause anthracnose on leaves and twigs (see Chapter 7) or ripening berries, to cause brown blight (see below). However, *Colletotrichum* strains isolated from leaves and twigs do not cause CBD symptoms when inoculated onto immature green berries. A condition known as 'stalk rot' has been reported from India (Muthappa, 1970). This causes the dehiscence of young berries and leaves and is related to die-back associated with *Colletotrichum gloeosporioides* – and not due to *C. kahawae*.

Pathogen characteristics

Morphology and growth in culture

Conidia are produced in profusion in a mucilaginous matrix from simple, short, conidiogenous cells at the bases of the acervular conidiomata produced on diseased berries. In culture, the fungus grows more slowly than *Colletotrichum gloeosporioides* (2–4 mm/day at 25°C on malt agar), and fresh monoconidial isolates from host tissues appear as dark greenish grey colonies with a profuse aerial mycelium on malt agar. No conidiomata are produced in culture; clusters of conidia are produced from the tips of simple conidiogenous cells produced directly from the mycelium. Simple, dark brown, ovate appressoria are commonly produced in mature cultures. Conidia are hyaline, unicellular, straight, generally cylindrical with rounded ends and are bigutulate; a proportion of conidia may be more ovate. Conidia are generally 12.5–19.0 × 4 μm, but some larger conidia exceeding 20 × 6 μm may occur in culture.

Older cultures revert to a pale grey to off-white colour after subculturing a few times, eventually turning a brownish colour and ceasing to sporulate (Waller *et al.*, 1993). The perfect state (*Glomerella cingulata*) has not been confirmed, although Hocking *et al.* (1967) observed perithecia producing ascopsores on inoculated green berries in Tanzania. Subsequent work was unable to confirm the teleomorph (Firman and Waller, 1977).

Colletotrichum kahawae and *C. gloeosporioides* cannot be distinguished by morphological features visible under the light microscope. However, the slow-growing, greyish green colonies of *C. kahawae* are readily distinguished from the fast-growing, off-white colonies of *C. gloeosporioides* (see Plate 22) and the pale pink, adpressed colonies of *C. acutatum* that are the other species found on coffee. *Colletotrichum kahawae* can also be distinguished easily from other *Colletotrichum* spp. on coffee in culture by its inability to utilize citrate or tartrate as a sole source of carbon. A change in pH of a basic mineral salt medium containing these carbon sources as acid salts can be readily observed with an appropriate indicator such as bromocresyl purple, when the carbon sources are metabolized (see Plate 23) (Bridge *et al.*, 2007).

Variability

All *Colletotrichum* isolates from coffee were previously referred to as *Colletotrichum coffeanum* Noack, but it was recognized from an early date that the pathogen was of a different form. This problem with nomenclature causes difficulties with earlier literature as it is often impossible to know whether the term '*C. coffeanum*' was actually used for the CBD pathogen or for *C. gloeosporioides* occurring on coffee.

Rayner (1952) recognized cultural differences and termed the pathogen *C. coffeanum* var. *virulans*, but it was Gibbs (1969) who first clearly differentiated four main strains, based on sporulation and colony type. Hindorf (1970) then expanded on this system of classification (see Table 8.1). Waller *et al.* (1993) demonstrated the biochemical differences between the CBD strain and other strains of *Colletotrichum* and, together with cultural, ecological and pathogenic differences, were able to differentiate the pathogen as a new species, *Colletotrichum kahawae* Waller & Bridge.

Molecular biological studies have demonstrated a close affinity between *C. kahawae* and *C. gloeosporioides.* Sreenivasaprasad *et al.* (1993) showed that there is no significant difference in the base sequences of the conserved ITS region of ribosomal DNA, indicating a very close evolutionary relationship. Waller and Bridge (2000) claimed that this and other evidence indicates that the pathogen has only recently evolved from the general *Colletotrichum* complex occurring on *Coffea* spp. in the central African region.

The process has been driven at least partly by the selection pressure following the arrival and widespread planting of the susceptible *C. arabica* in the region. On the basis of pathogenicity, morphological features and biochemical tests, Varzea *et al.* (2002) classified 31 isolates of *Colletotrichum* from Angola, Burundi, Cameroon, China, Ethiopia, Malawi, Kenya, Rwanda and Zimbabwe into three species: *C. kahawae* (24 isolates), *C. gloeosporioides* (six isolates) and *C. acutatum* (one isolate). Bridge *et al.* (2007) have shown that there are a range of molecular differences that differentiate *C. kawawae* from *C. gloeosporioides*, although few differences have been found within the species, again indicating an early stage in species evolution.

Some small behavioural variations do exist within the species *C. kahawae.* Rodriguez *et al.* (1991) reported that Malawian and Angolan isolates differed in morphology and aggressiveness from Kenyan isolates. Within the Kenyan population of the pathogen, strains resistance to fungicides appeared soon after these were used (Okioga, 1976), and differences in tolerance to prochloraz also occur (Mwang'ombe, 1994). Some variation in aggressiveness between Kenyan isolates has also been reported by Omondi *et al.* (2000), but no differences in virulence were found.

Molecular fingerprinting has tended to confirm that *C. kahawae* is genetically conserved. An RFLP study of 12 isolates from different African countries failed to detect any differences (Screenivasaprasad *et al.*, 1993). This result was supported by a study with 23 isolates that gave similar RFLP patterns (Beynon *et al.*, 1995). Further work with 26 isolates, including some showing pathogenic variation, using RFLPs – PCR-based DNA fingerprinting techniques using AFLPs and VNTR primers – showed slight differences in isolates from Cameroon and Burundi, but the overall variability within the group was small, indicating an early stage in speciation (Bridge *et al.*, 2007).

The study by Beynon *et al.* (1995) suggested that isolates could be

Table 8.1. Distinguishing characteristics of strains and species of *Colletotrichum* commonly isolated from coffee (modified from Hindorf, 1970, Firman and Waller, 1977 and Waller *et al.*, 1993).

Species	Sporulation	Cultural characters
Colletotrichum kahawae Waller & Bridge	Conidia 12.5–19.0 μm × 3.5–4.0 μm, borne on aerial conidiophores without acervuli	Slow-growing, profuse dark grey/green mycelium; older cultures variable, often pale and sterile without perithecia; cannot metabolize citrate and tartrate as a sole carbon source
C. acutatum Simmonds	Conidia average 12.1–3.4 μm, straight, fusiform with acute tips; borne directly on hyphae and later in acervuli	Slow-growing, profuse, pale, aerial mycelium, tinged pink with conidia; may produce setae; able to metabolize citrate and/or tartrate
C. gloeosporioides Penz. (white mycelial form)	Conidia average 12.8 × 3.7 μm, straight, cylindrical, borne directly on hyphae and later in acervuli	Fast-growing, profuse, white or pale grey, aerial mycelium, occasional setae and perithecia in older cultures; able to metabolize citrate and/or tartrate
C. gloeosporioides (acervuli form)	Conidia average 10.8 × 4.5 μm, short, straight, cylindrical and may be acute at tip; produced in large, black acervuli	Moderately fast-growing, sparse pale, aerial mycelium and acervuli, often with setae after 7–10 days; sometimes producing perithecia; able to metabolize citrate and/or tartrate
C. gloeosporioides (greenish mycelial form)	Rarely produces spores but, when it does, these are borne directly on mycelium	Slow-growing, pale greenish grey mycelium, not as dark as *C. kahawae*; associated with *G. cingulata* perithecia on bark; able to metabolize citrate and/or tartrate
Glomerella cingulata [Stonem.] Spauld. & Shrenk.	Perithecia globose to pyriform, dark brown, 80–100 μm diam.; asci containing eight ascopores that are hyaline, cylindrical to fusiform and *c.*12 mm long; may become septate on germination	White to grey mycelium, produces perithecia after 2–3 weeks; metabolizes citrate and/or tartrate

1

2

3

4

5

Plate 1. Adult *Monochamus leuconotus*.
Plate 2. Ring barking by *M. leuconotus* larva.
Plate 3. *M. leuconotus* larva in tunnel (courtesy N.Phiri).
Plate 4. Berry borer larva and damage to coffee bean.
Plate 5. Berry moth damage to berry cluster.

6

9

7

8b

8a

Plate 6. Green scale.
Plate 7. Star scale.
Plate 8. *Hemileia vastatrix* symptoms on underside **(a)** and upperside **(b)** of leaves.
Plate 9. Defoliation caused by severe rust infection.

10

11

12

13

14

Plate 10. Rust lesion with *Verticillium* hyperparasite (courtesy H. Evans).
Plate 11. South American leaf spot caused by *Mycena citricolor* (courtesy H. Evans).
Plate 12. *Mycena citricolor* lesions on stem.
Plate 13. Brown eye spot caused by *Cercospora coffeicola*.
Plate 14. Bacterial blight caused by *Pseudomonas syringae pv garcae*.

15

16

18

17

Plate 15. Koleroga (black rot) caused by *Corticium koleroga* (courtesy H.Evans).
Plate 16. Leaf Anthracnose caused by *Colletotrichum gloeosporioides*.
Plate 17. Twig die-back caused by *Ascochyta tarda*.
Plate 18. Symptoms of blister spot disease.

19

20

21

22

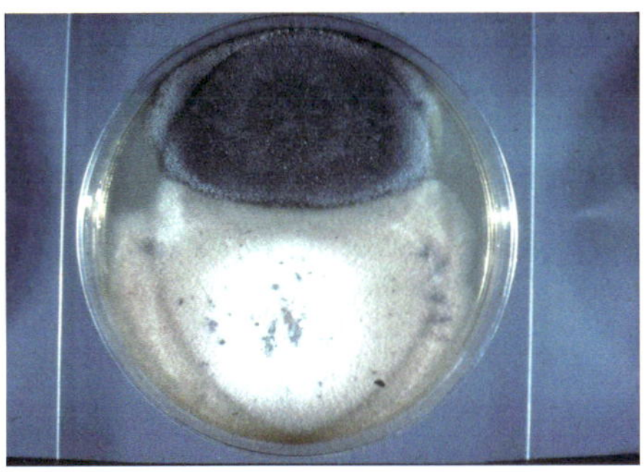

Plate 19. *Cephaleuros virescens* (algal leaf spot).
Plate 20. CBD, active lesions.
Plate 21. CBD, scab lesions.
Plate 22. *C. kahawae* and *C. gloeosporioides* in culture showing the darker slower growing *C. kahawae*.

25

23

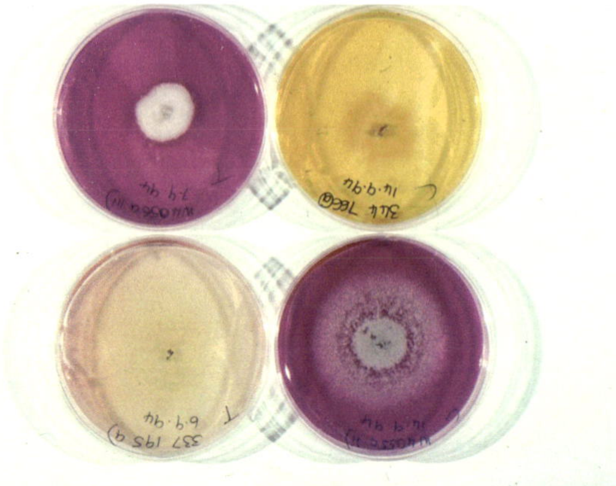

24

26

27

Plate 23. *C. kahawae* and *C. gloeosporioides* carbohydrate utilization. Pale plates; no change in pH as no utilization – *C. kahawae.*, Pink plates; pH raised as organic acid utilized – *C. gloeosporioides.*
Plate 24. Seedling pre-selection test for CBD resistance – seedling hypocotyl lesion.
Plate 25. Red blister disease (*Cercospora* berry blotch).
Plate 26. Tree with fusarium wilt (courtesy M. Rutherford).
Plate 27. Symptoms of fusarium wilt under bark of diseased tree (courtesy M. Rutherford).

28

29

30

31

32

Plate 28. Fusarium bark disease, stem cankers.
Plate 29. Sporophores of *Armillaria mellea*.
Plate 30. Root knot nematode damage caused by *Meloidogyne exigua* (courtesy J.Bridge).
Plate 31. Symptoms of overbearing dieback following severe rust infection.
Plate 32. Symptoms of 'hot and cold' disease.

33

Plate 33. Mineral deficiency symptoms (Adapted from Harrer, 1962 with acknowledgement to IICA as the original source)
a) nitrogen; **b)** potassium; **c)** phosphorus; **d)** magnesium; **e)** calcium; **f)** boron; **g)** iron; **h)** manganese; **i)** zinc

grouped into several sub-populations, through differences in vegetative compatibility. However, Gichuru *et al.* (2000) reported that all 42 isolates from Kenya and one from Malawi grouped in a single vcg and Varzea *et al.* (2002) found that the isolates of *C. kahawae* that they studied were self-compatible, forming a single vcg and separate from other species. Isolates of *C. kahawae* from Cameroon have been shown to have some differences from East African isolates (Muller *et al.*, 2004), and Bridge *et al.* (2007) found that there was a difference in vcgs.

Epidemiology

Infection and sporulation

Spore germination requires free water as rain, mist or dew. The optimum, maximum and minimum temperatures for spore germination in water and for lesion formation are, respectively, 22, 30 and 15°C (Nutman and Roberts, 1960a), but the range may be increased by 5°C by the presence of nutrient in the water. In the field, it appears that some infection can occur at temperatures as low as 10°C (Bock, 1956; Cook, 1975b), provided that coffee surfaces remain wet for long enough. Spore germination and appressorium formation can occur within 5 h at optimum temperature (Nutman and Roberts, 1960b), but take longer under field conditions.

The latent period of the disease on susceptible young berries is about 20 days, but can be much longer if infection occurs on berries that are approaching their resistant phase as they become fully expanded. Such infection may remain latent for several months, until the berry becomes susceptible again at ripening. Scab lesions may behave in a similar manner. The fungus may also persist for a time on berry surfaces as resistant, thick-walled appressoria.

Conidia of *C. kahawae* are produced in acervuli on the bark of young twigs and on berries. In humid conditions, spores are exuded on the lesions as pink, mucilagenous masses and the spores are dispersed by rain. A ripe berry can produce 20×10^3 spores/cm^2/h under optimum conditions for sporulation (Gibbs, 1969), and lesions begin to sporulate about 4 days from their appearance (Steiner, 1974). A saturated atmosphere is required for sporulation, but free moisture is optimum. Lateral dispersal of spores between berries occurs by rainwater flow and splash, but movement in rainwater is primarily downwards through the canopy (Waller, 1972). Damage to the berry by berry moth (*Prophantis smaragdina* Butler) (p. 76) or berry borer (*Hypothenemus hampei* (Ferrari) (p. 68) can increase susceptibility to CBD (Masaba and Waller, 1992).

Disease progress

Much of the work to resolve the epidemiology and control of the disease was undertaken in Kenya during the 1960s and 70s, and this has been reviewed by Firman and Waller (1977), although some parallel work was also carried out in Cameroon by Muller and colleagues. In early studies on the disease epidemiology, Nutman and Roberts (1961) found that the pathogen was a normal inhabitant of the immature bark of young twigs, especially in the zone of phellogen development, and they concluded that this was the main source of inoculum for berry infection.

Subsequent work by Gibbs (1971), based on the ability to culturally differentiate the pathogen from other forms of *Colletotrichum*, demonstrated that the pathogen formed only a small proportion (< 5%) of the *Colletotrichum* population in the bark, and that diseased berries were a far more potent source of pathogen inoculum. A single, actively diseased berry can produce 50 times more pathogen spores than the bark of a bearing stem over the same period.

However, the young bark of coffee twigs is the source of inoculum to initiate the epidemic at the start of the rainy season (when the sporulating capacity of the fungus in the bark is at its highest), providing there are no diseased berries from overlapping crops on the tree. Nevertheless, inoculum from diseased berries soon takes over as the major source, despite the fact that many diseased berries fall from the tree at an early stage of disease development.

The assessment of disease progress is complicated by the different effects of the disease on the crop. Whereas berries with active lesions can be readily recognized and recorded and are a potent source of inoculum, those with scab lesions may contribute neither to losses nor to disease progress unless they become active later. Furthermore, because a variable number of infected berries are constantly being shed, expressing disease as a percentage of berries present on the tree at any given time does not give an accurate estimate of the disease or of its effect on the crop. During berry ripening the disease has little effect on yield.

Disease progress has been most accurately assessed by expressing the number of healthy berries as a proportion of the total that have developed beyond the pinhead stage when physiologic shedding was minimal (Griffiths *et al.*, 1971). If accurate berry counts are conducted at an early stage, then crop loss can be estimated by the number of ripe berries harvested as a proportion of the initial count. This does not allow for berries lost to other causes and will only give an accurate estimate of yield loss to CBD when losses to the disease predominate over other causes, such as insect damage. Factors affecting the ratio of active to scab lesions and to dehisced berries are chiefly the resistance of the cultivar and the use of fungicides, and are considered further under the section 'control', below.

As all phases of the disease cycle, from infection through sporulation and spore dispersal, are dependent on water, disease progress can only occur during the rainy season when the crop is at a susceptible stage. The disease in less severe in dryer areas or when the expanding crop stage does not coincide with wet weather. Any changes in rainfall distribution that occur as a result of climate change will therefore influence the severity of CBD, with wetter conditions favouring the disease. Muller (1973) avoided the disease in Cameroon by the use of irrigation to produce early flowering, so that the berry expansion phase occurred during dry weather.

The disease is more severe at higher altitudes, largely because the higher rainfall, cooler conditions and longer periods of wetness favour disease development (Griffiths and Waller, 1971; Mulinge and Griffiths, 1974). The proportion of *C. kahawae* in the bark is also greater at higher altitudes (Mulinge, 1971). The disease is also more severe in older, multiple-stemmed trees where tall, dense heads provide a major source of inoculum to be washed onto the crop beneath, and removal of these heads by capping can reduce inoculum load and disease severity (Waller, 1972).

In equatorial countries with a bimodal rainy distribution, overlapping crops exacerbate the disease, as diseased berries may be carried over between crops and greatly enhance the early season inoculum produced from the bark, thus providing a major boost to the initial infection levels. The multiple-stem system of pruning permits free growth, with little control over cropping so that double cropping is usual, often with small, out-of season flowerings. These provide berries at different growth stages and enable a constant source of susceptible crop for infection and a constant source of inoculum (see Fig. 8.2).

Muller and Gestin (1967) showed that removal of out-of season flowerings could reduce disease development, although Gibbs (1971) did not find that disease levels were much reduced by removal of diseased berries from an earlier crop. The length of time diseased berries remain an active source of inoculum may be a crucial factor, as they can be rapidly colonized by saprophytes.

Although rainfall is essential for disease progress, it is wetness duration that is most significant (Cook, 1975). Heavy rainfall and irrigation can wash most inoculum from the trees, and rainfall events up to 10 mm have been shown to disperse the greatest number of spores (Waller, 1971, 1972). While *Colletotrichum* spores cannot be transported long distances on air currents, they can travel several metres in rain-splash droplets, and may be carried by passive vectors such as birds, insects, man and machinery. Transmission on unhulled coffee might occur, but the pathogen does not penetrate the testa of mature berries and transmission on hulled seed has not been demonstrated (Tagne and Mathur, 2003).

Control

Chemical control

Chemical control is essential in most areas affected by coffee berry disease, although resistant varieties exist and newer, resistant cultivars are being bred and grown in some areas. Much of the work on chemical control of CBD has been done in Kenya. Bordeaux mixture was first used against the disease in Kenya in the 1930s, and copper-based fungicides are still very effective. However, results were variable (Rayner, 1952) and depended on prolonged spraying with 50% copper formulations, applied at high volume.

Similar applications throughout the cropping season were needed to control the disease in Cameroon (Muller and Gestin, 1967). Copper fungicides

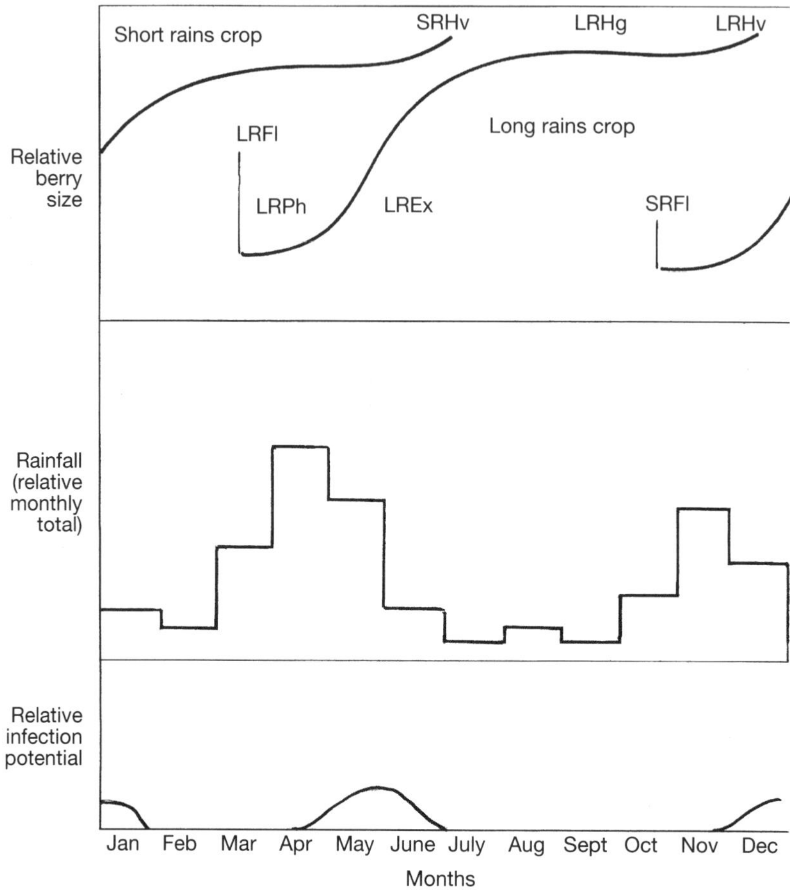

Fig. 8.2. Crop phenology and CBD susceptibility in Kenya. SRHv, short rains crop harvest; LRHv, long rains crop harvest; LRFl, long rains crop flowering; LRPh, long rains crop pinhead stage; LREx, long rains crop expansion phase; LRHg, Long rains crop hard green stage; SRFl, short rains crop flowering.

were already being used against rust and had a tonic effect by enhancing yields above those that could be attributed solely to disease control (see Chapter 7).

The first attempts to improve fungicidal control of CBD in Kenya centred on the use of early sprays to decrease the inoculum produced from the bark (Nutman and Roberts, 1961; Bock, 1963). Spray regimes emphasizing early sprays were adopted during the 1950s and 1960s, but it became apparent that these did not always provide acceptable levels of control. Sometimes early season sprays increased disease severity (Wallis and Firman, 1967; Griffiths and Gibbs, 1969; Gibbs, 1971). There were also some estates in Kenya where fungicides had never been sprayed against CBD or against rust and where the disease was of little consequence.

Furtado (1969) found that the 'CBD strain of *Colletotrichum*' was more abundant from the branches of trees on farms that had been sprayed than from trees on neighbouring farms that had never been sprayed. This iatrogenic effect of disease enhancement (Griffiths, 1981) appears largely due to the suppression of the natural microflora of coffee by fungicides. Elements of the microflora have been shown to compete with the pathogen, both to limit infection and sporulation and to stimulate natural host defence mechanisms (Waller and Masaba , 2006). This leads to severe disease outbreaks if fungcides are only applied early in the season (Firman and Waller, 1977), as both chemical and natural biological control mechanisms are suspended.

It is intriguing that CBD became a major problem in Kenya about the time that the practice of spraying copper fungicides for rust control was widely adopted. Early spraying gave good control in seasons when flowering was early, but poor control when it was late (Griffiths and Gibbs, 1969; Griffiths and Waller, 1971). It was then demonstrated that a spray schedule that protected the berries gave better control of CBD than sprays aimed at decreasing the sporulating capacity of the bark – which had the negative effect of increasing disease levels above those of unsprayed coffee (Griffiths *et al.*, 1971) (see Fig. 8.3).

Since then, monthly sprays to protect the crop during the rainy season has been the standard recommendation in Kenya. Subsequent research has been directed at methods of decreasing the number of sprays or amount of fungicide required, and on the efficacy of new fungicides.

Bock (1963) found that copper fungicides were more effective than dithiocarbamates, but captafol was found to be as good or better than copper (Vermeulen, 1968) and also had a 'tonic effect' on the crop similar to that of copper. Vine *et al.* (1973) screened 60 fungicides in the laboratory, of which nine performed well in the field and were recommended for CBD control in Kenya: 50% copper formulations, captafol, chlorothalonil, benomyl, thiophanate methyl, thiophanate, thiabendazole, dithianon and fentin hydroxide.

The benzimidazole systemic fungicides were very effective against the disease and were widely used on estates in the early 1970s as they provided a greater flexibility for application schedules, but resistance soon developed and control failed (Okioga, 1976; Ramos and Kamidid, 1982). The fungicide-resistant strains are very stable, and benomyl-resistant strains were shown to dominate the pathogen population, many years after the fungicide had been discontinued (King'ori and Masaba, 1991).

These failures provided an early warning that systemic fungicides should be used sparingly. It has become common practice to use them either alternating with, or in tank mixtures with, copper compounds (Masaba *et al.*, 1990). In Kenya, effective control of CBD and better yields than with organic fungicides alone were obtained with a tank mixture consisting of half the normal rate of the organic compound and a reduced rate of copper (Okioga, 1978).

Some proprietary mixtures are available, but there is some evidence that the tank mixture of two separate fungicides – such as anilazine or chlorothalonil – and copper is more effective (Masaba and Opilo, 1990). Fluzinam or prochloraz in a mixture with copper have also been recommended (Masaba *et*

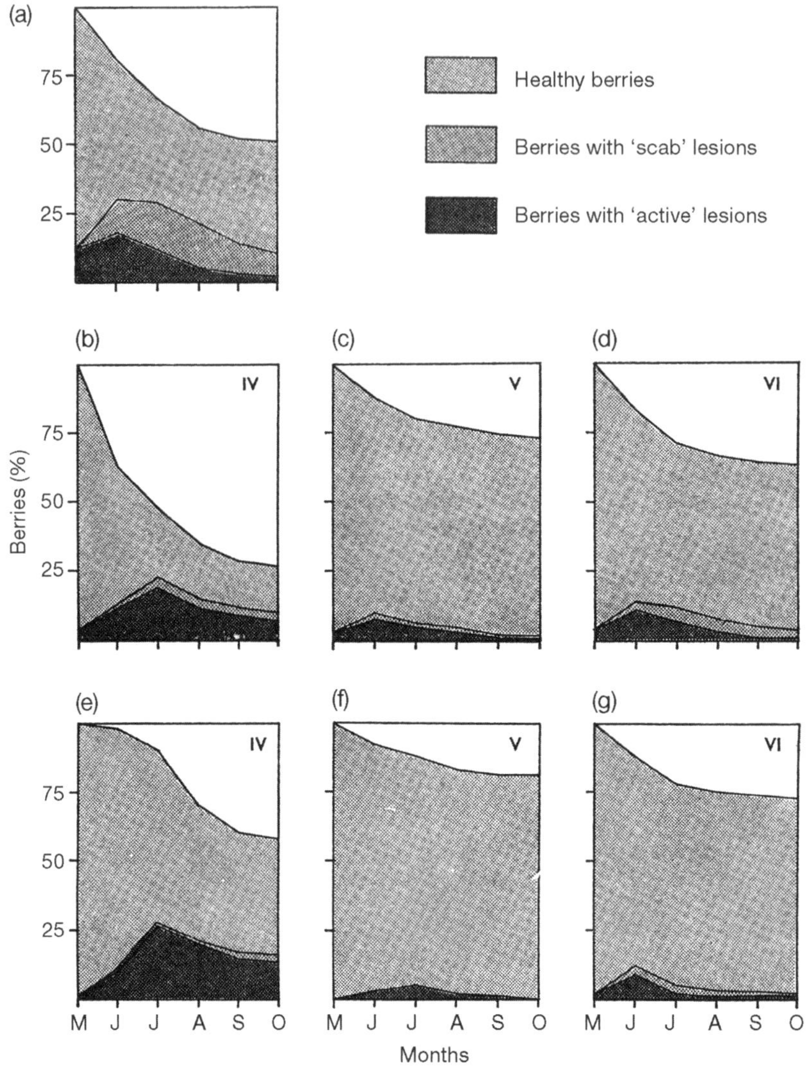

Fig. 8.3. CBD development and crop fate with different fungicide applications (from Griffiths *et al.*, 1971). (a), unsprayed control; (b), (c) and (d), sprayed with copper fungicide; (e), (f) and (g), sprayed with Captafol fungicide. IV, early season (pre-rains) stage; V, throughout rainy season; VI, during susceptible crop stages.

al., 1993). Resistance to prochloraz was detected in field trials after 2 years and developed quickly in cultures exposed to increasing concentrations of the fungicide (Mwang'ombe, 1994). More recently, triazole compounds such as cyproconazole and hexaconazole have shown promise and these will also control rust. Captafol, which probably gave the best control of the disease and set the standard, was withdrawn in the early 1990s due to health concerns.

All of the currently recommended products are class III WHO materials and pose no recognized health dangers. Small-scale farmers continue to use copper fungicides as they are relatively cheap compared to organic fungicides and also provide control of rust.

The need to control other disease such as rust and bacterial blight, as well as CBD, also influences the components of tank mixes. In situations where rust is a problem but CBD has not appeared in the crop, although present in the region, consideration should be given to rust control with ground-applied systemic fungicides. These should have less effect on surface microflora and be less likely to increase susceptibility to CBD through iatrogenic effects. Unfortunately, ground-applied fungicides cannot be used to control CBD as they are not well translocated to the berries. Fungicides recommended for CBD control are shown in Table 8.2.

Application schedules are aimed at protecting the susceptible stages of the crop during the rainy season, but both crop growth and erosion of fungicide deposits by rainfall can quickly reduce effectiveness. The target of bearing branches is also difficult to reach in heavily foliated coffee, and early attempts at spraying involved the use of high-volume applications (2000 l/ha) with hydraulic hand lances (Bock, 1963). Wallis and Firman (1967) found that good coverage of the target could be achieved with volumes below 1000 l/ha using knapsack sprayers, and later work showed that tractor-drawn or motorized knapsack mist-blowers using lower volumes were effective.

However, further research demonstrated that redistribution of fungicidal

Table 8.2. Some fungicides recommended for control of coffee berry disease.

Fungicide	Formulation
Inorganic contact	
Cuprous oxide	50%[a]
	75%[a]
Cupric oxychloride	50%[a]
Cupric hydroxide	50%[a]
Copper sulphate + lime (Bordeaux mixture)	25% (Cu)[a]
Organic contact	
Chlorothalonil	75%[a]
	50%[b]
Anilazine	75%[a]
	48%[b]
Dithianon	75%[a]
	50%[b]
Systemic	
Fluzinam	50%[b]
Mixtures	
Anilazine/copper	17/30[a]
Chlorothalonil/copper	25/30[a]

[a] Wettable powder.
[b] Suspension concentrate.

deposits in rainwater was a significant factor, especially as spores of the pathogen were also distributed by rainwater, and overhead spraying was shown to be as effective and much quicker that other methods (Periera *et al.*, 1973). This could be achieved by aerial spraying or using overhead booms that could cover eight rows at a time (Periera and Mapother, 1972). The importance of adequate fungicidal deposits in the tops of trees is now well recognized.

Resistant varieties

Harrar, Bourbon and SL types of *C. arabica* are particularly susceptible to CBD, as is Caturra. The SL varieties were popular in Kenya and elsewhere in Africa for their high yield and good quality but, with the arrival of CBD, they could no longer be grown, particularly at higher altitude, without continuous fungicide protection during berry development. The dense canopy of Caturra made it especially prone to CBD when grown in East Africa.

However, many varieties of *C. arabica* possess appreciable resistance to coffee berry disease but are not widely grown due to problems of quality or yield performance. Rume Sudan and Blue Mountain have long been known to have some resistance to CBD (Rayner, 1952). Firman (1964) screened 133 coffee cultivars in the field and laboratory for resistance. Rume Sudan exhibited the highest level and this seemed to be associated with internal resistance in the berry. Geisha 10 was moderately resistant and both Rume Sudan and Geisha produced a high proportion of scab lesions. Resistance was apparently associated with the narrow-leaved, bronze-tipped character and this was used in Congo to produce several CBD-resistant types known as 'Local Bronze' (Firman and Waller, 1977). Other sources of CBD resistance have been identified in Rwanda (Jackson hybrid) and Cameroon (Java variety).

Blue Mountain was recommended for use in Kenya as a CBD-resistant variety, but it did not yield well under shade and can be subject to heavy flower infection by CBD, although the berries are resistant (Bock, 1963). Although Rume Sudan is highly resistant to CBD, its poor quality makes it unsuitable for commercial production, but it is a source of resistance to CBD in hybrids. The cultivar K7 was widely grown in the middle- and lower-altitude coffee areas in Kenya, where it showed reasonable field resistance to CBD and to rust, combined with acceptable quality.

During the 1970s, an FAO-sponsored programme in Ethiopia selected a number of genotypes with field resistance from semi-natural coffee populations, and the resistance was found to be quantitative and polygenic, implying that it might be expected to be durable (Van der Graff and Pieters, 1983). Material from this programme is being developed and grown in Ethiopia.

The main source of resistance to CBD since the 1970s has been selections from Hybrido de Timor (HdT, see Chapter 7). Van der Vossen *et al.* (1976) developed a reliable and rapid screening method for CBD resistance, based on inoculation of the hypocotyls of 6-week old seedlings (see Plate 24). This method was also used to identify two genes controlling resistance in Rume Sudan and one in HdT (Van der Vossen and Walyaro, 1980), which were used in the Kenyan breeding programme.

Because of its resistance to rust, HdT became widely used in crossing programmes in Latin America. In Kenya, progeny of Colombian crosses between Caturra and HdT were selected for resistant to CBD as well as to rust, and hybridized with selections from crosses involving SLs and Rume Sudan. This produced the F_1 composite variety, Ruiru 11, with resistance to CBD and rust and other desirable characters (Van der Vossen and Walyaro, 1981). Ruiru 11 is a high-yielding, semi-dwarf type and it was estimated that its cultivation could cut the cost of production by 26%, due mainly to the elimination of fungicidal sprays (Roe and Nyoro, 1986). The demand for Ruiru 11 has been high in Kenya, but it has proved difficult to produce the hybrid seed, which requires hand-pollination of emasculated flowers, quickly enough to meet the demand (Masaba and Waller, 1992; Opile and Agwanda, 1993).

By the year 2000, there was no evidence that widespread cultivation of Ruiru 11 had selected for more virulent strains of *C. kahawae*. Variation in resistance among hybrid progeny of Ruiru 11 was attributed to differences in aggressiveness among isolates of the pathogen (Omondi *et al.*, 2001).

Other African countries have also opted to distribute Catimor-type cultivars. A number of Catimor populations have been planted in Malawi, for instance, although not all of these have CBD resistance. CBD resistance was identified in selections from Catimor 129, and a small amount of seed became available in the late 1990s under the variety name 'Nyika' (N. Phiri, Malawi, 1999, personal communication). New techniques to assist selection in ongoing coffee breeding programmes have been developed. RAPD markers have been identified that are associated with resistance in coffee to the disease and can be used in marker-assisted breeding (Agwanda *et al.*, 1997). Callus tissue and protoplasts can also be used as a basis for resistance selection (Nyange *et al.*, 1995).

Integrated disease management

Measures to enable adequate aeration of the coffee canopy such as pruning, shade control and adequate spacing will reduce humidity and wetness duration of berry surfaces and, to some extent, hinder the pathogen. This also enables better penetration and coverage of fungicides if these are used. The capping of the taller stems of multiple-stemmed coffee can also decrease the incidence of CBD, as these often provide a major source of inoculum for the crop. Waller (1972) found less disease in both sprayed and unsprayed capped multiple-stem trees than in their uncapped counterparts.

Removal of out-of-season berries also improves control (Muller and Gestin, 1967), and initiating flowering by irrigation during the dry season to avoid the susceptible stages of the crop coinciding with the rainy season may be applicable under some circumstances, as in Cameroon (Muller *et al.*, 2004).

Avoidance of overlapping crops in seasonally bimodal areas may have some effect, but is not an economic proposition. Possibilities of developing biological control of CBD exist. Natural control of the disease has been shown to occur through several effects of natural microflora on coffee. Some components such as *Fusarium stilboides* and *Colletotrichum gloeosporioides* are powerful antagonists to *C. kahawae*, are apparently endophytic and may induce systemic resistance to *C. kahawae* in unsprayed coffee, as evidenced by the greater proportion of scab lesions. Coffee that has been left unsprayed for several years regains some of its natural microflora and, under some circumstances, can yield as much as sprayed coffee (Waller and Masaba, 2006).

Other Berry Diseases

Brown blight

This is the term applied to anthracnose of the ripening berries and is caused by both *Colletotrichum gleoeporioides* and *C. kahawae* (see Chapter 7 and above). It is analogous to other anthracnoses of ripening fruits caused by *Colletotrichum*. The disease is common when coffee ripens during wet weather, but causes no yield loss. In some circumstances the dried, decayed pulp may stick to the testa causing problems in processing, including bean staining and loss of quality. The disease often arises through the activation of latent infections as the coffee berries ripen. Hocking (1966) studied the problem in Tanzania.

Berry blotch

Cercospora coffeicola can cause symptoms on the berries as well as on the leaves (see Chapter 7). The dark reddish brown, blotchy lesions may easily be confused with CBD, but they are typically surrounded by a yellow to red halo and not sunken – as is the case with CBD lesions – and may require control (Van der Vossen and Cook, 1975) (see Plate 25). Exposed fruit that is subject to sun scorch or under other forms of physiologic stress is particularly susceptible.

The disease on *C. canephora* berries causes small spots that become reddish and raised, when the disease is often called red blister disease. *Fusarium stilboides* causes a similar berry rot but it is distinguishable from CBD by typically affecting first the flower end of the berry, so that half of the berry is orange/red due to premature ripening and the other half is brown. It often colonizes berry blotch and brown blight lesions and may exacerbate the damage.

Warty berry

Caused by *Botrytis cinerea* Pers.ex Pers. (Teleomorph: *Sclerotinia fuckeliana* (de Bary) Fuckel.), this is a condition that only occurs in very damp, cool conditions. It is sometimes seen on shaded coffee at high altitudes, but rarely causes significant damage. As the name implies, affected berries develop warty, grey outgrowths and usually occur in sporadic clusters. The grey sporulation of the fungus is often evident on diseased berries. Where still, damp condition occur from flowering to the pinhead stage, the fungus can spread from dead flower remains to cause the disease, infecting clusters of pinheads (Baker, 1972).

Trachyspheria fructigena Tabor and Bunting can infect *C. liberica* berries under hot, humid conditions to cause a purple discoloration followed by pinkish sporulation of the fungus on the fruit surface. The problem requires control with copper sprays in Côte d'Ivoire (Tabor and Bunting, 1923; Mallamaire, 1934).

References

Agwanda, C.O., Lashermes, P., Trouslot, P., Combes, M.C. and Charrier, A. (1997) Identification of RAPD markers for resistance to coffee berry disease, *Colletotrichum kahawae*, in Arabica coffee. *Euphytica* 7, 1–248.

Baker, C.J. (1972) An outbreak of *Botrytis cinerea* on coffee in Upper Kaimbu. *Kenya Coffee* 36, 266–277.

Beynon, S.M., Coddington, A., Lewis, B.G. and Varzea, V. (1995) Genetic variation in the coffee berry disease pathogen, *Colletotrichum kahawae. Physiological and Molecular Plant Pathology* 11, 313–325.

Bock, K.R. (1956) Investigations of coffee berry disease – laboratory studies. *East African Agriculture Journal* 24, 405–506.

Bock, K.R. (1963) The control of coffee berry disease in Kenya. *Empire Journal of Experimental Agriculture* 31, 91–107.

Bridge, P.D., Waller, J.M., Davies, D. and Buddie, A. (2007) Variability of *Colletotrichum kahawae* in relation to other *Colletotrichum* species from tropical perennial crops; development of diagnostic techniques. *Journal of Phytopathology* {In press}.

Cook, R.T.A. (1975) The effect of weather conditions on infection by coffee berry disease. *Kenya Coffee* 40, 190–197.

Cramer, H.H. (1967) *Plant Protection and World Crop Protection*. Farbenkafbriken Bayer AG, Leverkeusen, Germany, 524 pp.

Firman, I.D. (1964) Screening coffee for resistance to coffee berry disease. *East African Agricultural Journal* 29, 192–194.

Firman, I.D. and Waller, J.M. (1977) *Coffee Berry Disease and other Colletotrichum Diseases of Coffee. Phytopathologcal Paper No. 20*. Commonwealth Mycological Institute, UK.

Furtado, I. (1969) Effect of copper fungicides on the occurrence of the pathogenic form of *Colletotrichum coffeanum. Transactions of the British Mycological Society* 53, 325–328.

Gibbs, J.N. (1969) Inoculum sources for coffee berry disease. *Annals of Applied Biology* 64, 515–522.

Gibbs, J.N. (1971) Some factors affecting the performance of spray programmes for the control of coffee berry disease. *Annals of Applied Biology* 67, 343–356.

Gichuru, E.K., Varzea, V.M.P., Rodrigues, C.J. and Masaba, D.M. (2000) Vegetative compatibility grouping of *Colletotrichum kahawae* in Kenya. *Journal of Phytopathology* 148, 233–237.

Griffiths, E. (1969) CBD: Kenya's biggest coffee problem. *SPAN* 12, 2.

Griffiths, E. (1981) Iatrogenic plant diseases. *Annual Review of Plant Pathology* 19, 69–82.

Griffiths, E. and Gibbs, J.N. (1969) Early season sprays for the control of coffee berry disease. *Annals of Applied Biology* 64, 523–532.

Griffiths, E. and Waller, J.M. (1971) Rainfall and cropping patterns in relation to coffee berry disease. *Annals of Applied Biology* 67, 75––91.

Griffiths, E., Gibbs, J.N. and Waller, J.M. (1971) Control of coffee berry disease. *Annals of Applied Biology* 67, 45–74.
Hindorf, H. (1970) *Colletotrichum* spp. isolated from *Coffea arabica* in Kenya. *Zeitschrift fur Pflanzenkrankenheiten und Pflanzenschutz* 77, 328–331.
Hindorf, H. (1974) Colletotrichun-Arten aus dem kaffeeanbaugebeit von Kiambu in Kenya. *Zeitschrift fur Pflanzenkrankenheiten und Pflanzenschutz* 81, 108–113.
Hocking, D. (1966). Brown blight (*Colletotrichum coffeanum* Noack) of arabica coffee in East Africa. *Annals of Applied Biology* 58, 409–421.
Hocking, D., Johanns, J.C. and Vermeulen, H. (1967) Ascospore production, discharge and infections by *Glomerella cingulata* causing coffee berry disease. *Nature* 214, 1144–1145.
King'ori, P.N. and Masaba, D. (1991) Distribution and importance of benomyl resistance in populations of *Colletotrichum coffeanum* in coffee. *Kenya Coffee* 56, 1071–1074.
Mallamaire, A. (1934) L'anneé phytopathologique en Côte d'Ivoire. *Agronomie Coloniale* 23, 114–119.
Masaba, D. and Opilo, V.L. (1990) The performance of formulated mixtures of copper and organic fungicides against coffee berry disease. *Kenya Coffee* 55, 825–831.
Masaba, D. and Waller, J.M (1992) Coffee berry disease: the current status. In: Bailey, J.A. and Jeger, M.J. (eds) *Colletotrichum: Biology, Pathology and Control*. CAB International, Wallingford, UK, pp. 237–249.
Masaba, D., King'ori, P.N. and Muthangya, P.M. (1990) Control of coffee berry disease using tank mixtures of copper and organic fungicides. *Kenya Coffee* 55, 949–956.
Masaba, D., King'ori, P.N. and Opilo, V.L. (1993) Effectiveness of new fungicides against coffee berry disease in Kenya. *Kenya Coffee* 58, 1517–1524.
McDonald, J. (1926) A preliminary account of a disease of green coffee berries in Kenya. *Transactions of the British Mycological Society* 11, 145–154.
Mulinge, S.K. (1970) Development of coffee berry disease in relation to stage of growth of the berry. *Annals of Applied Biology* 65, 269–276.
Mulinge, S.K. (1971) Effect of altitude on the distribution of the fungus causing coffee berry disease in Kenya. *Annals of Applied Biology* 67, 93–98.
Mulinge, S.K. and Griffiths, E. (1974) Effects of temperature on the *Colletotrichum* population on coffee bark. *Transactions of the British Mycological Society* 62, 610–614.
Muller, R.A. (1973) L'anthracnose des baies du caféier d'arabie (*Coffea arabica*) due à une forme virulent de *Colletotrichum coffeanum* Noack. *Café-Cacao-Thé* 17, 281–312.
Muller, R.A. and Gestin, A.J. (1967) Contribution à la mise au point de méthodes de lutte contre l'anthracnose de baies du caféier d'arabie (*Coffea arabica*) due à une forme de *Colletotrichum coffeanum* Noack au Cameroon. *Café-Cacao-Thé* 11, 157–178.
Muller, R.A., Berry, D., Avelino, J. and Bieysse, D. (2004) Coffee diseases. In: Wintgens, J.N. (ed.) *Coffee: Growing, Processing and Sustainable Production*. Wiley-Verlach, Weinheim. Germany, pp. 491–545.
Muthappa, B.N. (1970) Studies on the role of *Colletotrichum coffeanum* causing stalk rot of leaves and berries of arabica coffee. *Indian Coffee* 34, 263–264.
Mwang'ombe, A.W. (1994) Tolerance in isolates of *Colletotrichum coffeanum* to prochloraz-Mn in Kenya. *Journal of Phytopathology* 140, 114–122.
Nutman, F.J. and Roberts, F.M. (1960a) Investigations on a disease of *Coffea arabica* caused by a form of *Colletotrichum coffeanum*. I. Some factors affecting infection by the pathogen. *Transactions of the British Mycological Society* 43, 489–505.
Nutman, F.J. and Roberts, F.M. (1960b) Investigations on a disease of *Coffea arabica* caused by a form of *Colletotrichum coffeanum*. II. Some factors affecting germination and infection and their relation to disease distribution. *Transactions of the British Mycological Society* 43, 643–659.
Nutman, F.J. and Roberts, F.M. (1961) Investigations on a disease of *Coffea arabica* caused

by a form of *Colletotrichum coffeanum*. III. The relation between infection of the bearing wood and disease incidence. *Transactions of the British Mycological Society* 44, 511–521.

Nyange, N.E., Williamson B., McNicol, R.J. and Hackett, C.A. (1995) In vitro screening of coffee genotypes for resistance to coffee berry disease (*Colletotrichum kahawae*). *Annals of Applied Biology* 127, 251–261.

Okioga, D.M. (1976) Occurrence of strains of *Colletotrichum coffeanum* resistant to methyl benzimidazol-2-ylcarbamate (carbendazim) and chemically similar compounds. *Annals of Applied Biology* 84, 21–30.

Okioga, D.M. (1978) The role of copper fungicides in the control of coffee diseases. *Kenya Coffee* 43, 221–230.

Omondi, C.O., Ayecho, P.O., Mwang'ombe, A.W. and Hindorf, H. (2000) Reaction of some Coffea arabica genotypes to strains of *Colletotrichum kahawae*, the cause of coffee berry disease. *Journal of Phytopathology* 148, 61–63.

Omondi, C.O., Ayiecho, P.O., Mwang'ombe, A.W. and Hindorf, H. (2001) Resistance of Coffea arabica cv. Ruiru 11 tested with different isolates of *Colletotrichum kahawae*, the causal agent of coffee berry disease. *Euphytica* 121, 19–24.

Opile, W.R. and Agwanda, C.O. (1993) Propagation and distribution of cultivar Ruiru 11: a review. *Kenya Coffee* 58, 1496–1508.

Pereira, J.L. and Mapother, H.R. (1972) Overhead application of fungicide for the control of coffee berry disease. *Experimental Agriculture* 8, 117–122.

Pereira, J.L., Mapother, H.R., Cook, B.K. and Griffiths, E. (1973) Redistribution of fungicides in coffee trees. *Experimental Agriculture* 9, 209–218.

Ramos, A.H. and Kamidid, R.E. (1982) Determination and significance of mutation rates for *Colletotrichum coffeanum* from benomyl sensitivity to benomyl tolerance. *Phytopathology* 72, 181–185.

Rayner, R.W. (1952) Coffee berry disease. A survey of investigations carried out up to 1950. *East African Agricultural Journal* 17, 130–158.

Rodriguez, C.J., Varzea, V.M.P., Hindorf, H. and Mediros, E.M. (1991) Strains of Colletotrichum coffeanum Noack causing coffee berry disease in Angola and Malawi with characteristics different to the Kenya strain. *Journal of Phytopathology* 131, 205–209.

Roe, J.D.M. and Nyoro, J.K. (1986) Economic indications of introducing the new hybrid variety of arabica coffee. *Kenya Coffee* 51, 219–244.

Screenivasaprasad, S., Brown, A.E. and Mills, P.R. (1993) Coffee berry disease pathogen in Africa: genetic structure and relationship to the group species *Colletotrichum gloeosporioides*. *Mycological Research* 97, 995–1000.

Steiner, K.G. (1974) Die sporulation von *Colletotrichum coffeanum* auf kirschen von Coffea arabica. *Phytopathologische Zeitschrift* 79, 179–189.

Tabor, R.J. and Bunting, R.H. (1923) On a disease of cocoa and coffee fruits caused by a fungus hitherto undescribed. *Annals of Botany* 37, 152–157.

Tagne, A. and Mathur, S.B. (2003) Fungi associated with seeds of *Coffea arabica*. *Plant Genetic Resources Newsletter* 135, 44–46.

Tapley, R.G. (1964) Coffee berry disease in Tanganyika. *Tanganyika Coffee* 38, 45.

Van der Graaff, N.A. and Pieters, R. (1983) Resistance to coffee berry disease in Ethiopia. In: Lamberti, F., Waller, J.M. and Van der Graaff, N.A. (eds) *Durable Resistance in Crops*. Plenum Press, New York, pp. 317–334.

Van der Vossen, H.A.M. and Cook, R.T.A. (1975) Incidence and control of berry blotch caused by *Cercospora coffeicola* on arabica coffee in Kenya. *Kenya Coffee* 40, 58–61.

Van der Vossen, H.A.M. and Walyaro, D.J. (1980) Breeding for resistance to coffee berry disease in *Coffea arabica*. II. Inheritance of resistance. *Euphytica* 29, 77–791.

Van der Vossen, H.A.M. and Walyaro, D.J. (1981) The coffee breeding programme in Kenya: a review of progress made since 1971 and plan of action for the future. *Kenya Coffee* 46, 113–130.

Van der Vossen, H.A.M., Cook, R.T.A. and Murakura, G.N.W. (1976) Breeding for resistance to coffee berry disease caused by *Colletotrichum coffeanum* in *Coffea arabica* I. Methods of preselection for resistance. *Euphytica* 25, 733–745.

Varzea, V.M.P., Rodriguez, C.J. and Lewis, B.G. (2002) Distinguishing characteristics and vegetative compatibility of *Colletotrichum kahawae* in comparison with other related species from coffee. *Plant Pathology* 51, 202–207.

Vermeulen, H. (1968) Screening fungicides for control of coffee berry disease in Kenya. *Experimental Agriculture* 4, 255–261.

Vine, B.H., Vine, P.A. and Griffiths, E. (1973) Evaluation of fungicides for the control of coffee berry disease in Kenya. *Annals of Applied Biology* 75, 359–375.

Waller, J.M. (1971) The incidence of climatic conditions favourable to coffee berry disease in Kenya. *Experimental Agriculture* 7, 303–314.

Waller, J.M. (1972) Water-borne spore dispersal in coffee berry disease and its relation to control. *Annals of Applied Biology* 71, 1–18.

Waller, J.M. and Bridge, P.D. (2000) Recent advances in understanding Colletotrichum diseases of some tropical perennial crops. In: Prusky, D. Freeman, S. and Dickman, M.B. (eds) *Colletotrichum: Host Specificity, Pathology and Host–Pathogen Interaction*. APS Press, St Paul, Minneapolis, pp. 337–345.

Waller, J.M. and Masaba, D. (2006) The microflora of coffee surfaces and relationships to coffee berry disease. *International Journal of Tropical Pest Management* 52, 89–96.

Waller, J.M., Bridge, P.D., Black, R. and Hakiza, H. (1993) Characterisation of the coffee berry disease pathogen, *Colletotrichum kahawae* sp. nova. *Mycological Research* 97, 989–994.

Wallis, J.A.N. and Firman, I.D. (1967) A comparison of fungicide spray volumes for the control of coffee berry disease. *Annals of Applied Biology* 59, 111–122.

9 Wilt Diseases and Diseases of the Root and Stem

Fusarium Wilt

Pathogen: *Gibberella xylarioides* Heim & Saccas, Ascomycete, (anamorph: *Fusarium xylarioides* Steyaert; syn. *Fusarium oxysporum* forma *xylarioides* (Steyaert) Deassus).

The disease, also known as tracheomycosis, coffee wilt, 'sudden death' (French – trachéomycose du caféier, fusariose du caféier), was first reported from *Coffea liberica* in central Africa and then from *C. canephora* in the mid-20th Century. It was later found to occur on *C. arabica* coffee in Ethiopia and has recently caused serious damage to *C. canephora* in Democratic Republic of Congo and Uganda.

The primary hosts of the fungus are *C. arabica*, *C. canephora* and *C. liberica*, and probably other *Coffea* spp. It has also been isolated from rotting tomatoes in Nigeria (Onesirosan and Fatunla, 1976) and, recently, from banana in Uganda. The fungus has also been reported to be pathogenic to cotton seedlings (Pizzinatto and Menten, 1991).

History and distribution

A wilt disease affecting coffee was observed first in 1927, near Bangui in the Central African Republic on 'excelsa' coffee (*C. liberica* var. *dewevrei;* syn. *C. excelsa*) (Guillemat, 1946; Fraselle, 1950). The disease spread quickly during the 1930s and, by 1939, all excelsa plantations in the Central African Republic were affected and it had begun to affect *C. canephora* by 1945. In 1948–1949, a similar wilt disease decimated coffee plantations in Côte d'Ivoire growing *C. liberica* cv 'Indenie' and *C. canephora* cv 'kouillou' (Wrigley, 1988; Kebe, 1997).

From the mid-1940s to the mid-1950s, fusarium wilt reached epidemic proportions in the Belgian Congo, Central African Republic and Côte d'Ivoire, until the outbreak was contained in the 1960s by replanting with resistant cultivars. The disease was also reported from Cameroon around the same time as the epidemic in the Belgian Congo, but affecting only excelsa coffee (Muller, 1997). In 1958, similar symptoms were reported on *C. arabica* in Ethiopia (Lejeune, 1958), where coffee wilt is now considered to be endemic throughout the coffee-growing areas (Girma *et al.*, 2001; CMI map No. 464).

The most recent resurgence of Fusarium wilt seems to date from the 1980s, when outbreaks were reported in the Democratic Republic of Congo (Mfwidi-Nitu, 1994). Since 1986, the disease has spread through a corridor of robusta coffee plantations stretching from Isiro to Mambasa, Komanda and Beni. The disease appears to have spread from Democratic Republic of the Congo into south-west Uganda during the late 1980s or early 1990s. The first report in Uganda was received in 1993 from the Districts of Bundibugyo and Rukungiri, bordering Democratic Republic of the Congo and, by 2003, it was thought that all districts of Uganda growing robusta had become affected. The disease is now in north-west Tanzania. Rutherford (2006) provides a recent review of the current status.

Economic impact

Fusarium wilt was the most important disease of 'excelsa' and 'robusta' coffee and caused major losses to these coffees in the Central Africa Republic, Democratic Republic of the Congo and Côte d'Ivoire in mid-20th Century. Plantations of the robusta cvs 'bandama' and 'tuba' were completely destroyed, while some other cvs such as ' ebobo', 'nana' and 'robusta Congo', imported from the Belgian Congo, showed resistance to the disease. By 1945, many of plantations of *C. liberica* in Côte d'Ivoire had been completely destroyed and the disease was also affecting some cultivars of *C. canephora*, particularly the indigenous kouillou (*C. canephora* var. *kouillou*) (Muller, 1997).

However, by the 1960s the disease had been contained through the destruction of millions of trees and replanting with resistant cultivars, and was reduced to relatively minor significance. However, a recent major upsurge of the disease is causing considerable losses of robusta coffee in Democratic Republic of the Congo (Flood, 1996) and in Uganda (Flood and Brayford, 1997). Disease incidences in robusta coffee of over 90% were reported from the north-east of the Democratic Republic of the Congo (Flood, 1996).

A survey conducted in 1995/1996 revealed that tracheomycosis was present in 12 out of 27 coffee-growing districts in Uganda, with mean disease incidence per district ranging from 1 to 40%, and only robusta coffee being affected (Birikunzira and Lukwago, 1997). By 2003 it was thought that around 15 million trees had been lost, or about 5% of the country's total. There is a risk of the disease spreading to neighbouring countries. The government of Uganda and the Uganda Coffee Development Association, with financial support from the EU, embarked on an ambitious plan to replace the ageing robusta population with improved clones. However, although there are varying degrees of susceptibility among the six clones, none are resistant to wilt.

In Ethiopia, the disease affects *C. arabica*, but Pieters and Van der Graaff (1980) considered that it was not a problem in areas under traditional low-management systems, only reaching epidemic proportions where coffee was grown under intensive cultivation. In a survey conducted in 1996/1997 in Ethiopia, the mean percentage of wilted trees ranged from 45% at Gera to 69% at Bebeka (Girma *et al.*, 2001). Rutherford (2006) considers that this is now the most serious disease threatening coffee production in Africa, as it spreads into Tanzania and causes losses of 70% in Uganda.

Symptoms

Symptoms usually begin with generalized chlorosis of the leaves, which become flaccid and curled. Leaves dry up, turn brown, become very fragile and abscise. The crowns of diseased trees may be completely defoliated and branches develop a dark necrosis and dieback. Dieback often commences in one or more branches on one side of the coffee tree (Van der Graaff and Pieters, 1978), and gradually spreads downwards and progresses until the whole tree is leafless and dies (see Plate 26). The bark on the trunk is hypertrophied, with vertical or spiral cracks and cankers, especially in the collar region. Dark, violaceous ascomata can be often be seen in bark fissures of diseased trees. Under the bark the wood is stained with dark blue or brown streaks (see Plate 27) and the roots show a moist, black rot.

Internally, the main tracheids in diseased wood are heavily infected by mycelium. At the limit of spread the tracheids alone are affected but, where the infection is long established, the mycelium is found in the fibres surrounding the vessels and medullary rays. Wood parenchyma is rarely infected, but primary xylem and pith may be invaded. Tyloses develop and a yellow gum is observed. Mycelium is rarely present in the bark, reaching only the cortical medullary rays (Fraselle, 1950).

Young trees can be killed in a few days (sudden death syndrome), but mature trees can survive for up to 8 months after symptoms first appear. The period from infection to death of the tree is usually 2–3 months (Saccas, 1956).

A branch dieback of coffee caused by *G. xylarioides* was reported from parts of Swaziland and the Transvaal of South Africa (CSFRI, 1989). A disease of coffee stems, berries and seedlings in Zimbabwe was attributed to *G. xylarioides* (Clowes and Hill, 1981), but the symptoms are quite different from those of fusarium wilt observed in Central Africa and Ethiopia. The identity of the pathogen in Zimbabwe has been subsequently questioned (Ndimande, 1985), and the symptoms resemble those of coffee bark disease caused by the closely related *Fusarium stilboides* (see below) (Waller and Holderness, 1997).

Other *Fusarium* Diseases of Coffee

Fusarium solani (Mart.) Sacc. (*Nectria haematococca*) also causes a wilt of coffee accompanied by a dry root rot. The distinguishing feature of this disease is a purplish brown discoloration of the wood, seen in sections of the main roots and collar region. This disease lacks the blackish, darker discoloration seen under the bark in the collar region of trees affected by *G. xylarioides*, and stromata-bearing perithecia in bark fissures are also generally absent. *Fusarium stilboides* Wollenw. (*Gibberella stilboides*) causes coffee bark disease that also produces a discoloration beneath the bark and a progressive decline in vigour of the tree, but not the typical blackish discoloration of the lower vascular tissues associated with *G. xylarioides*.

Pathogen characteristics

Steyaert first identified the fungus causing coffee wilt in 1939, from samples sent from Bangui and described as *Fusarium xylarioides* Steyaert (1948) – a new species causing wilt of excelsa coffee, *Coffea liberica var. dewevrei*. Heim and Saccas (1950) concluded that the teleomorph of *F. xylarioides* should be referred to *Gibberella* and, as perithecia are readily formed by the fungus in nature and can be induced in culture, the causal organism of tracheomycosis in coffee is known by this name (Booth, 1971). Delassus (1954) later proposed that it be named *F. oxysporum* f. *xylarioides*. Earlier concepts of variation in *G. xylarioides* may have led to confusion with other *Fusarium* spp. on coffee. Earlier placement of the anamorph in the Lateritium section of the genus *Fusarium* now appears mistaken, as the fungus has been shown to be a member of the *Gibberella fujikuroi* (Sawada) Ito apud Ito & Kimura species complex (Geiser *et al.*, 2005).

Cultures of the anamorph on potato sucrose agar (pH 6.5) are pale beige, with sparse, white mycelium; a purple coloration later develops, accompanied by dark bluish-black, discrete stromata, some of which represent ascomatal initials. Microconidia are unicellular, allantoid, curved, 5–10 × 2.5–3.0 μm. Macroconidia are fusoid, falcate, bi- to tri-septate, 20–25 × 4–5 μm. Chlamydospores are oval to globose, smooth or roughened, 10–15 × 8–10 μm but may be sparsely produced and then mostly within macroconidia (see Fig. 9.1).

Perithecia are violaceous, embedded, single or in groups, in dark purple stromata; globose with a flattened base, 230–350 μm in diameter. Asci cylindrical, thin-walled, shortly pedicellate, 90–110 × 7.0–9.5 μm, with eight ascospores. Ascospores hyaline to straw-coloured, fusoid, uni- to tri-septate, finely roughened, 12.0–14.5 × 4.5–6.0 μm (Booth and Waterston, 1964a).

The fungus is heterothallic so that perithecia will form in culture only if compatible mating types are present, but they are abundant on diseased coffee. Previous reports of a morphologically distinct 'male' strain by Booth (1971) are now considered to be misplaced, as isolates of this strain have been shown to be quite different from *G. xylarioides* and related to *F. lateritium* (Rutherford, 2006). This has previously led to occasional confusion in distinguishing it from other species that occur on coffee as pathogens or saprobes (*F. stilboides* and *F. lateritium* Wollenw.).

There is some regional variation in the coffee species and cultivars reported to be affected by Fusarium wilt and the type of symptoms reported. Variation in

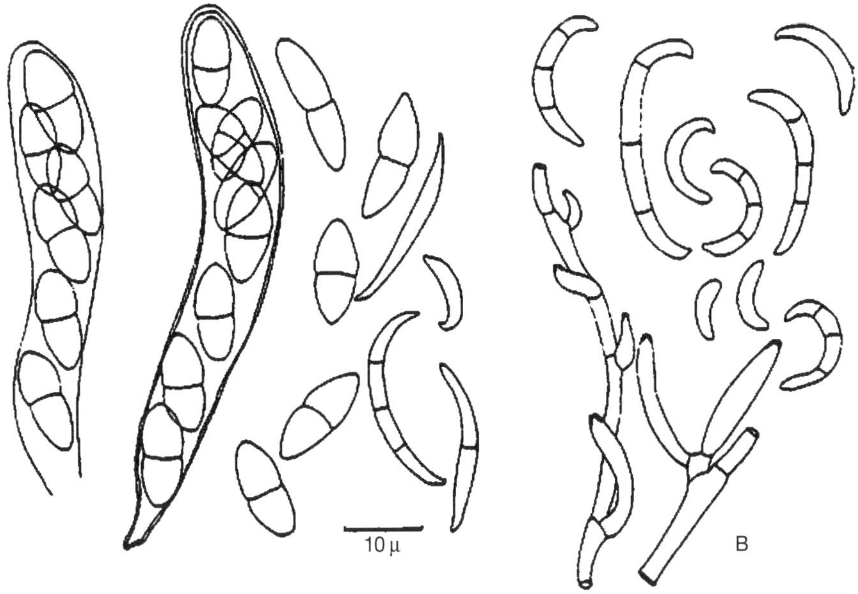

A, Asci, ascospores and conidia from coffee; B, conidia and conidiophores from culture

Fig. 9.1. *Gibberella xylarioides* (from Booth, 1971).

the pathogen is being studied as part of a project on biology and management of the disease, funded by the Common Fund for Commodities that began in 1997. The strain affecting *C. arabica* in Ethiopia is different from that in Uganda, where only *C. canephora* is affected. Isolates from the two regions do not cross-infect between the two host species and there are molecular differences as shown by RAPD-PCR markers (Adugna *et al.*, 2005). Molecular genetic variability within these two groups is very small, but old isolates from *C. excelsa* (*C. liberica var. dewevrie*) in central Africa are different from those recently obtained from *C. canephora* (Rutherford, 2006).

Nevertheless, the frequency of perithecial production in nature enables variability within populations of *G. xylarioides*, with the likelihood of pathogenic variation (Van der Graaff and Pieters, 1978; Muller, 1997), and this seems to be confirmed in recent studies. Adugna *et al.* (2005) propose that the fungus affecting *C. arabica* in Ethiopia be named *G. xylarioides* f.sp. *abyssiniae*, and that affecting *C. canephora* in central Africa be named *G. xylariodes* f.sp. *canephorae.*

Fusarium solani, the cause of dry root rot of coffee, is readily distinguishable from the *Fusarium* state of *G. xylarioides* primarily by its abundant microconidia borne on long, branched conidiophores. *Fusarium stilboides*, the cause of *Fusarium* bark disease, is also readily distinguishable from the *Fusarium* state of *G. xylarioides* by the carmine red colour produced in culture and by its longer, straighter macroconidia. Previous confusion with some *G. xylarioides* strains has occurred, but the pronounced curvature of *G. xylarioides* macroconidia should enable distinction (see Table 9.1).

Table 9.1. Morphological attributes of the main *Fusarium* species causing diseases of coffee.

	F. xylarioides *Gibberella xylarioides*	*F. stilboides* *Gibberella stilboides*	*F. oxysporum* *Gibberella?*	*F. solani* *Nectria haematococca*
Culture colour	Pale beige to pale violet or deep blue/purple to brown, blue/black stromata	Carmine red with white, becoming reddish brown	White to peach or salmon to grey to purple or violet	Beige/grey to blue to brown
Microconidia	Abundant, highly curved, elliptical to ovate	Not produced	Abundant, ovoid–ellipsoidal, cylindrical, from simple, short, lateral phialides	Abundant, cylindrical/oval from tips of long, sometimes branched conidiophores (phialides)
Macroconidia	Variable with mating type, sickle-shaped, curved and beaked, or almost straight with curved ends, from simple lateral branches; marked foot cell	Long, narrow cylindrical/ spindle shaped, from simple phialides or sporodochia	Spindle-shaped, often tri-septate, pointed at both ends	Inequilaterally fusoid, can vary in length with mating type, foot cell rounded or distinct
Chlamydospores	Rare or absent	Sparse, intercalary and on short branches	Globose, singly or in pairs, intercalary or on short lateral branches	Common, produced singly or in pairs on short branches or intercalary
Diagnostic characters	Differs from *F. udum* in pigmentation and perithecial state	Appearance in culture; spore form; coffee host	Short simple phialides (cf. *F. solani*) producing microconidia; presence of chlamydospores (cf. *F. moniliforme*)	Long phialides; branched, often elaborate micro-conidiophores (cf. *F. oxysporum*); macroconidial shape
Perithecia	Purple, frequent in culture and on host	Blue–black in culture	Not known	Orange/brown, sparse

Epidemiology

The life cycle of the fungus is not fully known, but it is considered to be an endemic saprophyte in tropical forest soils (Jacques-Felix, 1954). Although the fungus appears to be a typical soil-invading (*sensu* Garret, 1956) vascular wilt pathogen, Van der Graaff and Pieters (1978) considered that the fungus did not persist in soil because it only rarely produces resting structures (chlamydospores). Transmission can occur by conidia and ascospores that are spread by wind, rain and through human activities (harvesting, pruning, etc.) (Jacques-Felix, 1954).

Infection can occur through wounds in the aerial parts or superficial roots (Saccas, 1956; Muller, 1997), so that any agency causing wounds will aid the spread of the fungus. Kranz and Mogk (1973) noted that most dying and dead trees in Ethiopia had been wounded during weeding. Insects may also spread the disease from tree to tree (Wrigley, 1988).

The incubation period from first symptoms to death of the tree varies from days in young plants to 8 months in trees > 10 years old (Saccas, 1956), although most affected trees die 2–3 months after initial symptoms are observed.

Delassus (1954) reported that the disease spreads in contiguous plantations at a rate of 25 km/year, but that gaps between plants of a few hundred metres was sufficient to confine the disease to an affected plantation. This observation indicated that the disease spreads through dispersal of conidia and that ascospores must play a role in infection and spread of *Fusarium* wilt. In Ethiopia, the pathogen was isolated from all parts of diseased coffee trees except the seeds (Girma *et al.*, 2001), which did not transmit the pathogen. The relative importance of soil and airborne inoculum and the extent to which the fungus survives in the soil are also unknown, as are many other facets of the epidemiology of this disease.

Environmental factors may also influence infection, but the roles of soil type, climate and agronomic practice have yet to be evaluated (Waller and Holderness, 1997). Evidence from the earlier disease outbreaks suggests that older trees and those grown under poor management regimes are more susceptible to the disease. However, the high incidence of Fusarium wilt in new coffee plantations in Ethiopia has been attributed to intensive farming methods with frequent weeding, stumping and pruning, resulting in wounds that allow the entry of *F. xylarioides* (Girma, 1997). The reasons for the sudden upsurge and spread of the disease in the late 1990s remain unresolved.

Control

Cultural practices

Vigorous, well-managed coffee trees may be less prone to Fusarium wilt. Most of the coffee in Uganda is produced from robusta trees that are more than 40 years old, and this might explain the rapid spread of the disease. The distribution of new, improved clones and their careful planting and maintenance may therefore help the situation, until more resistant alternatives become available. Rotation may be of some benefit, although the persistence of the pathogen in the soil is unknown. Frequent inspection of the crop, along with burning of infected material and spraying the soil surface with 2.5% copper sulphate, has been advocated as an effective control measure (Saccas, 1956).

Replanting should not be undertaken until 6 months after the uprooting of infected trees to allow the viability of the soil inoculum to decline (Wrigley, 1988). Removal of bushes to reduce spread between plantations was effective in the Ivory Coast: gaps of a few hundred metres were enough to confine the disease (Delassus, 1954). Grafting of susceptible varieties to a more resistant Robusta variety was effective in French West Africa (Gaudy, 1956).

Resistant varieties

Several authors have reported varietal differences in resistance to the pathogen, and suggested the use of resistant varieties as a means of control (Fraselle, 1950; Delassus, 1954; Bouriquet, 1959; Porteres, 1959). Following the destruction of the coffee plantations by Fusarium wilt in Côte d'Ivoire during the period 1933–1950, the highly susceptible kouillou and indenie cultivars were replaced by cultivars of *C. canephora* (notably robusta), which were resistant and formed the basis of many of the West African breeding programmes (Kebe, 1997).

This was also the approach in the Democratic Republic of the Congo, replacing excelsa coffee with robusta, although even as far back as 1950 it was known that up to 10% of the robusta trees were infected (Muller, 1997). This may have led to the current epidemic in the Democratic Republic of the Congo and Uganda, where the present robusta varieties seem to be highly susceptible.

Nevertheless, some robusta trees have survived in the epidemic areas and material has been collected for evaluation. Other resistant robustas and arabica x robusta hybrids are being evaluated for Fusarium wilt resistance by the breeding programme at the Coffee Research Institute in Uganda, to provide a basis for a future control strategy. Muller (1997) listed some of the disease-resistant varieties and cultivars that were identified during the 1960s in Côte d'Ivoire and Guinea:

- *Coffea canephora*
  - Robusta ebobo
  - Robusta INEAC
  - Robusta lulla
  - Robusta Congo Belge
- *Coffea stenophylla*

Van der Graaff and Pieters (1978) reported that genotypes of *C. arabica* in Ethiopia also differed widely in resistance to *G. xylarioides*, and considered that these differences provided an excellent opportunity to control the disease with resistant varieties. They suggested that resistance in *C. arabica* was quantitative in nature and horizontal: no evidence of vertical resistance was seen.

Pieters and Van der Graaff (1980) reported two methods of screening for resistance: a seedling test that involved wounding seedlings with a knife dipped in spore suspension and a conidial germination test conducted directly on the bark of the tree. Both tests correlated significantly with field scores and with each other, and provided the basis for a screening programme to be used in a more extensive programme involving selection for resistance to coffee berry disease (*Colletotrichum kahawae*) and *G. xylarioides*. Screening for resistance to Fusarium wilt is a routine component of the coffee-breeding programme (Van der Graaff and Pieters, 1978).

The underlying resistance mechanisms remain unclear. Caffeine, which inhibits the pathogen, has been found in higher concentrations in tissues of *C. canephora* than in *C. liberica* (Rabechault, 1954), and a higher content of chlorogenic acid in the wood of resistant material (variety robusta) has also been reported (Bouriquet, 1959).

Phytosanitation

It is possible that considerable spread of the pathogen has occurred due to the local and long-distance haulage of wood from diseased trees. Since the earlier epidemics in Congo were contained, the general recommendation has been to uproot and burn affected trees and to avoid carrying wood from diseased trees out of the plantation. The relatively rapid spread throughout Uganda suggests that mechanisms of natural dispersal are efficient and that the enforcement of phytosanitory measures to prevent movement of wood are likely to have only limited effect.

Chemical and biological control

Jacques-Felix (1954) considered that treating wounds to the coffee tree with any disinfectant would limit the spread of the pathogen. Gaudy (1956) reported that spraying with copper oxychloride was effective in controlling the disease. Rabechault (1954) isolated four actinomycetes, one bacterium and *Corticium*, *Marasmius* and *Trichoderma* spp., all of which were inhibitory to *G. xylarioides*, but no methods of biological control are currently available.

Fusarium Bark Disease

Pathogen: *Gibberella stilboides* Gordon ex Booth, Ascomycete, (anamorph *Fusarium stilboides* Wollenweber. Syns *F. lateritium* var. *stilboides*; *F. lateritium* var. *longum*).

The disease is also known as Storey's bark disease, scaly bark, collar rot, fusariose du café (French), fusariosis del cafeto (Spanish). The pathogen is usually referred to as *Fusarium stilboides* by pathologists, as the teleomorph is not readily encountered on diseased material.

History, distribution and losses

Bark disease was first described by Storey (1932) in northern Tanzania, and it is widely distributed in Tanzanian coffee areas. The disease affects only *C. arabica* (Wrigley, 1988), although the fungus has a wide host range and it has been associated with a dieback of mandarin citrus in India. Wallace and Wallace (1955) studied the disease in Tanzania, followed by Siddiqi and Corbett (1963) in Malawi, where it has been most destructive. When much of the estate coffee in Malawi was being converted to multiple-stem pruning, the incidence of bark disease increased and it became epidemic between 1953 and 1955. Siddiqi and Corbett (1965) described bark disease as the main factor limiting coffee production in Malawi.

Bark disease is not widespread in Kenya, but can be a problem in hot, dry areas such as the Taita Hills and sporadically in other areas (Firman, 1964; Baker, 1970), and where there is damage by stem borers (Anon, 1987). The disease has been reported from India, Papua New Guinea and Dominica and is apparently widespread in many African countries: Ethiopia, Ghana, Liberia, Nigeria, Tanzania, Togo, Uganda, Zambia and Zimbabwe (EPPO, 2002), but the fungus has a much wider distribution.

Symptoms

The fungus attacks the bark meristematic tissues (phellogens), causing disruption in bark formation and necrosis of neighbouring tissues. One or more of the three symptoms in the syndrome of bark disease may be present on an affected coffee tree:

Storey's bark disease

A slightly sunken lesion with a water-soaked margin appears a few centimetres from the base of the green stem. The lesion is initially cinnamon to olive in colour, becoming brown to black and, under moist conditions, pink spore masses may appear (Storey, 1932). The lesion gradually extends and girdles the stem which, after a period that varies from a few days to a few months, leads to the wilting and death of the shoot, with the leaves usually remaining attached. The disease is particularly important on young suckers that grow from the trunk base and used to convert multiple-stem trees to a new cycle. The disease may be tolerated on older, mature stems and where the lesion does not completely encircle the stem, but these may remain constricted at the base and fail at a later date (see Plate 28).

Scaly bark

The bark rises up and becomes flaky, usually around a pruning scar. This form of the disease is more common on plants pruned on the single-stem system, where it can cause a dieback of new primaries. On young wood the symptoms are associated with cankerous areas around the base of branches and suckers, but affected mature branches survive and the disease may be difficult to detect, although a brown discoloration of the tissues beneath the bark is apparent if a little is scraped away. This is a more insidious form of the disease, may do little apparent damage but is more widespread than is sometimes realised (Baker, 1970). The infection may pass into the suckers to produce the Storey's bark symptom or progress into the collar rot symptom.

Collar rot

A cankerous lesion at the base of the stem similar to that described by Storey, but occurring on older plants causing unthrifty growth, chlorosis and sometimes wilting and dieback, and may eventually kill the tree. This is the least common form of the disease.

The fungus has been isolated from coffee berries in Malawi, apparently causing a rot of immature fruit (Siddiqi and Corbett, 1963) and in many other coffee-producing countries, where it occurs as a secondary invader of berries damaged by *Cercospora*, sun scorch, insects and overbearing. It also colonizes CBD lesions, where it is a strong competitor to *Colletotrichum kahawae* (see Chapter 7). The fungus is seed-borne and can cause a seedling blight.

Pathogen characteristics

Cultures (on potato sucrose agar) are carmine red with white, floccose, aerial mycelium that sometimes becomes a peach to brown colour. Older cultures of some isolates develop sporodochia that are 0.5–1.0 mm in diameter, becoming orange in colour with the development of a mass of macroconidia. Cultural mutants have little aerial mycelium and become slimy with the production of macroconidia, and may devleop a bluish black pigmentation. Aerial conidiophores have a loose, penicilliate form. Only macroconidia occur, and these are tri- to septi-septate, inequilaterally fusoid and pointed or beaked at the apex, measuring 20–82 μm in length according to the number of septa and 3–5 μm in width.

Perithecia can occur naturally and can be induced in culture with opposite mating strains on sterile straw. They are ovate to globose, 100–140 μm in height and 80–130 μm in diameter. Asci are 60–70 × 9–11 μm. Ascospores are elliptical, mainly one, but sometimes bi- or tri-septate and 12–18 × 4–5.5 μm. Chlamydospores are absent from the mycelium but sometimes form in the conidia (see Fig. 9.2). There is no information on pathogen variability, although with a sexual phase and the wide host range and distribution of the fungus, this is likely to occur (Booth and Waterston, 1964b; Booth 1971).

Epidemiology

Fusarium stilboides is seed-borne and a likely endophyte given its ubiquity in coffee tissues (Waller and Holderness, 1997; Waller and Masaba, 2006). It is

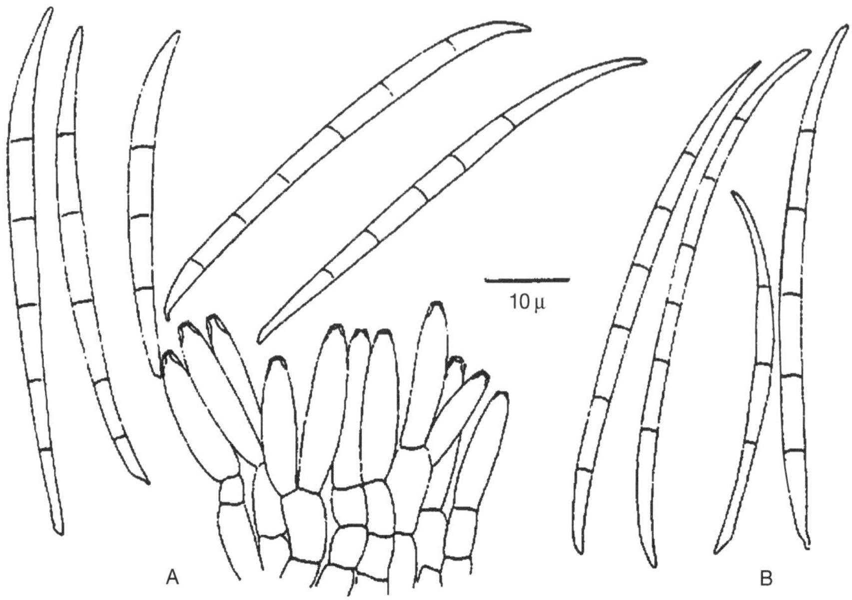

A, macroconidia and conidiophores from coffee; B, macroconidia from culture

Fig. 9.2. *Gibberella stilboides* (Booth and Wakerston 1964b).

not certain whether the fungus can survive in the soil, but it is associated with all aerial parts of the coffee plant and is normally non-parasitic. Sporulation occurs freely on the surface of damaged tissue under humid conditions Tiny, white sporodochia are often visible on old dead berries (cf. the pink acervuli of *Colletotrichum*). Conidia are dispersed in water by rain-splash, water films, farm implements and possibly by insects such as the yellow-headed borer (*Dirphya nigricornis* Olivier) (Anon, 1987) (p. 53). The pathogen is also seed-borne (Clowes *et al.*, 1989).

Plants are more susceptible if they are damaged, under stress, in marginal environments or poorly managed. The upsurge of the disease in Malawi was associated with conversion to multiple-stem coffee that involved stumping and regeneration through suckers in areas already conducive to the disease (Siddiqi and Corbett, 1963, 1968). Conditions favouring disease development are warm, wet weather, poor nutrient status and compaction of the soil, trees weakened by insect attack, wind-rock, mechanical or herbicide damage or physiological stress such as that caused by drought, over-bearing, root competition from weeds or intercrops (Anon, 1987).

Mulching close to the stem, excessive weed growth and intercropping can create humid conditions close to the stem that can favour the disease. Soils with a pH < 5.8 are also considered to be conducive to disease development (Baker, 1970). Increased drought stress that may occur in some areas as a result of climate change would increase the susceptibility of coffee to this disease.

Control

Cultural control and phytosanitation

The main objective in managing Fusarium bark disease is to avoid those conditions known to favour the disease, and much can be done by adopting appropriate cultural practices. Adequate fertilization, careful weeding to avoid damage, proper attention to mulching and pruning and control of stem borers are all important aspects of general husbandry. Since the epidemic of Fusarium bark disease in Malawi during the 1950s, the disease has been managed on coffee estates by good husbandry and sanitation. The production of disease-free seed (initially of Caturra and SL28) was the key to control, followed by careful phytosanitation to avoid introducing the disease into the nursery – and subsequently into the plantation. Benomyl has been recommended in Malawi and Zimbabwe for seed treatment to ensure that seed is free of *F. stilboides*, but safer chemicals such as copper formulations should now be used.

Regular scouting is required, both in the nursery and in the plantation, for early detection of infected plants that might act as a source of infection for the disease to spread to other plants, especially during pruning. Any infected plants should be uprooted and burned, with care being taken not to leave infected debris in the field. In plantations with a history of Fusarium bark disease, pruning should be kept to a minimum, and the trees replanted rather than stumped. Pruning should be undertaken in dry weather and the equipment disinfected between trees using a proprietary disinfectant or bleach solution. Application of protectant fungicides such as copper formulations to protect young suckers from infection by Storey's bark symptom and the treatment of wounds where diseased suckers or stems have been removed have been advocated.

Resistant varieties

A major selection programme was undertaken in Malawi by Siddiqi and Corbett during the 1960s to search for cultivars resistant to Fusarium bark disease (Siddiqi and Corbett, 1965; Siddiqi, 1980). It was discovered that the cultivars Geisha and Agaro had some resistance to Fusarium bark disease, although the seed can still be infected (Clowes *et al.*, 1989). Because of the difficulty of ensuring careful phytosanitation under smallholder conditions, it was decided to recommend the resistant cultivars for smallholders in the north of the country, while the estate sector opted for the higher-yielding cultivars that were susceptible, but protected by cultural controls and phytosanitation as described above. No information is currently available on the susceptibility of Catimor cvs to Fusarium bark disease, but if these prove to be susceptible, the disease might become important again in northern Malawi.

Ceratocystis Wilt

Pathogen: *Ceratocystis fimbriata* Ellis & Holst. Ascomycete (Syns: *Ceratostomella fimbriata* (Ellis & Halst.) J.A. Elliott; *Endoconidiophora fimbriata* (Ellis & Halst.) R.W. Davidson; *Ophiostoma fimbriatum* (Ellis & Halst.) Nannf.).

Ceratocystis wilt is also known as *Ceratocystis* canker, South American wilt, mal de machette, Llaga macana (Spanish), chancre du caféier (French).

Distribution and host range

Ceratocystis wilt is generally a minor disease of coffee, although the fungus is of worldwide distribution (CMI distribution map No. 91) and causes important diseases of some root and tree crops; in the tropics it is more important on cacao. *Ceratocystis fimbriata,* the type species of the genus, was originally described on sweet potato (*Ipomoea batatas*). A fungus attacking *Coffea* in Indonesia was described by Zimmerman in 1900 as *Rostrella coffea*, and this species was later synonymized with *C. fimbriata* (Pontis, 1951), although no careful comparisons have been made.

The literature on the disease on coffee is limited, as its significance on coffee is mainly limited to parts of South and Central America (Pontis, 1951). In Colombia it is regarded as one of the most important coffee diseases (Marin *et al.*, 2003), and it was important in Guatemala and Surinam (Van Emden and Van Suchtelen, 1959; Snyder *et al.*, 1960). Much of the information on the disease and the pathogen is derived from reports on other crops. CABI (2005) list some 30 genera as hosts, which includes many tropical tree crops such as *Theobroma*, *Mangifera*, *Cocos*, *Anacardium* and *Acacia.*

Symptoms

Ceratocystis wilt is primarily a vascular disease that infects through wounds. Symptoms first appear as a chlorosis and wilting of the leaves on affected branches. Wilted leaves typically become dry and curled rather suddenly, but remain attached to the tree for several weeks. On the surface of the trunk or branches, sunken cankers may develop over areas of irregularly shaped necrotic lesions. The cankers are clearly defined and spread and extend vertically, often involving the roots where cankers are near the stem base, as is often the case. Cankers are more often seen on older trees and where the bark is at least 3 years old.

The fungus causes dark reddish brown to purple to deep brown or black staining in the xylem that may extend several metres from the roots, up the trunk of the tree and into branches. Mycelium and spores enter wounds and move through the xylem in water-conducting cells and into ray parenchyma cells, which can be seen as a staining pattern in cross-sections of diseased wood.

Pathogen characteristics

Morphology

The fungus grows readily on most agar media. Mycelium is hyaline at first, later turning dark greenish brown. Within a few days there are usually abundant conidiophores that produce chains of hyaline conidia (endoconidia), characteristic of the anamorph genus *Chalara*. Endoconidia are cylindrical and may vary in size from 11–16 μm long by 4–5 μm wide. Specialized conidiophores give rise to thick-walled, pigmented chlamydospores (aleurioconidia), probably a survival spore; these are characteristic of the anamorph genus *Theilaviopsis*. Chlamydospores are typically 9–16 μm long and 6–13 μm wide, borne singly or in short chains. Endoconidia may also darken and become thick-walled, thus resembling chlamydospores. All of these spore types may be produced on and within the substratum.

The teleomorph of the fungus is readily produced on the surface of the host or in culture if thiamine is present. Most field isolates are self-fertile. Ascomata are dark brown to black, consisting of globose perithecia 130–200 μm in diameter with a long, thin neck up to 800 μm in length, through which the ascospores are exuded. The opening at the tip of the neck has eight to 15 ostiolar hyphae ranging in length from 50 to 90 μm.

Ascospores are small, hyaline and hat-shaped, 4.5–8.0 μm long by 2.5–5.5 μm wide and accumulate in a sticky matrix at the tip of the ascomatal neck, where they appear as a cream to pink ball or coil. Cultures produce a fruity odour. *Ceratocystis fimbriata* usually grows best at temperatures from 18 to 28°C and is able to produce ascospores within a week (see Fig. 9.3) (Hunt, 1956; Morgan-Jones, 1967).

Pathogenic variation

Pathogenic specialization occurs on the wide variety of annual and perennial plants attacked by *C. fimbriata*. There are several apparently host-specialized strains that are sometimes called 'types', 'races' or 'forms' (Wellman, 1972; Harrington, 2000), and many of these may prove to be distinct species. Webster and Butler (1967) considered such types as members of a single, highly variable species. There have been detailed studies of the genetics, cytology and morphology of the fungus, particularly of North American strains (e.g. Baker *et al.*, 2003). Most isolates are homothallic, but out-crossing does occur.

Cross-inoculation studies have established the host specificity of some of these types. For example, isolates from *Mangifera* (Ribeiro and Coral, 1968), *Ipomoea*, *Platanus* (Baker and Harrington, 2000), *Crotalaria*, *Cajanus* and *Acacia* (Coral *et al.*, 1984) did not infect *Theobroma*. Isolates from *Ipomoea*, *Hevea* (Olson and Martin, 1949) and *Coffea* (Pontis, 1951) were host-specific when inoculated into the other hosts. Isolates from *Platanus*, *Prunus* and *Theobroma* were not pathogenic to *Ipomoea*, and isolates from *Ipomoea*, *Prunus* and *Theobroma* were not pathogenic to *Platanus* (Baker and

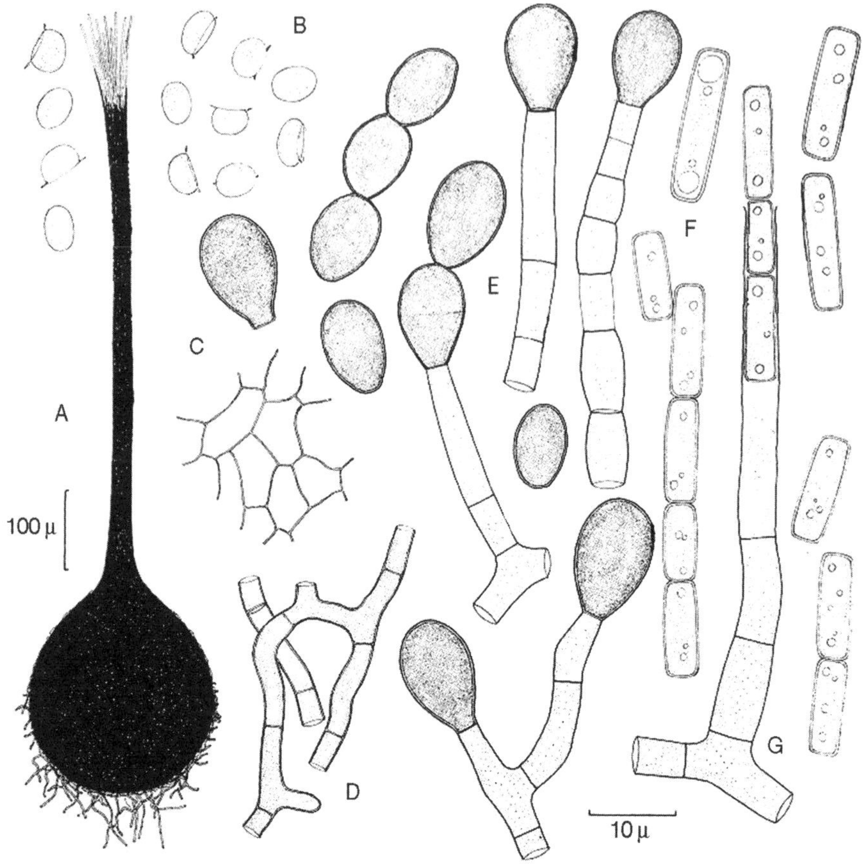

A, Perithecium; B, ascospores; C, surface of perithecium wall; D, mycelium; E, chlamydospores; F, conidia; G, conidiophore

Fig. 9.3. *Ceratocystis fimbriata* (Morgan-Jones, 1967).

Harrington, 2000). Kojima and Uritani (1976) also reported that isolates from *Coffea* failed to infect *Ipomoea*. Each host-specific type of *C. fimbriata* appears to have a distinct geographic distribution, although the total number of types and the geographic and host boundaries of each of them have not been fully determined.

Isolates of *C. fimbriata* from coffee in different parts of Colombia were found to be highly variable. There was wide variation in pathogenicity, with some killing 90% of inoculated plants and others < 5% (Marin *et al.*, 2003).

Epidemiology

Most knowledge of the epidemiology of *C. fimbriata* comes from studies on crops other than coffee. Infection typically occurs through fresh wounds that may be natural or man-made. Cultivation practices such as pruning, and wounds arising from soil cultivation close to tree bases provide infection courts, and the fungus can be carried on machetes or pruning tools (Teviotdale and Harper, 1991). Mango trees may be infected through the roots from soil-borne inoculum (Rossetto and Ribeiro, 1990), and the fungus may also be transmitted through root grafts, as shown in *Platanus* (Accordi, 1986). The disease can be exacerbated by abiotic stresses such as drought, waterlogging, etc., and Snyder *et al.* (1960) found that it was most severe in the humid upland areas of Guatemala.

The fungus may be dispersed as fragments of mycelium, conidia, chlamydospores or ascospores. Inoculum may reach an open wound by being blown in the wind, by rain-splash, by being carried by insects that visit the wound or by carriage on implements. After infection, there may be extensive mycelial growth within a plant before symptoms appear, and growth in the tissues produces the characteristic dark colour. Sporulation of the anamorph state occurs within the host tissue. Perithecia are produced on the outside of the host.

Ceratocystis fimbriata produces a strong, fruity odour that has been assumed to be an adaptation for dispersal by insects. These are attracted to diseased plants and can become covered with sticky spores if the fungus is sporulating. Ambrosia beetles (especially *Xyleborus* and *Hypocryphalus* species) are attracted to diseased plants (such as *Theobroma*, *Mangifera* and *Eucalyptus*) and produce large amounts of wood dust (frass) when creating breeding galleries in the trunk and branches. This is pushed outside the tree as the galleries are excavated, and contains spores and fragments of mycelium that may be blown in the wind (Iton, 1960) or in rain-splash. Nitidulid beetles that feed on fungi and plant sap may be important vectors (Moller and DeVay, 1968).

In India, an increase in the disease on coffee was associated with attempts to detect the presence of white stem borer larvae (*Xylotrechus quadripes* (Chevrolat) (p. 48) that involved scraping away the outer layers of the bark. Bhat *et al.* (2002) have shown that infection and subsequent spread is greatest on fresh wounds and in the wet season.

Chlamydospores are probably the most common survival units outside the woody host, because they are thick-walled and durable and they probably facilitate survival in soil (Accordi, 1989), wood particles and insect frass (Iton, 1960).

Control

Careful attention to cultivation practices is particularly important to avoid wounds on the trunk base that might easily be infected with *Ceratocystis*. Removal and destruction of diseased trees should also be undertaken in order to remove sources of inoculum. Disinfecting machetes and pruning tools between plants may help control the spread of the disease.

Protection of wounds with fungicides can also help to prevent infection, and this method has been used with some success to treat tapping panels of *Hevea* (Chee, 1970).

There are differences in resistance to the disease among *Coffea* species, with *C. canephora* and *C. liberica* being more resistant than *C. arabica*. Resistance has been attributed to the higher chlorogenic acid levels (Echandi and Fernandez, 1962), and resistance in *C. arabica* to polyphenol levels (Zuluaga *et al*., 1971). Resistant selections have been made in Colombia (Castilla, 1982), where outputs from the breeding programme are tested for resistance (Izquierdo, 1988). Host–plant resistance has also been used successfully with mango (Ribeiro *et al*., 1995) and cocoa (Simmonds, 1994).

Hymenomycete Root Rots

Pathogens: *Armillaria* spp. (syns: *Armillariella*, *Clitocybe*) (*A. mellea* (Fr.) Karst; *A. fuscipes* Petch); *A. heimii* Pegler (syns: *Armillariella elegans* Heim., Bres; *Armillaria tabescens* (Scop. Ex Fr.) Singer).

Hymenomycete or agaric root rots, also known as pourridiés agarics (French) and podredumbre agaricos (Spanish), are sporadic but serious and widespread causes of mature coffee tree death. They are widely distributed throughout the world, but are more common in cooler parts of the tropics such as montane areas, and hence tend to be more prevalent on *C. arabica*. These fungi have a very wide host range and are particularly important as the cause of root rots of woody plants; their taxonomy has undergone several revisions (CAB International, 2005).

They are recognized by the characteristic pale yellow/brown toadstools, usually produced in clusters around the base of many tree species (see Plate 29) (Pegler and Gibson, 1972). *Armillaria heimii* (= *Clitocybe elegans*) has caused serious damage in shade-grown coffee in Madagascar and Cameroon (Blaha, 1978; Coste, 1992) and is the commonest African species (Pegler, 1977; Mohammed and Guillaumin, 1993), although *A. mellea* is found in plantations in Kenya and Uganda.

Symptoms

On coffee, early symptoms are a general decline, with chlorosis and leaf-fall eventually leading to death of the tree. The disease can be recognized by the longitudinal splitting of affected root, collar and sometimes stems. Fan-shaped sheets of creamy white mycelia can be observed under the bark, and flattened brown, cord-like rhizomorphs may also be present. Sporophores are produced during the wet season at the base of trees in an advanced stage of the disease. Black rhizomorphs can be found on the roots and in the surrounding soil of disease trees, and it is through these structures that the disease spreads between trees.

Epidemiology

The disease usually occurs where land has been previously cleared for planting or shade trees removed. Old stumps or larger roots left in the soil act as food bases for the fungus, from which it can develop rhizomorphs that spread through the soil to infect coffee; spread between adjacent coffee trees can subsequently occur.

Rayner (1959) considered that *Armillaria* was significant only above 1650 m in Kenya, where it could spread from old shade tree stumps, but it did not produce sporophores on coffee. When forest or shade trees are to be removed to plant coffee, they should first be ring-barked 1 year in advance of felling to exhaust the tree roots of nutrient, so that any fungus that is present is starved. Stumps and large roots should be thoroughly uprooted and burned. When a coffee tree is identified with this disease, a trench, 30 cm wide × 60 cm deep, should be dug around the base to isolate it. The tree is then uprooted and burned, the soil treated with lime and the land fallowed for 6 months or planted with a resistant crop – such as a legume – before replanting.

The presence of the fungus in the soil can often be detected using the 'trap stick' method. This involves planting short stakes (about 40 × 3 × 3 cm) of a woody plant particularly susceptible to *Armillaria* – such as *Albizzia* or *Leuceana* – in the soil that will soon become infected and show characteristic signs of the fungus if it is present in the soil (Blaha, 1978).

Control

There seems to be little that can be done once these diseases appear, except uproot and burn the tree with as much of the root system as possible. Care should be taken to avoid infection in the first place, by good field sanitation and early isolation and destruction of infected plants. Chemical eradication from the soil using fumigants is no longer recommended, but fungicides applied to roots or trunks may be effective (Adaskaveg *et al.*, 1999). *Trichoderma harzianum* is a strong antagonist to *Armillaria* (Dubos *et al.*, 1978) but attempts to use this in the field on tea have met only with limited effects (Otieno, 1998).

Other hymenomycete fungi

Similar hymenomycete fungi can also occur on coffee, causing similar diseases:

- *Phellinus noxius* (Corner) G. Cunn. (syn: *Fomes noxius*) causes brown root rot, characterized by a yellow–brown encrustation of mycelium and soil particles covering affected roots. It is more important on tree crops in the humid tropics such as cocoa, rubber and palms. It produces grey–brown, bracket-shaped sporophores on dead tree trunks. Occurs in Africa and Asia.

- *Rigidoporus microporus* (Fr.) Overeem (syn: *R. lignosus*; *Fomes lignosus*) causes a white root rot, characterized by whitish rhizomorphs and mycelium covering the roots. It is particularly important on rubber and occurs more in warm, humid areas throughout the tropics. Sporophores are thin, brownish brackets that occur in tiers on the stems of diseased trees.
- *Poria* spp. cause red root disease of tea, and have been reported from coffee in India.
- *Ganoderma* spp. are most significant on palms and on tea, but occasionally affect coffee.

Other Diseases of Stems and Roots

Black root rot

Black root rot (llaga negra, Spanish) is caused by *Rosellinia* spp. *Rosellinia bunodes* (Berk. & Br.) Sacc. is the most widespread on coffee, whereas *R. pepo* Pat. is largely confined to Central America and the Caribbean. *Rosellinia arcuata* Petch is less frequent but more important on tea in Asia. All species are pathogens for a wide range of woody perennials in tropical or subtropical humid forest ecologies. Tea, citrus, cassava, nutmeg, cocoa and potato are all primary hosts for *Rosellinia*.

The external symptoms are foliar wilting and eventual death of affected stems. The roots are discoloured internally and covered with black, branching strands. A sheet of white mycelium forms around the collar, which gradually turns purplish black and extends above the soil line in damp conditions. Perithecia form at the base of the stem and on dead woody debris in the soil, usually occuring in clusters. Ascospores are dark, aseptate, filiform, 80–120 $\times$ 5–9 μm.

The anamorph is *Dermatophora* state, consisting of a synnematum some 2 mm tall with a branching head bearing conidia 4–10 $\times$ 2–5 μm (Sivenasan and Holliday, 1972) (see Fig. 9.4). The fungus survives on forest floor litter, and infected plant debris is the source of inoculum (Fernandez-Borrero and Lopez, 1964). Adequate field sanitation to remove such material is usually sufficient to prevent the disease from causing damage.

In Guatemala, treatment of early infection of young trees with 0.25% copper fungicide is recommended. Clearance from planting sites of tree debris and soil treatment with sulphur have been recommended in Colombia (Castaño, 1953). Older trees should be removed and burned and the hole treated with fungicide. *Xylaria thwaitesii* also causes black root rot, a similar disease, also associated with rubber plantations. *R. necatrix* Prill. causes a white root rot, which has also been reported from coffee.

Santavery root disease

This is a vascular wilt disease caused by *Fusarium oxysporum* f. sp. *coffeae* that has been reported from India and Central America (Wellman, 1954). In

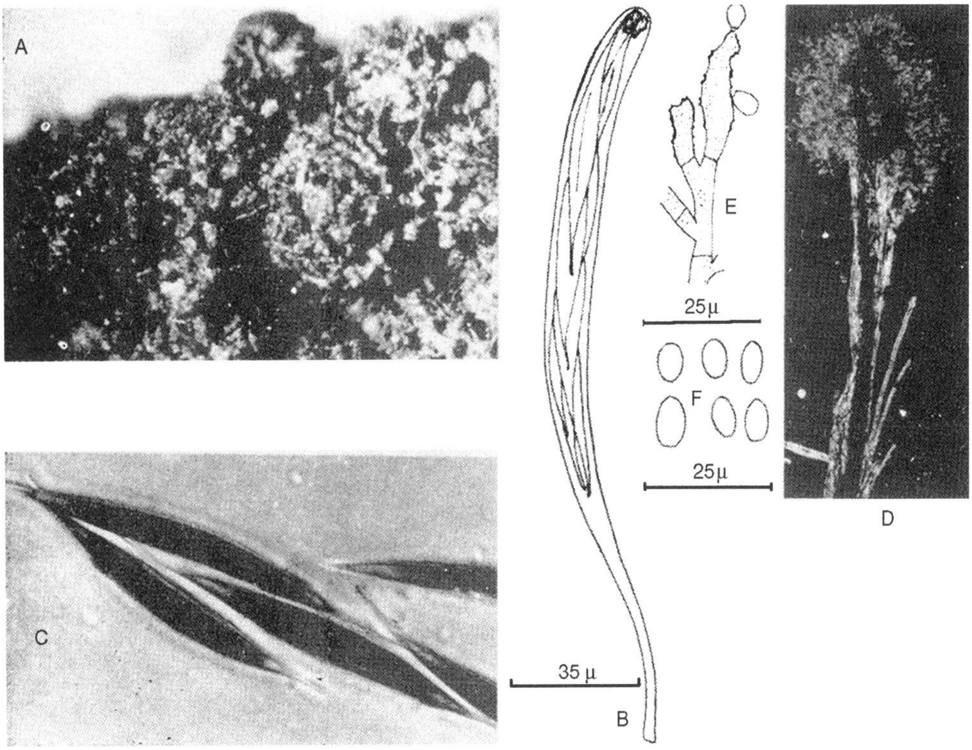

A, Perithecia x 13; B, ascus; C, ascospores x 400; D, synnemata x 100;
E, conidiogenous branch; F, conidia.

Fig. 9.4. *Rosellinia bunodes* (Sivanesan and Holliday, 1972).

the initial stages of the disease the leaves wilt followed by defoliation and die-back. A pinkish vascular discoloration is apparent. Attempts to halt the disease can be made as soon as leaf-wilting becomes apparent, by drenching the soil around the base of the plant with 3 l of a solution containing 8 g/l of carbendazim, or carboxin at 4 g/l. The application of organic manure can help prevent infection in plantations prone to santavery disease (Govindarajan, 1988; CCRI, 1999).

However, *F. oxysporum* is a common root- and soil-inhabiting fungus and frequently invades roots that have been damaged by nematodes or other causes, and can exacerbate the damage (see Chapter 10). Some strains are antagonistic to nematodes.

Dry root rot

Fusarium solani (Mart.) Sacc. (Teleomorph: *Nectria haematococca* Berk. and Br. Ascomycete) is responsible for a root rot of arabica coffee in East Africa. The characteristic symptom is a purpleish brown discoloration of the wood in the crown area. The coffee tree is predisposed to infection by drought and, although not common, the attack is often lethal (Baker, 1972). *Fusarium solani* is a common soil- and root-inhabiting fungus and is frequently isolated from damaged roots, often together with *F. oxysporum* (see above).

Phloem necrosis

Caused by *Phytomonas leptovasorum* Stahel, a plant-parasitic protozoan that is apparently limited to parts of northern South America and the Caribbean (Van Emden and Van Suchtelen, 1959). The disease is often associated with high water tables that may predispose plants to infection. Affected plants wilt and eventually die, showing a characteristic dark necrosis of the phloem tissues. The disease is most prevalent on *C. liberica* (Vermeulen, 1968). Plant-inhabiting *Phytomonas* spp. have a wide host range and can live as harmless endophytes in the vascular tissues of some plants. Plant-inhabiting phytomonads have been reviewed by Dollet (1984).

Root and collar rot complex

A range of other organisms has been associated with collar rot of coffee plants at various times in various countries. In Java, collar rot is attributed to *Helicobasidium compactum* (Boedijn) Boedijn, and occurs on coffee grown next to tea. *Helicobasidium compactum* has also been reported from Guatemala and El Salvador (Schieber and Zentmeyer, 1967) where it causes a collar canker, the characteristic symptom of which is the production of a brownish mycelial cushion at the base of the stem. *Phomopsis coffea* Bond.-Mont. is a minor pathogen associated with stem canker on coffee, and may act in unison with *Gibberella stilboides* (Firman, 1964). Collar rot has been associated with *Rhizoctonia solani* Kühn., *Rosselinia bunodes* and *Botryodiplodia theobromae* Pat. in India (CCRI, 1999). *Phytophthora* spp. have been associated with root disease in Brazil (Zentmeyer, 1976).

Seedling diseases

Damping-off of seedlings in the nursery is a common problem. The usual symptom is one of cotyledons failing to emerge and the seedling collapsing due to a root or collar rot, but seedlings may fail to emerge from the soil or be attacked at a later stage, when the cotyledons become blighted. The disease is usually caused by *Rhizoctonia solani*, and seedlings are predisposed to infection by excessive soil moisture. *Fusarium stilboides* can also cause a seedling blight and is seed-borne.

Other fungi previously associated with seedling death include *Myrothecium roridum* (see Chapter 7), *Corticium rolfsii* Curzi, *Aspergillus niger* Tiegh. and *Pythium* spp. Improved drainage of seed beds and reduced irrigation frequency can prevent most problems. Seedlings can be protected by treating seedbeds with fungicide, dipping seedlings in fungicide before potting or by applying an approved fungicidal seed dressing. Copper fungicides or mancozeb are effective (see also Chapter 12).

References

Accordi, S.M. (1986) Diffusione di *Ceratocystis fimbriata* f. *platani* attraverso le anastomosi radcali. *Informatore Fitopatologico* 36, 53–58.

Accordi, S.M. (1989) Sopravvivenza nel terreno di *Ceratocystis fimbriata* f.sp. *platani*. *Informatore Fitopatologico* 39, 57–62.

Adaskaveg, J.E., Förster, H., Wade, L., Thompson, D.F. and Connell, J.H. (1999) Efficacy of sodium tetrathiocarbonate and propiconazole in managing *Armillaria* root rot of almond on peach rootstock. *Plant Disease* 83, 240–246.

Adugna, G., Hindorf, H., Steiner, U., Nirenberg, H.I., Dehne, H.W. and Schellander, K. (2005) Genetic diversity in the coffee wilt pathogen (*Gibberella xylarioides*) populations: differentiation by host specialization and RAPD analysis. *Zeitschrift fur Pflanzenkrankheiten und Pflanzenschutz* 112, 134–145.

Anon (1987) Recognition and control of fusarium bark disease. Technical circular No. 20. *Kenya Coffee* 52, 169–171.

Baker, C.J. (1970) Coffee bark disease in Kenya. *Kenya Coffee* 35, 226–230.

Baker, C.J. (1972) *Fusarium solani* associated with a wilt of *Coffea arabica* in Kenya. *East African Agricultural and Forestry Journal* 337, 137–140.

Baker, C.J. and Harrington, T.C. (2000) Host-specialized *Ceratocystis fimbriata* on cacao, sweet potato and sycamore (abstr.). *Mycological Society of America Newsletter* 51, 15.

Baker, C.J., Harrington, T.C., Krauss, U. and Alfenas, A.C. (2003) Genetic variability and host specialisation in the Latin American clade of *Ceratocystis fimbriata*. *Phytopathology* 93, 1274–1284.

Bhat, S.S., Rajamani, R. and Nandagopal, N. (2002) Manifestation of *Ceratocystis fimbriata* on coffee in relation to age of the bark wound and season. *Journal of Coffee Research* 30, 8–13.

Birikunzira, J.B. and Lukwago, G. (1997) The status of coffee wilt disease in Uganda. In: *Proceedings of the First Regional workshop on the Coffee Wilt Disease (Tracheomycosis)*, Kampala, Uganda, July 1997. National Agricultural Research Organisation (NARO) and the European Union, pp. 36–42.

Blaha, G. (1978) Un grave pourridié du *C. arabica* au Cameroon: *Clytocybe elegans* Hiem. *Café-Cacao-Thé* 22, 203–216.

Booth, C. (1971) *The Genus Fusarium*. Commonwealth Agricultural Bureaux, Farnham, UK.

Booth, C. and Waterston, J.M. (1964a) *Gibberella xylarioides*. *CMI Descriptions of Plant Pathogenic Fungi and Bacteria No. 24*. CAB International, Wallingford, UK.

Booth, C. and Waterston, J.M. (1964b) *Gibberella stilboides*. *CMI Descriptions of Plant Pathogenic Fungi and Bacteria No. 30*. CAB International, Wallingford, UK.

Bouriquet, G. (1959) Plant diseases and pests in some African territories. *FAO Plant Protection Bulletin* 7, 61–63.

CABI (2005) *Crop Protection Compendium*. CD ROM. CAB International, Wallingford, UK.

Castilla, Z.J. (1982) Production de una selección resístanse a llaga macana *Ceratocystis fimbriata* (Ell. and Halst.) Hunt con relacion a las variedades Tipica y Bourbon. *Cenicafe* 33, 53–66.

Castaño, J.J. (1953) Algunas observaciones sobre la llaga negra radicular del cafeto. *Boletin Informatif Colombia* 4, 28–34.

CCRI (1999) Diseases of coffee in India and their management practices. *Indian Coffee* 63, 14–16.

Chee, K.H. (1970) Evaluation of fungicides for control of mouldy rot (*Ceratocystis fimbriata*) of *Hevea brasiliensis*. *Plant Disease Reporter* 54, 897–899.

Clowes, M.J. and Hill, R.K. (1981) *Coffee Handbook* (2nd edn). Zimbabwe Coffee Growers Association, Harare, Zimbabwe, 198 pp.

Clowes, M.J., Nicoll, W.D. and Shelley, R.S. (eds) (1989) *Coffee Handbook for Malawi*. Tea Research Foundation of Central Africa, Mulanje, Malawi, 224 pp.

Coral, F.J., Bovi, O.A. and Ribeiro, I.J.A. (1984) Action of *Ceratocystis fimbriata* Ell. & Halst., isolated from various hosts, on cocoa seedlings. *Bragantia* 43, 245–248.

Coste, R. (1992) *Coffee, the Plant and the Product*. Macmillan, London, 327 pp.

CSFRI (1989) A new disease of coffee in Southern Africa? *Citrus and Sub-Tropical Fruit Research Institute Information Bulletin* No. 205, 13.

Delassus, E. (1954) La tracheomycosis de Caféier. *Bulletin Scientifique. Ministère des Colonies. Section Agronomie Tropical* 5, 345–348.

Dollet, M. (1984) Plant diseases caused by flagellate Protozoa (*Phytomonas* spp). *Annual Review of Plant Pathology* 22, 115–132.

Dubos, B., Guillaumin, J.J. and Schubert, M. (1978) Action du *Trichoderma viride* Pers., apporté avec divers substrata organiques, sur l'initiation et la croissance de rhizomorphs d'*Armillaria mellea* (Vahl.) Karst. dans deux types de sols. *Annales de Phytopathologie* 10, 187–196.

Echandi, E. and Fernandez, C.E. (1962) Relation between chlorogenic acid content and resistance to coffee canker incited by *Ceratocystic fimbriata*. *Phytopathology* 52, 544–546.

EPPO (2002) PQR database (version 4). European & Mediterranean Plant Protection Organisation, Paris.

Fernandez-Borrero, O. and Lopez, S. (1964) Llagas radiculares negra (*Rosellinia bunodes*) y estrella (*Rosellinia pepo*) del cafeto. I, Patogenicidad y influencia de la clase de inoculo en la infeccion. *Cenicafe* 15, 126–144.

Firman, I.D. (1964) Fusarium disease of coffee in Kenya. *Kenya Coffee* 29, 353–355.

Flood, J. (1996) *A Study of the Tracheomycosis or Vascular Wilt Disease of Coffee in Zaire*. Report presented to the Zairean Coffee Organisation (OZACAF), August 1996, 13 pp.

Flood, J. and Brayford, D. (1997) Re-emergence of Fusarium wilt of coffee in East and Central Africa. *Proceedings of the 17th International Conference on Coffee Science*, July, 1997. Association Scientifique Internationale du Café (ASIC), Paris.

Fraselle, J. (1950) Observations preliminaires sur une tracheomycosis de *Coffea robusta*. *Bulletin Agricole du Congo Belge* 41, 361–372.

Garret, S.D. (1956) *Biology of Root-infecting Fungi*. Cambridge University Press, Cambridge, UK.

Gaudy, M.R. (1956) Contribution du techniques, scientifique ou développement de l'agriculture en Afrique Occidentale Francaise. *Journal of the West African Science Association* 2, 172–197.

Geiser, D.M., Lewis-Ivey, M.L., Hakiza, G., Juba, J.H. and Miller, S.A. (2005) *Gibberella xylarioides* (anamorph: *Fusarium xylarioides*), a causative agent of coffee wilt disease in Africa, is a previously unrecognised member of the *G. fujikuroi* species complex. *Mycologia* 97, 191–201.

Girma, A. (1997) Status and economic importance of fusarium wilt disease of arabica coffee in Ethiopia. In: *Proceedings of the First Regional workshop on the Coffee wilt Disease (Tracheomycosis)*, Kampala, Uganda, July 1997. National Agricultural Research Organisation (NARO) and the European Union, pp. 46–54.

Girma, A., Halluka, M. and Hindorf, H. (2001) Incidence of tracheomycosis, *Giberella xylarioides* on arabica coffee in Ethiopia. *Zeitschrift fur Pflanzenkrankenheiten und Pflanzenschutz* 108, 136–142.

Govindarajan, T.S. (1988) A review on the incidence of root diseases on coffee and their management. *Journal of Coffee Research* 18, 16–28.

Guillemat, J. (1946) Quelques observations sur la traceomycose du *Coffea excelsa*. *Revue de Botanique Appliqueé et d'Agriculture Tropicale* 26, 542–550.

Harrington, T.C. (2000) Host specialization and speciation in the American wilt pathogen *Ceratocystis fimbriata*. *Fitopatologia Brasileira* 25, 262–263.
Heim, R. and Saccas, A. (1950) La tracheomycose des *Coffea excelsa* et Robusta des plantations de L'Oubangui-Chari. *Revue de Mycologia. Supplément Colonial* 15, 97.
Hunt, J. (1956) Taxonomy of the Genus *Ceratocystis*. *Lloydia* 19, 1–59.
Iton, E.F. (1960) Studies on a wilt disease of cacao at River Estate. II. Some aspects of wind transmission. In: *Annual Report on Cacao Research*, 1959–1960. Imperial College of Tropical Agriculture, University of the West Indies, St Augustine, Trinidad, pp. 47–58.
Izquierdo, J.E. (1988) Comportamiento de genotipos de cafetos ante *Ceratocystis fimbriata*. *Ciencia y Tecnica en la Agricultura, Café y Cacao* 10, 53–59.
Jacques-Felix, H. (1954) La carbunculariose. *Bulletin Scientifique. Ministère des Colonies. Section Agronomie Tropical* 5, 296–344.
Kebe, B.I. (1997) Coffee wilt disease in Côte d'Ivoire. In: *Proceedings of the First Regional Workshop on the Coffee Wilt Disease (Tracheomycosis)*, Kampala, Uganda, July 1997. National Agricultural Research Organisation (NARO) and the European Union, pp. 55–56.
Kojima, M. and Uritani, I. (1976) Possible involvement of furanoterpenoid phytoalexins in establishing host–parasite specificity between sweet potato and various strains of *Ceratocystis fimbriata*. *Physiological Plant Pathology* 8, 97–111.
Kranz, J. and Mogk, M. (1973) *Giberella xylarioides* on arabica coffee in Ethiopia. *Phytopathologische Zeitschrift* 78, 365–366.
Lejeune, P. (1958) *Rapport au Governement Imperial d'Ethiopie sur le Production Caféiere*. FAO /58/3/1881, Rome.
Marin, M., Castro, B., Gaitan, A., Preisig, O., Wingfield, B.D. and Wingfield, M.J. (2003) Relationships of *Ceratocystis fimbriata* isolates from Colombian coffee-growing regions based on molecular data and pathogenicity. *Journal of Phytopathology* 151, 395–405.
Mfwidi-Nitu, P. (1994) The recrudescence of tracheomycosis of robusta coffee in Zaire. *Africa Coffee Bulletin* 40, 9–12.
Mohammed, C. and Guillaumin, J.J. (1993) *Armillaria* in tropical Africa. In: Isaac, S., Frankland, J.C., Watling, R. and Whalley, A.J.S. (eds) *Aspects of Tropical Mycology*. Cambridge University Press, Cambridge, UK, pp. 207–217.
Moller, W.J. and DeVay, J.E. (1968) Insect transmission of *Ceratocystis fimbriata* in deciduous fruit orchards. *Phytopathology* 58, 1499–1508.
Morgan-Jones, G. (1967) *Ceratocystis fimbriata*. *CMI Descriptions of Pathogenic Fungi and Bacteria No. 141*. Commonwealth Mycological Institute, Kew, UK.
Muller, R.A. (1997) Some aspects of past studies conducted in western and central francophone Africa on coffee tracheomycosis (Côte d'Ivoire, Cameroon and Central African Republic). In: *Proceedings of the First Regional Workshop on the Coffee Wilt Disease (Tracheomycosis)*, Kampala, Uganda, July 1997. National Agricultural Research Organisation (NARO) and the European Union, pp. 15–26.
Ndimande, B.N. (1985) A review of the current knowledge of Fusarium bark disease of coffee in Zimbabwe. In: Clowes, M. St J., Logan, W.J.C. (eds) *Advances in Coffee Management and Technology in Zimbabwe 1980–1985*. Coffee Grower's Association, Harare, Zimbabwe. pp. 118–119.
Olson, E.O. and Martin, W.J. (1949) Relationship of *Ceratostomella fimbriata* from the *Hevea* rubber tree and sweet potato. *Phytopathology* 39, 17.
Onesirosan, P.T. and Fatunla, T. (1976) Fungal fruit rots of tomatoes in Southern Nigeria. *Journal of Horticultural Science* 51, 473–479.
Otieno, W. (1998) *Armillaria-Trichoderma* interactions and management of Armillaria root rot of tea (*Camellia sinensis* (L.) O. Kuntze). *Tea* 19, 11–16.
Pegler, D.N. (1977) *A preliminary agaric flora of East Africa*. Kew Bulletin Additional Series VI. HMSO, London, 615 pp.

Pegler, D.N. and Gibson, I.A.S. (1972) *Armillariella mellea CMI Descriptions of Pathogenic Fungi and Bacteria No. 321*. Commonwealth Mycological Institute, Kew.

Pieters, R. and Van der Graaff, N.A. (1980) Resistance to *Giberella xylarioides* in *Coffea arabica*: evaluation of screening methods and evidence for horizontal resistance. *Netherlands Journal of Plant Pathology* 86, 37–43.

Pizzinatto, M.A. and Menten, J.O.M. (1991) Pathogenicidade de oito especies de *Fusarium*, isoladas de sementes, as platulas de algodoeiro. *Phytopathologica* 17, 124–134.

Pontis, R.E. (1951) A canker disease of the coffee tree in Colombia and Venezuela. *Phytopathology* 41, 178–184.

Porteres, R. (1959) Valeur agronomique des caféiers des types Kouilou et Robusta cultivars en Côte d'Ivoire. *Café Cacao Thé* 3, 3–13.

Rabechault, H. (1954) Sur quelques facteurs de resistance du caféier a la tracheomycose. *Bulletin scientifiques. Ministère des Colonies. Section Agronomie Tropical* 5, 292–295.

Rayner, R.W. (1959) Root rot of coffee and ring barking of shade trees. *Kenya Coffee* 24, 361–365.

Ribeiro, I.J.A. and Coral, F.J. (1968) Preliminary studies on the fungus *Ceratocystis fimbriata* Ell. and Halst., causing wilt of mango (*Mangifera indica* L.) and cocoa (*Theobroma cacao* L.). *Bragantia* 27, 87–89.

Ribeiro, I.J.A., Rossetto, C.J., Donadio, L.C., Sabino, J.C., Martins, A.L.M. and Gallo, P.B. (1995) Mango wilt. XIV Selection of mango (*Mangifera indica* L.) rootstocks resistant to the mango wilt fungus *Ceratocystis fimbriata* Ell. & Halst. *Acta Horticulturae* 370, 159–161.

Rossetto, C.J., Ribeiro, I.J.A. (1990) Mango wilt. XII. Recommendations for control. *Revista de Agricultura* 65, 173–180.

Rutherford, M.A. (2006) Current knowledge of coffee wilt disease, a major constraint to coffee production in Africa. *Phytopathology* 96, 663–666.

Saccas, A.M. (1956) Recherches experimentales sur la tracheomycose des caféiers en Oubangui-Chari. *L'Agronomie Tropicale* 11, 7–38.

Schieber, E. and Zentmeyer, G.A. (1967) Collar canker of coffee trees in Guatemala. *Plant Disease Reporter* 51, 267–269.

Siddiqi, M.A. (1980) The selection of arabica coffee for fusarium bark disease resistance at Bvumbwe. *Kenya Coffee* 45, 55–60.

Siddiqi, M.A. and Corbett, D.C.M. (1963) Coffee bark disease in Nyasaland I. Pathogenicity, description and identity of the causal organism. *Transactions of the British Mycological Society* 46, 91–101.

Siddiqi, M.A. and Corbett, D.C.M. (1965) Fusarium bark disease of coffee. Report on research carried out in Malawi. *East African Agricultural and Forestry Journal* 31, 11–15.

Siddiqi, M.A. and Corbett, D.C.M. (1968) Coffee bark disease in Malawi II. Properties of the causal organism and conditions favouring the disease. *Transactions of the British Mycological Society* 51, 129–135.

Simmonds, N.W. (1994) Horizontal resistance to cocoa diseases. *Cocoa Growers' Bulletin* 47, 42–52.

Sivanesan, A. and Holliday, P. (1972) *Rosellinia bunodes. CMI Descriptions of Plant Pathogenic Fungi and Bacteria* No. 531. Commonwealth Mycological Institute, Kew.

Snyder, W.C., Trujillo, E.E., McOnie, K.C. and Berger, C.A. (1960) Trunk and branch canker of coffee trees in Guatelamala. *Plant Disease Reporter* 44, 566–567.

Steyaert, R.L. (1948) Contribution a l'étude des parasites des vegetaux du Congo Belge. *Bulletin de la Societé Royale de Botanique du Belque* 80, 11–58.

Storey, H.H. (1932) A bark disease of coffee in East Africa. *Annals of Applied Biology* 19, 173–183.

Teviotdale, B.L. and Harper, D.H. (1991) Infection of pruning and small bark wounds in almond by *Ceratocystis fimbriata*. *Plant Disease* 75, 1026–1030.

Van der Graaff, N.A. and Pieters, R. (1978) Resistance levels in coffee arabica to *Gibberella xylarioides* and distribution pattern of the disease. *Netherlands Journal of Plant Pathology* 84, 117–120.

Van Emden, J.H. and Van Suchtelen, N.J. (1959) Ziefvanteziekte en kaffiekanker. *Surinaamse Landbouw* 7, 111–112.

Vermeulen, H. (1968) Investigation into the cause of phloem necrosis of *Coffea liberica* in Surinam, South America. *European Journal of Plant Pathology* 74, 202–218.

Wallace, G.B. and Wallace, M.M. (1955) The bark disease of coffee. *East African Agricultural and Forestry Journal* 21, 25–27.

Waller, J.M. and Holderness, M. (1997) Fusarium diseases of coffee. In: *Proceedings of the First Regional workshop on the Coffee wilt Disease (Tracheomycosis)*, Kampala, Uganda, July 1997. National Agricultural Research Organisation (NARO) and the European Union, pp. 27–36.

Waller, J.M. and Masaba, D.M. (2006) The microflora of coffee surfaces and relationships to coffee berry disease. *International Journal of Pest Management* 52, 89–96.

Webster, R.K. and Butler, E.E. (1967) A morphological and biological concept of the species *Ceratocystis fimbriata*. *Canadian Journal of Botany* 45, 1457–1468.

Wellman, F.L. (1954) The *Fusarium* phase of the root disease complex in coffee. *Phytopathology* 44, 509 (abstr.).

Wellman, F.L. (1972) *Tropical American Plant Diseases*. Scarecrow Press, Metuchen, New Jersey, 200 pp.

Wrigley, G. (1988) *Coffee*. Longman, London and New York, pp. 339–342.

Zentmeyer, G.A. (1976) Phytophthora plant destroyer. *Biosience* 26, 686–689.

Zuluaga, V.J., Valencia, A.G. and Gonzalez, J. (1971) Contribucion al estudio de la natureleza de la resistencia del cafeto a *Ceratocystis fimbriata* (Ell.and Halst.)Hunt. *Cenicafe* 22, 43–68.

10 Nematodes

Introduction

Nematode populations may build up to damaging levels in perennial crops, and coffee is no exception. Around the world, a great diversity of nematode genera and species has been identified from the root zone of coffee. Many of these are potentially pathogenic and several are known to cause yield reductions. There is a paucity of information on yield affects for most of these, although there is some information from Latin America on root-knot nematodes, and breeding for resistance has been carried out in Brazil, Colombia and Costa Rica (Bertrand *et al.*, 2001). The subject of nematodes on coffee has been well reviewed by Campos and Villain (2005).

Species distribution and host range

The most important nematodes on coffee worldwide are the root-knot nematodes (*Meloidogyne* spp.) and the root lesion nematodes (*Pratylenchus* spp.), together with *Radopholus similis* (Cobb) Thorne and *Rotylenchulus reniformis* Linford and Oliveira (Campos and Villain, 2005). The nematodes isolated in greatest numbers from *C. arabica* in Brazil were, in order, *Helicotylenchus* sp., *Meloidogyne exigua* Goeldi, *Aphelenhcus* sp., *Criconemella* sp., *Xiphinema* sp., *Ditylenchus* sp. and *Pratylenchus* sp. (Dias *et al.*, 1996). A survey of coffee nurseries in Costa Rica found that the main nematodes were *Meloidogyne* spp. and *Pratylenchus* spp. (Figueroa and Vargas, 1989).

The following nematodes are listed as those extracted from coffee roots in Colombia: *M. exigua*, *M. incognita* (Kofoid and White) Chitword, *M. javanica* (Treub) Chitwood, *Pratylenchus coffeae* (Zimmermann) Filipjev & Schuurmans Steckhoven, *Helicotylenchus erythrinae* (Zimm.) Golden, *Xiphenema* sp. and

Criconemoides sp. (Villalba *et al.*, 1988). From soil associated with the root zone of coffee plants in Colombia, *Helicotylenchus* was again by far the most populous plant parasitic genus, followed by *Meloidogyne*, *Pratylenchus*, *Xiphinema* and *Criconemoides* (Villegas-Garcia and Arango-Bernal, 1990).

In Malawi, Saka and Siddiqi (1979) list the plant parasitic nematodes isolated from arabica coffee as *Helicotylenchus dihystera* (Cobb) Sher, *Pratylenchus coffeae*, *Scutellonema validum* Sher and *Xiphenema* sp. A combination of ectoparasitic nematodes (*Hemicricenomoides* sp., *Nothocricenomoides* sp. and *Helicotylenchus* sp.) was reported to cause crinkle leaf disorder in India (Kumar, 1988). Kumar and Samuel (1990) listed *Meloidogyne*, *Hemicricenomoides* and *Pratylenchus* as the most important nematode genera on coffee in India.

The mixture of crops found in different coffee-farming systems influences the balance of nematode genera found on coffee. For instance, the practice of intercropping banana with coffee must influence the constituents of the nematode population on coffee, and might explain the high numbers of *Helicotylenchus* reported on coffee in Colombia and parts of Brazil. In a survey of bananas in two coffee districts in Colombia, the plant-parasitic nematode genus extracted from roots in greatest number was *Helicotylenchus*, followed by *Pratylenchus*, *Meloidogyne* and *Radopholus*.

In Hawaii, when coffee was planted following sugarcane, populations of *Criconemella* sp. and *Pratylenchus zeae* declined, whereas those of *Rotylenchulus* and *Meloidogyne* increased (Schenck and Schmidt, 1992). However, it was noted that coffee is a poor host for *Rotylenchulus*, and the populations had probably increased on weed species. Results from a survey of coffee plantations on five of the Hawaiian Islands showed that spiral (*Helicotylenchus* sp. and *Scutellonema* spp.) and pin nematodes (*Pratylenchus* spp.) were the most widely distributed, being found associated with coffee soils on all the islands. Reniform (*Rotylenchulus* sp.) nematode was found on three out of the five islands, but burrowing nematode from only one. *Meliodogyne incognita* was rarely found, but *M. koanensis* Eisenbert, Bernard & Schmidt, although restricted to the main island, was present in 34% of the samples and is considered to be a highly damaging species (Hue *et al.*, 2004).

There is little information about the effect of simultaneous invasion of coffee roots by more than one nematode genus. One report presents evidence for a competitive interaction between *Pratylenchus coffeae* and *Meloidogyne* spp. (Herve *et al.*, 2005).

Symptoms

With the exception of root galls produced by the root-knot nematodes (*Meloidogyne* spp.), symptoms of nematode attack tend to be rather general, and the causal nematode can rarely be determined from symptoms alone. The affected plant may become unthrifty, the leaves yellow and may be shed, but such symptoms can be confused with nutrient deficiency.

Pathogen characteristics

Most plant-parasitic nematodes cannot be seen easily with the naked eye, ranging from 0.2–2.5 mm in length, although some Longidorids may reach 10 mm. Even the larger nematodes can be difficult to see until they are separated from the soil. Identification to genus level usually requires their extraction from roots or soil, placing them in water in a Petri dish or similar vessel and examination under a dissecting or inverted microscope. Soil extracts usually contain large numbers of non-parasitic nematodes that can be distinguished by the absence of a feeding stylet. More detailed examination is usually required to identify a nematode to species level, such as the distinctive perineal patterns on mature *Meloidogyne* females. Electrophoretic and DNA-based techniques are being used increasingly to distinguish nematode species (e.g. Randig *et al.*, 2004).

Root-knot Nematodes

Species distribution and host range

Several species of *Meloidogyne* are known to attack coffee. Campos and Villain (2005) list *M. exigua*, *M. incognita*, *M. coffeicola* Lordello and Zamith and *M. paranaensis* Carneiro, Carneiro, Abrantes, Santos & Almeida as the species that are most common and/or most damaging, with another 13 species listed as the less widely distributed and generally less damaging group (Table 10.1).

Among the four most damaging species of *Meloidogyne* on coffee, *M. incognita* is the most widely distributed worldwide, having been reported from Africa, India and Latin America. *Meloidogyne exigua* (see Fig. 10.1) has been found in all the major coffee-producing states in Brazil, but has not been found outside of Latin America. In Central and South America, *M. exigua*, *M. coffeicola* and *M. incognita* have been identified as the most widespread species; *M. paranaensis* has been reported only from Brazil, but can be as damaging there as *M. incognita* (Castro and Campos, 2004).

Conflicting results on species identification with respect to esterase phenotype, perineal pattern and reactions of host differentials suggest that *M. paranaensis* may also occur in Guatemala and Colombia (Campos and Villain, 2005). The species *M. africana* Whitehead, *M. decalineata* Whitehead, *M. kikuyensis* De Grisse and *M. megadora* Whitehead are confined to the African continent.

Extensive nematode surveys were conducted in Brazil between 1999 and 2005. In 2002, almost 70% of the coffee produced in Brazil came from the States of Parana, São Paulo and Minas Gerais, the latter state accounting for over 50% of coffee production. In coffee plantations in São Paulo and Parana, *M. exigua*, *M. coffeicola*, *M. incognita* and *M. paranaensis* were found to occur in separate or mixed populations. All four nematode species have been recorded on the same plantation (Otoboni *et al.*, 2003).

Table 10.1. Main parasitic nematodes found on coffee, and countries where they have been reported.

Nematode	Countries where recorded
Root-knot nematodes[a]	
Meloidogyne incognita	Brazil, Tanzania, Jamaica, Venezuela, Guatemala, Côte d'Ivoire, India, Costa Rica, El Salvador, Hawaii, Nicaragua, Cuba
M. exigua	Brazil, Guatemala, Dominican Republic, Nicaragua, Costa Rica, Puerto Rico, El Salvador, Venezuela, Bolivia, Honduras, Peru, Colombia, Martinique
M. coffeicola	Brazil
M. javanica	Brazil, Tanzania, Kenya, Congo, El Salvador, Cuba, Sao Tome, India
M. hapla	Brazil, Tanzania, Kenya, Congo, El Salvador, Guatemala, India
M. africana	Kenya, Congo
M. decalineata	Tanzania, São Tomé
M. kikuyensis	Tanzania
M. arenaria	Jamaica, Cuba, El Salvador
M. megadora	Angola, Uganda
M. inornata	Guatemala
M. oteifae	Congo
M. thamesi	India
M. paranaensis	Brazil, Guatemala
M. arabicida	Costa Rica
M. konaensis	Hawaii
M. mayaguensis	Cuba
Root lesion nematodes	
Pratylenchus coffeae	India, Guatemala, Indonesia, Costa Rica, Dominican Republic, El Salvador, Puerto Rico, Brazil, Tanzania, Congo, Matinique, Madgascar, Barbados, S.E. Asia.
P. brachyurus	Costa Rica, Brazil, Peru, West Africa.
P. goodeyi	Tanzania
P. loosi	Sri Lanka
P. pratensis	India
Burrowing nematode	
Radopholus similis	India, Indonesia
Other nematodes	
Rotylenchulus reniformis	India, Hawaii, Philippines, South Pacific, Brazil
Helicotylenchus spp.	India, Brazil, Malawi
Criconemella sp.	Brazil
Cricenomoides spp.	Colombia, India
Ditylenchus sp.	Brazil
Xiphenema spp.	Brazil, Malawi
Scutellonema sp.	Malawi

[a] Modified from Campos and Villain (2005).

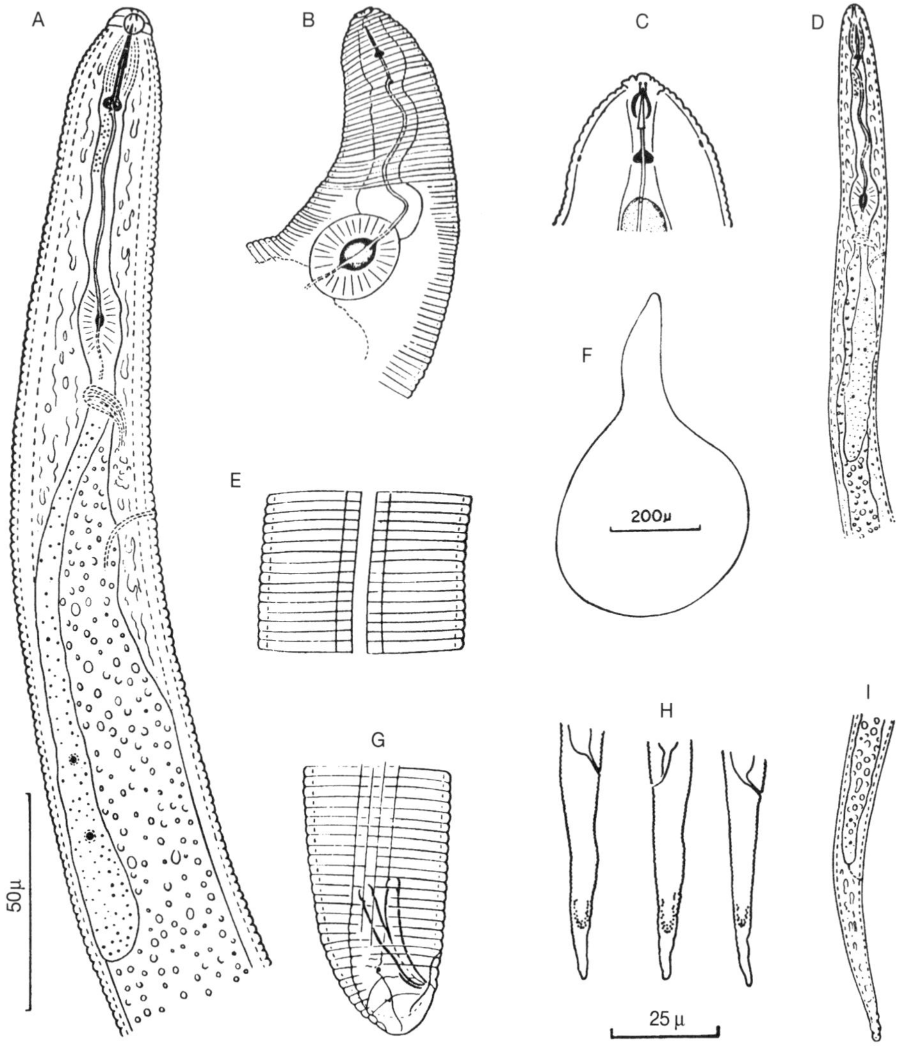

Fig. 1. *Meloidogyne exigua*. A. Male oesophageal region. B. Female oesophageal region. C. Female anterior end. D. Second-stage larva, oesophageal region. E. Male lateral field. F. Whole female. G. Male posterior end. H, I. Larval tails. (C, H. from Whitehead, 1968: the rest from Lordello & Zamith, 1958, courtesy Proceedings of the Helminthological Society of Washington.)

Fig. 10.1. *Meloidogyne exigua* (Eain, 1974).

In Parana State 34% of 657 samples contained *Meloidogyne* species, of which 44% were *M. paranaensis* and 17% were *M. incognita* (Krzyzanowski *et al.*, 2001). In São Paulo State, *M. incognita*, *M. exigua*, *M. paranaensis* and *M. javanica* were identified as 37%, 21%, 13% and 1% of plant parasitic nematodes, respectively (Lordello, 2002). *Meloidogyne exigua* is widespread in Minas Gerais, and was found in 22% of soils from coffee plantations in the State (Campos, 2002).

Surveys in Brazil suggest that over time, the incidence of *M. incognita* has increased, whereas that of *M. coffeicola* has decreased, but *M. exigua* remains the most important species in Minas Gerais State (Campos *et al.*, 1990). These two species can be distinguished from the morphology of the mature female; *M. coffeicola* is larger than the other species (up to 1300 μm) and more elongated in shape (Lordello and Zamith, 1960). *Meloidogyne incognita* has a wide host range among crop and weed species and *M. exigua* has also been reported from a number of common weeds, as well as from cocoa (Bridge *et al.*, 1982).

Meloidogyne coffeicola seems to be native to forest species and moved to coffee as the Brazilian forest was cleared, and has therefore been found on few agricultural hosts (Jaehn *et al.*, 1980). *Meloidogyne hapla* Chitwood causes slight swelling of the root of arabica coffee in Tanzania (Bridge, 1984) but, in Brazil, it was reported to cause typical galls, together with necrosis and lateral root formation (Lordello, 1982).

Recent identification of a new species among specimens previously identified as *M. incognita* – and given the name *M. paranaensis* – has shown the value of polymorphic DNA and esterase phenotype for distinguishing *Meloidogyne* species that may have similar perineal patterns (Carneiro *et al.*, 2004). Specific RAPD (random amplified polymorphic DNA) markers have been identified for the three most important root-knot nematode species in Brazil, and transformed into SCAR (sequence characterized amplified region) markers. This was shown to be a fast and reliable method of reliable identification to species level, based on DNA from egg masses (Randig *et al.*, 2004). Carneiro *et al.* (2004) reported that esterase phenotypes were species-specific and more reliable than perineal patterns for distinguishing *Meloidogyne* species.

Species identification using RAPD markers was consistent with results from perineal patterns and esterase phenology, and showed that *M. exigua* and *M. mayaguensis* were closely related, as were *M. arenaria* (Neal) Chitwood and *M. javanica*. A low level of intraspecific variability was detected in *M. exigua*, *M. incognita* and *M. paranaensis*, but higher levels were found in *M. arenaria*. Hernandez *et al.* (2004a) used esterase phenotype to identify *Meloidogyne* species from coffee plantations in Central America and Brazil. *Meloidogyne exigua* was found in Costa Rica, Nicaragua and Honduras, *M. arenaria* was found in El Salvador and *M. incognita* in Brazil.

Information on the species and their distribution in Africa and elsewhere outside Latin America is scarce, as fewer surveys have been undertaken. In the highlands of Tanzania, *M. decalineata* was the predominant species and *M. kikuyensis* was also reported from this area (Swai, 1981), while *M. africana* is widespread in Kenya and Congo (Whitehead, 1959) and *M. megadora* has been found in Angola and Uganda (Whitehead, 1968a).

Bridge (1984) added *M. decalineata* to the list from Tanzania. *Meloidogyne oteifae* Elmiligy occurs in Congo (Elmiligy, 1968). According to Campos and Villain (2005), *M. mayaguensis* Rammah and Hirschmann is widespread in Africa, having been reported from several West African States and from South Africa. *Meloidogyne thamesi* (Neal) Chitwood has been reported from India (Kumar, 1984).

Damage and losses

Affected roots show swelling, malformation and sometimes galls on younger roots (see Plate 30). The mature coffee plant becomes debilitated but is rarely killed. The affect of root-knot nematode is more marked on seedlings, producing the typical root galls and the plant becomes chlorotic and stunted. Such plants should not be transplanted as they will yield poorly. Mature trees attacked by root-knot nematodes may exhibit premature fruit-fall, nutrient deficiency, defoliation, die-back and stunted growth, and can lose up to 20% of their yield potential under optimal growing conditions. If the plants are growing under stress, the nematode can cause complete crop failure (Wrigley, 1988).

The exact symptoms seen on the roots differ between species: *M. exigua* forms typical round galls on young roots and the egg sac is produced under the epidermis. Lateral root formation is stimulated around the galls (Mendes, 1977), and necrotic areas on galled roots are associated with secondary pathogens, causing root decay. *Meloidogyne coffeicola* does not form galls, but the egg sac is extruded on the outside of the root through cracks made by the expanding female. This species can be very destructive to coffee plantations due to the loss of main and feeder roots and invasion of secondary pathogens (Schmidt, 1969).

Meloidogyne incognita causes swellings that may resemble galls, but does not form typical galls on mature plants. Instead, the nematode causes cracking of the root cortex as giant cells expand, and this is associated with root death.

More typical galls are formed on young seedlings in the nursery. *Meloidogyne paranaensis* does not cause root galls but, instead, causes the cortical root tissue to split and crack; egg masses are produced within the root tissue. Root damage is sufficient to cause a general decline in the tree, and heavy infestation can result in death of the tree (Campos and Villain, 2005). *Meloidogyne paranaensis* was recorded in Minas Gerais from coffee plantations showing symptoms of early decline, and the roots of affected plants were peeled and roughened with few side rootlets (Castro and Campos, 2004).

Meloidogyne africana and *M. decalineata* cause small galls from 1–5 mm in diameter. Affected seedlings become stunted, with a proliferation of rootlets behind the root tip (Whitehead, 1959). Lordello and Fazuoli (1980) also reported galls caused *by M. decalineata* in nurseries on both arabica and robusta seedlings, but also reduced growth and leaf yellowing in the plantation. *Meloidogyne oteifae* was reported to cause moderate-sized galls on *C. canephora* (Elmiligy, 1968). *Meloidogyne africana* causes poor growth of *C. arabica* sedlings in Kenya (Whitehead, 1959) and on *C. canephora* in Congo (Whitehead, 1969a).

Root-knot nematodes have caused huge economic losses on coffee in Latin America, notably in Brazil and Colombia. Some areas of Brazil that grew coffee in the past have been forced to grow other crops due to nematode infestation and, by law, any seedlings found in a nursery with root knot must be destroyed (Campos *et al.*, 1990). Barbosa *et al.* (2004) estimated that, in the best-managed plantations, a yield loss of 45% could be attributed to *M. exigua*. However, in poorly managed plantations, several other factors caused yield loss, and these factors were much more important than root-knot nematodes. *Meloidogyne incognita* has been responsible for the total destruction of coffee plantations within the space of 5 years in Brazil's São Paulo State, and is considered to be more pathogenic than *M. exigua* to *C. arabica* (Moraes and Lordello, 1977).

Root-knot nematodes have been shown to increase infection by many vascular wilt fusaria, and there is one report of a similar association between *M. incognita* and *Fusarium oxysporum* f.sp. *coffeae* (Negron and Acosta, 1989). *Meloidogyne arabicida* Lopez & Salazar has also been associated with *F. oxysporum* in Costa Rica (Bertrand *et al.*, 2000). Damage to the root system caused by root-knot nematodes can also decrease the uptake of soil-applied granular fungicide and decrease the control of coffee leaf rust.

Control

The main way by which root-knot and other nematodes are introduced into previously uninfested coffee fields is through the transplantation of seedlings that had become infested in the nursery. The first rule of nematode control is exclusion, and strict measures are enforced in some states in Brazil to ensure that seedlings are certified as nematode-free before transplantation. Nursery sites must be chosen carefully to ensure there is no risk of nematode infestation from the soil at the site, or from the water used to irrigate the nursery. New coffee plantations should not be established on sites where old coffee trees are present or where run-off water from nearby infested fields might reach the new site.

Root-knot nematodes can be controlled in the nursery and, to some extent, in the plantation on young trees that are not severely damaged by application of granular nematicides such as aldicarb, carbofuran and phenamiphos (Huang *et al.*, 1983). These are applied to wet soil in a furrow along both sides of the plant row and are then incorporated by hand or machine. Typical application rates are in the range 1.6–6.0 g a.i./ha. However, these substances are very toxic, being classified as WHO Class I substances and given red status by the Common Code for Coffee Community (see Chapter 15). They cannot be recommended for general use and may not be registered in some countries. Soil solarization or steam sterilization of soil used in pots are alternatives to the use of chemicals. Soil solarization was shown to decrease the numbers of *M. incognita* (race 2) and other nematodes, resulting in enhanced plant growth (Cuadra *et al.*, 1999a).

Once root-knot nematodes are introduced into a coffee plantation, they build up over time on the coffee roots, and build-up is generally more rapid on sandy soils and less rapid in soils with a high content of organic matter. Cover crops were found to slow the build-up of root-knot nematodes. *Desmodium ovalifolium* decreased the population of *M. incognita*, although *Rotylenchulus reniformis* populations increased (Herrera, 1997).

When a plantation becomes infested with root-knot nematodes, there may be no alternative but to destroy the crop and plant a suitable non-host. However, for *M. incognita*, the decline in nematode population may be slow. In one study, 27% of the original population level was still found after 6 months without host plants (Jaehn and Rebel, 1984). A break of more than 6 months without coffee plants to feed on is sufficient for the removal of *M. exigua* from the soil (Moraes and Lordello, 1977).

A rotation involving 1 year with cotton, soybean or maize is therefore sufficient to allow coffee to be replanted where the main species is *M. exigua* (Moraes *et al.*, 1977). *Meloidogyne coffeicola* also has a low capacity for persistence in the soil. However, only *Arachis hypogaea* and *Ricinis communis* were found to be immune to Brazilian populations of *M. incognita*, and therefore suitable for rotation with coffee where this nematode occurs (Carneiro and Carneiro, 1982).

Outside of the nursery, nematicides are of limited value. In plants heavily infested with *M. incognita* there is usually too much root damage to obtain an economic return from chemical application. If the infestation is less severe, it may be possible to increase the productivity of an infested plantation with judicious use of nematicides. Lordello *et al.* (1990) reported that a 30% increase in yield was obtained from nematicidal treatment of a plantation infested with *M. exigua*.

In parts of Brazil where the soils are heavily infested with root-knot nematodes, grafting *C. arabica* scions onto *C. canephora* root stocks has been recommended and shown to provide effective control. A root-knot nematode-resistant *C. canephora* (cv. C.2258) was obtained from Costa Rica and, after further selection for root-knot nematode resistance, a highly resistant line (LC2258) was obtained, onto which was grafted either Mundo Novo or Catuai Vermelho (Fazuoli, 1986, cited by Campos *et al.*, 1990). Further selection within cv. 2258 has increased the level of resistance to root-knot nematodes. The improved line to be used as a rootstock for *C. arabica* has been named Apoatã, and is resistant to most populations of *M. incognita* and *M. paranaensis* and immune to *M. exigua* (Fazuoli *et al.*, 2002).

Breeding for resistance

There seems to be little variation in resistance to nematodes as regards *C. arabica*. *Coffea arabica* cvs Catuai, Mundo Novo and Bourbon Amarelo, together with *C. liberica var. dewevrei* (*C. excelsa*) and *C. canephora* cvs Guarini and Larentii, were reported to be susceptible to at least one race of *M. incognita* (Moraes *et al.*, 1973). *Coffea canephora* however, has proved to be a good source of resistance to several root-knot nematodes. Extensive work has been conducted in Brazil on breeding and selection for resistance to *M. exigua*, with a number of robusta lines having been identified as good sources of resistance (Fazuoli and Lordello, 1978; Lehman and Lordello, 1982).

Collections of wild coffee species have been screened for resistance to *Meloidogyne* spp. in Latin America. Accessions resistant to *M. exigua* have been identified in several *C. canephora* accessions and in *C. racemosa*. Resistance to *M. arabicida* was detected in 25% of wild *C. arabica* tested and in all *C. canephora* accessions. Using 15 isolates of *Meloidogyne* spp. from

Central America and Brazil, Hernandez *et al.* (2004b) inoculated two *C. arabica* cultivars, Catuai and Sarchimor, and two wild accessions from Ethiopia, ET15 and ET28. All but one isolate multiplied on Catuai. An isolate of *M. incognita* from Brazil did not multiply on Sarchimor or on the two wild accessions. *Meloidogyne arenaria* multiplied at a high rate on Sarchimor, at a low rate on ET15 and did not multiply on ET28. *Meloidogyne exigua* was the most pathogenic, multiplying on all four test hosts.

Root invasion by *M. exigua* on the resistant cv. Iapar 59, which contains the recently identified Mex-1 resistance gene, was compared with invasion on a susceptible Caturra cultivar. There was decreased penetration and retarded development of the nematode in the resistant cultivar compared to Caturra. A minimum of 11 galls per plant developed on the Caturra, compared to only one or two per plant on Iapar (Anthony *et al.*, 2005).

Several lines from *C. canephora* and *C. congensis* are resistant to race 3 of *M. incognita*, while some of the Icatu (*C. arabica* x *C. canephora*) and Sarchimor (*C. arabica* cv. Villa Sarchi x Timor Hybrid) lines have shown moderate resistance (Goncalves and Ferraz, 1987). Similarly for *M. exigua*, Bertrand *et al.* (2001) found that the frequency of resistant plants in populations of *C. arabica* was low, but there was a high frequency of resistant individuals in *C. canephora* and in a segregating population of crosses with H. de Timor (both Catimors and Sarchimors).

A *C. arabica* hybrid has been created by crossing nematode-resistant Sarchimor with some Ethiopian lines, to combine resistance to both *M. exigua* and CLR. Some of these selections are showing a high level of resistance to nematodes and to rust, as well as excellent cup quality (Montagnon *et al.*, 2002). Work continues in Latin America to develop suitable root-knot nematode-resistant rootstocks and arabica hybrids (Anthony *et al.*, 2003).

Bridge (1984) reported that most of the coffee varieties and lines he tested in Tanzania – including some robusta lines – were susceptible to root-knot nematodes. Two species (*C. liberica var. dewevrei* and 'semperflorens' coffee) and four breeding lines were found to be nematode-free. The contradictory report by Moraes *et al.* (1973), that *C. liberica var. dewevrei* was susceptible, indicates that differences in pathogenicity exist between races and populations of *M. incognita*.

Root Lesion Nematodes

Distribution, symptoms and damage

Root lesion nematodes (*Pratylenchus* spp.) are migratory endoparsites, feeding on the root cortex. Affected roots become brown, then yellow, and channels formed by the nematodes moving through the cortex can be seen easily in longitudinal sections of the affected root. Adult trees become chlorotic, with short branches and generally retarded growth. Often, roots in the top 5 cm remain unaffected by these nematodes, while the growth of deeper roots is restricted. *Pratylenchus coffeae* can also affect seedlings in the nursery that become chlorotic and stunted (Wrigley, 1988). Where soil populations of *P. coffeae* are high, newly transplanted trees first show leaf yellowing, followed by stunting of the shoot, wilting and eventual death of the tree (Whitehead, 1969b). Affected roots show discoloured lesions that rot as nematodes migrate to fresh roots.

Pratylenchus spp. are found in all the main coffee-growing areas of the world, the two main species affecting the crop being *P. coffeae* (see Fig. 10.2) and *P. brachyurus* (Godfey) Filipjev & Schuurmans Steckhoven. Until the late 1960s, *P. brachyurus* was the only root lesion nematode known to attack coffee in Latin America. Later, *P. coffeae* was identified from coffee in a number of Latin American countries (see Table 10.1). Lesion nematodes are more common than root-knot nematodes in Guatemala, and were reported to be among the pests with the greatest negative impact on the economy of coffee production in Guatemala (Villain *et al.*, 2000).

In São Paulo, Brazil, surveys of root lesion nematodes showed that *P. brachyurus* was the most frequently isolated species, occurring in 18% of root samples, but population densities were low. *Pratylenchus coffeae* occurred in 5% of root samples but, where it occurred, pest density was higher than for *P. brachyurus* and it caused more severe damage. Damage to coffee by *P. brachyurus* was seen mainly in plantations with graminaceous cover crops (Kubo *et al.*, 2004).

In the State of Pernambuco in north-eastern Brazil, *P. coffeae* was reported to be the cause of severe root damage to coffee grown in a field in which yams had been grown previously (Moura *et al.*, 2002). More than 70% of young trees < 2 years old were killed by the nematode. *Pratylenchus coffeae* also occurs in Tanzania (Bridge, 1984), Madagascar, India and Indonesia (Whitehead, 1969b) and on several of the Caribbean Islands (Campos and Villain, 2005).

The lesion nematodes attacking coffee in Central America and Brazil form a species complex, and the taxonomy of individual species is under review (Campos and Villain, 2005). Two isolates collected on coffee in Guatemala are morphologically close to *P. coffeae* but do not interbreed, and one is much more pathogenic on *C. arabica* than the other (Villain *et al.*, 2002). Another species, widely distributed across ecologiocal zones in Guatemala, is morphologically distinct from *P. coffeae* (Villain, 2000). In Brazil, some *Pratylenchus* populations collected from coffee are closer taxonomically to *P. jaehni* (Inserra *et al.*, 2001; Siciliano-Wilcken *et al.*, 2002).

Pratylenchus coffeae is the main lesion nematode and the most destructive nematode on coffee in India (Palanichamy, 1973), South East Asia and Africa (Whitehead, 1968b, 1969a; Bridge, 1984). In India, *P. coffeae* was shown to cause crop losses of between 5 and 75% (Sreenivasan *et al.*, 1999). Some 3000 ha of arabica coffee were affected by lesion nematode in the states of Karnataka, Kerala and Tamil Nadu. Giribabu and Saha (2003) confirm that *P. coffeae* is the most important coffee nematode in Tamil Nadu. In Java it became a major pest of coffee (Whitehead, 1968b).

Pratylenchus goodeyi Sher & Allen is found on coffee in parts of Africa. In Cameroon, 40% of coffee root samples – from plantations where coffee was intercropped with banana – contained substantial numbers of *P. goodeyi* (Jacobsen *et al.*, 2004). Bridge (1984) reported *P. goodeyi* from Tanzania. As

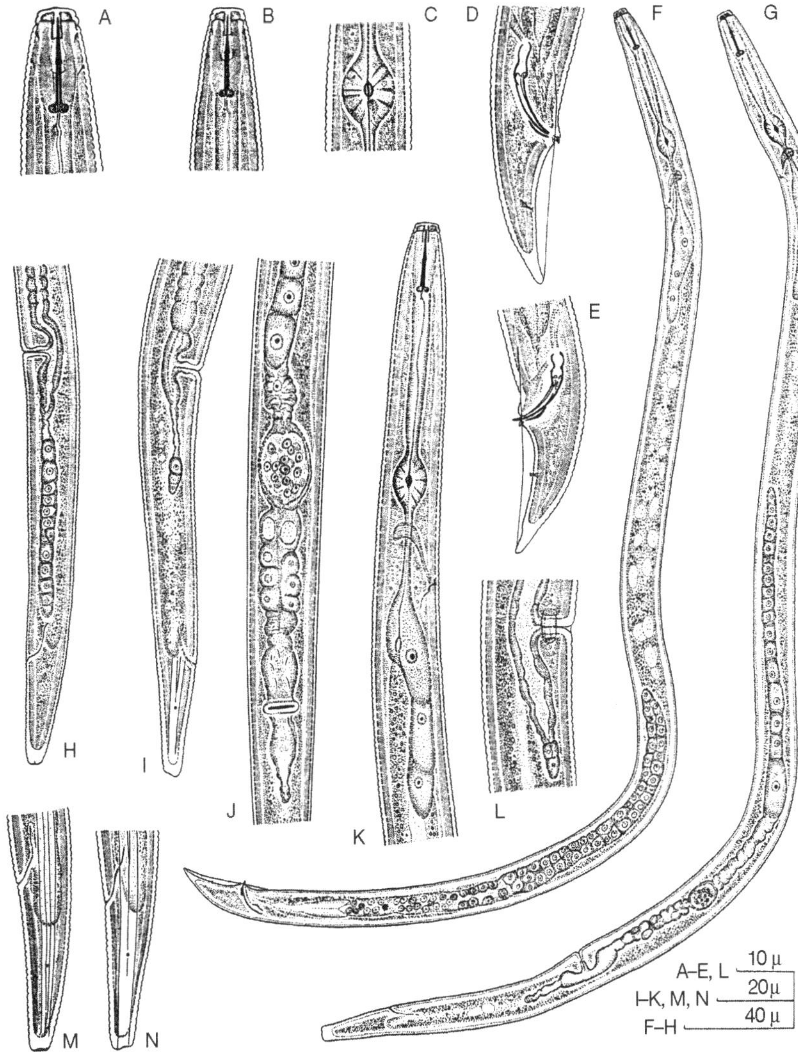

Pratylenchus coffeae (Zimmermann). B–F. Male; remainder female. A–G, I–L. Topotypes; H, M & N, from coffee soil, South India. A & B. Head ends. C. Median oesophageal bulb. D, E, H, I, M & N. Tail ends. F. Entire male. G. Entire female. J. Reproductive region. K. Oesophageal region. L. Vulval region. J. Ventral view; remainder lateral views. (Topotypes—courtesy of M. W. Allen and P. A. A. Loof.)

Fig. 10.2. *Pratylenchus coffeae* (Siddiqi, 1972).

yet, there is no information on damage or yield loss caused by other *Pratylenchus* spp. recorded from coffee, including *P. goodeyi*.

Biology and host range

The eggs of *Pratylenchus* spp. are laid in the root cortex; hatching takes place with *P. coffeae* in 6–8 d at 28–30°C and the life cycle is completed in 27 d

(Siddiqi, 1972). Most *Pratylenchus* spp. have a wide host range and *P. coffeae* can be a major pest on banana, citrus, sweet potato and *Solanum* spp. *Pratylenchus goodeyi* is an important pest of banana in Africa, and *P. loosi* Loof on tea (Luc *et al.*, 2005). *Pratylenchus coffeae* has a wide host range (Nickle, 1984), and *P. brachyurus* infects a wide range of crops around the world.

The grass species common in coffee plantations in South America, *Melinis minutiflora* and *Hyparrhenia rufa*, are good hosts for *P. brachyurus* (Lordello, 1972). In Hawaii, *P. brachyurus* and *P. zeae* Graham have been reported on coffee planted after sugarcane but populations decline beyond 3 years after planting, indicating that coffee is a poor host for these two species. Differences between isolates of *P. coffeae* with respect to host preference have been reported, indicating that there may be some physiological specialization in the species (Kumar and Viswanathan, 1972).

Control

As with root-knot nematodes, control of lesion nematodes in the nursery is of paramount importance, and the nursery site and soils used in pots should be checked regularly for nematodes. In coffee plantations that are already seriously infected with *Pratylenchus*, the best solution may be to switch to the cultivation of robusta cultivars that have good tolerance to *P. coffeae*, or to grow arabica on robusta root stocks. Villain *et al.* (2000) reported that grafting onto *C. canephora* provided efficient control of root lesion nematodes, resulting in significantly higher yields than from ungrafted plants. *Pratylenchus coffeea* was found to have a wide host range, and *C. arabica* was not among the best hosts. There was also evidence of differences in host preference between strains (Silva and Inomoto, 2002).

The granular nematicide formulations aldicarb, carbofuran, fensulfothion, oxamyl and fenamiphos all provided good control of *P. coffeae* in El Salvador (Abrego, 1974) and in Indonesia (Wiryadiputra, 1987), but are all WHO class 1 highly toxic materials.

Sanchez and Viesca (1989) reported that decomposed coffee pulp and the nematophagous fungus, *Paecilomyces lilacinus*, both provided control of *P. coffeae* that was similar in efficacy to the nematicide, fenamiphos. In Guatemala, treatment of *C. arabica* with terbufos suppressed populations of root lesion nematodes only until the second year after planting. However, this was sufficient to reduce mortality rates in ungrafted seedlings (Villain *et al.*, 2000).

There is some variation between cultivars of *C. arabica* in their ability to support reproduction of lesion nematodes. In Guatemala, among coffee cultivars grown in field soil infested with both *Pratylenchus* spp. and *M. exigua*, a significantly larger population of lesion nematode was found on cv. CR-95 than on IAPR-59, while the root-knot nematode was able to reproduce on CR-95 but not on IAPR-59 (Alpizar, 2004).

Burrowing Nematode

The burrowing nematode (*Radopholus similis*) (see Fig. 10.3) is a major pest of citrus and banana, and also attacks coffee. This nematode is also a migratory endoparasite, causing tunnelling in the root cortex similar to that caused by lesion nematodes. The outside of the affected roots may be warty and abnormally thickened. Mature trees may tolerate attack by *R. similis*, but the build-up in nematode population makes it impossible to successfully replant *C. arabica*, and the less susceptible *C. canephora* must be grown instead (Bally and Reydon, 1931, cited by Wrigley, 1988).

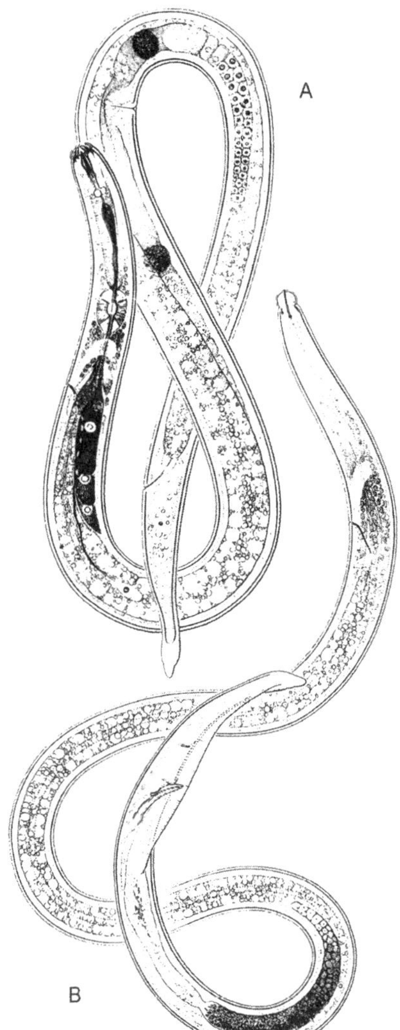

A, Female; B, Male.

Fig. 10.3. *Radopholus similis* (Orton Williams and Siddiqi, 1973).

A burrowing nematode identified as *R. arabocoffeae* (Trinh *et al*., 2004) was reported as a new problem on coffee cv. Catimor in Vietnam. This species was reported to be easily distinguishable morphologically from *R. similis*. The affected cv. Catimor was susceptible to all three nematodes affecting coffee in this part of Vietnam: *P. coffeae*, *R. duriophilus* and *R. arabocoffeae*, but especially to the latter species of burrowing nematode.

Other Nematodes

A number of other nematodes have been associated with coffee roots, but their distribution tends to be localized and little is known about their economic impact. Among these, *Rotylenchulus reniformis* reduces the growth of coffee trees if the inoculum density is greater than ten per 50 cm^3 of soil (D'Souza and Screenivasan, 1965). *Helicotylenchus* spp. have been associated with coffee cultivation in South America (Dias *et al*., 1996), India (Kumar, 1988) and Africa (Saka and Siddiqi, 1979), but there is no information on their importance in relation to yield loss.

General Nematode Control

There is no way economically to control nematodes on mature coffee trees. The adverse affects of nematodes on tree health and crop production can be alleviated by good fertilizer and irrigation management but, once trees have become unproductive, uprooting is the only solution. Prevention is much better than control where nematodes are concerned, and great care should be taken to prevent the introduction of nematodes into the plantation on infected seedlings transplanted from the nursery. It may be worthwhile to have the soil tested for nematodes before a site is chosen for a coffee plantation or nursery. Soil used in pots for raising seedlings must also be sourced from a site known to be free of plant-parasitic nematodes, particularly root-knot nematodes. If the soils local to the nursery site are known to be nematode infested, then soil used in pots must first be sterilized.

In the past in Brazil, soil used in nurseries was routinely treated with a granular nematicide or methyl bromide. As the use of methyl bromide is being phased out and granular nematicides are highly toxic and expensive, steam sterilization is a less costly and less hazardous alternative method of sterilizing soil. These methods of soil treatment will have a general effect on nematodes and other soil micro-organisms. Water sources used for watering nursery seedlings may also be contaminated with nematodes.

The alternatives to chemicals for nematode control include incorporation of organic residues (e.g. Sanchez and Viesca, 1989), soil solarization (e.g. Cuadra *et al*., 1999b) and the use of legume cover crops (Herrera, 1997). Bridge (1984) suggested that, in northern Tanzania, some species of *Meloidogyne* were endemic in the forest soil, infecting wild Rubiaceae. Elevated nematode populations in some coffee plantations were attributed to the use of copper sprays for many years to control coffee berry disease and rust. The accumulation of copper in the soil decreased the population of soil micro-organisms that normally kept the nematode population in check.

References

Abrego, L. (1974) Evaluating the selectivity of nematicides for control of *Pratylenchus coffeae* in coffee nurseries. *Nematropica* 4, 17.

Alpizar, E. (2004) Evaluación de la dinamica poblacional de *Meloidogyne exigua* y *Pratylenchus* sp. en tres genotipos de café (*Coffea arabica*). *Boletin PROMECAFE* 99, 6–8.

Anthony, F., Topart, P., Astorga, C., Anzueto, F. and Bertrand, B. (2003) La resistencia genetica de *Coffea* spp. a *Meloidogyne paranaensis*: identificacion y utilizacion para la caficultura latinamericana. *Manejo Integrado de Plagas y Agroecologia* 67, 5–12.

Anthony, F., Topart, P., Martinez, A., Silva, M. and Nicole, M. (2005) Hypersensitive-like reaction conferred by the Mex-1 resistance gene against Meloidogyne exigua in coffee. *Plant Pathology* 54, 476–482.

Barbosa, D.H.S.G., Vieira, H.D., Souza, R.M., Viana, A.P. and Silva, C.P. (2004) Field estimates of coffee yield losses and damage thresholds by *Meloidogyne exigua*. *Nematologia Brasileira* 28, 49–54.

Bertrand, B., Nuez, C. and Sarah, J.L. (2000) Disease complex in coffee involving *Meloidogyne arabicida* and *Fusarium oxysporum*. *Plant Pathology* 48, 383–388.

Bertrand, B., Anthony, F. and Lashermes, P. (2001) Breeding for resistance to *Meloidogyne exigua* in *Coffea arabica* by introgression of resistance genes of *Coffea canephora*. *Plant Pathology* 50, 637–643.

Bridge, J. (1984) *Coffee Nematode Survey of Tanzania. Report of a Visit to Examine Plant Parasitic Nematodes of Coffee in Tanzania.* Commonwealth Institute of Parasitology, St Albans, UK, 22 pp.

Bridge, J., Page, S.L.J. and Waller, J.M. (1982) *Plant-parasitic Nematodes and Diseases of Crops in the Santa Cruz Department of Bolivia.* UK Overseas Development Administration Report, 60 pp.

Cain, S.C. (1974) *Meloidogyne exigua CIH Descriptions of Plant-parasitic Nematodes No. 49.* CAB International, Wallingford.

Campos, V.P. (2002) *Coffee Nematode Survey in Minas Gerais State, Brazil.* Research report of the grant by PNP and D-Café, EMBRAPA, Brazil.

Campos, V.P. and Villain, L. (2005) Nematode parasites of coffee and cocoa. In: Luc, M., Sikora, R.A. and Bridge, J. (eds) *Plant Parasitic Nematodes in Subtropical and Tropical Agriculture*, 2nd edn. CAB International, Wallingford, UK, pp. 529–579.

Campos, V.P., Sivapalan, P. and Gnanapragasam, C. (1990) Nematode parasites of coffee, cocoa and tea. In: Luc, M., Sikora, R.A. and Bridge, J. (eds) *Plant Parasitic Nematodes of Tropical and Subtropical Agriculture.* CAB International, Wallingford, UK, pp. 387–430.

Carneiro, R.G. and Carneiro, R.M.D.G. (1982) Seleção preliminary de plantas para rotação de culturas em areas infestadas por *Meloidogyne incognita*. *Nematologia Brasiliera* 6, 141–148.

Carneiro, R.M.D.G., Tigano, M.S., Randig, O., Almeida, M.R.A. and Sarah, J.L. (2004) Identification and diversity of *Meloidogyne* spp. on coffee from Brazil, Central America and Hawaii. *Nematology* 6, 287–298.

Castro, J.M.C. and Campos, V.P. (2004) Ocorrencia de *Meloidogyne paranaensis* em cafeeiro na região do sul de Minas Gerais. *Fitopatologia Brasileira* 29, 227.

Cuadra, R., Aguilera, C. and Perez, J.A. (1999a) Effect of soil solarization on nematodes in coffee nurseries. *Revista de Proteccion Vegetal* 14, 23–26.

Cuadra, R., Perez, J.A., Machado, J. and Vazquez, J. (1999b) Effect of Nemacur, Terracur and

Furadan on root-knot nematodes in coffee plantations. *Revista de Proteccion Vegetal* 14, 111–115.
Dias, W.P., Liberato, J.R. and Fonseca, A.F.A da (1996) Survey of plant parasitic nematodes associated with the coffee crop in Espirito Santo State of Brazil. *Revista Ceres* 43, 808–812.
D'Souza, G.I. and Screenivasan, C.S. (1965) A note on the reniform nematode *Rotylenchulus reniformis* on arabica coffee in South India. *Indian Coffee* 29, 11–13.
Elmiligy, I.A. (1968) Three new species of the genus *Meloidogyne* Goeldi. *Nematologica* 14, 577–590.
Fazuoli, L.C. and Lordello, R.R.A. (1978) Fontes de resistencia em especies de cafeeiro a *Meloidogyne exigua. Sociedade Brasileira de Nematologia* 3, 49–52.
Fazuoli, L.C., Medina-Filho, H.P., Goncalves, W., Guerreiro-Filho, O. and Silvarolla, M.B. (2002) Coffee improvement: arabica varieties from the Instituto Agronomico de Campinas. In: Zambolin, L. (ed.) *State of the Art Technologies in Coffee Production*. Editora UVF, Vicosa, Brazil, pp. 162–215.
Figueroa, M.A. and Vargas, G.E. (1989) Survey of nematodes in coffee nurseries. *Investigacion Agricola* 3, 26–29.
Giribabu, P. and Saha, M. (2003) Studies on nematode communities at different depths, altitudes and periods of the year in coffee plantations. *Annals of Plant Protection Sciences* 11, 360–363.
Goncalves, W. and Ferraz, L.C.B. (1987) Resistencia do cafeeiro a nematoides. II Testes de progenies e hibridos para *Meloidogyne incognita. Nematologica Brasiliera* 9, 125–142.
Hernandez, A., Fargette, M. and Sarah, J.L. (2004a) Characterisation of *Meloidogyne* spp. (Tylenchida: Meloidogynidae) from coffee plantations in Central America and Brazil. *Nematology* 6, 193–204.
Hernandez, A., Fargette, M. and Sarah, J.L. (2004b) Pathogenicity of *Meloidogyne* spp. isolated from Central America and Brazil on four genotypes of *Coffea arabica. Nematology* 6, 205–213.
Herrera, S.I. (1997) Effect of legume cover on the control of phytoparasitic nematodes of coffee. *Memoirs of the XVIII Latin American Symposium of Coffee Production*, Costa Rica, 1997. IICA Miscellaneous Publication No. A1 SC-97-05, pp. 387–391.
Herve, G., Bertrand, G., Villain, L., Licardie, D. and Cilas, C. (2005). Distribution analyses of *Meloidogyne* spp. and *Pratylenchus coffeae sensu lato* in coffee plots in Costa Rica and Guatemala. *Plant Pathology* 54, 471–475.
Huang, S.P., Resende, I.C., Souza, P.E. de and Campos, V.P. (1983) Effect of aldicarb, ethoprop and carbofuran on control of coffee root-knot nematode, *Meloidogyne exigua. Journal of Nematology* 15, 510–514.
Hue, N.V., Serracin, M., Schmidt, D.P. and Bittenbender, H.C. (2004) Nutrient and nematode status of coffee soils from orchards in Hawaii. *Communications in Soil Science and Plant Analysis* 35, 2023–2036.
Inserra, R.N., Duncan, L.W., Troccoli, A., Maia dos Santos, J., Kaplan, D.T. and Volva, N. (2001) *Pratylenchus jaheni* sp. n. from citrus in Brazil and its relationship with *P. coffeae* and *P. loosi. Nematology* 3, 653–665.
Jacobsen, K., Fogain, R., Mouassom, H. and de Waele, D. (2004) Musa-based cropping systems of the Cameroon Highlands: a case study of the West and North-west Provinces of Cameroon, with emphasis on nematodes. *Fruits-Paris* 59, 311–318.
Jaehn, A. and Rebel, E.K. (1984) Sobrevivencia do nematoide de galhas *Meloidogyne incognita* em substrato infestado para producão de mudas de cafeeiros sadios. *Nematologica Brasileira* 8, 319–324.
Jaehn, A., Rebel, E.K. and Lordello, L.G.E. (1980) A origem do nematoide *Meloidogyne coffeicola. Nematologica Brasileira* 4, 159–161.
Krzyzanowski, A.A., Figueiredo, R., Santiago, D.C. and Favoreto, L. (2001) Identification of species and races of *Meloidogyne* on coffee estates in Parana State. *Symposium of Coffee Research of Brazil, II*, Vitoria, Espirito Santo, Brazil, p. 81.

Kubo, R.K., Oliveira, C.M.G., Antedomenico, S.R., Monteiro, A.R., Ferraz, L.C.B. and Inomoto, M.M. (2004) Ocorrencia de nematoides do genero *Pratylenchus* em cafezais do Esdado de Sao Paulo. *Nematologia Brasileira* 28, 159–165.

Kumar, A.C. (1984) Resistance in coffee to *Meloidogyne* spp. and occurrence of intersexes in *M. thamsei*. *Nematologica* 30, 108–110.

Kumar, A.C. (1988) Nematode problem in coffee and its management. *Indian Coffee* 52, 12–19.

Kumar, A.C. and Samuel, S.D. (1990) Nematodes attacking coffee and their management – a review. *Journal of Coffee Research* 20, 1–27.

Kumar, A.C. and Viswanathan, P.R.K. (1972) Studies on physiological races of *Pratylenchus coffeae*. *Journal of Coffee Research* 2, 10–15.

Lehman, P.S. and Lordello, L.G.E. (1982) *Meloidogyne exigua*, a root-knot nematode of coffee. *Nematology Circular No. 88*, Florida Department of Agriculture and Consumer Service Division of Plant Industries, 4 pp.

Lordello, L.G.E. (1972) Nematode pests of coffee. In: Webster, J.M. (ed.) *Economic Nematology*. Academic Press, London and New York, pp. 268–282.

Lordello, L.G.E. (1982) Nova ocorrencia do *Meloidogyne hapla* em cafeeiro. *Revista de Agricultura Piracicaba* 57, 6.

Lordello, L.G.E. (2002) *Coffee Nematode Survey in São Paulo State, Brazil*. Research Report of the grant by PNP and D-Café, EMBRAPA, Brazil.

Lordello, L.G.E. and Zamith, A.P.L. (1960) *Meloidogyne coffeicola* sp. n., a pest of coffee trees in the state of Parana, Brazil. *Revista Brasileira de Biologia* 20, 275–279.

Lordello, J.G.E., Lordello, A.I.L., Martins, A.L.M. and Pereira, J.C.V.N.A. (1990) Treatment with cafezal in an area infested with *Meloidogyne exigua*. *Nematologia Brasileira* 14, 18–19.

Lordello, R.R.A. and Fazuoli, L.C. (1980) *Meloidogyne decalinelata* parasita cafeeiro na ilha de São Tomé. *Revista de Agricultura* 55, 238.

Luc, M., Sikora, R.A. and Bridge, J. (2005) *Plant Parasitic Nematodes of Tropical and Subtropical Agriculture*. CAB International, Wallingford, UK.

Mendes, B.V. (1977) Observaçoes histopatologicas de raizes de cafeeiros parasitadas por *Meloidogyne exigua*. *Nematologia Brasileira* 2, 207–229.

Montagnon, C., Leroy, T., Bertrand, B., Charmetant, P. and Dufour, M. (2002) Recent coffee genetic improvements results. *Plante Recherche Développement (Special Issue: Research and Coffee Growing)*, 84–94.

Moraes, M.V. and Lordello, L.G.E. (1977) Estudo de tres populaçoes de nematoides nocivos ao cafeeiro. *Nematologia Brasileira* 2, 249–255.

Moraes, M.V., Lordello, L.D.G., Piccinin, O.A. and Lordello, R.A. (1973) Susceptibilidada de cafeteeros ao ataque por um nematoide. *Nematologica Mediterranea* 2, 107–110.

Moraes, M.V., Lordello, L.G.E., Reis, A.J., Thomaziello, R.A. and Goncalves, W. (1977) Rotation for the rehabilitation of coffee lands infested with *Meloidogyne exigua*. *Nematologia Brasiliera* 2, 257–265.

Moura, R.M., Pedrosa, E.M.R. and Prado, M.D.C. (2002) Incidencia de *Pratylenchus coffeae* causando severa nematose em cafeeirono nordeste. *Fitopatologia Brasileira* 27, 649.

Negron, J.A. and Acosta, N. (1989) The *Fusarium oxysporum* f. sp. *coffeae – Meloidogyne incognita* complex in 'Bourbon' coffee. *Nematropica* 19, 161–168.

Nickle, W.R. (1984) *Plant and Insect Nematodes*. Marcel Dekker, New York and Basel, Switzerland, 925 pp.

Orton Williams, K.J. and Siddiqi, M.R. (1973). *Radopholos Similis. CIH Descriptions of Plant-parasitic Nematode No. 27*. CAB International, Wallingford.

Otoboni, C.E.M., Santos, J.M. and Otoboni, A.M.M.B. (2003) Occurrence of species of *Meloidogyne* in the same coffee estate. *Abstracts from the Brazilian Congress of Coffee Research, XXIX*, Araxa, Minas Gerais, Brazil, pp. 277–278.

Palanichamy, L. (1973) Nematode problems on coffee in India. *Indian Coffee* 37, 99–100.

Randig, O., Carneiro, R.M.D.G. and Castagnone-Sereno, P. (2004) Identificaçao das principais especies de *Meloidogyne* parasites do cafeeiro no Brasil com marcadores SCAR-café em multiplex PCR. *Nematologia Brasileira* 28, 1–10.

Saka, V.W. and Siddiqi, M.A. (1979) Plant-parasitic nematodes associated with plants in Malawi. *Plant Disease Reporter* 63, 945–948.

Sanchez, V.R. and Viesca, R.S. (1989) Evaluation of four methods of control of phytoparasitic nematodes on coffee. *Revista Cafetalera* 303, 15–22.

Schenck, S. and Schmidt, D.P. (1992) Survey of nematodes on coffee in Hawaii. *Journal of Nematology* 24, 771–775.

Schmidt, C.T. (1969) *Meloidogyne coffeicola*, a serious root nematode problem in Brazilian coffee. *FAO Plant Protection Bulletin* 17, 56–57.

Siciliano-Wilcken, S.R., Inomoto, M., Ferraz, L.C. and Oliveira, C.M. (2002) Morphometry of *Pratylenchus* populations from coffee, banana, ornamental plants and citrus in Brazil. *Fourth International Congress of Nematology, Programme and Abstracts*, 8–13 June 2002, Tenerife, Canary Islands, Spain. *Nematology* 4, 356.

Siddiqi, M.R. (1972) *Pratylenchus coffeae. CIH Descriptions of Plant-parasitic Nematodes* No. 6. CAB International, Wallingford.

Silva, R.A. and Inomoto, M.M. (2002) Host range and characterization of two *Pratylenchus coffeae* isolates from Brazil. *Journal of Nematology* 34, 135–139.

Sreenivasan, T.N., Sundarababu, R. and Sankar, V. (1999) Nematode menace in coffee plantations. *Indian Coffee* 63, 22.

Swai, I.S. (1981) Root-knot nematodes, *Meloidogyne* species in Tanzania. In: *Proceedings of the Third Research Planning Conference on Root-knot Nematodes*, November 1980, Ibadan, Nigeria, pp. 28–30.

Trinh, P.Q., Nguyen, C.N., Waeyenberge, L., Subbotin, S.A., Karssen, G. and Moens, M. (2004) *Radopholus arabocoffeae* sp. n. (Nematoda: Pratylenchidae), a nematode pathogenic to *Coffea arabica* in Vietnam, and additional data on *R. duriophilus*. *Nematology* 6, 681–693.

Villain, L. (2000) Biology and characterisation of the parasite complex of the genus *Pratylenchus* associated with *Coffea* spp. PhD Thesis, ENSAR, Rennes, France.

Villain, L., Molina, A., Sierra, S., Decazy, B. and Sarah, J.L. (2000) Effect of grafting and nematicide treatments on damage by root-lesion nematodes (*Pratylenchus* spp.) to *Coffea arabica* L. in Guatemala. *Nematropica* 30, 87–100.

Villain, L., Baujard, P., Anzueto, F., Hernandez, A. and Sarah, J.L. (2002) Integrated protection of coffee plantings in Central America against nematodes. *Plante Recherche Développement (Special issue: Research and Coffee Growing)*, 118–133.

Villalba, D.A., Baeza, C.A. and Gil, L.V. (1988) Nematodos fitoparasiticos del cafeto en Colombia. *Tecnologia del Cultivo del Café*. Federacion Nacional de Cafeteros de Colombia, Bogota, Colombia, pp. 170–173.

Villegas-Garcia, C. and Arango-Bernal, L.G. (1990) *Nematodes in Banana Harton Enano (Musa AAB)*. Technical Bulletin No. 150, October 1990, Cenicafe, Chinchina, Colombia, 5 pp.

Whitehead, A.G. (1959) The root-knot nematodes of East Africa. *Nematologica* 4, 272–278.

Whitehead, A.G. (1968a) Taxonomy of *Meloidogyne* (Nematoda: Heteroderidae) with descriptions of four new species. *Transcations of the Zoological Society, London* 31, 263–401.

Whitehead, A.G. (1968b) Nematodea. In: Le Pelley, R.H. (ed.) *Pests of Coffee*. Longmans, Green & Co., London and Harlow, UK, pp. 407–422.

Whitehead, A.G. (1969a) The distribution of root-knot nematodes *Meloidogyne* spp. in tropical Africa. *Nematologica* 15, 315–333.

Whitehead, A.G. (1969b) Nematodes attacking coffee, tea and cocoa and their control. In: Peachy, J.E. (ed.) *Nematodes of Tropical Crops*. Technical Communication No. 40, Commonwealth Agricultural Bureaux, St Albans, UK, pp. 238–250.

Wiryadiputra, S. (1987) Control of coffee parasitic nematodes with granular systemic nematicides. *Pelia Perkebunan* 3, 100–107.

Wrigley, G. (1988) *Coffee*. Longman, London.

11 Nutrient Deficiencies and Physiological Disorders

Introduction

As a perennial crop that may remain in the field for up to 10 years in the case of dwarf types such as Caturra, and up to 40 years for the more traditional varieties, the coffee tree is demanding of soil fertility. Supplementary fertilizer is required to ensure good establishment and growth of the tree when it is transplanted from the nursery and, thereafter, to maintain yields without over-bearing, which leads to dieback and a tendency to biennial bearing.

Assuming that all other conditions for healthy growth of the tree are met, the response to added nitrogen decreases as shade density increases. Flower production and yield potential from the coffee tree is greatest when it is grown in full sun, but regular replacement of nutrients is essential and may need to be combined with soil moisture conservation, and possibly irrigation.

When particular nutrients are in short supply, the plant may show signs of specific nutrient deficiencies. Deficiency symptoms may also appear as a result of some diseases. Imbalance of nutrients and adverse environmental conditions can also lead to a number of other disorders (non-pathogenic diseases) that result from interference with the physiology and growth of the plant. These are more evident when coffee is grown in conditions removed from its natural equatorial forest habitat or in zones at the extremes of the tropics.

Specific cultural practices may be able to alleviate some of the problems associated with growing coffee in sub-optimal conditions, and some cultivars may be more suited to some conditions than others, but there remain finite limits to where coffee can be grown economically and without physiological disorders.

Diagnosis

Some nutrient deficiency symptoms, such as iron deficiency, may be readily identified by visual inspection. This is based primarily on the type and distribution of leaf symptoms, e.g. chlorosis occurring on the youngest or mature leaves, whether or not it is uniformly distributed in the leaf, and if it is associated with colour changes or necrosis. Often, it will be necessary to confirm any diagnosis based on visible symptoms with foliar analysis, and this should be backed up with soil analysis. Foliar analysis is usually based on sampling the third or fourth pair of leaves from the terminal shoot of lateral branches (Robinson, 1969). Table 11.1 gives optimum levels of nutrients in leaves according to Clowes and Hill (1981).

Younger leaves may give better results if a deficiency of calcium, copper, sulphur or boron is suspected (Malavolta *et al*., 1983). Four pairs of leaves, free of disease or insect damage, are collected from bearing laterals from each of ten trees. Soil and dust may need to be rinsed from the leaves before they are placed in a polythene bag. They must be stored in a freezer unless analysis is to take place within 24 h (Clowes and Hill, 1981). Colour illustrations of mineral deficiencies in coffee can be found in Haarer (1962), CAB (1974), CRF (1978), Valencia (1979) and see Plate 11.1. Brief descriptions of symptoms are given in Table 11.2.

Nutrient deficiency symptoms shown by the plant are often due to deficiencies in the soil, but may also be due to poor uptake by plant roots. This may be due to other soil factors such as drought or waterlogging, which are often associated with nitrogen deficiency symptoms, or to a high soil calcium content preventing adequate iron uptake and causing lime-induced chlorosis.

Table 11.1. Optimum range for nutrients in fourth pair of leaves taken from bearing laterals (from Clowes and Hill, 1981).

Nutrient	Optimum range
Macronutrients (%)	
Nitrogen	2.5–3.0
Phosphorus	0.1–0.2
Potassium	1.5–2.5
Sulphur	0.1–0.2
Calcium	0.7–1.5
Magnesium	0.2–0.4
Micronutrients (ppm)	
Boron	30–40
Copper	5–40
Iron	50–150
Manganese	50–150
Zinc	10–30

Table 11.2. Identification of nutrient deficiency symptoms in coffee.

Nutrient	Symptoms
Nitrogen	Chlorosis is uniform and appears first on young leaves and usually on the bearing branches, particularly in full sun and in dry conditions; leaves may be smaller and leaf production and growth decreased
Phosphorus	Begins with older leaves turning blue–green, yellow patches develop on older leaves becoming bronze, red or purple; affected leaves hang downwards from lateral branches and may be prematurely shed; root development retarded, which causes reduced growth in young trees
Potassium	Symptoms appear first on mature leaves – chlorotic patches along the edge join and become necrotic; also intervenal chlorosis becoming bronze; leaves may be twisted along the midrib; leaves become necrotic and are shed
Manganese	Uniform chlorosis giving the leaf a pale olive green appearance to youngest leaves at the end of the branch, while nearby older leaves remain dark green; leaf production and growth may be retarded but no reduction in leaf size; may be accompanied by a coarse mottling and veins remaining dark green
Iron	Begins on the youngest leaves which are pale green progressing to yellow, with veins remaining dark green; affected leaves are normal in shape and size; in advanced stages the leaf may become completely pale yellow to creamy white
Calcium	Trees appear generally chlorotic due to bleaching around the leaf margin; the bleached edges grow more slowly than the rest of the lamina, resulting in a cupped appearance
Magnesium	Evident on older leaves which are olive brown, becoming bronze with main veins remaining green; the discoloration forms regular patches between the main veins, giving a herring-bone pattern
Zinc	Leaves yellow–green with main veins remaining dark green; leaves are smaller than normal with short internodes giving a 'rosette' appearance that distinguishes zinc deficiency from other forms of interveinal chlorosis
Boron	Appears on older leaves which show chlorosis of the tip, becoming necrotic, while the rest of the leaf remains dark green; young leaves may be malformed due to failure of the leaf tip to develop; the midrib becomes corky and some leaves become narrow, with a leathery texture; growing points of primary and secondary branches die, producing a fan effect

Impaired root or vascular function in the plant due to nematode, root rots or wilt-inducing fungi may also cause poor nutrient uptake.

All diseases result in some physiological disturbance to the plant and, equally, adverse environmental conditions predispose plants to a number of diseases, and plant tissues damaged by non-pathogenic agents are invariably colonized by opportunistic pathogens that are not the primary cause of the condition. This close linkage between disease and environment can confuse the accurate diagnosis of the primary cause of some of these problems.

Nutrient Deficiencies

Nitrogen

Provided coffee trees are well managed and soil miconutrients are in balance, then available nitrogen has the greatest effect on yield. However, the response is poor or absent in coffee grown under heavy shade. Nitrogen deficiency symptoms are most likely to occur on unshaded trees, due to a combination of low leaf nitrogen and high light intensity and in situations where there is intense weed competition. The yield potential is greatest under high light intensity, increasing the amount of nitrogen required to replace that removed when the crop is harvested. Nitrogen deficiency symptoms appear first on the leaves of new shoots that become yellow or pale green (see Plate 33a).

Symptoms will become most marked under dry conditions, and severe symptoms may include a reduction in leaf area and shortened internodes. Equally, a high water table resulting in impaired root function may also cause typical nitrogen deficiency chlorosis. Nitrogen is translocated from older leaves to the new leaves and berries and, under heavy fruiting and inadequate fertilization, loss of nitrogen from older leaves causes premature defoliation. When a heavy crop is expected, additional nitrogen must be applied to prevent defoliation and associated die-back. Dry season chlorosis due to poor nutrient uptake by roots growing in dry soil can be ameliorated by the use of mulch.

Nitrogen should normally be applied after the onset of the first rains or in irrigation water. Ammonium sulphate may be used if the soil pH is > 6.5, otherwise urea is suitable or calcium ammonium nitrate is required for acid soils (pH < 5.2). Depending upon rainfall and cropping patterns, nitrogen applications may be split to reduce leaching and to ensure that the trees are well supplied during berry development.

Potassium

Potassium is the second most important nutrient for coffee trees in terms of frequency of response, and the high level of uptake requires the addition of potassium fertilizer to the soil to prevent the development of deficiency. Demand for potassium is particularly high during berry maturation, and low levels contribute to leaf loss associated with over-bearing. Developing fruits are a major sink for the element and, if in short supply, it will be translocated from the leaves to the fruits. Potassium is also important for maintaining basic levels of disease resistance.

The main symptom of deficiency is marginal chlorosis on older leaves that eventually becomes necrotic (see Plate 33b), and affected leaves are shed. Mulching and the use of animal manure can reduce the requirement for potassium. As with nitrogen deficiency, symptoms of potassium deficiency are likely when the trees are heavily bearing; this can be treated with applications of potassium sulphate or potassium chloride (muriate of potash).

Phosphorus

Phosphorus is immobilized in many tropical soils due to the formation of phosphates with low solubilities. However, under normal growing conditions, coffee trees are able to take up the phosphorus they require even when soil levels are low. Phosphorus deficiency symptoms are, therefore, uncommon. Nevertheless, available phosphorus is essential for flower development and fruit set (Cannel and Kimeu, 1971). Newly planted coffee is also given phosphate fertilizer to encourage good root development.

Deficiency symptoms have occurred on certain soils when high demand coincides with dry soil conditions. Symptoms appear on the older leaves that take on autumnal colours with lemon yellow patches, possibly with purple tinges, turning bronze (see Plate 33c). The younger leaves may be darker than normal, having a blue–green hue and hang downwards. Severe symptoms lead to defoliation of fruiting branches and contribute to symptoms of over-bearing. The use of organic manures is the best way to maintain soil phosphorus levels, but deficiency symptoms can be treated by applications of single superphosphate. If soils are $<$ pH 5.6, triple superphosphate is required.

Magnesium

Magnesium deficiency is not uncommon on coffee trees, particularly when grown on soils that are rich in potassium. First signs of the deficiency appear on the older leaves during fruit development as chlorosis between the veins, forming a herringbone pattern (see Plate 33d). The condition may be exacerbated by the use of mulches based on plants such as elephant grass that have a high potassium content. Magnesium deficiency appearing late in the season can be treated with foliar sprays of Epsom salts that may be applied with copper sprays, if these are being used for disease control. Ground application of dolomite worked into the topsoil around the tree may be required to treat symptoms appearing early in the season.

Calcium

On acid soils liming is required to raise the pH and prevent the development of calcium deficiency. Below pH 5.0, lime at the rate of 0.5–1.0 t/ha is required pre-planting. If pH is $<$ 4.5 then 1–2 t/ha may be required. Dolomitic lime is a good source of calcium, if magnesium is also lacking in the soil. Red dolomitic soils and brown granites usually require lime to counteract manganese deficiency (Clowes and Hill, 1981).

Calcium deficiency appears as pale bronzing or bleaching at the leaf margins and tips, initially of the younger leaves (see Plate 33e), giving the tree a generally chlorotic appearance. It is different from nitrogen deficiency in that the chlorosis is marginal and the bleached edges and tips grow more slowly than the green centre of the lamina, giving a cupped shape to the leaves. Also, the main veins split and become corky and the apical meristem may be killed, resulting in die-back. There is considerable genetic variation within *C. arabica* for sensitivity to low calcium.

Boron

Boron deficiency is most common on light soils that are low in organic matter. Other contributing factors are dry soil conditions or excessive rainfall and pH > 6.5, which may be caused by over-liming (Malavolta *et al.*, 1983). Deficiency symptoms are marked by foliar chlorosis that begins at the leaf tip and spreads to affect about half the lamina. Corky tissue develops along the midrib on the underside of affected leaves. Terminal buds may die and the laterals develop to from a fan (see Plate 33f). Leaves may be reduced in size and distorted. Fan-branching may also be caused by damage to terminal buds by Antestia. The condition can be treated with foliar sprays of boric acid or borax applied to the soil. Care must be taken to avoid boron toxicity.

Copper

Copper deficiency is uncommon, except on certain soils rich in organic matter, especially when liming to raise pH has been heavy and on sandy, calcareous soils. Where copper sprays are used for disease control, copper deficiency is even less likely. Symptoms may be seen as narrowing of the leaf, intervenal chlorosis and shoot die-back. Copper toxicity may be a problem where frequent copper sprays are used, coffee seedlings being particularly sensitive. Symptoms of copper toxicity are production of dark, water-soaked spots in the leaf lamina.

Iron

Iron deficiency symptoms are common when root function is impaired by dry soil conditions or the effects on the root system of over-bearing and subsequent dieback. Usually the symptoms disappear when it rains, unless the underlying cause is low or high pH. On iron-deficient coffee trees, the leaves appear pale green from a distance and, on closer inspection, a characteristic pattern of the network of green veins can be seen against the pale green or yellow lamina (see Plate 33g). The effect is more pronounced in younger leaves, but the whole tree is a paler green than normal.

Iron deficiency also affects the quality of coffee beans as it imparts an amber colour. Persistent symptoms can be treated by applying either iron sulphate to the soil or foliar sprays of iron chelate. Alternatively, the pH may need to be raised by liming, or lowered with ammonium sulphate or sulphur.

Manganese

Manganese is an important micronutrient for coffee, and severe deficiency decreases the yield and quality of green coffee. Light, alkaline soils high in organic matter are particularly associated with manganese deficiency (Malavolta *et al.*, 1983). When mild symptoms occur, the young leaves become pale in colour between the main veins, contrasting with the normal dark green of older leaves on the shoot. In more severe symptoms, the youngest pair of leaves is uniformly light green and diffuse, interveinal chlorosis may appear on some of the older leaves (see Plate 33h). The symptoms differ from iron deficiency in the absence of a pronounced network of dark green veins. Manganese deficiency can be corrected with foliar sprays of manganese sulphate. Manganese can become toxic on soils $<$ pH 4.5, but coffee is rarely grown on such soils.

Sulphur

Sulphur deficiency is very uncommon. The main symptom is a general chlorosis of the foliage that is often more intense near the main veins and on younger leaves. Mottling is less evident than with nitrogen deficiency. Growth is apparently unaffected. Sulphur deficiency is usually corrected by the use of one of the many sulphur-containing fertilizers such as ammonium sulphate or superphosphate.

Zinc

Zinc deficiency is not associated with particular soil types but with low soil zinc in general, and is often seen on regrowth after over-bearing. Deficiency symptoms may be induced by soil compaction due to restricted root development. Other predisposing factors are over-liming or over-fertilization with potash and phosphate. In Latin America, zinc and boron deficiencies often occur together, making diagnosis based on visual inspection more difficult. Zinc deficiency is more common on 'Mundo Novo' than on other cvs (Malavolta *et al.*, 1983).

First symptoms appear on terminal shoots as chlorosis between the main veins, with the lamina being narrower that normal (see Plate 33i). Severe deficiency may cause shortening of the internodes and reduction in the size of leaves and berries. Healthy outer leaves contrast with primary shoots in the centre of the tree, which exhibit shortened internodes and pale green, narrow leaves. Symptoms can be corrected by foliar sprays of zinc oxide.

Physiological Disorders

Over-bearing dieback

Coffee is especially prone to this physiological dieback because of its basic physiology. Physiological shedding of berries seldom occurs after the pinhead stage and the plant is committed to carrying a crop of expanding and maturing berries that absorb nutrients from other parts of the tree (see Chapter 1). Large reserves of dry matter and nutrients are utilized by the developing berries (Cannel and Kimeu, 1971). A range of conditions can lead to dieback and include lack of the major nutrients, excessive cropping due to full sun exposure, poor root function due to drought, nematode damage, weed competition and defoliation due to rust or leaf-eating pests.

Typical symptoms of dieback usually commence on bearing branches, with leaves becoming chorotic, often with marginal necrosis. Leaves are progressively shed until only the terminal cluster remains and berries become dull with signs of premature ripening such as yellow–orange coloration. The beans in such berries are either empty or incompletely filled. Thereafter, all leaves are lost and both stems and berries become necrotic (see Plate 31). Various studies have shown that there is a depletion of starch from stem tissues and that roots as well as shoots are affected, often accompanied by secondary invasion of opportunistic pathogens such as *Colletotrichum gloeosporioides* (see Chapter 7) and *Fusarium oxysporum* (see Chapter 9). Application of shade, nitrogenous fertilizers and crop thinning have been shown to alleviate the condition, known as Lyamungu dieback in Tanzania (Burdekin, 1964; Burdekin and Baker, 1964).

Star flowers

Under certain conditions, flowers may be deformed, either opening prematurely to produce small, rudimentary flowers less < 3 mm in length with imperfectly formed green floral parts or having small, erect petals that often remain green and known as star flowers. Both types remain sterile and are apparently produced during periods of high temperatures and when anthesis is stimulated by limited rainfall. The condition is due to an imbalance of hormones (Kumar, 1982).

Berry squeeze

This is basically a form of physiological berry shedding that occurs at the pinhead stage. Clusters of pinheads fail to develop and become chlorotic, then necrotic and are eventually shed. It is more commonly encountered on stems that are carrying overlapping crops. On alkaline soils, nutrient imbalance may also contribute to this condition.

Hot and cold disease

This is commonly encountered on coffee grown without shade at high altitudes or latitudes. Young shoots are exposed to low night temperatures and high levels of daytime insolation, and this interferes with the development of young tissues. Symptoms are distorted leaves, often curled and thickened and with a marginal chlorosis (see Plate 32). Internodes may also be shortened and lateral buds develop giving a rosette appearance in severe cases, and it can be confused with zinc deficiency that also produces a rosette appearance.

Leaves eventually develop marginal, necrotic patches and may become invaded by opportunistic pathogens such as *Colletotrichum gloeosporioides* or *Ascohyta tarda* (see Chapter 7) that can lead to dieback. Thorold (1945) considered that there was a linkage between hot and cold disease and Elgon dieback (subsequently shown to be due to *Pseudomonas syringae* pv. *garcea*), as resistance to both was apparent in 'Kapretwa series A' selections he made in the Mount Elgon regions.

Climatic damage

Hail damage can be encountered on coffee at high altitudes, especially if grown without shade. Symptoms are mostly apparent as small tears in the leaf lamina. These wounds can allow entry of pathogens such as *Pseudomonas syringae* pv. *garciae*. Wind also causes mechanical damage to plant shoots, but an important effect on coffee is root and collar damage caused by wind-rocking and subsequent predisposition to root and stem diseases. Suitable windbreaks can be provided by rows of suitable trees planted between coffee blocks (see Chapter 13), and are an essential feature of coffee plantation in some areas.

Lightning strike is sometime responsible for the death of one or a small group of trees, and may be initially misdiagnosed as a problem caused by a root rot or vascular pathogen. The presence of a scorch line or vertical split down the trunk is often indicative. Sun-scorch is often seen on exposed berries of unshaded coffee under hot dry conditions and is often associated with premature ripening due to other forms of stress. Dark patches appear on berry surfaces that soon become dry and necrotic, and usually colonized with *Fusarium stilboides*. Berry infection by *Cercospora coffeicola* may also be associated with sun-scorch damage (see Chapter 8).

Frost is a threat to coffee grown in some regions, and its incidence determines the southern limit of production in South America and Africa and limits the altitude for growing coffee in tropical areas. Frost has caused major damage to coffee plantations in parts of Brazil. Incidence of frost is influenced both by meteorological conditions and by topography. Exposed ground or plant surfaces cool during the night as they lose heat through radiation into space. This is greatest during periods of clear, still, dry weather and can result in ice crystals forming inside the tissues of exposed plant organs, with lethal effects. Frost may form under humid conditions, but under these conditions is often limited to the formation of ice crystals on plant surfaces, causing relatively little damage.

Cold, dense air that results from surface cooling flows down slopes and can accumulate in mountain valleys and hollows in upland plateaux to produce frost pockets. This may result to frost damage to the lower parts of the plant. The effect of frost is a collapse and subsequent necrosis of affected tissues, giving a scorched effect. If sharp frost penetrates woody tissues, plants suffer major damage and may require stumping for regeneration. Various measures can be undertaken to mitigate the effects of frost in frost-prone areas. Shade and raised humidity provided by ponds can help avoid radiation frost, and measures to assist air flow through coffee on slopes can avoid air frost damage.

Other Conditions

Minor malformations of unknown aetiology such as: (i) 'crinkle leaf', which causes deformation of the lamina of young leaves on suckers; and (ii) 'stem pitting', evident as deep grooves beneath the bark of swollen stems, have occasionally been reported from East Africa; these are possibly related to 'hot and cold disease'.

Various forms of mechanical damage can be caused by domestic and wild animals grazing shoots, damaging bark and sometimes digging up roots or young plants or, very occasionally, by birds and fruit bats eating ripening berries.

References

Burdekin, D.A. (1964) 'Lyamungu dieback' of arabica coffee in Tanzania I. Symptoms, distribution and experimental treatments. *Annals of Applied Biology* 53, 281–289.

Burdekin, D.A. and Baker, R.M. (1964) 'Lyamungu dieback' of arabica coffee in Tanzania II. Relation of starch reserves to Lyamungu dieback. *Annals of Applied Biology* 54, 107–113.

CAB (1974) *An Annotated Bibliography of Colour-illustrated Mineral Deficiency Symptoms in Tropical Crops*. Technical Communication No. 34, Commonwealth Bureau of Horticulture and Plantation Crops, CAB International, Wallingford, UK, 85 pp.

Cannel, M.G.R. and Kimeu, B.S. (1971) Uptake and distribution of macro-nutrients in trees of *Coffea arabica* in Kenya as affected by seasonal climatic differences and presence of fruits. *Annals of Applied Biology* 68, 213–230.

Clowes, M.St J. and Hill, R.H.K. (1981) (eds) *Coffee Handbook*, 2nd edn. Zimbabwe Coffee Growers Association, Harare, 198 pp.

CRF (1978) *An Atlas of Coffee Pests and Diseases*. Coffee Board of Kenya, Nairobi, 146 pp.

Haarer, A.E. (1962) *Modern Coffee Production*. Leonard Hill, London, 495 pp.

Kumar, D. (1982) Preliminary investigations into some flowering abnormalities of coffee in Kenya. *Kenya Coffee* 7, 6–25.

Malavolta, E., Carvalho, J.G. and Guimaraes, P.T.G. (1983) Effect of micronutrients on coffee grown in Latin America. *Journal of Coffee Research* 13, 64–77.

Robinson, J.B.D. (1969) Effects of environment and cultural conditions on nitrogen and phosphorus in coffee leaves. *Experimental Agriculture* 5, 301–309.

Thorold, C.A. (1945) Elgon dieback disease of coffee. *East African Agriculture and Forestry Journal* 10, 198–206.

Valencia, A.G. (1979) *Mineral Deficiencies in Coffee and Methods of Correction*. Technical Bulletin No. 1, National Coffee Research Centre, Chinchina, Colombia, 16 pp. (translated from original Spanish).

IV Integrated Crop Management

There are numerous definitions of integrated crop management (ICM) and it is probably more of a concept than a system, but central to the concept is husbandry of the land. Agricultural practices that are compatible with the principles of ICM are sustainable and protect and improve the most important basic resource, the soil.

A suitable definition is quoted in the 1996 British Agrochemicals Association Handbook *Integrated Crop Management: A Complete Training and Resource Pack*, as follows: 'A management system which employs controlled inputs to achieve sustained profitability with minimum environmental impact, but with sufficient flexibility to meet natural and market challenges economically.' This definition recognizes that the aim of agriculture is to produce crops profitably over the long term and that market demand influences supply and profitability. Planting coffee is a long-term investment and the trees may remain productive for 8–30 years, depending on the cultivar and quality of crop and land management.

The interrelationship between crop management practices is more apparent in coffee than perhaps for some other crops, beginning with nursery management and proper crop establishment in the field. Soil fertility and shade management interact to affect yield as well as susceptibility to pests and diseases. The importance of markets and the impact that world overproduction has had on the livelihoods of small producers has been painfully evident in the new millennium. Coffee producers, particularly smallholders in developing countries, are able to invest in sustainable crop management practices when they are sure of a market for their crop and a fair return on their effort and investment. This principle has been recognized by some large coffee-marketing companies who have seen an opportunity to market speciality coffees by working with growers to promote sustainable cultivation practices.

The aim of integrated crop management is to provide optimal crop husbandry that minimizes the need for crop protection interventions, and any additional crop protection required should be based on the principles of ICM. This section provides an overview of the components of ICM for coffee as a basis for sustainable crop protection, but is not intended to be a detailed guide to coffee production methods.

12 Nursery Management, Transplantation and Crop Maintenance

Preliminary Considerations

The development of a coffee plantation is a long-term endeavour and a coffee grower will want the trees to remain healthy and productive for as long as possible. An important aspect of sustainable crop management is an understanding of how the various activities of crop husbandry interact with pests and pathogens. In order to manage pests and diseases in a rational and sustainable fashion, knowledge of how the crop itself should be managed is essential. This chapter gives a brief account of general coffee husbandry. The general texts on coffee referred to in Chapter 1 (e.g. Clarke and Macrae, 1988; Wrigley, 1988; Cambrony, 1992; Coste, 1992) provide more detail, and there are many local handbooks on coffee husbandry published by individual coffee authorities, some of which are mentioned in the bibliography at the end of this chapter.

The starting point for new plantations is site selection. Bearing in mind the ecological requirements for coffee outlined in Chapter 1, there are limits to the climatic and edaphic conditions under which coffee can be grown. Although various agricultural practices can mitigate the effects of adverse environmental conditions and avoid the stresses that can exacerbate diseases, the economic cost of these measures must be considered. Furthermore, in an era of climatic change, conditions that are currently marginal could become intolerable. The broader socio-economic factors involved with coffee production – such as labour availability, processing facilities and marketing arrangements – also need to be taken into account.

Healthy soil is the fundamental basis for all sustainable agricultural systems. Soil is the primary resource base and its quantity and quality must be maintained and, if possible, improved for agriculture to be sustainable and profitable. Coffee is usually grown on forest land where soils are sufficiently deep and the dry periods not too long. However, the ecology of such land is fragile, especially when natural vegetation is removed. The physical process of land clearance can damage the soil structure and lead to much wider environmental problems. In particular, soil is exposed to erosion, and the natural nutrient recycling that is a feature of natural vegetation systems is halted.

Measures have to be taken to protect the soil from erosion and to replenish nutrients by the application of fertilizer. Where the land slopes by not more than about 10%, coffee should be planted on the contour, but steeper slopes should be terraced. Perennial vegetation barriers such as grass, bananas, etc. planted at intervals along contours or along the edges of bunds (embankments) can retain soil and moisture. Vetiver grass is often used to stabilize terrace edges. The subject of site selection and land preparation for establishment of new coffee plantations is covered in detail by Descroix and Wintgens (2004) and in several general texts on coffee, and will not be dealt with further. However, it is necessary to realize that the environmental conditions under which coffee is grown have an underlying effect on the health of the plants and therefore on the impact of pests and diseases.

Propagation

Seed selection and preparation

The first task in propagation is to select the seed both of the chosen cultivar and from a healthy source. *Coffea arabica* is most commonly propagated by seed, as being mostly autogamous, selected lines breed true to type. An important exception to this is that seed from hybrid plants will not be true to type, as they are genetically heterogeneous and characters will continue to segregate at succeeding generations to produce variable offspring. *Coffea canephora* can also be propagated by seed but, due to its allogamous nature, material raised from seed is very variable and vegetative propagation is usual.

Seed of an appropriate cultivar should preferably be obtained from the relevant local coffee authority. Many factors influence the choice of cultivar such as quality, productivity, disease resistance and cultivation system, but ultimately local conditions and marketing considerations are the major determinants. If a farmer is collecting seed of *C. arabica*, then it should be taken from the most healthy and heavily bearing trees, avoiding those with obvious symptoms of diseases. Ripe, but firm, healthy berries are selected for seed extraction. These are washed and the seed is squeezed out by hand and all the skin removed before washing again in clean water. Any seeds that float or show signs of discoloration or damage should be discarded.

The seed may be fermented for a few hours or simply dried in a dry and shady place; the testa (parchment) should remain intact. Seed from mechanically pulped berries or that has had a long fermentation period may have reduced viability. Fresh seed should be used as stored seed loses viability relatively rapidly, so that after 6 months' storage under ambient conditions germination will be $< 50\%$. Airtight storage at cool temperatures will maintain viability for longer periods.

Arabica seed germinates in 6–10 weeks depending on seed age, planting depth and temperature, with fresher seed taking a bit longer than older seed. Viability can be increased by soaking the seed in water for 24 h. Several coffee pathogens can be carried in berry tissues and can contaminate seed material but, provided seeds are properly prepared from healthy berries, there is little evidence of problems with seed transmission of pathogens. As an added insurance against seed-borne diseases, the seed can be treated with an inexpensive seed dressing fungicide where local regulations permit this.

Vegetative propagation

Nodal cuttings must be taken from healthy, young orthotropic stems, such as suckers produced from the main stem. Bending the stem over will stimulate the production of orthotropic shoots. The leaves are trimmed to about half size and the cuttings planted up to the leaf bases in a moist, well-drained soil in propagation beds or boxes. High humidity needs to be maintained above the cuttings until they have rooted (after about 10 weeks). They are then normally transplanted to polybags to grow and harden-off before field transplanting, as described below. There is little evidence of transmission of pathogens provided the cuttings are taken from healthy stems on healthy trees.

This is the normal method for propagation of robusta; it can also be used for arabica, but cuttings root less readily. Coffee can also be propagated by grafting, and arabica scions can be cleft- or approach-grafted onto robusta seedling stock, which has been used as a method for nematode control. This and other methods can be also used to graft new scion material onto older stumps. Micropropagation from meristemic tissue and somatic embryogenesis has also been used in propagating breeding material.

Nursery practices

Traditionally, coffee was sown direct into the field, usually with several seeds at each station and up to four seedlings allowed to grow on, often producing clumps of trees as in the Brazilian 'cova' system. Directly sown seeds need adequate protection from weed competition and climatic extremes, usually provided by erecting a small shelter of branches above the seed station. Seedlings are now usually raised in nurseries for later transplantation, and this system has its own requirements for pest and disease management.

The nursery site must be carefully chosen. It should be close to a water supply, sheltered from weather extremes and with a nearby source of fertile soil for pot-filling, unless soil is to be imported to the site. Nursery design varies and may be simple or more complex – and therefore more expensive to set up. Provision of shade that can be decreased as the seedlings grow is the most costly part, requiring the erection of posts and a framework to support shading by thatch, plastic strips, etc. Smallholders in Africa usually rely on shade provided by trees or roughly constructed grass tents. Whatever the level of sophistication, some method of protecting the young seedlings from the sun will be required (see Fig. 12.1).

The soil to be used for potting should be reasonably free-draining. If it has a high clay content it will be necessary to mix it with coarse sand before use. The potting soil can be supplemented with phosphate fertilizer, and lime as required. An important plant protection issue at this stage is to ensure that the potting soil is not infested with nematodes, as may be the case if the site has been used for crops in the recent past. Sites where banana is or has been grown should be avoided for topsoil collection, as they are especially prone to nematodes that also attack coffee. Natural water sources close to cultivated land may contain plant-parasitic nematodes, particularly if not free flowing.

Ideally, the soil from a prospective nursery site should first be sent for analysis to determine the fertilizer requirements and also, if possible, its nematode status. If the seedlings become infested with nematodes in the nursery, the infestation will be carried to the field, contaminating the site. Nematode populations will increase as the coffee tree grows, interfering with root function and making the tree prone to nutrient deficiency and dieback.

Soil for nurseries can also be sterilized by heat such as steam, by solar heating under a polythene sheet (solarization) or by chemical agents to kill weed seeds, pests and diseases, including nematodes. Methyl bromide,

Fig. 12.1. Coffee nursery (courtesy of T. Ragsdale).

formerly widely used for soil sterilization, is now prohibited or severely restricted by most countries in accordance with international conventions. Other chemicals may be used for this purpose, such as dazomet and pencycuron, in accordance with local regulations. However, healthy coffee plants normally have symbiotic vesicular arbuscular mycorrhizal fungi within their roots that assist uptake of nutrients, particularly phosphorus. Using sterilized soil would prevent early colonization of seedling roots by these beneficial organisms. Inoculating seedlings or growing them in soil known to be infested with vesicular arbuscular mycorrhizal fungi had been shown to be beneficial (Siqueira *et al.*, 1995).

Seed may be pre-germinated on a bed of moist sand covered with hessian sacks, or simply between layers of sack that are kept moist but not waterlogged. When germination begins the seed has to be examined each day, and seed with root tip just visible carefully removed for potting. This allows viable seed to be selected when the root tip is just appearing, minimizing the risk that it is broken or bent during potting.

Seed is often sown 1–2 cm deep in prepared seedbeds, composed of sandy soil with added compost and usually raised some 10–20 cm above soil level to ensure adequate drainage. Seedlings are transplanted to containers just before or just after cotyledons emerge ('soldier' or 'butterfly' stages). Although it is possible to transplant seedlings straight to the field, they are usually grown on and hardened-off under nursery conditions in containers such as black plastic polybags (15 × 30 cm) or black plastic sleeves. During transplantation, the roots are more or less bare and can be inspected, so that only those seedlings with strong, straight taproots are planted.

Coffee seedlings require shade during the early stages of growth. The depth of shade is gradually decreased and the space between containers increased to prevent etiolation as the young plants are hardened off. Transplantation can be carried out any time from 6 to 12 months after potting, depending on the growth rate of the plants and local practice. Generally, it is best to transplant before the ten-leaf stage and before roots appear at the base of the container. Plants left in containers too long will develop distorted root systems. Care during transplantation is needed, especially to avoid damage to the root system, over exposure and weed competition, in order to produce vigorous young plants capable of tolerating minor damage caused by pests and diseases.

The main diseases of coffee seedlings in the nursery are damping off (mainly *Rhizoctonia solani*, *Fusarium* spp. and *Pythium* spp.) and brown eye spot (*Cercospora coffeicola*). Damping-off is largely a consequence of overwatering and/or overshading that predisposes seedlings to infection by the pathogens involved. Seedlings are predisposed to brown eye spot by high humidity due to poor ventilation and overshading. Both diseases can be largely avoided by using clean soil and attention to watering and shading regimes. They can also be controlled with copper-based fungicides applied to the soil against damping-off and to the foliage against leaf spot (see also Chapter 9).

The main insect pests in the nursery are green scale, mealybugs, leaf miner and thrips, all of which can be controlled with insecticide sprays of e.g.

fenitrothion. Leaf-eating caterpillars can sometimes cause damage and should be controlled with an appropriate insecticide before their populations increase. Root- and collar-feeding insects such as mole crickets and cutworms can be a problem, as can twig boring beetles that bore into the stems of small nursery plants.

Transplantation

Transplantation is carried out in the wet season, and young plants are normally transported to the field in their containers, from which the root ball and soil can be easily removed by cutting away the plastic. Only healthy, vigorous coffee seedlings that appear true to type should be transplanted. One of the most important requirements of the seedling is that the taproot is not bent, as this will produce a weak tree that cannot sustain a full crop, and which is susceptible to pests and diseases. One of the main objectives at transplantation is to ensure that the plant is able rapidly to develop a strong root system.

Soil preparation

Coffee plants will become established and remain productive for many years in the sites into which they are transplanted, so it is essential that appropriate attention is given to their initial placement. Hence, adequate site preparation is necessary before transplantation. On newly cleared land, one of the main disease problems is that posed by basidiomycete soil-borne pathogens, which are endemic in natural forest systems and can readily spread from old stumps and roots remaining in cleared land. Such debris needs to be removed and destroyed before transplantation.

Alternatively, ring-barking of trees a year or so before removal starves the roots and reduces their effectiveness as an inoculum source. Perennial weeds, particularly grasses and sedges, also need to be eradicated. A leguminous cover crop such as *Mucuna* is sometimes grown for a season and ploughed in before planting coffee. Wherever land is cleared for coffee cultivation, care should be taken to ensure that the soil is protected from erosion as far as possible.

Transplants can be placed in holes or trenches. Where the coffee is to be grown at close spacing in the row to produce a 'hedgerow' system with plants less than 1 m apart, a trench may be preferred, but the usual system is to plant in individual holes. These need to be 60 cm diameter × 60 cm deep, with the topsoil separated from the subsoil, so that the topsoil can be placed around the roots at the base of the hole and the subsoil placed at the top.

Holes should be prepared about 1 month before transplantation and partially filled with excavated soil mixed with any required fertilizer. This will vary with soil type at each location, but usually requires phosphorus for root establishment in the form of triple super-phosphate, or single super-phosphate if sulphur levels in the soil are low. Acid soils (pH 4.5–6.5) are most suitable for coffee but, at pH 5.0 and below, lime is also required. Soils that are low in organic matter benefit from the addition of compost or well-rotted animal manure. All the fertilizer and manure should be well mixed with the excavated soil before it is returned to the planting hole.

Once the soil around the transplanted seedling has been firmed in, the plant should be well watered and a heavy mulch applied around the coffee trees, but kept clear from the stem. This will help to decrease soil loss and compaction caused by heavy tropical rainstorms impacting on the soil surface. It is preferable to provide some shelter for young plants, and temporary light shade can be provided by inter-row planting of *Cajanus*, *Tephrosia*, *Crotolaria* or *Sorghum* during the first year.

Spacing

An initial consideration when transplanting coffee is the system under which it will be subsequently grown, as this affects the positioning and spacing of the transplants (see Fig. 12.2). Traditional coffee in forest shade is grown at about 1200 plants/ha, but unshaded coffee is usually at a density of about 2000 plants/ha for robusta and 3000 for arabica. Higher densities of planting are possible using semi-dwarf cultivars, and various configurations have been used giving up to 10,000 plants/ha in very intensive systems. Strip or contour 'hedgerow' planting uses single or double rows of plants at 1.0–1.5 m spacing, with 3–5 m avenues between rows (see Fig. 12.3). This allows easy mechanization and interplanting of other crops (see below).

Planting in partially cleared forest where some trees are left for shade cover is the traditional method, requiring relatively wide and uneven spacing to avoid competition with shade trees – which may be large. Coffee can also be planted

Fig. 12.2. Young coffee plantation.

with other perennial crop plants providing shade, thus giving a mixed cropping system that provides a variety of products. There are many examples of this in smallholder situations, e.g. bananas/robusta in Uganda, coconuts/robusta in parts of Indonesia.

In the mixed perennial tree crop gardens typical of parts of Sri Lanka, South India and Indonesia, coffee may form only a small proportion of total components. Spacing of trees in this system has to accommodate the positioning of other trees and avoid root competition, which can be a problem with some tree crops. Coffee may also be planted with selected shade trees in more intensively managed systems, where shade trees can be planted uniformly or on contours to give more even shading to regularly spaced coffee. The various aspects of shade for coffee in discussed in the next chapter.

Coffee grown in full sun is the most intensive system and has been encouraged by the high yields that can be obtained with high levels of fertilizer input and careful management. This system is particularly suited to modern semi-dwarf cultivars of arabica derived from Caturra, etc. Although coffee trees provide some protection for the soil once fully established, the removal of shade trees has resulted in major erosion problems in Guatemala, for instance, where coffee is grown on steep hillsides. Some form of artificial shade may be required to protect the newly transplanted seedlings. Shade was removed from many of the older coffee plantations in the mid-20th century.

Fig. 12.3. Contour-planted coffee.

The main crop protection concerns at transplantation, in addition to the avoidance of soil-borne pathogens and perennial weeds, are to ensure that the seedlings are free of plant-parasitic nematodes and of diseases such as Fusarium bark disease, tracheomycosis or ceratocytis wilt, and that roots and stems are not damaged. In areas affected by stem borers, a systemic insecticide applied to the soil in the pots will protect the young coffee tree from egg-laying on the bark of the trunk for the first 2–3 years (J. Biscoe, personal communication, Malawi, 2000).

Crop Maintenance

Fertilization

Fertilizer is required to maintain optimal yields from coffee plants, but the quantity and type of fertilizer required differs with soil type and crop management practice. Coffee grown at wide spacing under shade will require less fertilizer than coffee grown at close spacing without shade. Failure to match fertilizer input with expected yield can lead to overbearing, dieback and predispose the crop to diseases such as Cercospora leaf spot and rust.

Mulching, cover plants and shade provide significant amounts of organic matter and nutrients, especially nitrogen (N) from leguminous plants. Organic matter also has the important function of improving soil structure, increasing cation exchange capacity and alleviating mineral toxicity in some situations. In general, young coffee plants have a high requirement for N and phosphorus (P), while in mature bearing plants the greatest need is for N and potassium (K). If levels of available K are adequate, mature coffee plants respond only to applied N. In soils with low cation exchange capacity, a complete fertilizer containing NPK may be required. Willson (1985) provides more details of coffee nutrition.

Soil analysis is helpful in planning macronutrient fertilizer requirements in the coffee plantation, while foliar analysis may be needed for micronutrients and for the diagnosis of suspected mineral deficiencies in the crop (see Chapter 11). Foliar analysis provides information on the nutritional status of plant tissues that is dependent both on how much is available in the soil and on the ability of the plant to take it up.

Ideally, fertilizer should be given in four split applications but, in practice, the heavy labour requirement limits the applications to two per season. Fertilizer should be applied during the rainy season, when demand through crop growth is greatest and when levels in the soil may be depleted through leaching. It should be applied above the rooting zone of the tree within the drip-line of the canopy. The aim should be to maintain or improve soil fertility during the crop cycle, for which the minimum requirement is to replace nutrients removed by each harvest and used in growth of the tree.

It has been estimated that each 100 kg of arabica cherry harvested removes 34 kg of N, 7 kg of P and 40 kg of K from the land (Catani and De Moraes, 1958); similar figures for robusta are provided by Carvajal (1984).

There is then a need to add the nutrients that will be taken up by vegetative growth and lost by removal through pruning, leaching, etc. before assessing the gross fertilizer usage of coffee. While there may be minor differences in nutritional requirements of arabica and robusta, in general, fertilizer recommendations are similar for both (see Table 12.1).

Soil pH

This can provide a useful indicator of fertilizer requirements and likely deficiencies. Coffee grows well in soils with a neutral to acid pH, but at pH (calcium chloride) < 5.0, lime will be required to raise the pH. Red dolomites and brown granites often require lime to reduce the likelihood of manganese toxicity; aluminium toxicity may also be present in soils with low pH. Soils with a pH of 4.5–5.0 require 0.5–1.0 t of lime. Below pH 4.5 this increase to 1–2 t/ha. Lime can be simply ground limestone or dolomitic limestone that contains magnesium. If the soil contains < 0.2 meq magnesium, then half the lime applied should be dolomitic (Clowes *et al.*, 1989). At pH levels > 7.0, iron availability may be impaired.

Nitrogen (N)

In coffee grown intensively for maximum yield without shade, regular applications of nitrogen are essential to prevent over-bearing and to prolong the productive life of the tree. Under less intensive production systems under shade, there may be little response to N. Nitrogen may be applied as urea, sulphate of ammonia or calcium ammonium nitrate. Urea contains the highest percentage of N and is the most readily available source. Sulphate of ammonia should not be used on already acid soils < pH 6.0 unless there is sulphur deficiency. Calcium ammonium nitrate is best used on more acid soils < pH 5.0. Where N is limiting crop growth it may be translocated away from older eaves to supply berries and young shoots. This can lead to the shedding of older leaves.

Table 12.1. Amount of fertilizer (kg) required to provide 1 kg of each nutrient (from Clowes *et al.*, 1989).

Fertilizer	N	P_2O_5	K_2O	S
Compound (e.g. 15:5:20)	6.67	20.0	5.0	10.0
Calcium ammonium nitrate (28% N)	3.57			
Urea (46% N)	2.17			
Sulphate of ammonia (21% N, 24% S)	4.76			4.17
Diammonium phosphate (18% N, 46% P)	5.56	2.17		
Single super phosphate (19% P, 12.4% S)		5.26		8.0
Triple super phosphate (45% P, 2.5% S)		2.22		40.0
Muriate of potash (60% K)			1.67	
Sulphate of potash (0% K, 18% S)			2.22	5.56

N, nitrogen; P, phosphorus; K, potassium; S, sulphur.

Leaf retention is vital for maximising yields, and if N uptake is prevented due to the soil being too dry, supplementary N may have to be applied as a foliar spray on heavily bearing trees. Mulching can help to keep the soil moist and minimize drought-induced N deficiency. Nitrogen fertilizer should be applied after the start of the rains and again as the fruit begins to develop. In areas with a dry season, at the very start of the rains the crop benefits from natural N flush.

Phosphorus (P)

Although P is often chemically immobilized in many tropical soils, the coffee tree is very efficient at taking it up, even when P levels in the soil are low. Phosphorus deficiency symptoms are rare, except where a heavy crop is developing under drought or waterlogged conditions. Phosphorus is always recommended at planting to encourage root development. Phosphorus can be applied as single super phosphate or triple super phosphate. Single super phosphate is usually preferred as it contains 12% sulphur, which can be an important nutrient. Diamonium phosphate provides both N and P. Vesicular arbuscular mycorrhizal fungi in coffee root systems have been shown to assist uptake of P (Vast *et al.*, 1996).

Potassium (K)

Coffee has a high demand for K, particularly when berries are developing. Where the element is in short supply as the crop develops, leaf shedding and symptoms of overbearing may develop. Deficiencies are usually seen only on heavily bearing trees. Potassium interacts with other elements and excess can cause magnesium deficiency. For optimum response to K, the ratio with N should be in balance. Potassium is applied as muriate of potash (K chloride) or as sulphate of potash (K sulphate). Mulching can provide much of the K needs of the plant.

Compound fertilizers

NPK may be provided by compound fertilizers containing varying ratios of the macronutrients and usually contain sulphur as well. The choice of which compound to use will depend on the most economic way of providing the required combination of nutrients.

Micronutrients

Magnesium (Mg) deficiency can occur in coffee grown on soils with a high K content and on light alkaline soils, especially where there is a high level of organic matter. This can be corrected with a foliar application of Mg sulphate or ground application of dolomitic lime. Boron (B) deficiency is associated with soils low in organic matter. Boron can be added to the crop in supplements such as 'Solobor' (20.5% B), which is mixed with water (500 g/100 l) and applied as a full-cover spray, equivalent to 3–5 kg Solobor/ha. Zinc (Zn) deficiency is associated with soil compaction. Zinc can be applied as Zn oxide (250 g/100 l water), equivalent to 2.0–2.5 kg Zn oxide/ha. Iron (Fe) deficiency is quite common due to over-bearing or high soil pH. Where the problem is serious, long-term correction will require the application of sulphate of ammonia to lower the pH.

Organic fertilizers

The application of organic fertilizers such as manure, compost and mulch helps to retain soil moisture and protects the surface soil from erosion by heavy rainfall. As they are incorporated into the soil, they improve soil structure and facilitate root development. There is also the effect that the enhanced microbial status of soils with adequate organic matter content may have on inhibiting the growth of root pathogens. However large the quantity of mineral nutrients in the soil, they can only be taken up by the plant if the roots are healthy and active and there is sufficient moisture for translocation.

Mulching and application of cattle manure is often the best way to provide both the P and K and some of the N required, while also maintaining levels of soil organic matter. Waste products, husks or parchment from primary processing are a good source of mulch. Increasing the organic matter content of the soil also increases the effectiveness of mineral fertilizers.

The aim of fertilizer application should be to maintain the required balance of nutrients in the soil and to maintain soil structure. It may be necessary to invest in soil improvement for some time before the benefit is seen as increased yields. Coffee is unlikely to respond to fertilizer in the short term if low yields are due to poor crop management, poor soil structure or old trees. Once other crop and soil management issues have been addressed and the yield is at least 500 kg/ha without fertilizer, then a yield response representing a cost benefit to fertilizer application may be possible.

Effect on diseases and pests

Apart from the direct effects which deficiencies have (see Chapter 11), poor nutrient status predisposes coffee trees to several diseases such as rust and brown eye spot, causing defoliation and exacerbating over-bearing effects. Diseases themselves may cause poor nutrition if they damage roots or interfere with translocation. Unthrifty trees are also more attractive to white stem borer.

Mulching

Mulching consists of covering the topsoil with organic residues to protect the soil from erosion by heavy rainfall, to retain moisture and to decrease surface soil temperature (see Fig. 12.4). This is particularly beneficial in unshaded coffee and during dry periods. Mulching also reduces weed establishment, helps to maintain soil structure by adding organic matter, enhances the microbial activity in the soil and adds nutrients as it decomposes.

Fig. 12.4. Mulched coffee.

Mulching is practised more in Africa than in Latin America, but increasing land shortage in some areas has led to a decrease in mulching, as sources of mulch become scarce. Cutting, carrying and applying mulch is labour-intensive and is unpopular where labour is scarce or expensive relative to the returns from coffee growing. With the trend towards intensive cultivation of dwarf coffee varieties without shade, there is an argument in favour of mulching. This is particularly so in areas with a defined dry season where soil temperatures may be sufficiently high to cause root damage and the topsoil becomes dry, limiting nutrient uptake.

Soil temperatures in the warmer coffee areas of East Africa can exceed 30°C, resulting in rapid loss of soil moisture and rapid decay of soil organic mater. Root damage occurs at temperatures of 35°C and above. Mulching can reduce the temperature at the soil surface by 4–5°C. Above a soil temperature of 25°C, the rate of humus decomposition exceeds its formation. Mulching helps in two ways: by decreasing soil temperature and by adding organic matter.

Mulching is highly beneficial in tropical zones with long dry seasons and torrential rainstorms by preventing loss of topsoil and fertilizer, increasing rainfall penetration and reducing moisture loss through evaporation. It was observed in Kenya that 230 mm of rainfall penetrated to 1.4 m under a 15 cm mulch of Napier grass (*Pennisetum purpureum*), but to only 6 cm in bare soil trampled by pickers (Pereira and Jones, 1950).

Mulching replaces nutrients removed by the coffee crop. Results from Tanzania show that annual application of a mulch of dried banana leaves (at 25 t/ha) more than replaced the nutrients removed at harvest (Robinson and Hogwood, 1965). Materials vary in their effect: in Ghana, mulching was shown to increase yields by up to 150%, the best sources of mulch being *Tripsacum laxum* (Guatemala grass) and *Chromolaena odorata* (Siam weed), which improved yields considerably more than either coffee husk or banana trash (Afrifa *et al.*, 2003). Mulch supplies a disproportionate amount of K, which may antagonize the uptake of Mg in some soils. If this occurs, a foliar spray of Mg sulphate might be required. The use of sisal residues as mulch in Kenya resulted in a rise in soil pH, with a resultant nutrient imbalance.

A wide range of materials can be used as mulch, including by-products of coffee processing and trash from other crops. Material shedding seeds should be avoided as a source of mulch, and live grass stems may take root at the nodes. Banana leaves are particularly useful and convenient if grown as an intercrop with the coffee bushes. Dried grasses and weeds can also be used, and the larger perennial grasses such as Napier and Guinea grass (*Panicum maximum*) are suitably bulky, but about 1 ha of grass crop is needed for mulching 1 ha of coffee.

Crop residues such as maize and sorghum stover are also used. Mulch is usually applied at the end of the rainy season and should be at least 10 cm deep but leaving a gap around stem bases. Mulching alternate rows each year is also effective. Inert plastic-based materials used for weed control and moisture retention in horticultural crops are generally not economic for use in coffee.

It has been suggested that increased damage from leaf miner in Kenya was associated with the increased use of mulching in the 1950s. This may be due to increased survival of pupae under the mulch. However, the use of broad-spectrum insecticides also increased in Kenya at the same time and this might have had a greater effect on leaf miner than did mulching, by decreasing its natural enemy populations. Mulch reduces the incidence and severity of many diseases, both because it reduces moisture and nutrient stress – which are predisposing factors for many diseases – and the improved microbial status of soils under mulch increases antagonistic effects operating against soil-borne pathogens. Possibly the greatest danger from mulch is fire: a tinder-dry layer of dead stems and leaves at the height of the hot dry season is easily ignited.

Weed control

Coffee feeder roots are in the upper layers of the soil, so the plant is quite sensitive to weed competition. Furthermore, cultivation techniques to remove weeds can damage surface roots. These effects are most noticeable on robusta because of its shallower root system. Perennial grasses and sedges are the most damaging; rhizomes can intertwine with coffee roots and these plants compete strongly for available moisture and nutrients, especially N; *Cyperus* spp. may also be allelopathic and actively inhibit coffee root growth.

Removal of perennial weeds before planting is essential. This can be done with mechanical cultivators or by hand fork hoe, as used by smallholders. In shaded or densely planted coffee, weeds are seldom significant as they are shaded out. In open coffee, some method for controlling weeds is necessary. Mulch will prevent weed growth and cover crops or intercrops will control weeds between rows of widely spaced coffee (see below). Mechanical control by hand-slashing or hoeing is laborious, and care is needed to avoid damage to coffee roots close to the surface and damage to coffee stem bases, a common problem allowing entry of *Fusarium*, *Ceratocystis* and other stem pathogens.

Herbicides such as paraquat (contact) and glyphosate (systemic) are widely used for weed control on coffee. However, weeding around young trees needs to be done by hand, as the trees at this stage are susceptible to herbicide damage. In plantations over 3 years old, a combination of hand-weeding close to the tree and herbicide between the rows can be used. Removal of weeds can create an erosion hazard, especially on sloping land. It may not be beneficial to clean weed outside an area equivalent to the drip line when coffee is irrigated, but weeds should be kept in check by hand or mechanical slashing before they are able to set seed. After weeding around the trees, replacing the mulch will suppress weed regrowth and reduce erosion of the exposed topsoil. Cover crops are sometimes used to decrease weed establishment and prevent erosion (see below).

Cover crops and intercrops

Leguminous cover crops are sometimes used to good effect in coffee plantations and serve many of the same functions as mulch, with the added advantage of N fixation. They can be used in unshaded, widely spaced coffee, but also provide a temporary check on topsoil erosion when the soil is bare in a new plantation. The disadvantages are mainly the labour required to keep the growth in check and the competitive effect of the cover crop on the coffee. In drier areas, competition for moisture is a critical factor and the main benefits of added N and weed suppression occur in areas with adequate moisture throughout the year. The weed-suppressant effect may be important where labour for weeding is scarce or expensive.

Cover crops are more suited to robusta areas, but choice of species is critical. Numerous procumbent, spreading legumes are used as ground cover, e.g. *Indigofera endecaphylla*, *Calopogonium mucunoides*, *Vigna oligosperma*, *Crotolaria anagyroides*, *Mucuna* spp., *Arachis pintoi* and *Mimosa invisa*.

Intercroppping can take many forms. Permanent perennial intercrops play an important role both as shade and in providing a diversified income; annual crops such as beans are also important in providing an alternative source of income, and are often grown between rows of young or stumped coffee. Intercrops such as pigeon pea and sorghum can provide a degree of temporary shelter and shade for young coffee.

The benefits of intercropping include weed control and erosion prevention, and the coffee may also benefit from fertilizers applied to the intercrop. However, the intercrop may compete with the coffee for moisture and nutrients, and care has to be exercized to ensure that the cultivation of intercrops does not damage coffee roots close to the surface. Cover crops and intercrops generally enhance pest and disease management by improving soil microbial status and providing a habitat for a more diverse fauna, including the natural enemies of pest species.

Irrigation

Irrigation may be used to supplement rainfall in areas with a long dry season or prone to periods of drought, more commonly encountered in non-equatorial areas. Irrigation is more applicable to arabica cultivation, and seldom used for robusta. Irrigation is used in parts of East Africa, India, Brazil and Vietnam, but mainly in areas with a fairly level topography. In sun-grown coffee, especially the high-yielding dwarf varieties, high yields and the avoidance of over-bearing requires high fertilizer input.

To maximize the benefits of fertilizer, the soil must be moist throughout the period of flower and fruit development, so that irrigation is required in areas experiencing dry conditions during this time if die-back is to be avoided.

Irrigation of young, establishing coffee trees encourages root growth in the upper soil layers and may hinder development of deeper axial roots unless sufficient irrigation water is applied to penetrate below the root zone and the soil is allowed to dry out between irrigations. This can make the plants more susceptible than non-irrigated trees to drought if irrigation ceases to be available.

Irrigation can be used to control flowering and, in Cameroon, early flowering initiated by irrigation during the dry season allowed the young expanding berries to escape CBD. At the end of the dry season, irrigation can be used to initiate a uniform flowering, which is more amenable to pest and disease control operations than is a staggered crop.

Overhead irrigation is often used on large plantations, but this can erode fungicide deposits and make control of diseases such as rust and CBD more difficult. The most efficient way to apply irrigation water is directly to the root zone by drip and trickle irrigators or mini-sprinklers. However, these techniques require more technical and capital input. Care should be taken to avoid over-irrigation, which leaches N from the soil.

Pruning

The objectives of pruning are: (i) to control the size and shape of the tree; (ii) to maintain a healthy open canopy; and (iii) to control the cropping level to avoid over-bearing and biennial bearing. Pruning removes the less productive parts of the tree and opens the canopy to sunlight to encourage flowering and fruit-set. The principles of coffee pruning are based on the type of growth and cropping pattern that the tree exhibits; vertical extension is provided by orthotropic branches, whereas the crop is borne on recently mature plagiotropic branches growing laterally from the nodes of the orthotropic stem. These then produce a succession of secondary and tertiary plagiotropic stems that in turn bear fruit after one season.

There are two basic types of pruning: free growing – usually multiple stem – and capped – usually single stem, but with several variations on these (see Fig. 12.5). In the free-growing system, orthotropic stems with associated series of plagiotropic stems grow more or less unchecked, apart from some thinning of older, crowded plagiotropic stems until they become too tall, when they are cut down and replaced by new, young orthotropic stems growing from the stump.

This is the simplest form of pruning and, for semi-dwarf cultivars such as Caturra and Catimor, all that is required is regeneration by stumping or replanting after 7–10 years. High yields may be possible without pruning, provided the crop is well managed and fertilized to prevent over-bearing. For standard cultivars a multiple-stem system is usually adopted, whereby a succession of three to four stems is raised from each plant, and replaced as they become too large, by new orthotropic shoots raised in succession from the base of the tree.

With the capped pruning system, vertical growth of the orthotropic stem is stopped at about 2 m by removal of the terminal shoot and any succeeding shoots. A system of plagiotropic branching established by careful pruning, in which the development of a succession of new, lower-order shoots is induced to provide the required bearing wood.

Opening the canopy by thinning plagiotropic stems increases air flow and discourages diseases such as web blights and CBD that require surface moisture. Antestia bug is less attracted to open canopies. Low branches that touch the ground should be cut back, as these allow ants to climb into the canopy and attend to pests such as scales and mealy bugs. However, pruning creates wounds and can allow development of bark and stem diseases caused by *Fusarium*, *Ceratocystis* and similar pathogens.

Arabica coffee trees are prone to biennial bearing, brought on by the tendency for arabica to over-bear. In a year when the crop is heavy, the tree sacrifices the following year's bearing wood to support the current year's developing crop. As a consequence, the following season's crop is small and vegetative growth is enhanced, to give another heavy crop the following season.

To break this pattern of biennial bearing, a suitable ratio of leaf to crop must be maintained by pruning to reduce the risk of die-back of the primaries resulting from over-bearing. This also avoids predisposition to the range of diseases exacerbated by physiologic stress and maintains bean quality by ensuring that the tree is able to adequately fill the crop that it is carrying. Robusta is not affected by this condition, and so requires less pruning than Arabica.

Pruning should be carried out immediately after harvest. Correct pruning requires training and some knowledge of coffee tree morphology – and of its

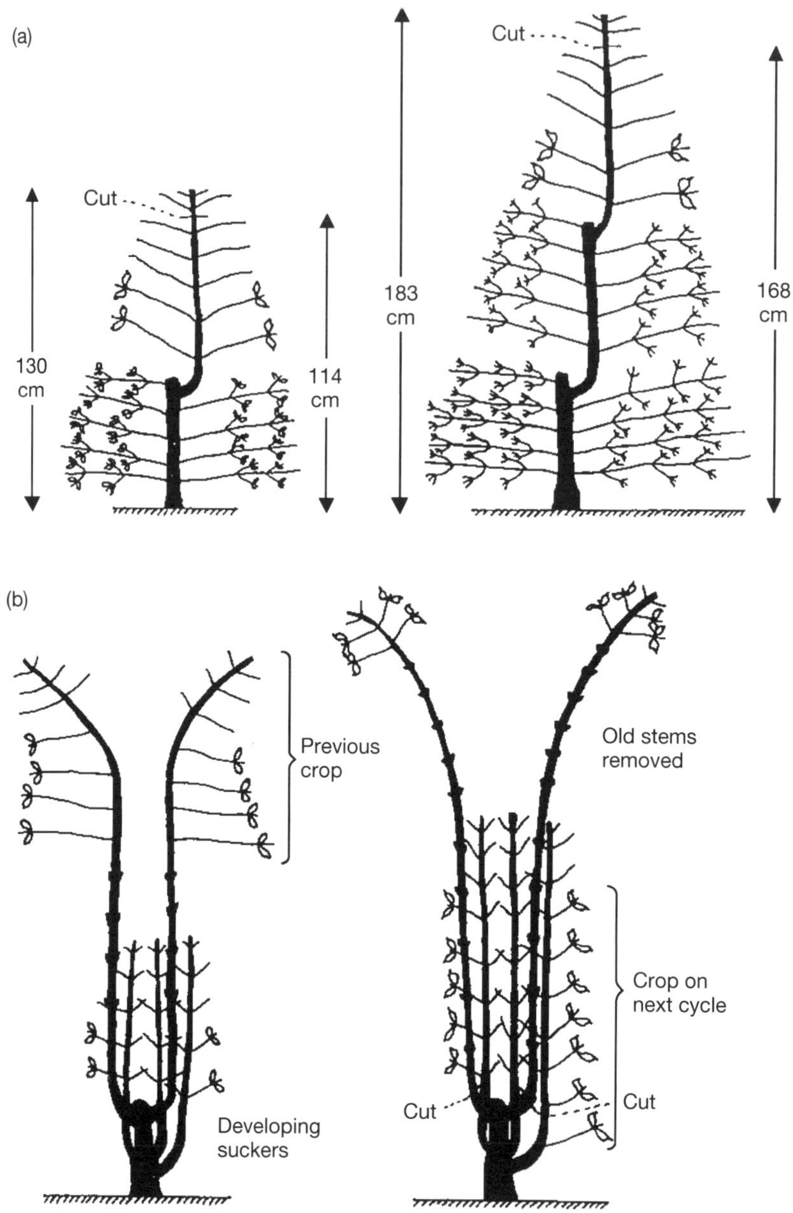

Fig. 12.5. Diagram of capped single-stem (a) and multiple-stem coffee (b) (adapted from Fernie in Robinson, 1974).

cropping potential under the relevant management system. The various pruning systems are described in detail by Lambot and Bouharmont (2004) and in most local handbooks describing the practical aspects of coffee agronomy – and in general coffee texts referred to below.

References

Afrifa, A.A., Ofori-Frimpong, K., Appiah, M.R. and Halm, B.J. (2003) Effect of mulching on soil nutrients and yield of Robusta Coffee. *Tropical Agriculture* 80, 105–109.

Cambrony, H.R. (1992) *Coffee Growing*. Macmillan Press Ltd., London, 119 pp.

Carvajal, J.F. (1984) *Cafeto – cultiva y Fertilizacion*. Institto Internacional de la Potasa, Bern, Switzerland.

Catani, R.A. and De Moraes, F.R.P. (1958) A composico quimica de cafeeiro. Quantidade e distribucao de N, P_2O_5, CaO e MgO em cafeeiro de la 5 anos de idade. *Revista de Agricultura, Brazil* 33, 45–52.

Clarke, R.J. and Macrae, R. (eds) (1988) *Coffee. Vol. 4, Agronomy*. Elsevier, London.

Clowes, M., Nicoll, W.D. and Shelley, R.S. (eds) (1989) *Coffee Manual for Malawi*. Tea Research Foundation of Central Africa, 224 pp.

Coste, R. (1992) *Coffee: the Plant and the Product*. Macmillan Press Ltd., London, 328 pp.

Descroix, F. and Wintgens, J.N. (2004) Establishing a coffee plantation. In: Wintgens, J.N. (ed.) *Coffee: Growing, Processing and Sustainable Production*. Wiley-Verlach, Weinheim, Germany, pp. 178–245.

Lambot, C. and Bouharmont, P. (2004) Pruning. In: Wintgens, J.N. (ed.) *Coffee: Growing, Processing and Sustainable Production*. Wiley-Verlach, Weinheim, Germany, pp. 284–307.

Pereira, H.C. and Jones, P.A. (1950) The maintenance of fertility in dry coffee soils. *East African Agricultural Journal* 15, 174–179.

Robinson, J.B.D. (1974) (ed.) *A Handbook on Arabica Coffee in Tanganyika*. Tanganyika Coffee Board, Moshi, Tanzania.

Robinson, J.B.D. and Hogwood, P.H. (1965) Effects of organic mulch on fertility of a lactosolic coffee soil in Kenya. *Experimental Agriculture* 1, 67–80.

Siqueira, J.O., Saggin Jn., O.J., Colozzi, F.A. and Oliveira, E. de (1995) Influencia do substrata de formacao e de micorriza no crescimento de muda de cafeeiro transplantadas. *Pesquisa Agropecuaria Brasileira* 30, 1417–1425.

Vast, P., Zasoski, R.J. and Bledsoe, C.S. (1996) Effects of vesicular arbuscular mycorrhizal inoculation at different soil P availabilities on growth and nutrient uptake of *in vitro* coffee (*Coffea arabica* L.) plants. *Mycorrhiza* 6, 493–497.

Willson, K.C. (1985) Mineral nutrition and fertilizer needs. In: Clifford, M.N. and Willson, K.C. (eds) *Coffee: Biochemistry and Production of Beans and Beverage*. Croome Helm, Kent, UK, pp. 135–156.

Wrigley, G. (1988) *Coffee*. Longman Scientific and Technical, London, 637 pp.

13 Shade Management, Conservation and Biodiversity

Introduction

Coffee originated as an understorey shrub in the forests of the Ethiopian highlands and is adapted to growing in shade. Most of the world's coffee was originally grown under the shade of natural forest or trees planted for that purpose. However, coffee flower initiation is stimulated by light and higher yields are obtained when coffee is grown without shade provided that appropriate management strategies are adopted.

Much of the shade was removed from the larger coffee plantations in parts of East Africa and Brazil in the mid-20th century and new areas were planted without shade. From the 1970s, the semi-dwarf cultivars such as Caturra and Catuai that were better adapted for growth in full sun became widely planted in South America, and there was an increasing trend to remove shade trees and convert to high-input systems to obtain higher yields. Intensive management was also compatible with the requirements for the higher returns to offset the costs of the control of coffee rust, which was spreading through South and Central America.

The destruction of shade trees in coffee plantations gathered pace in the 1980s, and much forest and its associated biological diversity has been lost in the process (Muschler *et al.*, 1997). By the year 2000, over 50% of the land planted to coffee in Mexico, Colombia and the Caribbean had been converted to production in full sun. Nevertheless, the value of shade was still recognized as essential in exposed areas, and as an integral part of cultivation systems and for smaller coffee holdings where the intensive management required for sun-grown coffee was not feasible.

Because areas suitable for coffee cultivation are often forested hillsides, land clearance for coffee farming has become an important conservation issue. Many coffee-growing areas are significant centres of forest biodiversity, and trees used as coffee shade can help to preserve this. Although the introduction of farming into previously natural areas will inevitably impact on wildlife and displace larger animals, retaining some of the indigenous trees and growing coffee under their shade minimizes habitat destruction and species loss (Perfecto *et al.*, 1996). With falling coffee prices in the second half of the 1990s and concerns about environmental degradation, calls to reverse the trend towards sun-grown coffee were heard, and research undertaken to assess the value of shade trees for the sustainability of coffee culture and for their contribution to biodiversity.

Systems of Shade-grown Coffee

The simplest way to plant shade-coffee is to cultivate the ground under natural forest trees, and this represents a sustainable way of managing native forest. Alternatively, new shade trees can be planted when the coffee plantation is established, and these may be indigenous or introduced species. The shade might be of a temporary nature to protect only the young trees, permanent or a combination of both, with quick-growing trees to provide shade for the transplanted seedlings until the canopy of the more permanent trees is established.

Rustic coffee

The most ecologically sustainable method of planting coffee is in partially cleared indigenous forest where the understorey and some trees are removed but selected trees are left as shade cover. In its simplest form this is a low-input system requiring little initial investment, but yields are low. This method has been used traditionally in Latin America and Asia, provides the best protection for biodiversity and causes least soil erosion – often referred to as sustainable, eco-friendly coffee.

Polyculture or mixed-shade coffee

In order to conserve biodiversity and protect the soil, the next best option is to grow coffee under planted shade trees. In polyculture, shade is provided by a mixture of indigenous trees similar to those found in nearby natural forest, mixed with fruit trees (see Fig. 13.1), often found to provide sustainable livelihoods for smallholders who can benefit from timber, firewood and fruit, in addition to the coffee harvest.

There are various forms of polyculture, from traditional systems – similar to the rustic management system with some modification to the natural vegetation, such as the addition of fruit trees that may actually increase canopy diversity – to commercial polyculture systems, usually on a larger scale, more intensively managed and often with less shade diversity. Temporary shade cover can be provided by quickly establishing tall plants (banana, pigeon pea,

Fig. 13.1. Coffee under tall shade, India.

tall sorghum) that will also reduce soil erosion while the more permanent shade trees establish. The main problem from a crop health point of view is that roots of many trees will compete with coffee, and resultant stress will predispose the plants to various diseases.

Reduced-shade culture

This substitutes the natural forest with introduced shade trees planted at regular spacing and often of a single or few species such as *Inga*, *Erythrina*, *Gliricidia* or *Grevillea*. The forest is cleared, trees sold for timber or otherwise disposed of before coffee is planted and new shade trees established. The range of shade trees commonly grown is discussed below. There is a high risk of soil loss due to erosion when the forest is cleared and before the coffee and shade trees become established, especially on hillsides. Temporary shade must be provided while the trees become established.

Effects of Shade

Climatic buffering

The microclimate in coffee plantations grown under shade trees can be very different from that in full-sun plantations. The main effects are: (i) altered temperature and light intensity; and (ii) reduced wind velocity. Shade trees have the effect of decreasing the maximum temperature and raising the minimum temperature. This may be important in protecting the coffee bushes at high altitudes, for instance where there are large diurnal variations in air temperature that can induce 'hot and cold' disease. Temperatures at the top of the coffee canopy may be as much as 7°C lower during the day and 3°C higher at night in shaded coffee compared to open-field sites. Shading also protects coffee trees from frost damage. In southern Brazil, shade trees such as *Mimosa scabrella* are vital for frost protection (Calamori *et al*., 1996).

Decreased light intensity under shade regulates production by restricting flower initiation, so that there is a better physiological balance between cropping level and plant nutrition. This prevents problems with biennial bearing, over-bearing and dieback, thus extending the productive life of the coffee bushes. For lower-input systems there is the added benefit that fertilizer rates may be reduced or in some smallholder systems or dispensed with altogether, without running into problems of die-back. However, the lower yield obtained under shade is one of the primary reasons why much coffee is grown in full sun, where yields can be doubled under ideal conditions and with intensive management.

Protection from wind and storm damage is an important function of shade trees in some areas. In addition to decreasing direct mechanical damage caused by high wind, protection from cold or hot drying winds is also important for healthy coffee plants. Lower wind speeds in shaded coffee result in less evapotranspiration and assist with soil moisture conservation. Trees may be planted in strips to serve as windbreaks, but the species selected needs to be tall and resistant to wind damage. Protection from heavy rain and hail can also be a significant benefit.

Soil and water retention

Where coffee plantations are established on hillsides, shade trees help to stabilize the topsoil and prevent erosion. Soil compaction and run-off under heavy rain are reduced and percolation into deeper soil layers assisted. The protective effect is greatest when the coffee plantation is established under already existing trees, rather than clear-felling the land before replanting. Sometimes it may be possible to retain at least some of the indigenous forest trees within the coffee plantation. Early experiments with shade removal in Guatemala failed due to the combined effects of soil erosion and inadequate fertilizer application for prevention of over-bearing.

The organic matter content of soils in shaded coffee is generally higher than in unshaded coffee plantations. There is a reduced rate of oxidation of organic matter, as the soil is protected from direct sun radiation and temperatures are lower. Falling leaves from shade trees also provide a small amount of mulch, and leguminous shade species contribute nitrogen to the soil. Not only is the moisture content of soils under shaded coffee greater than that under sun coffee, but organic matter content is also higher (Rice, 1991). The general effects of trees on soil fertility have been fully documented by Schroth and Sinclair (2003).

Pest and diseases

Weed growth is greatly reduced under shade, and a number of pests and diseases are also less damaging. Soto-Pinto *et al.* (2002) have reported the effects of shade on berry borer, rust and weeds in Mexico. In India, the stem borer *Xylotrechus quadripes* is a less serious pest in shaded coffee as egg-laying is inhibited, and this may also be the case with a similar pest in eastern and southern Africa, *Monochamus leuconotus.* Although preferring a dense canopy on individual trees, Antestia bug is also less serious in shaded coffee. Leaf miner incidence is less in shaded coffee but berry borer is favoured by excessive shading, and twig borer on robusta is more prevalent under shade.

Reduced incidence of rust in shaded coffee is one aspect of the effective integrated management of this disease, and is related to decreased physiologic stress, but hyperparasitic fungi also appear to be more prevalent in rust pustules under shaded coffee. Cercospora leaf spot is another disease that is seldom problematic under shade.

However, the more humid environment under shade can exacerbate some diseases, especially when shade becomes too dense. These include the web blight fungi, *Corticium koleroga* and *Erythricium salmonicolor*, and South American leaf spot caused by *Mycena citricolor*. Some shade tree species can also be hosts of these fungi and thus provide very effective inoculum sources for infecting coffee beneath them. Root diseases caused by root pathogens of shade trees such as *Armillaria* and *Ganoderma* can also spread from infected shade trees and the remnants of old forest trees.

Coffee quality

There is mounting evidence indicating that the quality of coffee produced under shade is greater than that of sun-grown coffee. As growth is slower and flowering is physiologically controlled, beans are larger and tend to be denser, without 'lights' (incompletely filled beans). Furthermore, there is less damage to berries caused by sun scorch, Cercospora berry blotch and associated problems. Depending on the premium paid, both for quality and other ecological attributes of shade coffee (see later), these advantages can compensate economically for the lower yields obtained under shade (Muschler, 2004).

Disadvantages

The environmental advantages of shade for coffee are quite clear but there are some disadvantages for production, the most important of which is suppression of yield. Shade trees themselves also require management. As the trees grow and their canopies close, shade levels can become too dense, resulting in excessive yield suppression and the creation of a dark, moist microclimate that favours diseases.

The canopies of shade trees therefore require pruning or thinning-out to maintain optimal shade levels, but as the trees increase in height this becomes an ever more demanding task. Shade trees may also compete with coffee for water and nutrients, so care is needed in selection and spacing. This can be a major factor in polyculture, where some fruit trees commonly grown in these systems are shallow-rooted. Another factor to be considered is the damage to coffee caused by fallen boughs of shade trees and during the felling of timber trees. The effect of shade on pest and disease levels differs with the particular pest or pathogen and with shade levels but, overall, moderate shade is beneficial for their control.

Choice of Shade Tree

Shade should aim to provide a homogeneous canopy and allow adequate air circulation above the coffee. The optimum level of shade is generally bout 30–40% light interception. Suitable shade trees should be deep-rooted to avoid competition with coffee roots and to aid recirculation of nutrients and moisture from the deeper layers of soil. They should also be locally adapted, long-lived, resistant to wind damage and tolerant of lopping. Leguminous trees have the advantage of fixing atmospheric nitrogen that can become available to coffee (see Fig. 13.2). Species producing large amounts of seed and those with a deciduous tendency should be avoided.

Fig. 13.2. Coffee under *Inga* shade, Central America.

Until recently in Central America, shade was regarded as necessary for successful cultivation of arabica, and species of *Inga*, *Erythrina*, *Albizia* and *Pithecolobium* are used as shade trees; *Mimosa scabrella* is widely used in Brazil and *Gliricidia* is used as shade cover for both cocoa and coffee. In India, a number of leguminous trees such as *Acrocarpus*, *Adenanthera*, *Albizia* and *Dalbergia* and non-legumes such as *Ficus* and *Atrocarpus* have been used as traditional shade trees in coffee plantations. In addition, *Grevillea robusta* (see Fig. 13.3) and *Erythrina lithosperma* have been introduced, and these are widely used in other parts of Asia and Africa.

In Indonesia, *Leucaena* spp. have been the shade tree of choice, but some species produce an abundance of seed that can cause a weed problem when they germinate, and attacks by the *Leuceana* psyllid can cause substantial defoliation. In East Africa, some indigenous trees such as *Ficus thoningii* and *F. natalensis* have traditionally been used to shade robusta plantations. A number of trees have been introduced to Africa to provide shade for coffee planations, such as *G. robusta* and *Leucaena* spp. *Gliricidia* is also used to provide quick shade, but this tends to suffer from seasonal defoliation. Other species used include *Casuarina* (in South East Asia), *Croton* and *Cordia alliodora* as a source of timber. Prunings from shade trees can be used as mulch or cattle fodder.

Fig. 13.3. Coffee with *Grevillea* shade, East Africa.

The shade trees themselves have insect pests, which may become a source of infestation for coffee. In Puerto Rico, the ant *Myrmelachista ambigua ramulorum* hollows out branches of old shade trees such as *Inga vera* and *Inga laurina*, where it tends scale insects. From there the ants invade the coffee, eating out longitudinal channels in the pith of the fruit-bearing branches, which become weakened and may break off at harvest time. Shade trees may also be a source of leaf-eating caterpillars, which are not normally pests of coffee but may move onto coffee when their food supply becomes depleted.

The composition of shade trees in coffee plantations may be diverse, as in the polyculture systems referred to above, consisting of trees that provide timber and firewood as well as fruit trees. Common mixed perennial crop systems include robusta coffee with bananas, especially the tall African types grown in Uganda (see Fig. 13.4), and with coconuts in Indonesia. These can give adequate shade cover, with the added advantage of providing a food crop and a supply of mulch.

The mixed perennial crop gardens common in Sri Lanka, parts of South India, Indonesia and parts of Mexico are examples where coffee forms a component of the system. Mixed cropping with coffee is common in smallholder plantations in India where crops such as cardamom, black pepper and citrus provide a diverse source of income (see Fig. 13.5). Shade trees are used to support the black pepper vines (Korikanthimath *et al.*, 1997).

Fig. 13.4. Coffee with banana shade, Uganda.

Fig. 13.5. Coffee under areca palm with pepper and vanilla, Kerala, India.

In Venezuela, smallholders grow a mixture of timber and fruit trees such as banana, citrus and guava (Escalante, 1995). Many insect pests of coffee are shared by tree crops such as citrus and guava, so intercropping can be a mixed blessing from the pest control point of view, the intercrops providing a source of not only pest species but also their natural enemies. In addition to the conservation of biodiversity, inclusion of timber and fruit trees in the coffee plantation provides income diversity that can be an important livelihood strategy. The income from the non-coffee components of the plantation may be substantial and can exceed the income from coffee when coffee prices are low.

Shade-grown Coffee and Biodiversity

'Modernization' or 'Technification' of coffee growing, which began in Latin America but has spread to other parts of the world, was driven by the need to maximize productivity by using a more intensive management system. This was based primarily on the use of high-yielding semi-dwarf cultivars that responded well to fertilizer and could be planted at a higher density than the traditional taller coffee varieties such as Mundo Novo. Exposure to full sun stimulated flowering, and fertlizer usage enabled a heavier crop to be carried. Therefore, shade was either greatly reduced or removed altogether.

In the early days of the trend towards intensification of coffee-growing methods, Caturra and Catuai were the preferred varieties and became widely planted. However, when coffee rust arrived these were found to be susceptible and, more recently, Catimor cultivars and similar hybrids that have been selected for rust resistance have been favoured.

'Technification' generally achieved the main objective of increased yields and often increased profitability, but at the cost of increased fertilizer use, soil erosion and loss of biodiversity (Perfecto *et al.*, 1996). The sustainability of sun-grown coffee is also in doubt due to the heavy requirement for fertilizer, soil erosion caused by tree removal and the need to replant every 10 years or so. With increasing management costs and falling coffee prices, the economics of such coffee has become less persuasive.

Trees

Species richness generally has been found to decrease with increasing intensification of coffee cultivation, where the intensification gradient is defined as a reduction in the number of species of shade tree and of shade cover. The species diversity of shade trees can be large in some systems. In the Corg district in the western Ghats of India, 129 tree species were recorded in 12 plantations. In Central America, it is not unusual for traditional smallholder coffee plantations to have at least 25 species of associated fruit and timber trees, and a similar range of species occurs in the perennial crop gardens of Sri Lanka.

Seventy-five woody species were recorded in polyculture systems in Chiapas, Mexico, of which 40% were native species and most of the rest were introduced species of fruit trees or shrubs (Soto-Pinto *et al.*, 2001). It has been suggested that there is a significant role for shade-grown coffee based on indigenous vegetation, as a diversity refuge for woody plants that would be expected to support a diverse fauna, with the added value that growers can market their coffee into biodiversity-friendly niche markets.

Vertebrates

Effects on the biodiverstiy of other groups of associated organisms are also apparent in shaded coffee. Particular attention has been paid to birds in Central America, because of the significance of the ecological corridor linking the two halves of the continent and its relevance to bird migration. In Costa Rica, it was found that conversion from forest to coffee grown without shade trees diminished nesting sites for 'forest-interior' bird species, but numerous forest-edge and open-country species were able to nest in and around coffee plantations (Lindell and Smith, 2003).

In eastern Chiapas, Mexico, much of the remaining forest vegetation is managed for coffee production. Where shade is provided, it is either indigenous pine-oak woodland or a planted canopy, dominated by *Inga* spp. Both of these systems were found to harbour a large and diverse bird population. The combined species list contained 180 species, with 105 species recorded in each plantation type. The diversity of bird species exceeded that found in other habitats nearby. In the region, only moist tropical forest has greater abundance of bird species (Greenberg *et al*., 1997).

While the numbers of species of ants and butterflies decreased with the decrease in shade cover, birds declined at one site but increased at another (Pefecto *et al*., 2003). Bird species diversity was correlated with distance from forest fragments and not with habitat type, suggesting that leaving the remnants of indigenous forest among coffee plantations may help to maintain biodiversity.

A reduction in mammalian populations has also been associated with the removal of shade cover in coffee plantations in Veracruz, Mexico, where a 43% loss of diversity was recorded. A high diversity in the shade stratum was recommended for protection of the mammalian population, with species such as *Inga jinicuil*, *Musa* spp., *Citrus* spp., *Persea americana* and *Mangifera indica* (Gallina *et al*., 1996).

Invertebrates and microorganisms

The biodiversity of invertebrates and microorganisms is of particular relevance to pests and pathogens. There is ample evidence that pest and pathogen populations are constrained by the biodiversity present in natural ecosystems above and below ground and that monoculture, where biodiversity has been shown to be much reduced, favours expansion of these populations to the extent that control interventions are needed. The large diversity of these organisms associated with a varied flora supports a wealth of natural biological control mechanisms that are essential to the sustainability of crop production.

Despite the major significance of these groups of organisms in ecosystem function, there has been only limited investigation of their influence in coffee.

The nutrition of coffee plants also benefits from the microbial diversity associated with shade trees. Leguminous shade trees fix nitrogen through *Rhizobium* bacteria in their roots, and free-living, nitrogen-fixing bacteria inhabit the phyllosphere of tropical trees. There is also evidence that vesicular arbuscular mycorrhizal fungi are more prevalent in agroforestry soils than in coffee monoculture soils (Cardosa *et al.*, 2003).

During the early 1990s, the coffee agro-ecosystem in the Central Valley of Costa Rica underwent a major transformation to capital-intensive monoculture, where all shade trees were eliminated. Loss of shade trees was found to be associated with a significant loss of arthropod species living in the coffee bushes (Perfecto *et al.*, 1997). As well as indicating a general decline in biodiversity, a decrease in arthropod species has implications for natural control of insect pests in coffee plantations. The complex multitrophic interactions of arthropod populations and how they can be influenced to suppress pest species in shaded coffee has been reviewed by Staver *et al.* (2001).

Economics

Vaast and Harmand (2002) postulated that agroforestry systems of coffee cultivation are economically less risky and more sustainable than coffee monocultures. Shade-grown systems help to diversify farmers' sources of income through timber and fuel wood production and decrease pressure on the forests. There has been a trend to replace indigenous forest with coffee monocultures shaded by single species such as trees of the species *Inga*. However, Peeters *et al.* (2003) have shown that the income derived from tree wood is no greater in the Inga system than in the more diverse system, while coffee yields were similar. They conclude that there is no benefit to be derived from the additional labour required to partially clear the natural forest for planting of *Inga* trees.

Niche Markets for Shade-grown Coffee

Since the early 1990s, there have been numerous projects around the world promoting the conservation of biodiversity by maintaining or returning to shade-grown coffee under indigenous trees. The UK government's Darwin Initiative, for instance, funded one such project in El Salvador, where 80% of the country's forests are associated with shade coffee plantations. The World Bank and the Global Environment Facility have sponsored the Rainforest Alliance (RA) and local partners to evaluate the environmental and livelihood implications of sustainable coffee cultivation.

Forested coffee farms were found to be vital to the survival of 188 avian, 31 mammalian, 26 reptilian and 326 plant species. Smallholders have benefited through reduced costs by using organic mulches and biological control and have been able to obtain premium prices for their crop by selling to organizations marketing 'eco-friendly' products.

Once the increasing trend towards sun-grown coffee became a conservation issue, several conservation groups, especially those concerned with shrinking avian populations, began to campaign for the retention and restoration of shade-grown coffee. This offered an opportunity for coffee-marketing companies to exploit niche markets for speciality coffee by promoting eco-friendly or biodiversity-friendly coffee. This was also a new market opportunity for smallholders in Latin America and elsewhere, who still grew their coffee under shade, and would be able to sell their coffee at a time when supply was exceeding demand on the world markets. One such project was launched in El Salvador, specifically with the aim of encouraging shade cultivation of coffee by promoting biodiversity-friendly coffee, supported by a certification scheme (Gobbi, 2000).

The *Conservation Principles for Coffee Production*, published in 2001, was commissioned jointly by three environmentalist groups: Conservation International, the RA and the Smithsonian Migratory Bird Centre (SMBC). This was a comprehensive set of guidelines for producing environmentally friendly coffee, and its publication reflected widespread concern about the destruction of shade trees on coffee plantations. The principles outlined conditions and practices that apply to farms and processing facilities in coffee-growing regions throughout the world, and provided a foundation for conservation-based certification programmes.

The principles were divided into seven main areas of concern:

- Ecosystem and wildlife conservation
- Soil conservation
- Water conservation and protection
- Energy conservation
- Waste management
- Pest and disease management
- Sustainable livelihoods.

Certification schemes for 'Bird-friendly coffee' and 'Eco-OK' generally recommend 11–12 different tree species, preferably indigenous, non-deciduous trees and forming at least two strata, with emergent crowns of the tallest being at least 15 m high.

Organizations such as the RA and the SMBC have developed partnerships with multinational coffee-marketing companies to promote sustainable coffee-growing practices. In 2003, two of the world's largest coffee brokers, Neuman Kaffee Gruppe and the Volcafe Group, signed agreements with the RA to promote environmentally and socially sustainable coffee production. The RA aims to bridge the gap between certified sustainable coffee farms and coffee consumers around the world. The two traders are currently working with sustainable coffee projects in Central America, Peru, Ethiopia and Uganda.

The agreements with the RA cover consultation on best practice and standards in producing countries and marketing agreements in consuming countries. On coffee farms, the agreements will promote coffee grown sustainably under forest cover, rather than sun-grown, monoculture plantations. The RA provides technical support to farmers within their projects to incorporate environmentally sustainable practices into their farm management strategies.

The RA also works with Kraft foods, who buy 10% of global coffee production, to develop a range of coffees produced from sustainable sources. Kraft has committed to purchase > 2000 t each year from farms in Brazil, Colombia, Mexico and Central America that have been certified as sustainably managed by RA. Kraft also supports training of local specialists to assist farmers in achieving certification. Monitoring and verification is provided by RA and its partner organizations such as the Sustainable Agriculture Network, a coalition of non-profit conservation groups in El Salvador, Guatemala, Mexico, Nicaragua, Panama, Costa Rica, Brazil and Colombia.

In the UK, the Royal Society for the Protection of Birds (RSPB) also sees the coffee crisis and loss of shade coffee as a threat to biodiversity in general, and more specifically to bird populations. The RSPB teamed up with Birdlife International (BI) to promote sustainable land use in coffee areas and has proposed the establishment of a global fund to support sustainable production of commodity crops. The idea of a 'sustainability fund' is supported by the Global Alliance on Commodities and Coffee and by the UN (RSPB, 2003).

Organic Coffee

Organic coffee is another important niche market in which shade is an essential ingredient necessary to enable sustainable coffee production without the inputs prohibited under regulations for organic coffee. The International Federation of Organic Agriculture Movements (IFOAM), which issued *Guidelines for Organic Coffee, Cocoa and Tea* in 1996, has set the standards for organic production. These include a range of measures to ensure conservation of the environment and its biodiversity and the sustainable, pollution-free production and processing of coffee. There are a number of different regulatory systems that vary between regions (the EU differs from the USA, for example), and some 25 certification authorities across the world. Hence organic coffee production (or production for other niche markets) requires significant attention to bureaucratic procedures. The subject has been reviewed by van Elzakker (2001).

References

Calamori, P.H., Androcioli, F.A. and Leal, A.C. (1996) Coffee shade with *Mimosa scabrella* for frost protection in southern Brazil. *Agroforestry Systems* 33, 205–214.

Cardosa, I.M., Boddington, C., Jansen, B.H., Oenema, O. and Kuyper, T.W. (2003) Distribution of mycorrhizal fungal spores in soil under agroforestry and monocultural coffee systems in Brazil. *Agroforestry Systems* 58, 33–43.

Escalante, E. (1995) Coffee and agroforestry in Venuzuela. *Agroforestry Today* July–December, 5–7.

Gallina, S., Mandujano, S. and Gonzales, R.A. (1996) Conservation of mammalian biodiversity in coffee plantations of Central Veracruz, Mexico. *Agroforestry Systems* 33, 13–27.

Gobbi, J.A. (2000) Is biodiversity-friendly coffee financially viable? *Ecological Economics Amsterdam* 33, 267–281.

Greenberg, R., Bichier, P. and Sterling, J. (1997) Bird populations in rustic and planted shade coffee plantations of Eastern Chiapas, Mexico. *Biotropa* 29, 501–514.

Korikanthimath, V.S., Ravindra, M., Rajendra, H. and Hosmani, M.M. (1997) Coffee, cardamom, black pepper and mandarin mixed cropping system – a case study. *Journal of Spices and Aromatic Crops* 6, 1–7.

Lindell, C. and Smith, M. (2003) Nesting bird species in sun coffee, pasture and understory forest in southern Costa Rica. *Biodiversity and Conservation* 12, 423–440.

Muschler, R.G. (2004) Shade management and its effect on coffee growth and quality. In: Wintgens, J.N. (ed.) *Coffee: Growing, Processing and Sustainable Production*. Wiley-Verlach, Weinheim, Germany, pp. 391–418.

Muschler, R.G., Bonnemann, A. and Huttl, R.F. (1997) Potentials and limitations of agroforestry for changing land-use in the tropics: experiences from Central America. *Forest Ecology and Management* 91, 61–73.

Peeters, L.Y.K., Soto-Pinto, L., Perales, H., Montoya, G. and Ishiki, M. (2003) Coffee production, timber and firewood in traditional and Inga-shaded plantations in Southern Mexico. *Agriculture Ecosystems and Environment* 95, 481–493.

Perfecto, I., Rice, R.A., Greenberg, R. and Vandervoort, M.E. (1996) Shade coffee: a disappearing refuge for biodiversity. *BioScience* 46, 598–608.

Perfecto, I., Vandermeer, J., Hanson, P. and Cartin, V. (1997) Arthropod biodiversity loss and the transformation of a tropical agro-ecosystem. *Biodiversity and Conservation* 6, 935–945.

Perfecto, I., Mas, A., Dietch, T. and Vandermeer, J. (2003) Conservation of biodiversity in coffee agroecosystems: a tri-taxa comparison in southern Mexico. *Biodiversity and Conservation* 12, 1239–1252.

Rice, R.A. (1991) Observaciones sobre la transicion en el sector cafetalero en Centroaomerica. *Agroecologia Neotropical* 2, 1–6.

RSPB (2003) *Sweet Like Chocolate? Making the Coffee and Cocoa Trade Work for Biodiversity and Livelihoods*. Report of the Royal Society for the Protection of Birds, Sandy, UK, 22 pp.

Schroth, G., and Sinclair, F.L. (eds) (2003) *Trees, Crops and Soil Fertility: Concepts and Research Methods*. CAB International, Wallingford, UK.

Soto-Pinto, L., Romero, A.Y., Caballero, N.J. and Segura, W.G. (2001) Woody plant diversity and structure of shade-grown coffee plantations in Northern Chiapas, Mexico. *Revista de Biologia Tropical* 49, 3–4.

Soto-Pinto, L., Perfecto, I. and Caballero, N.J. (2002) Shade over coffee: its effects on berry borer, leaf rust and spontaneous herbs in Chiapas, Mexico. *Agroforestry Systems* 55, 37–45.

Staver, C., Guharay, F., Monterroso, D. and Muschler, R.G. (2001) Designing pest-suppressive multistrata perennial crop systems: shade-grown coffee. *Agroforestry Systems* 52, 151–170.

Vaast, P. and Harmand, J.M. (2002) The importance of agroforestry systems for coffee production in Central America and Mexico. In: *Plantations, Recherche, Développement et Cafeiculture*. Centre de Cooperation Internationale en Recherche Agronomique pour le Développement, Montpellier, France, pp. 34–43.

van Elzakker, B. (2001) Organic coffee. In: Baker, P.S. (ed.) *Coffee Futures: a Sourcebook for Critical Issues Confronting the Coffee Industry*. CABI-FEDERACAFE, USDA-ICO, Chinchiná, Colombia, pp. 74–81.

14 Postharvest and Processing Pests and Microbial Problems

Introduction

Ripe coffee fruit ('cherry') is processed after harvesting to remove the seeds ('beans'). Most coffee is dry-processed – the whole fruit is dried either in the sun or mechanically, and the remains of the outer layers of the fruit are removed by hulling to leave 'green' coffee beans. Wet-processing is undertaken for the higher-quality ('washed') arabicas (see Fig. 14.1). This involves removing the fruit epi- and mesocarp layers by mechanical pulping and washing/fermenting processes and drying the beans, complete with testa, to produce parchment coffee. The testa (parchment) is removed by milling or hulling to produce green coffee.

Comprehensive accounts of coffee processing can be found in the general texts referred to in the bibliography to Chapter 1. Damage caused by pests and diseases can affect these processes in several ways. Diseased fruits and those containing damaged beans may interfere with extraction of the beans and may reduce the quality of the final product, and the processes themselves may enable microorganisms to interfere with the quality of the resultant coffee. Damage to green coffee may occur after processing and during storage or transportation, particularly by storage pests. Large quantities of green coffee are stored in producing countries, especially after good harvests and when prices are low (see Fig. 14.2).

Damaged Coffee Fruits at Harvest

For high-quality coffee, only healthy red berries should be harvested but, if all fruit is stripped from the tree, berries need to be picked over by hand before processing. Green (unripe) berries are difficult to process adequately and spoil the flavour. Over-ripe berries are darker with a purplish tinge and can also impart an undesirable flavour. Prematurely ripe berries have a yellow or dull

Fig. 14.1. Coffee pulpery, Malawi.

Fig. 14.2. Large coffee warehouse, Paraná, Brazil.

orange colour and are a characteristic sign of over-bearing, and often contain deformed or light beans, which degrade the quality of the final product. Such berries often show sun scorch damage and are frequently infected by *Cercospora coffeicola* and/or *Fusarium stilboides*. Prematurely ripe berries or others in which the beans are deficient may be removed as 'floaters' during the initial phase of wet-processing.

Brown blight of ripening fruit caused by *Colletotrichum* spp. can be a problem during wet weather. Although this does not infect the coffee bean, it can make pulping during wet processing difficult, as the diseased pericarp sticks to the testa. This can result in a discoloured parchment coffee, which may result in a downgrading of quality. Fruits infected by coffee berry disease earlier in the season may remain on the tree as black, mummified fruit. Black berries may also result from pest damage and subsequent secondary infection by fungi and bacteria. Pest damage by the antestia bug causes discoloured strips across the beans ('zebra' beans), and berry borer damage causes holes in beans (see Plate 4).

Full details of the pests and diseases that affect coffee berries are given in Chapters 4 and 8. Damaged or diseased berries are often infected by fungi and bacteria that can grow in the tissues of healthy berries during processing before the coffee is dry; this can result in the production of undesirable flavours and mycotoxins that impair the quality of the final product.

Microbial Problems during Drying/Fermenting

In dry processing, growth of contaminating fungi and bacteria can occur in the fruit tissues before drying is complete. These may produce off-flavours and mycotoxins, so that rapid drying to constrain this is desirable. 'Monsooned coffee' is a particular type of coffee produced in India and owes its apparent special flavour to being kept under moist conditions for a long period. The microbial population on monsooned coffee is different from that on conventional coffee, and the associated fermentation produces higher concentrations of certain biochemicals that impart the characteristic flavour. Production of mycotoxins such as ochratoxin A (see below) on this coffee would occur unless stored adequately (Ahmed and Magan, 2003).

The 'fermentation' process that occurs after pulping in wet-processing is largely a process of autolysis of the soft pulp remaining on the bean caused by pectolytic enzymes already present in the tissues and released during the pulping process. However, as this is not a sterile process, yeasts, other fungi and bacteria already present on the berry surface, in the environment generally around coffee processing plants and in water used in the process, will grow in the pulped coffee and contribute to the fermentation process. If this stage is prolonged, various microbial products such as propionic acid – which imparts an onion flavour, butyric acid and other secondary metabolites of microbial activity – including mycotoxins – may be produced in quantities large enough to be detected in the final product.

Over-fermentation of coffee beans produces 'stinkers' which, as the name implies, impart a bad flavour to the coffee. 'Potato taste' has been shown to be due to bacterial growth in damaged beans, particularly those damaged by the antestia bug (Bouyjou *et al.*, 1999). Recent surveys of the mycobiota of harvested coffee fruits have been prompted by the need to prevent ochratoxin contamination.

A wide range of common mould fungi is found on coffee at harvesting and during processing, and these occur naturally in the soil, on plant debris and in coffee tissues. *Fusarium*, *Cladosporium*, *Aspergillus*, *Penicillium*, *Rhizopus*, *Mucor* and yeasts were found on harvested coffee and during the processing stages in India. The toxigenic *A. ochraceus* was found to be less common on dry processed coffee and most frequently associated with damaged beans (Panneerselvam *et al.*, 2000).

Generally, hydrophilic moulds predominate during the wet stages of processing, with the more xerophilic genera such as *Aspergillus*, *Penicillium* and *Cladosporium* becoming predominant as the coffee dries out. Hence these can be more problematic during coffee storage and over 90% of stored coffee samples in Brazil have been found to be contaminated by *Aspergillus* species, with *A. niger*, *A. ochraceus* and *A. flavus* being most predominant (Batista *et al.*, 2003; Martins *et al.*, 2003).

Mycotoxins

Mycotoxins are secondary metabolites produced by a number of fungi under certain conditions, having toxic affects on animals – including humans. Several have been shown to occur in coffee, but most are insignificant and/or are destroyed during roasting. Ochratoxin A, however, is of some concern as this is a virulent toxin to kidney function, is a carcinogen and can survive roasting; International bodies have set limits on the levels of ochratoxin A that foods should contain. There is no current evidence that ochratoxin is causing a human health problem, and the proportion of ochratoxin in the human diet contributed by coffee is small (a few per cent) in comparison to that provided by major food items such as cereals (Paterson *et al.*, 2001).

Nevertheless, in October 2004, the Standing Committee on the Food Chain and Animal Health of the European Union set limits for ochratoxin A levels in coffee: 5.0 μg/kg in roasted coffee beans and ground roasted coffee and 10 μg/kg in soluble coffee. A study undertaken by the Institute for Scientific Information on Coffee in conjunction with the International Coffee Organization estimated that a 3 ppb limit could result in a reduction in exports of up to 600,000–700,000 t of coffee worldwide (Duris, 2002). To put this into perspective, the export of green coffee from the whole of the African continent in 2003 amounted to 555,636 t (FAO statistical database).

Ochratoxin A is produced by *Aspergillus ochraceus* and *A. carbonarius* in the tropics, and by *Penicillium verrucosum* in temperate climates. However, not all isolates of these fungi produce the toxin and production is also affected by the environmental conditions under which the fungus is growing. Essentially, the growth of these moulds and production of ochratoxin requires a warm, moist environment (20–30°C and water availability level > 0.8 (Paterson *et al.*, 2001) – equivalent to a bean moisture content of about 16%). Growth is also more prevalent on damaged beans and on the outer tissues of the fruit.

Recent work, including that sponsored by an ICO/CFC project, aims to provide more information on the problem of ochratoxin production in coffee and how it can be managed. Indications are that some important aspects of mould control include: (i) rapid drying of processed coffee to a moisture content < 16%; (ii) avoidance of local sources of mould contamination such as contact with soil postharvest and removal of coffee-processing debris; and (iii) removal of damaged beans and subsequent storage under dry conditions. Paterson *et al.* (2001) have provided a review of the subject.

Storage Pests and Diseases

If kept under proper conditions, stored coffee is not particularly prone to damage from pests and diseases. Most problems arise if the moisture content is too high or if the beans are damaged in any way. In February 2002, the International Coffee Council adopted a resolution establishing a Coffee Quality Improvement Programme and set out standards for the export of coffee.

This requires that arabica should not have more than 86 defects per 300 g sample (New York green coffee classification/Brazilian method or equivalent), robusta not more than 150 defects per 300 g (Vietnam, Indonesia or equivalent) and that both types should have a moisture content between 8 and 11% using the ISO 6673 method of measurement (Coffee and Cocoa International Conference, London, 2002).

At a moisture content of < 11%, problems are minimal. At around 13%, moulds such as *Rhizopus* and *Aspergillus* can begin to develop and the coffee bean weevil may become a problem. With moisture contents > 18%, wet rots caused by yeasts and bacteria can become troublesome (Mabbett, 2002). These organisms can be a source of off-flavour in processed coffee, but a more important consideration is that certain fungi produce mycotoxins that can contaminate the product.

Coffee bean weevil

Araecerus coffea (Fabricius), [Coleoptera: Anthribidae]; synonyms: *Araecerus fasciculatus* (De Geer)

Morphology

The adult is a small, dark brown beetle 2–5 mm long. The head and thorax are liberally coated with whitish scales and the abdomen mottled, with whitish or yellowish patches. The larva is legless, with a fat, yellowish white, curved body and shining brown head and is 4–6 mm long when fully mature (see Fig. 14.3).

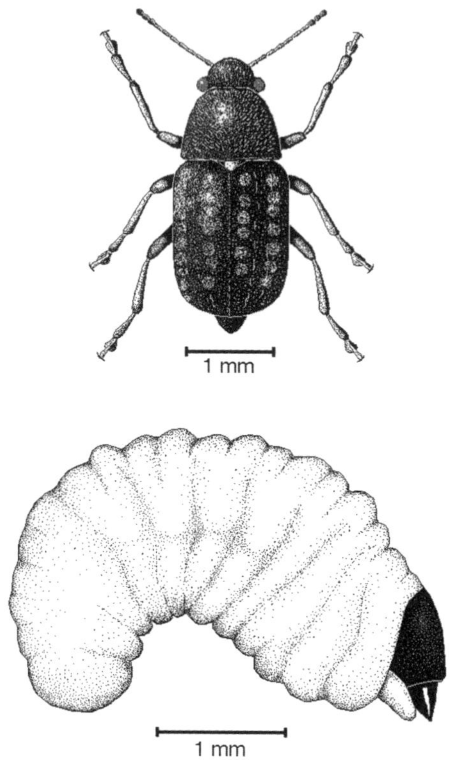

Fig. 14.3. Adult and larva of *Araecerus coffea.*

Distribution and status

There is some confusion over the naming of this species, which for many years has been known as *Araecerus fasciculatus*. De Geer described *fasciculatus* in 1775, so this name would appear to have priority over *coffea*, named by Fabricius in 1801. However, *coffea* was chosen as the type of species of *Araecerus* when the genus was set up by Schönherr in 1823, and it is the opinion of Valentine (1998) that *Araecerus coffea* (Fabricius) is the correct name.

The beetle is almost cosmopolitan as a storage pest, although it does not breed and cannot survive winter in temperate climates. It attacks a wide range of stored products and has been recorded from over 100 hosts, but is most troublesome as a pest of coffee, cocoa, cassava and nutmeg. In coffee, the pest may infest the fruit whilst still on the tree and be carried into store after harvest. The larva feeds within the bean, producing holes not unlike those of coffee berry borer in size. In Brazil, losses in store during a 6-month period amounted to 30% (De Figueiredo, 1957).

Specific records from coffee in the literature include the following countries, but doubtless the pest is present wherever coffee is grown:

- Central America: Costa Rica, El Salvador, Guatemala, Nicaragua
- Caribbean: Cuba, Haiti, Trinidad and Tobago
- South America: Brazil, Colombia, Peru, Venezuela
- Atlantic Ocean: São Tomé and Príncipe
- West Africa: Ghana, Ivory Coast, Senegal, Sierra Leone, Togo
- Central Africa: Democratic Republic of Congo
- Eastern Africa: Kenya, Tanzania, Uganda
- Indian Ocean: Madagascar
- Asia: Indonesia, Java and Sumatra, Malaysia, Taiwan
- Pacific Ocean: Australia, Hawaii, New Caledonia

During the First World War, enormous stocks of produce accumulated in Java (Roepke, 1926), and it was found that *A. coffea* did most damage to stocks of arabica and liberica and practically none to robusta.

Life cycle

The life cycle of the insect in the field was investigated in Brazil by Autuori (1931). A single egg is laid in each berry and this hatches in 6–9 d. The larvae at first feed in the pulp of the berry for 10–15 d before feeding in the bean for another 25–30 d. They then pupate within the bean for 6–9 d. The insect cannot survive in well-dried stored coffee, requiring a moisture content in excess of 12–13%.

Natural enemies

Most of the natural enemies of *A. coffea* have been recorded from Colombia by Cabal-Concha (1956). They include three wasps parasitic on the larvae: *Anisopteromalus calandrae* (Howard), *Cephalonomia gallicola* (Ashmead) and *Plastanoxus* sp., and two mites that are predators on the eggs, *Cheyletus* sp. and *Monieziella* sp. Another parasitic wasp, *Apanteles araeceri* Wilkinson, attacks larvae in Indonesia (Wilkinson, 1928).

Control

The main line of defence against this pest is to ensure a low moisture content in the produce. Chemical control should be avoided if possible but, if necessary, fumigation is the best option.

Coffee berry borer

Hypothenemus hampei (Ferrari) [Coleoptera: Scolytidae].

Although damage by berry borer takes place primarily in the field (see Chapter 4), nevertheless the infestation will persist in stored beans unless measures are taken to prevent it. Kalshoven (1950–1951) states that beetles can survive and multiply for up to 1 year in beans enclosed in closed containers. It is the ability to survive in parcels of fresh coffee beans which has led to its dissemination around the world, and there are many recorded instances of *H. hampei* being intercepted in quarantine.

Indeed, the type specimen of *H. hampei* described by Ferrari in 1867 was found in stored produce in France. Between 1985 and 2000, *H. hampei* was intercepted by the US Department of Agriculture (USDA) on 65 occasions (Haack, 2001), and there may have been many more because 821 interceptions of unspecified *Hypothenemus* were also recorded.

Like other coffee storage pests the beetle will not breed if the moisture content of the beans is low, below 13.5% for Arabica and below 12.5% for Robusta (Waterhouse, 1998).

A further danger from *H. hampei* is that it may act as a vector for ochratoxin A-producing fungi, and the same is probably true for *A. fasciculatus*, although this has not yet been tested. A survey of *H. hampei* beetles emerging from coffee beans in West Africa (Vega and Mercadier, 1998) showed 5.3% to be carrying spores of *Aspergillus ochraceus* in Uganda and 17.4% in Benin.

Other beetles of stored products

Ahasverus advena (Waltl), *Lasioderma serricorne* (Fabricius), *Stegobium paniceum* (Linnaeus), *Tribolium castaneum* Herbst, *T. confusum* Duval. [Coleoptera: Silvanidae, Anobiidae, Tenebrionidae].

There are scattered records of familiar storage beetles associated with stored coffee, but they are not normally of great importance. *Ahasverus advena* is primarily a fungus feeder and is generally associated with mouldy produce, and is therefore a good indicator of damp storage conditions. The Tobacco Beetle, *L. serricorne*, has been recorded from coffee beans in Haiti (Anon, 1929) and Surinam (Reyne, 1919). It is a serious pest of cocoa beans in the tropics, and the infestation of coffee may well have been the result of the coffee having being stored in proximity to bagged cocoa beans.

Cross-infestation is probably also responsible for attacks by the grain weevils, *T. castaneum* and *T. confusum*, which are ubiquitous in warehouses in both tropical and temperate countries. The Anobiid, *S. paniceum*, was intercepted in the USA in coffee imported from Colombia which showed signs of feeding damage and appeared to have been infested prior to shipment (De Ong, 1948).

Moths of stored products

Cadra cautella (Walker), *Corcyra cephalonica* (Stainton) [Lepidoptera: Pyralidae].

As with the storage beetles, attacks by these two moths are probably induced by proximity to other stored produce – in particular, cocoa, but reports of infestation are not common. *Cadra cautella* has been recorded as infesting stored coffee in Malaysia (Yunus and Ho, 1980) and Papua New Guinea

(Bourke *et al.*, 1973), and *C. cephalonica* from Brazil (Bitran and Oliveira, 1978) and Malaysia (Yunus and Ho, 1980). Measurements of weight loss due to *C. cepahlonica* in Brazil amounted to only 0.6–1.54% of total mass after 6 months' storage and 0.84–2.13% after 8 months' (Bitram and Oliveira, 1978).

Moulds of stored products

Aspergillus niger Tiegh., *A. flavus* Link., *A. ochraceus* Wilhelm, *A. carbonarius* (Bainier) Thom.

These are the common mould species most likely to be encountered on stored coffee. They all produce masses of dry spores in chains from globular heads of conidiogenous cells carried on erect conidiophores (see Fig. 14.4) and have characteristic colours. *Aspergillus niger* and *A. carbonarius* are black or dark brown, *A. flavus* and *A. ochraceus* are yellow. However, because of the wide range of species in this genus, accurate identification requires careful mycological analysis (Sampson and van Reenen-Hoekstra, 1988). Although primarily xerophytic fungi, they generally require moisture levels of above a water activity level of 0.8 (equivalent to about 16% moisture content for stored coffee beans) and temperatures of 15–30°C for growth.

Aspergillus flavus can produce aflatoxins, while *A. ochraceus* and *A carbonarius* can produce ochratoxins on nutrient-rich media such as foodstuffs. They are cosmopolitan in occurrence, are primarily saprobes, can grow on a wide range of substrates and can be found in most natural environments. *Aspergillus niger* is recognized as a pathogen of living plant tissues and has been associated with a collar rot of coffee seedlings (see Chapter 9).

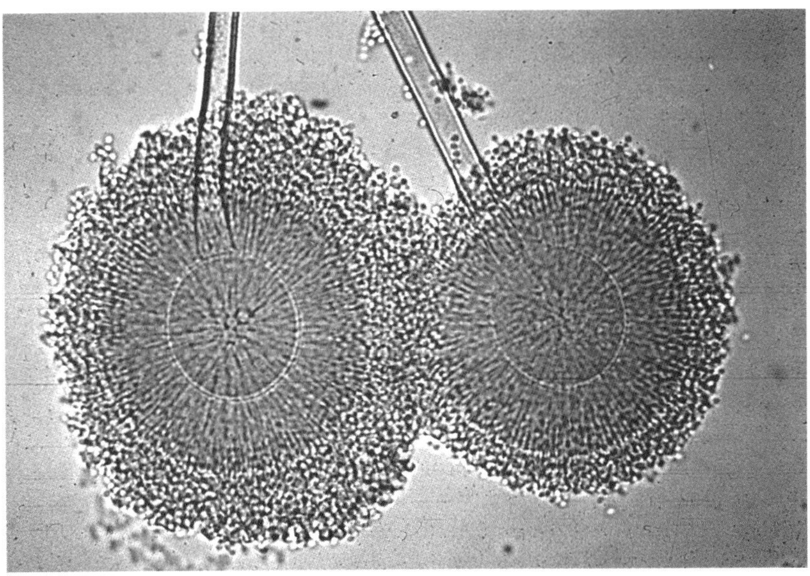

Fig. 14.4. *Aspergillus ochraceus.*

References

Ahmad, R. and Magan, N. (2003) Monsooned coffee: impact of environmental factors on fungal community structure, enzyme production profiles and ochratoxin. In: Bryson, R.J., Kennedy, R., Magan, N. and Scudamore, K.A. (eds) *Mycotoxins in food production systems*. Bath, UK, 25–27 June, 2003. *Aspects of Applied Biology* 68, 161–168.

Anon (1929) Department of Entomology and Zoology. *Report of Technical Services of the Department of Agriculture, Haiti, 1928–1929* No. 17, 157–166.

Autuori, M. (1931) Dados biologicos sobre o *Araecerus fasciculatus* De Geer. *Revista de Entomologia* 1, 52–61.

Batista, L.R., Chalfoun, S.M., Prado, G., Schwan, R.F. and Wheals, A.E. (2003) Toxigenic fungi associated with processed (green) coffee beans (*Coffea arabica* L.). *International Journal of Food Microbiology* 85, 293–300.

Bitran, E.A. and Oliveira, D.A. (1978) *Corcyra cephalonica* (Stainton, 1865) (Lepidoptera, Galleriidae) como praga de cafe beneficiado armazenado. *Archivos do Instituto biologico* 45, 153–168.

Bourke, T.V., Fenner, T.L., Stibick, J.N.L., Baker, G.L., Hassan, E., O'Sullivan, D.F. and Li, C.S. (1973) *Insect Pest Survey for the Year Ending 30th June 1969*. Entomology Branch, Department of Agriculture, Stock and Fisheries, Port Moresby, Papua New Guinea, 57 pp.

Bouyjou, B., Decazy, B. and Fourny, G. (1999) L'elimination du 'gout de pomme de terre' dans le café Arabica du Burundi. *Plantations, Recherche, Developpement* 6, 107–112.

Cabal-Concha, A. (1956) Biologia y control del gorgoja del café: *Araecerus fasciculatus* de Geer, (Fam: Anthribidae) en Barranquilla, Colombia. *Revista de la Facultad Nacional de Agronomia* 18, 30–31, 49–72.

De Figueiredo, E.R. (1957) O controle do 'caruncho' das tulhas. *Biologico* 23, 197–200.

De Ong, E.R. (1948) Damage to coffee by the drug store beetle. *Journal of Economic Entomology* 41, 124–125.

Duris, D. (2002) Coffee and ochratoxin contamination. *Food Safety Management in Developing Countries, Proceedings of the International Workshop, CIRAD-FAO, 11–13 December 2000*, 5 pp.

Haack, R.A. (2001) Intercepted Scolytidae (Coleoptera) at U.S. ports of entry: 1985–2000. *Integrated Pest Management Reviews* 6, 253–282.

Kalshoven, L.G.E. (1950–1951) *De Plagen van de Cultuurgewassen in Indonesie*. Van Hoeve, Gravenhage/Bandoeng, 1065 pp.

Mabbett, T. (2002) Storing up problems? *Coffee and Cocoa International*, September, 2 pp.

Martins, M.L., Martins, H.M. and Gimeno, A. (2003) Incidence of microflora and of ochratoxin A in green coffee beans (*Coffea arabica*). *Food Additives and Contaminants* 20, 1127–1131.

Panneerselvam, P., Velmourougane, K., Shanmukhappa, D.R., Gopinandan, T.N. and Naidu, R. (2000) Incidence of toxigenic moulds in cured and uncured coffee samples. *Journal of Coffee Research* 28, 40–48.

Paterson, R.R.M., Baker, P.S. and van der Stegen, G.H.D. (2001) Ochratoxin A in coffee. In: Baker, P.S. (eds) *Coffee Futures: a Sourcebook of some Critical Issues Confronting the Coffee Industry*. CABI-FEDERACFE-USDA-ICO, Chinchina, Colombia.

Reyne, A. (1919) Verslag van den entomoloog. *Verslagen Departement van den Landbouw in Suriname 1918*, 21.

Roepke, W. (1926) Vorratsschädlinge auf Java. *Mitteilungen Der Gesellschaft fur Vorratsschutz* 2, 50–53.

Sampson, R.A. and van Reenen-Hoekstra, E.S. (1988) *Introduction to Food-Borne Fungi*. CBS, Baarn, Netherlands, 299 pp.

Valentine, B.D. (1998) A review of Nearctic and some related Anthribidae (Coleoptera). *Insecta Mundi* 12, 251–296.

Vega, F.E. and Mercadier, G. (1998) Insects, coffee and ochratoxin A. *Florida Entomologist* 81, 543–544.

Waterhouse, D.F. (1998) *Biological Control of Insect Pests: Southeast Asian Prospects*. Australian Centre for International Agricultural Research, Canberra, Australia, 548 pp.

Wilkinson, D.S. (1928) A revision of the Indo-Australian species of the genus *Apanteles* Pts I & II. *Bulletin of Entomological Research* 19, 79–105, 109–146.

Yunus, A. and Ho, T.H. (1980) List of economic pests, host plants, parasites and predators in West Malaysia (1920–1978). *Bulletin of the Malaysian Department of Agriculture* 153, 538 pp.

15 Integrated Pest Management and Pest Management Technologies

Introduction

Pest management aims toward prevention of pest damage to crops, and there are a wide range of activities and technologies that can be used for this. Some involve no more than adopting optimum cultivation practices, while others are direct interventions that target different phases of the crop–pest system. Measures can be grouped according to whether they are aimed (i) directly at the pest or (ii) at the crop to increase its resistance or tolerance to pest damage.

Measures aimed at the pest include:

- Exclusion or eradication of the pest from a country or region: this depends on legislative plant health measures at national or regional level, e.g. plant quarantine regulations.
- Avoidance: a range of practices are aimed at avoiding or reducing the sources of pests and pathogens, such as using clean soil in nurseries and attention to site selection.
- Direct destruction: either by physical methods such as removal of pests and diseased tissues or by the application of pesticidal chemicals aimed at killing pest or pathogen propagules.
- Use of natural methods: such as the use semio-chemicals to disrupt insect behaviour (pheromones) or the use of natural enemies. These may be encouraged by the avoidance of pesticides and through ecological measures. Biological control agents can also be selected and introduced as in classical biological control, or used as biopesticides.

Measures aimed at the crop include:

- The breeding/selection and use of inherently tolerant or resistant cultivars is a major aspect of disease control.
- Cultural factors to encourage crop health, avoid physiological predisposition

to diseases and to enable physiological tolerance of mild pest attack and rapid recovery.

These measures may be integrated into a holistic programme for the most economical or ecologically efficient methods for pest management, and this forms the basis of integrated pest management.

Pest Management Options

Exclusion or eradication

The first line of defence against pest organisms is to exclude them from areas where they do not occur. National and regional plant health regulations are aimed at preventing the introduction of exotic pests into new areas. They usually limit or prohibit the importation of specified plants or plant parts from countries where exotic pests and diseases are known to occur. Import permits may specify the need for an international phytosanitary certificate guaranteeing the health of the imported material, and inspections or treatments may be required. Importation of material intended for planting is very closely controlled as this represents the greatest risk of such introductions, and such material is often kept under quarantine conditions.

The various regulations and administrative procedures that need to be followed are laid down by the International Plant Protection Convention that is mediated through FAO, and nearly all countries are members of this convention. There are a number of regional phytosanitary organizations that coordinate activities between geographically related countries, each having their own national plant health institutions that undertake the necessary plant health procedures such as inspections, treatments, quarantine and containment activities. For a more detailed review of this major topic of crop protection see Ebbels (2003).

Pest organisms tend to be most diverse in the centres of origin or diversity of crop species, as many of them have co-evolved with their host or in the same habitat. Most of the more serious diseases and pests of coffee originated with the species in Africa and several, such as CBD, Fusarium wilt and the white stem borer, still only occur there. Coffee berry borer and rust have spread from Africa to reach most other major coffee-growing areas during the last century, and other pest organisms had doubtless spread with coffee material before that. Hence the importation of coffee seed or planting material from Africa to other parts of the world is very restricted.

When exotic pest organisms become established in new areas, measures can be undertaken to eradicate them or to contain their spread. When coffee rust was first detected in Papua New Guinea, eradication was successfully achieved as the affected bushes were in a locality isolated from the main coffee areas (Shaw, 1968). However, when rust was first detected in Brazil, it was already distributed over a wide area of Minas Gerais State, and it soon spread southwards to other states through airborne spores.

The first outbreak in Central America was restricted to small, well-defined areas in Nicaragua, and attempts were made to eradicate this by destroying affected trees and those in the immediate area (Schuppener *et al.*, 1977). These measures, which cost US$20 million over 2 years, succeeded in limiting the disease to the Carazo area for a few years, but did not eradicate it (Waller, 1979), and it eventually spread to neighbouring countries.

Eradication of disease is both costly in terms of operations and compensation, and is only effective in exceptional circumstances. When the disease became established in Papua New Guinea for the second time, it had already spread over too large an area of coffee for eradication to be contemplated.

Avoidance

The most obvious way to avoid pests and diseases is to grow the crop in areas where they do not occur or at times when the susceptible phase of the crop does not coincide with active stages of the pest organism. Although more relevant for the growing of annual crops where sites can be changed between seasons, this strategy is applicable to the management of some coffee pests and diseases that are dependent upon external factors during part of their life cycles.

Both rust and CBD are limited in their range by climatic conditions. The cooler temperatures that are prevalent above about 1700 m in the tropics are unfavourable for the development of rust epidemics, as infection is reduced on cool, wet leaves and the generation time of the fungus is prolonged, so that the rate of spore production is slower. By contrast, CBD thrives better in the more prolonged, wet condition at altitudes above about 1500 m. Thus rust can be avoided at higher altitudes in many areas, but this favours CBD, which is a problem for arabica growers in Africa. Climatic change will affect the incidence and severity of these diseases, with rust likely to become more prevalent at higher altitude and CBD more prevalent at lower elevations.

A number of serious coffee pests are also more prevalent at lower altitudes, including the African white stem borer, leaf miner and berry borer. Potential infection by soilborne pathogens can be avoided by removal of sources of pathogen. This includes such methods as ring-barking of shade or forest trees a year before removal and the extraction and destruction of large roots and stumps from sites to be planted to coffee. Use of clean soil for nurseries will also avoid potential problems with soilborne pathogens and nematodes.

Modification of the ecology of the coffee plantation to make conditions less favourable to pests and diseases can also be classed as a type of avoidance: control of shade and tree density are obvious examples of this. Shade inhibits the Asian white stem borer that prefers to lay its eggs on sunny stems, but berry borer favours shaded coffee. Removal of loose bark and smoothing of coffee stems can also restrict the egg-laying of stem borers. Pruning to achieve a more open coffee canopy assists the drying of wet coffee foliage and stems, and so reduces infection by many plant pathogens. Opening the canopy to light also helps to control antestia bug, which prefers dense canopies within the coffee tree. Lower branches that touch the ground should be removed, as these create bridges allowing the attendant ants to reach colonies of mealybugs and scale insects.

The control of cropping patterns can be used as a method for the avoidance of some problems. Double cropping or continuous cropping in areas that have bimodal rainy seasons or are irrigated enables pests and pathogens that affect berries to spread from one crop to the next in a continuous sequence. Provision of a natural break in the cropping pattern by restricting flowering through relevant pruning techniques will halt the continuous build-up of berry pests and pathogens.

Irrigation may also be used as a technique to stimulate out-of-season flowering, so that susceptible stages of crop development avoid the wet season. This has been used in Cameroon to avoid CBD (Muller, 1973).

Chemical and physical control

Physical methods

Direct removal and destruction of pests can form an important part of an integrated pest management programme. Phytosanitary methods are important in the integrated management of many of the more slowly spreading diseases. These include the pruning out and destruction of branches affected by pink disease (*Erythricium salmonicolor*) or koleroga (*Corticium koleroga*), and the removal and destruction of whole trees affected by root or wilt diseases caused by Hymenomycete or *Fusarium* pathogens.

Mechanical removal and destruction can be an effective measure for some pests. The larger insects or larvae such as caterpillars can be removed by hand, and destruction of stem borer larvae by pushing a wire down their entry holes in coffee stems has been practised. Removal of dead berries infested by the berry borer at the end of the season is a recognized way of limiting infestation by this pest.

Pesticides

Chemicals were first used on coffee for the control of coffee rust in India, where Bordeaux mixture was found to be effective against the disease, and it soon became standard practice in arabica-growing areas in Asia and Africa in the 1920s. The development of effective insecticides and organic fungicides enabled a wider range of products to be used against coffee pests and diseases in the middle part of the last century. Persistent chlorinated hydrocarbon pesticides became widely used for control of the major pests, and the development of systemic organophosphates increased the potency and applicability of chemical control of pests.

New and more potent products were developed because pest populations soon developed resistance to older materials. Towards the latter part of the

20th century, the deleterious effect that these substances had on beneficial organisms, on other non-target organisms in the environment and on man became widely accepted and the more toxic substances were withdrawn from the market or banned from use on coffee and other crops. The sale and use of pesticides is now strictly controlled by law in virtually all countries, and chemical control is becoming a much more limited part of pest management.

Measures to limit the production and use of certain pesticides have been agreed at international conventions. The Stockholm Convention on Persistent Organic Pollutants was negotiated by more than 100 countries in 2001, with the objective of stopping the production and use of persistent pesticides and industrial chemicals harmful to man and the environment. The International Code of Conduct on the Distribution and Use of Pesticides was set out under international agreement by FAO in 2003 and, although voluntary, provides guidelines that have been incorporated in the legislation of many coffee-producing countries. The 'Codex Alimentarius' is a joint FAO/WHO programme that develops and updates international food safety standards, including recommended minimum residue levels for pesticides in food.

The WHO has classified pesticides according to their toxicity as follows:

- Ia and Ib: extremely or highly hazardous – should not be used
- II: moderately hazardous – use should be avoided if possible
- III: slightly hazardous
- Other: unlikely to present acute hazard under normal use.

The Pesticides Action Network (PAN) lists pesticides that are: (i) acutely toxic; (ii) cholinesterase inhibitors; (iii) known/probable carcinogens; (iv) known groundwater pollutants; or (v) known reproductive or developmental toxicants as 'bad actors' and 'dirty dozen'. More specifically to coffee, the Common Code for Coffee Community (4C) – a European-based initiative of producer organizations, coffee trade and industry, trade unions and NGOs – has categorized pesticides used for the control of the most common coffee pest problems using a red, yellow and green system as follows:

- Red: highly toxic (e.g. WHO Classes Ia and Ib) and not compatible with sustainable coffee production
- Yellow: moderately hazardous (e.g. WHO Class II) and should be used only as part of a monitoring system
- Green: might be used as part of an integrated pest management strategy.

Several marketing and consumer organizations restrict or prohibit the used of the more toxic pesticides on crops intended for markets in which they have an interest. The Fairtrade standard, for example, prohibits the use of Classes Ia, Ib and PAN 'dirty dozen' compounds.

Table 15.1 lists pesticides commonly used for the main coffee problems and their ratings, and a full list of the chemicals in the WHO categories can be found at http://www.inchem.org/documents/pds/pdsother/class.pdf

Of the many pesticides developed and used on coffee over the last 70 years, Bordeaux mixture and its related copper-based derivatives have endured for disease management as being effective, safe and cheap. Modern pesticides

Table 15.1. Examples of conventional and IPM alternatives for control of coffee pests and diseases.

Pest/disease	Pesticide	IPM alternative
Berry borer	Endosulphan[a] Fenitrothion Chlorpyriphos	Entomopathogen: *Beauvaria bassiana*; Parasitoids: *Cephalonomia stephanoderis*, *Prorops nasuta*, *Phymastichus coffea*; spot-spraying of insecticides may be needed; removal of unpicked and fallen berries
Leaf miner[b]	Fenitrothion Cypermethrin, etc. Imidacloprid	Many natural enemies – scouting may show that the pest does not reach the damage threshold Applied to soil, less effect on natural enemies
Antestia bug	Fenitrothion	Pruning to thin the canopy, although the pest is more of a problem in coffee grown without shade
Berry moth	Fenitrothion	Spot-spraying; need for careful scouting and to treat at first sign of damage before the insects become established within the berry clusters
Mealy bug	Chlorpyrifos Fenitrothion, etc.	Best controlled by preventing attendant ants from reaching the colonies, by removing branches that touch the ground and trunk-banding with insecticide
Scale insects	Imidacloprid	Soil-applied systemic will allow build-up of natural enemies
Stem borers	Fipronil[a] Chlorpyriphos	Applied to the trunk only, therefore IPM compatible Cleaning of trunk and coat with botanical repellents to prevent egg-laying
Rust	Copper or systemics	Resistant cultivars; use of shade to ensure the trees are adequately fertilized in relation to bearing; ground-applied systemic fungicide, e.g. triadimefon
CBD	Copper and/or other	Resistant cultivars; prevention of over-shading; adequate pruning

[a] Potentially dangerous chemicals – mammalian, avian or piscine toxins (WHO Class II) restricted or banned.

[b] Soil-applied systemics aldicarb and carbofuran, formerly used for leaf miner, are highly toxic (WHO Class I) and not recommended.

(that are acceptable for use on coffee) are effective at low doses and are generally formulated in powders or concentrates that can easily be diluted in water as a carrying agent. Those for soil application are usually formulated as granules. Recently, some simple compounds that are not toxic but can induce resistance to diseases in plants have been coming into use. However, costs in relation to farm-gate prices for coffee are high particularly when costs of application are added.

Pesticide application

Pesticides are toxic and can pose dangers to farmworkers, consumers and the environment. They must always be stored, used and disposed of in accordance with the manufacturer's instructions and local government regulations. How pesticides are used is as important as their chemical basis in the management of pests and diseases. The method of application, dosage and timing must all be considered.

Safe methods of application are particularly important for operators, as they are frequently contaminated both by direct contact via handling, spillage, etc. and through spray drift, so that protective clothing – including overalls, boots, gloves, etc. – should always be used, however uncomfortable. Face masks and goggles may also be needed where there is danger of much spray drift when treating trees.

Pesticides readily contaminate the environment and harm non-target organisms. This occurs through excessive application (spraying to 'run-off'), bad targeting, spillage and dumping of excess or waste spray liquid. Disposal of pesticides must be undertaken according to local regulations to prevent contamination of soil, groundwater and watercourses.

Timing of applications also needs to be carried out in relation to the relevant phase of the pest population build-up or disease epidemiology and the harvest time, in order to avoid residues being transferred to the processed crop. Optimal control of insect pests often requires application at particular times during the life cycle, e.g. when particular instars are most susceptible. Insecticide should be applied at the most appropriate time in the life cycle of the target pest. To be effective in controlling leaf miner for instance, insecticide should be applied about 7 d after a peak in moth populations, when eggs can be targeted. Once the larvae are under the epidermis, they are protected to some extent from the insecticide.

For optimal disease control, fungicides need to be applied to protect the susceptible stages of the crop at times when conditions are favourable for infection. Poor timing can exacerbate pest and disease problems by, for example, killing more natural enemies than pests (Nyambo *et al.*, 1970) or by interfering with other natural biological control systems, as may occur with CBD control (Waller and Masaba, 2006).

Pesticide application is costly. Hand-operated machines are labour-intensive, especially in hilly terrain requiring up to five man-days/ha on steep slopes in remote locations. Petrol-driven machines are expensive and beyond the capital resources of many smallholders; they also require fuel and maintenance. Hydraulic pumps, either hand- or machine-operated, form the most basic system of pesticide application and vary greatly in design.

The familiar knapsack sprayer with tank, lever-operated pump, lance and nozzle is perhaps the most widely used by millions of smallholder farmers across the world and, provided the pump operates efficiently, there are no leaks and the spray nozzle is effective, it can be one of the best spraying systems. The main advantage is that the spray can be hand directed to ensure adequate coverage of the foliage in the tree canopy, but it is labour-intensive and slow.

However, these sprayers can be effective at low volumes (100 l/ha) (Wallis and Firman, 1962) and are often the only machines that can be used in coffee grown on slopes. Pre-pressurized knapsack sprayers operate through air pressure pumped into a pressure-retaining tank that forces the spray liquid out through the lance and nozzles; these have the added advantage that pumping is not needed during movement through the coffee plantation.

Hydraulic systems used on large estates vary from petrol-driven standing pumps with long hoses and hand-directed lances to a variety of tractor-drawn or tractor-mounted machines (see Fig. 15.1), some of which can cover eight rows of coffee in one pass using a system of overhead nozzles (see Fig. 15.2). These machines can only be used in well-spaced coffee on relatively flat land.

Most hydraulic systems depend on the application of large volumes of spray liquid and high-pressure nozzles to generate the right spectrum of spray droplet to obtain complete coverage of foliage in the tree canopy. Ceramic or hard plastic cone nozzles are used for application of insecticides and fungicides and are graded according to operating pressure. The large volumes required, weight of machines and maintenance of the pumps and nozzles can be problematic in many coffee-producing countries. Collection and transport of water can be particularly difficult in mountainous terrain.

Air-assisted machines blow the spray droplets into the tree canopy, use lower volumes of liquid and have some of the advantages of hydraulic systems. Spray droplets are usually produced by an 'air-shear' effect on a simple tubular nozzle. Knapsack mist-blowers (see Fig. 15.3) are particularly efficient for coffee spraying as they use low volumes and can be hand-directed.

Spray droplets can also be produced by the centrifugal effect of a spinning disc. This produces very uniform droplets and is the basis of controlled droplet application (CDA) systems that can give effective coverage at ultra-low volumes (ULV). Some ULV/CDA sprayers have been successfully used on coffee, but the fine droplets are subject to drift and lack penetration of the canopy.

Fig. 15.1. Tractor-drawn air-assisted sprayer.

Fig. 15.2. Overhead hydraulic multirow sprayer.

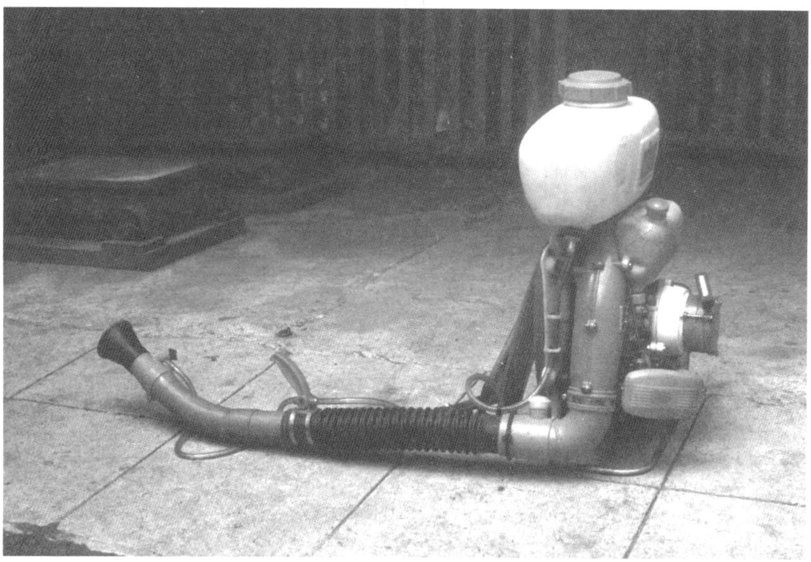

Fig. 15.3. Knapsack mist-blower.

Systems that combine CDA with air assistance to blow the droplets into the canopy are better adapted for coffee, and such a machine (Motax®, Micron Sprayers Ltd., Bromyard, UK) was devised for control of coffee rust in Colombia. This machine incorporates a back-mounted axial fan blowing air over a spinning disc droplet generator and gives good overall leaf coverage with minimal operator contamination (see Fig. 15.4). It gave control at volumes

Fig. 15.4. 'Motax' sprayer.

down to 20 l/ha and reduced labour requirements for coffee-spraying in the mountainous terrain to a quarter of that needed for conventional knapsack spraying.

Aerial spaying has been used for rust and/or CBD control over large, level areas of coffee, and thermal fogging – in which a fog of pesticide is generated, usually by a hot exhaust system and drifts into the coffee canopy – has also been tried. Dusting using hand machines was at one time used for berry borer control in South America. Soil application of chemicals has had limited use for control of soilborne nursery problems such as *Rhizoctonia* and nematodes, and systemic compounds for translocation following root uptake such as triadimenol for coffee rust control, but soil application risks environmental contamination and should be avoided where possible.

More detailed accounts of pesticide application technology are given by Matthews and Hislop (1993) and Pfalzer (2004).

Natural methods of pest control

Populations of pest organisms are moderated in natural ecosystems by a range of other organisms that either compete with them or prey on them. Pathogens are affected by the competing effects of similar organisms, and by antagonists that may produce antibiotic chemicals or induce a resistant reaction in the host plant. These mechanisms limit the development of CBD to some extent on unsprayed coffee and operate in the soil, where a rich microflora stimulated by organic matter can limit the development of root pathogens.

Parasites and predators of arthropod pests are particularly important in providing a natural biological control of almost all insect pests. These natural mechanisms are easily disturbed by agricultural systems that reduce biodiversity – and especially by the use of pesticides. Appropriate attention to ecological factors, such as increasing the floral biodiversity of coffee plantations through use of shade trees, etc. in coffee plantations, can increase the populations of the natural enemies of pests and thus help to restore natural biological control systems (see also Chapter 13).

Natural enemies can be reared or cultured in isolation and used to augment natural populations or as a biological pesticide. The fungus *Beauveria bassiana*, an important insect pathogen, has been used against a wide variety of insect pests, a specially formulated preparation of fungal spores usually being applied as a spray. Similarly, preparations of *Trichoderma* spp. are sometimes used as an additive to soils for control of soilborne pathogens.

Where pests have been introduced into new areas, they may multiply unchecked by natural enemies and cause major damage. In these situations, natural enemies reared in captivity can be released into areas where the pest has been introduced in an attempt to restore the natural balance between the pest and its natural enemies. This classical biological control intervention has been used against green scale of coffee in Papua New Guinea and berry borer in Latin America.

Some aspects of the natural behaviour of insect pests, such as mating, are controlled by pheromones. Minute traces of these volatile chemicals are often used to attract males, and some of these substances have been identified and synthesized. They can then be used to disrupt mating behaviour by attracting males to traps, thus inhibiting mating. This method has been used against the coffee white stem borer in India.

Resistant cultivars

The use of disease resistance in the host plant is the main method for the control of most crop diseases. Selection and breeding of plants with resistance to disease is an important aspect of research and development, with funding support from both national and international sources. Resistance to disease is a property common to all plants, and particular species are susceptible to very few of the many pathogenic organisms that are known to exist: most plants are non-hosts to most pathogens. Plants possess a range of different types of resistance that may be classified by mechanisms, genetics or effects, and are often interdependent.

All plants possess a general or background level of resistance that operates even against specialized pathogens, although this may not be particularly evident against pathogens such as virulent races of coffee rust during the height of an epidemic. This non-specific or field resistance is usually only partial, and influenced by factors that affect the physiological health of the plant. Other specific resistance mechanisms operating at the cellular level are directed at biotrophic pathogens that typically rely on a hypersensitive response by the plant cell to kill the pathogen at the earliest stage of infection.

Unspecialized pathogens, such as necrotrophs or opportunistic invaders of weakened plant tissues, usually stimulate the resistance mechanisms of a healthy plant sufficiently to prevent infection. If, however, the plant is under physiological stress through environmental or mechanical damage, for example, the resistance mechanisms may be insufficient to contain the infection.

By contrast, biotrophic pathogens are more specialized and tend to have very restricted host ranges. They are able to infect the plant by direct parasitism of cells without stimulating the plant's defence reactions, and often cause diseases that are more damaging.

Most non-specific resistance is under polygenetic control and, although not easily manipulated by conventional plant breeding techniques, is not readily overcome by genetic changes in pathogens and is therefore usually durable (Johnson, 1984). This type of resistance generally operates after infection and restricts the pathogen's subsequent development in the host plant. It usually causes longer incubation and latent phases and suppressed or restricted sporulation, all of which delay or prevent the epidemic progress of the disease. Because of this it may only be evident when crops are grown under field conditions, but is effective against all races of a pathogen – a characteristic of 'horizontal' resistance (Van der Plank, 1963).

By contrast, single or few genes in the host usually control pathogen-specific resistance, which often imparts immunity – or at least a high degree of resistance – and is thus easy to assess and can be manipulated by conventional plant-breeding techniques. A series of virulence factors in the pathogen interacts with corresponding resistance factors in the host, giving the classic 'gene-for-gene' mechanism typical of diseases such as rust, and selection pressure imposed on the pathogen population often results in the emergence of new, virulent races – a typical characteristic of 'vertical' resistance.

The lack of durability frequently exhibited by such resistance is a major problem in perennial crops such as coffee, where change of cultivar to deploy new effective resistance against changing pathogen virulence is not a practical option within the time frame of most agricultural enterprises. Resistance to coffee rust is mainly of the major 'vertical' type, but there is a significant element of general resistance effected by host physiology as well, whereas resistance to CBD and Fusarium diseases is apparently mainly of a polygenic nature. Rodriguez and Eskes (2004) provide further details of resistance to rust and CBD.

Disease resistance is one of only several factors incorporated into plant-breeding programmes, the main thrust of which is to achieve a suitable combination of many desirable properties. Typically, sources of resistance need to be identified – usually from wild populations of the crop species or related species – which can be incorporated by hybridization and back-crossing with selections having other desirable properties, during which individual progeny are carefully selected for incorporation at each stage. For perennial crops such as coffee, this is a long-term programme to which must be added the need to achieve durability of resistance in order to have a corresponding longer-term effect.

Following advances in molecular biology, it is now possible to identify genetic sequences associated with resistance or other characters and to detect these in breeding progeny through the use of DNA probes or PCR; this marker-assisted selection considerably shortens the time required for progeny screening (Nelson and Leung, 1994). Other biotechnological advances such as embryo rescue techniques and pollen culture allow a wider range of hybrid plants to be produced.

There are examples of apparently single gene resistance that have proved relatively durable, such as the S_H3 gene from liberica coffee in the arabica cultivar S795 that remained effective against rust for more than a decade in India. However, generally polygenic quantitative resistance or horizontal resistance is the most durable (Simmons, 1991). Gene pyramiding aims to incorporate a range of different resistance factors into plants, making it more difficult for simple genetic changes in the pathogen to be effective against them. Modern hybrid cultivars – such as the catimor group – incorporate gene pyramiding against coffee rust.

Races of *Hemileia vastatrix* which are able to overcome many host resistance factors lose 'avirulence' genes and, as a result, are apparently less fit and do not develop into large epidemics. Other factors – such as the use of multilines and variety mixtures with different resistance genes and integration with cultural measures which may hamper the development and spread of virulence within the pathogen population – can assist durability and, especially, the way in which the resistance is deployed within the cropping system. Although apparently durable resistance may not be complete, it can still form an important component of integrated disease management strategies.

More recently, exotic sources of resistance to pests and diseases may be incorporated by direct genetic modification using genetic engineering techniques, without recourse to traditional breeding techniques. The gene from the insect-pathogenic bacterium *Bacillus thuringiensis* that produces an insect toxin has been incorporated into many crop species to genetically engineer resistance to insect pests. This has been done for both arabica and robusta coffee to produce varieties resistant to leaf miner. Caffeine-free coffee can also be produced using genetic modification.

Cultural factors

As indicated above, disease resistance can be affected by the plant's physiology and it also affects tolerance to damage by both pests and diseases. The physiological health of the plant is closely related to cultural conditions, and these also affect environmental conditions in the coffee plantation that can play a part in the avoidance of pests and disease (see above). Cultural factors are therefore an important element of integrated pest management.

Stress caused by over-bearing, where excessive cropping leads to carbohydrate depletion of plant leaves and other tissues, is a major factor in reducing the efficacy of resistance to rust, and can be prevented by adequate nutrition, pruning or provision of shade. Similarly, hot, dry conditions predispose plants to Fusarium bark disease and Fusarium dry root rot.

Most stem diseases also rely on wounds for infection, so that careful cultivation practices that avoid stem base wounds are necessary. Mulching to conserve soil moisture, suppress weed competition and add organic matter to the soil is an important measure to improve plant health, especially on young, unshaded coffee or in drier localities.

Integrated Pest Management

Integrated pest management (IPM) is a decision support system for the selection and use of pest control tactics, singly or harmoniously coordinated into a management strategy that takes into account the interests of and impact upon producers, society and the environment (Kogan, 1998). Although there are many definitions of IPM, this one seems suited to the context of integrated crop management (ICM), encompassing the needs of producers and consumers.

The British Agrochemicals Association (1995) defines ICM as 'A management system which employs controlled inputs to achieve sustained profitability with minimal environmental impact, but with sufficient flexibility to meet natural and market challenges economically'. This defines ICM as an approach that can provide the basis for increased crop production in high-input or low-input agriculture. While off-farm inputs are discouraged, the approach is sufficiently flexible to allow for pesticides to be used when necessary, and encompasses the needs of producers and consumers.

The aim of IPM is to protect crops from damaging pest levels (in the wider sense, including diseases and weeds) with targeted use of pesticides where necessary. If the integration of sound crop husbandry and non-chemical methods is sufficient to keep pest population levels below their damage thresholds, then chemical pesticides may not be required. Unlike in organic farming systems, chemical pesticides are not proscribed in IPM if profitability is threatened by pest attack, provided they can be used safely and without damaging, non-target effects on the environment.

The principles of integrated control are also used in organic farming systems but, in this case, pest and disease management is based on sustainable crop husbandry and, if pesticides are required, their use must conform to the requirements of the certifying authority. Smallholders are more likely to adopt pest management technologies if they are perceived as part of good crop husbandry as a whole, rather than as a separate exercise that requires additional expenditure of time, labour or cash.

IPM as originally conceived was intended for use by relatively well-educated and progressive farmers to restrict pesticide use, primarily to decrease the rate at which resistance to the pesticide developed in the target population. The successes of IPM are fewer in the developing world among less well-educated and resource-poor farmers. Nyambo *et al.* (1996) have reviewed both the need for and constraints upon the adoption of IPM practices by small holder coffee farmers in East Africa. They identify the need for national IPM policies to provide a framework for action, better extension and linkages from farmer to researcher and a holistic approach to pest management.

Implementation of IPM may require an understanding of pest life cycles, of cause and effect processes within these cycles and an ability to identify and distinguish pest species and beneficial species. The typical knowledge transfer systems in developing countries are usually inadequate for the task of educating farmers to the point where they have sufficient understanding of their agro-ecosystems to be willing to implement the control measures. However, the farmer field school approach, where groups of farmers receive intensive training on their own crops, is having an impact in a number of areas. Successful IPM adoption has usually occurred where there is a predominant pest that has a major impact on crop yield and where, in the absence of pesticides, natural enemies keep damage below the economic threshold, e.g. brown plant hopper in rice as a vector of tungro disease – a major viral disease of rice.

It should be noted that prior to the introduction of the powerful modern pesticides, during and after the Second World War, coffee farmers had very little choice but to use alternative measures of control that today are being advocated as elements of IPM control strategies, and the pre-war scientific literature deserves scrutiny for such measures. On the other hand, it is evident that some of the measures recommended, such as the widespread practice of hand-collecting of pests, were very labour-intensive and often ineffective.

The first line of defence in IPM programmes, where possible, should be the use of resistant or tolerant varieties and is mostly applicable to diseases. Next in line are the common farming practices that have major pest and disease control benefits. Measures such as rotation and altered planting and harvesting dates often require no additional cost to the farmer, but are more applicable to annual crops.

For perennial crops such as coffee, initial site selection and preparation before planting is crucial, as this can determine the subsequent likelihood or otherwise of particular pest and disease problems. Factors such as soil type, drainage, mesoclimate, the need to plant in shade and accessibility will all influence the future requirements for the maintenance of the general health of the crop.

Seedling health

The starting point for IPM in coffee plantations is to ensure that the crop is properly managed in a way that does not encourage pest and disease build-up. Avoidance is always better than cure. This begins even before the nursery, by selecting unblemished seeds taken from trees that themselves are free of disease.

This applies primarily to Fusarium bark disease. In Malawi, where Fusarium bark disease threatened coffee production during the 1960s, the estate sector opted for phytosanitation to combat the disease, by ensuring that seedlings were disease free before transplanting. In the smallholder sector, where it is more difficult to apply rigorous standards of phytosanitation, planting of resistant varieties was the option chosen for bark disease management.

Care must be taken in the nursery to keep seedlings healthy and to ensure that diseases and pests – particularly nematodes – are not introduced into the plantation at transplantation. This requires attention to sources of soil or to the treatment of soil, and the preferred option here is to use solarization, where moist soil is covered with polythene sheeting in full sun to provide heat, which destroys pest organisms. Copper fungicide may be needed to control *Cercospora* on nursery seedlings.

Crop management

As outlined earlier, there is a wide range of crop management practices that will help to avoid pest attack and disease infection and to boost the plant's natural resistance and ability to tolerate damage. Most crop management operations, such as adequate attention to soil fertility and organic matter, weeding, mulching, shade and a suitable pruning regime, are designed to benefit the overall health and performance of the crop. Others, such as crop sanitation measures to remove diseased trees or branches and the clean harvesting of berries if berry borer is a problem, may also be required to maintain the health of the crop. These activities all require labour inputs but are an important part of IPM.

The choice and the degree to which these measures are implemented depends upon a wide variety of factors, the most significant of which is the local status of particular pests and diseases. Often there is a trade-off between different practices, and this is influenced by economics.

It is apparent from several of the earlier chapters in this book that the provision of shade has many ecological advantages, but shade trees themselves require management and shade suppresses flowering. Some shade trees that produce marketable products can provide some diversity of cash income that may offset other costs. However, in times of high coffee prices, this might prove to be disadvantageous. Mulching is another theoretically favourable practice, but locating sources of mulch, cutting, transportation and spreading it are often difficult and expensive.

IPM of plant diseases requires a basic understanding of the interactions between host, pathogen and environment, often depicted as the 'disease triangle'.

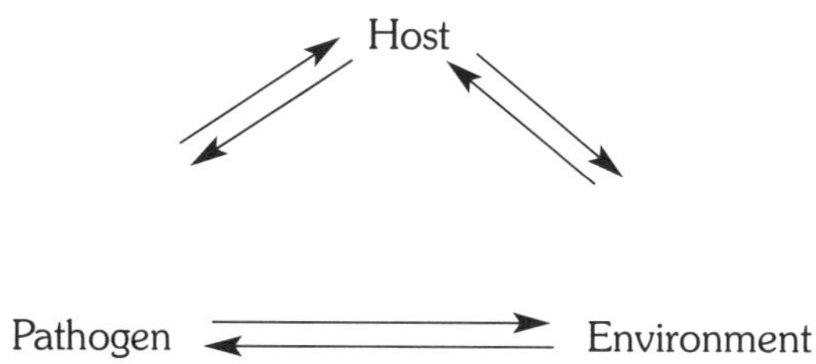

A similar scheme is applicable to pest interactions, except that the effect of natural enemies predominates as the main factor in determining population levels and the resultant damage to crops.

Pest thresholds

It is difficult to implement a true IPM system without some type of pest scouting. At its most simple, this may involve walking through the crop on a regular basis, examining the crop for damage and looking for pests. The next step is to use some form of recording system to record pest or visible damage levels, so that it is readily apparent when they have reached damaging levels. The damage threshold or the action threshold is the level of damage, or of pest populations, that starts to cause economic loss and, when this is reached, action to control the pest should be taken to avoid economic loss. Damage thresholds for the common problems on coffee are shown in Table 15.2.

Below this level, control measures will be wasted, as the benefit gained will be less than the cost of applying those measures. Far above the damage threshold and the pest management interventions may be rather 'revenge killing' than crop protection, or significant crop loss may already have occurred, once again decreasing the cost benefit of the intervention.

Table 15.2. Action thresholds for coffee pests and diseases.

Pest/Disease	Action threshold
Rust	5% of leaves affected by even a single rust pustule; examination of underside of leaves
CBD	Any evidence of disease on young berries
Antestia	A single insect (nymph or adult) per tree; scouting in early morning or late evening
Berry borer	5% of berries infested on the tree (not regarded as workable in Colombia) (See Baker, 1999)
Leaf miner	Number of moths per tree that are damaging changes with the age of the tree: 1 year = 2, 2 year = 5, 3 year = 10, > 4 year = 20. Select a sample of trees in the early morning and shake them, estimating moth numbers as they fly off. Spray about 1 week after first moths observed if the threshold population is reached. In some Latin American countries the spray threshold is 30% of leaves damaged
Berry moth	Need to be able to scout for moths and eggs. Alternatively, if damage on the buds or young berries is seen, spray immediately. Once the webbing is seen on the berry clusters, a lot of damage has already been done

Action thresholds are reached as pest and disease damage is advancing, so that the measures applied must stop this advance immediately, and this is often the point at which recourse to pesticides may be justified. This is particularly important for diseases where rates of increase are logarithmic and where latent periods often mean that there is a significant amount of disease in the crop that is not yet symptomatic. Only systemic fungicides that can eradicate latent infection will halt epidemics at this stage.

Scouting is best carried out in the early morning or late evening, when the insects are most active. Based on previous experience of the pest and disease problems in an area, the scouting schedule will target certain key pests. The trees should be examined for insect pests first, as they will be disturbed when foliage is examined for disease.

Where antestia is a problem, this is the first pest to scout for, by carefully moving around the tree and inspecting within the canopy. The next step is to shake the tree and estimate the number of leaf miner or other moths that fly out and to examine young branches and leaves for evidence of other pests and diseases. Finally, buds and young berries should be examined for evidence of disease or damage by pests such as berry borer, berry moth and antestia.

The number of trees to be sampled and the sampling design depends to some extent on the size of the plantation and its variability, and might range from ten trees in a small plot to 40 in a large plantation. Generally, trees should be selected at random to avoid any bias in observations. If the plantation is very large and consists of separate areas, or areas with different ecologies, these may have to be separately scouted. Scouting should ideally be conducted each week from flower initiation to harvest. Bushes may need to be examined at other times for damage caused by pests such as stem borers.

Observations and records of natural enemies, parasitized or diseased pests can also be made at this time by those sufficiently expert in this aspect. Very often, high levels of pest incidence are associated with high levels of parasitized or diseased individuals, in which case the population is already in decline and application of pesticides will do more harm than good.

Natural enemy populations

The concept of 'iatrogenic' pest and disease problems is now well established – that human intervention, usually in the form of pesticide spraying, can cause an upsurge of pest and disease species previously suppressed by ecological balance. Insect pest populations are normally kept below damage thresholds due to parasitism and predation by other organisms that are natural enemies of the pest species, and therefore known collectively as 'beneficials'.

There are a wide range of such organisms, as detailed in appendix A, and they are the most important factor in limiting pest 'outbreaks'. In a similar but less dramatic way, pathogens are suppressed by the competitive effects of microorganisms that exist on plant surfaces or in plant tissues as harmless endophytes. Soil-inhabiting, plant-parasitic nematodes can also be checked by competition from non-pathogenic species and by direct parasitism by nematophagous fungi.

There are recorded examples of all of these naturally suppressive systems becoming unbalanced through the use of pesticides used to control coffee pests and diseases. Berry moth, looper caterpillars, scale insects and leaf miner are all pests that have known natural enemies that would be adversely affected by broad-spectrum insecticides. The frequent use of insecticidal sprays is likely, in the long term, to have an undesirable effect on pest populations in coffee plantations, both through encouraging the development of resistance to the chemical used but also by causing a depletion of the natural biological control systems that limit pest populations.

The use of broad-spectrum organophosphate insecticides was partly responsible for the upsurge in leaf miner populations in Kenya and Tanzania during the 1980s, due to their effect on natural enemies and the development of pesticide resistance (Bardner and Mcharo, 1988; Nyambo *et al.*, 1996). The use of fungicides has been shown to exacerbate CBD if applied to the wrong schedule, and can result in disease being more severe than before spraying had begun (Griffiths *et al.*, 1971). This and similar effects have been shown to be due to the disruption by fungicides of natural biological control mechanisms mediated by the coffee microflora (Waller and Masaba, 2006). Some evidence has been presented that copper fungicide residues accumulating in the soil under coffee trees have upset the microbial balance in the soil in favour of plant-pathogenic nematodes (Bridge, 1984).

Pesticides as part of IPM

What is clear from this evidence and that drawn from similar experiences with other crops is that pesticides should be used only when necessary (not routinely) and used within an IPM framework, so that chemical use is properly scheduled with minimal non-target effects. It is a misconception, not exclusive to smallholders in developing countries, that pesticidal spraying can be used to increase yields. This is not the case: they cannot be used to compensate for poor crop management. Pesticides may be used to protect the yield that has been achieved by sound crop husbandry from losses caused by pests and pathogens when this cannot be avoided by other means.

Pesticides became widely and routinely used because they were seen as offering an easy insurance against pest attacks and, in some cases, as a labour-saving device. They still have an important role to play in IPM, but only when used judiciously after cultural measures have been implemented, possibly supplemented with biological control agents where these are available commercially. Spot-spraying of localized upsurges of a pest is one of the more sensitive uses of pesticides, as the target is defined and limited.

Routine applications of fungicides may still be needed for the control of endemic disease such as CBD and rust, where resistant cultivars have not been deployed, but this can only be justified where the crop is properly managed and fertilized. Ground-applied systemic fungicides – such as granular formulations of triadimefon – have given good control of rust and also of other leaf spot diseases (e.g. Villaraga, 1987). Although this would have less effect on beneficial phylloplane microflora than would be the case with a full cover spray, ground application of pesticides has deleterious effects on soil microbiota.

Soil application of aldicarb can control leaf miner, but this is a WHO Class I substance and poses risks to both the environment and those using it. In Brazil, it has sometimes been necessary to combine a single application of aldicarb with one spray of ethion, to keep insect populations below the damage threshold. The foliar spray was carried out only when 30% of leaves had been attacked by leaf miner and where there was no predation by wasps (Reis and Souza, 1996). Systemic pesticides are not translocated into the berry sufficiently to provide control of insect pests that feed in the berries – or on berry diseases.

Biological control

Biological control agents have been bred and introduced into areas where particular exotic pest species have become established and increased to damaging levels in the absence of their native natural enemies. This classical biological control approach has been used against berry borer in many countries. Parasites were introduced to Indonesia in 1924, Brazil in 1929 and Sri Lanka in 1938 (Le Pelley, 1968), and more recently to other Latin America countries and to some extent in India, under a recent international development programme (Baker, 1999). In the highlands of Kenya, the coffee mealybug, *Planococcus kenyae*, which up to 1939 had been a devastating pest, was much reduced in severity by the introduction of a parasite from Uganda (Le Pelley, 1968).

Usually no single agent is sufficiently effective to control the pest by itself, and several species acting in concert are more effective. It may also be necessary to integrate with other measures. In India, the use of endosulphan sprays has been successfully integrated with cultural measures to control berry borer. Use of the insecticide is strictly controlled in this IPM system, being permitted only once per season and applied only to the trees on which the pest outbreak is first noticed (Reddy-Samba Murthy and Rao, 1999).

It may be possible to confine pesticide use to spot-spraying if scouting is done thoroughly and regularly, a biological control agent is used and care is taken to remove all berries that remain on the trees and on the ground after harvest. It should also be noted that where endosulphan has been frequently used to control berry borer, resistance to the insecticide has developed in the insect population (Brun and Suckling, 1992). This substance is not recommended by most agencies (see Table 15.1).

In Guatemala, the parasitoid *Cephalonomia stephanoderis* was produced on-farm and was able to provide adequate control of berry borer at low levels of infestation (6%) at a cost equivalent to chemical control (Campos and Garcia, 1997). More effective control might be possible in future by combining *C. stephanoderis* with another parasitoid, *Phymastichus coffea*. Research efforts in 2004 were aimed at improving mass-rearing methods for these insects.

In Colombia, it was found that the parasitoid *Prorops nasuta* could be used in combination with insecticides, provided the parasitoid was released first and 20 d had elapsed before insecticides were applied (endosulphan, chorpyirfos, fenitrothion and pirimiphos-methyl) (Mejia *et al.*, 2000). The entomopathogenic fungus *Beauvaria bassiana* has also shown promise for borer control, but it has proved difficult to produce high-quality product at acceptable cost. In India, spot-spraying with endosulphan is combined with phytosanitary measures, particularly important amongst these being the removal of unharvested berries.

In order to preserve the natural enemy population, routine or indiscriminate use of insecticides should be avoided. Fenitrothion, for instance, is a useful insecticide, effective against several coffee pests but, as with the fungicides, in order to get a positive cost benefit from its use, the crop must also be well managed and application based on scouting. Proper timing of insecticide sprays requires some form of scouting to monitor pest infestations. In São Paulo, Brazil, for instance, it was found that naturally occurring parasitoids kept populations of leaf miner (*P. coffeella*) below the damage threshold, without the need for insecticide spraying (Gravena and Yamamoto, 1994).

Basic rules for IPM:

- Begin with careful nursery management.
- Ensure suitable site selection and preparation.
- Transplant strong healthy seedlings.
- Implement sound soil and nutrient management.
- Avoid over-shading.
- Mulch the trees, especially when young.
- Prune trees as required.
- Manage the plantation to encourage beneficial insects by retaining a

diverse flora of flowering plants between the coffee rows and/or the field margins.

- Carry out suitable crop sanitation measures, such as removal of old berries on the tree or on the ground if berry borer or CBD is a recurring problem.
- Inspect [scouting] trees weekly for diseases and pests or pest damage.
- Resort to pesticides only if there is danger that economic damage will otherwise occur and when there will be a cost benefit to their use. If possible, spot-spray in the first instance.

While it is possible to draw up basic guidelines for the implementation of IPM (see above) in coffee plantations, it is not possible to suggest a generalized IPM system for coffee, as each system has to be tailored according to the pest spectrum and to the needs and capabilities of coffee farmers in a particular area. The biological and socio-economic variables may be such that IPM systems have to be specific to individual farmers, in which case the best approach is to offer a 'basket' of interventions that can be selected by farmers and integrated into a system that meets their needs.

Usually, IPM systems are required where insect pest control is heavily reliant on chemical insecticides. The aim is then to design a system that keeps the main pest(s) below their damage threshold with the minimal use of insecticide. For instance, this might require the combination of biological control agents with the use of insecticide for the control of berry borer. Research in Brazil has shown that the insecticides alpha-cypermethrin and thiamethoxam may be integrated with application of the entomopathogenic fungus *Beauvaria bassiana*, without significant effect on growth or sporulation of the protective fungus (Oliveira *et al.*, 2003).

As concerns about the health and environmental impact of insecticide spraying become more widespread, public attention is driving the agenda to reduce reliance on pesticides. While legislation for safe use of pesticides becomes more rigorous and the more toxic pesticides are withdrawn from the market, there is an urgent need for practical IPM systems. Endosulphan for instance, has been widely used on coffee in Asia and Latin America, particularly for control of berry borer, but its use is now strictly regulated in many countries and banned in others.

References

BAC (1995) *Integrated Crop Management: a Complete Training and Resource Pack*. British Agrochemicals Association, Peterborough, UK.

Baker, P.S. (1999) *The coffee berry borer in Colombia. Final Report of the DFID-Cenicafe – CABI BioScience IPM for Coffee Project*. CABI BioScience, Ascot, UK, 144 pp.

Bardner, R. and Mcharo, E.Y. (1988) Confirmation of resistance of the coffee leaf miner (*Leucoptera meyricki*) to organophosphate insecticide spray in Tanzania. *Tropical Pest Management* 34, 52–54.

Bridge, J. (1984) *Coffee Nematode Survey of Tanzania. Report of a visit to examine plant parasitic nematodes of coffee in Tanzania, February/March 1984*. Commonwealth Institute of Parasitology, St Albans, UK, 22 pp.

Brun, L.O. and Suckling, D. (1992) Field selection for endosulphan resistance in coffee berry borer in New Caledonia. *Journal of Economic Entomology* 85, 325–334.

Campos, A.O. and Garcia, A. (1997) Control biologico de broca del cafe (*Hypothenemus hampei*). Aplicacion de manejo comercial en finca. *Boletin Promecafe* 76, 8–12.

Ebbels, D.E. (2003) *Principles of Plant Health and Quarantine*. CAB International, Walllingford, UK.

Gravena, S. and Yamamoto, P.T. (1994) Manejo de pragas do cafeeiro com endosulfan na região de Espirito Santo do Pinhal. *Cientifica Jaboticabal* 22, 123–131.

Griffiths, E., Gibbs, J.N. and Waller, J.M. (1971) Control of coffee berry disease. *Annals of Applied Biology* 67, 45–74.

Johnson, R. (1984) Analysis of durable resistance. *Annual Review of Phytopathology* 22, 309–378.

Kogan, M. (1998) Integrated pest management: historical perspectives and contemporary developments. *Annual Review of Entomology* 43, 243–270.

Le Pelley, R.H. (1968) *Pests of coffee*. Longmans, London, 590 pp.

Matthews, G.A. and Hislop, E.C. (1993) *Application Technology for Crop Protection*. Oxford University Press, New York.

Mejía, M.J.W., Butillo, P.A.E., Orozoco, H.J. and Cháves, C.B. (2000) Efecto de cuatro insecticidas de *Beauvaria bassiana* sobre *Prorops nasuta* (Hymenoptera: Bethylidae) parasitoide de la broca del café. *Revista Colombiana de Entomologia* 26, 117–123.

Muller, R.A. (1973) L'anthracnose des baies du caféier d'arabie (*Coffea arabica*) due à une forme virulent de *Colletotrichum coffeanum* Noack. *Café-Cacao-Thé* 17, 281–312.

Nelson, R.J. and Leung, H. (1994) The use of molecular markers to characterize pathogen populations and sources of disease resistance. In: Teng, P.S., Heong, K.L. and Moody, K. (eds) *Rice Pest Science and Management*. International Rice Research Institute (IRRI), Manila, Philippines, pp. 173–192.

Nyambo, B., Masaba, D.M. and Hakiza, G.J. (1996) Integrated pest management of coffee for small-scale farmers in East Africa: Needs and limitations. *Integrated Pest Management Reviews* 1, 125–132.

Oliveira, de C.N., Neves, P.M.O.J. and Kawazoe, L.S. (2003) Compatibility between the entomopathogenic fungus *Beauvaria bassiana* and insecticides used in coffee plantations. *Scientia Agricola* 60, 663–667.

Pfalzer, H. (2004) Spraying equipment for coffee. In: Wintgens, J.N. (ed.) *Coffee: Growing, Processing and Sustainable Production*. Wiley-Verlach, Weinheim, Germany, pp. 565–589.

Reddy-Samba Murthy, A.G. and Rao, L.V.A. (1999) Incidence of coffee berry borer in non-conventional coffee area of Karnataka. *Indian Coffee* 63, 15–16.

Reis, P.R. and Souza, J.C. (1996) Manejo integrado do bicho-mineiro, *Perileucoptera coffeella* (Guérin-Meneville) (Leucoptera: Lyonetiidae), e seu reflexo na produção de café. *Anais da Sociedade Entomológica do Brasil* 25, 77–82.

Rodriguez Jr, C.J. and Eskes, A.B. (2004) Resistance to Coffee Leaf Rust and Coffee Berry Disease. In: Wintgens, J.N. (ed.) *Coffee: Growing, Processing and Sustainable Production*. Wiley-Verlach, Weinheim, Germany, pp. 551–564.

Schuppener, H., Harr, J., Sequeira, F. and Gonzales, A. (1977) First occurrence of the coffee leaf rust *Hemileia vastastrix* in Nicaragua, 1976, and its control. *Café Cacao Thé* 21, 197–202.

Shaw, D.E. (1968) Coffee rust outbreaks in Papua from 1892 to 1965 and the 1965 eradication campaign. *Department of Agriculture, Stock and Fisheries, Port Moresby, Research Bulletin, Plant Pathology Series* 2, 20–25.

Simmons, N.W. (1991) Genetics of horizontal resistance to disease of crops. *Biological Reviews* 66, 198–241.

Van der Plank, J.E. (1963) *Plant diseases: Epidemics and Control*. Academic Press, New York.

Villaraga, A.L.A. (1987) Dosage and frequency of application of granular systemic fungicide to control coffee leaf rust in Kenya. *Ascolfi Informa* 13, 50–57.
Waller, J.M. (1979) The recent spread of coffee rust (*Hemileia vastatrix* Berk. & Br.) and attempts to control it. In: Ebbels, D.L. and King, J.E. (eds) *Plant Heath – the Scientific Basis for the Administrative Control of Plant Disease and Pests*. Blackwell, Oxford, UK.
Waller J.M. and Masaba, D.M. (2006) The microflora of coffee surfaces and relationship to coffee berry disease. *International Journal of Tropical Pest Management* 52, 89–96.
Wallis, J.A.N. and Firman, I.D. (1962) Spraying Arabica coffee for the control of leaf rust. *East African Agricultural and Forestry Journal* 28, 89–104.

Appendix A Natural Enemies and Other Insects Associated with the Main Pest Species

Note that this list includes only natural enemies actually recorded on coffee and therefore is not necessarily a complete list of the enemies that may attack the pest. Each entry comprises, in order, genus, family, location(s) discovered and relevant reference(s).

Lepidoptera

Archips occidentalis (see p. 100)

Parasites

Apanteles angaleti Muesebeck, Braconidae, Kenya, (Evans *et al.*, 1968)
Apanteles coffea Wilkinson, Braconidae, Zimbabwe, (Kutywayo, 1989)
Apanteles evansi Nixon, Braconidae, Kenya, (Nixon, 1971)
Apanteles sp., Braconidae, Kenya, (Evans, 1970)
Ascogaster cava De Saeger, Braconidae, Kenya, (Evans, 1970)
Ascogaster sp. near *cava* De Saeger, Braconidae, Zimbabwe, (Kutywayo, 1989)
Ascogaster sp., Braconidae, Kenya, (Evans, 1970)
Brachymeria olethria (Waterston), Chalcididae, Kenya, (Evans, 1970)
Brachymeria sp. near *dunbrodyensis* (Cameron), Chalcididae, Kenya, (Evans, 1970)
Brachymeria sp., Chalcididae, Kenya, (Evans, 1970)
Distatrix ugandaensis (Gahan), Braconidae, Kenya, (Evans, 1970), Zimbabwe, (Kutywayo, 1989)
Euplectromorpha obscurata Ferriere, Eulophidae, Uganda, (Ferriere, 1941)
Exorista xanthaspis Wiedemann, Tachinidae, Kenya, (Evans, 1970), Zimbabwe, (Kutywayo, 1989)
Goryphus sp., Ichneumonidae, Kenya, (Evans, 1970)

Itoplectis glabra (Morley), Ichneumonidae, Kenya, (Evans, 1970)
Itoplectis suada (Tosquinet), Ichneumonidae, Kenya, (Evans, 1970)
Palexorista sp., Tachinidae, Kenya, (Evans, 1970)
Pristomerus sp., Ichneumonidae, Kenya, (Evans, 1970)
Sjostedtiella rufescens (Tosquinet), Ichneumonidae, Kenya, (Evans, 1970)
Trichogrammatoidea nana (Zehntner), Trichogrammatidae, Kenya, (Evans, 1970)
Venanides sp., Braconidae, Zimbabwe, (Kutywayo, 1989)

Hyperparasites

Aphanogmus fijiensis (Ferriere), Ceraphronidae, Kenya, (Evans, 1970)
Mesochorus sp. near *breviscapus* Kerrich, Ichneumonidae, Kenya, (Evans, 1970)
Pediobius sp. near *foveolatus* (Crawford), Eulophidae, Kenya, (Evans, 1970)

Ascotis selenaria reciprocaria (see p. 97)

Parasites

Afromelanichneumon transvaalensis rubicundus Heinrich, Ichneumonidae, Kenya, (Abasa and Mathenge, 1972)
Charops spinitarsis Cameron, Ichneumonidae, Zimbabwe, (Kutywayo, 1989)
Congochrysosoma sp., Tachinidae, Zimbabwe, (Kutywayo, 1989)
Cryptus australis nigropictus Cameron, Ichneumonidae, Kenya, (Abasa and Mathenge, 1972)
Exorista bombycis (Louis), Tachinidae, Kenya, (Wheatley, 1963)
Palexorista quadrizonula (Thomson), Tachinidae, Kenya, (Crosskey, 1970), Zimbabwe, (Kutywayo, 1989)
Pediobius sp. near *foveolatus* (Crawford), Eulophidae, Kenya, (Crowe and Greathead, 1970)

Predators

Macrorhaphis acuta Dallas, Pentatomidae, Kenya, (Wheatley, 1963), Zimbabwe, (Kutywayo, 1989)
Rhynocoris segmentarius (Germar), Reduviidae, Zimbabwe, (Kutywayo, 1989)

Cephonodes hylas (see p. 106)

Parasites

Actia heterochaeta Bezzi, Tachinidae, Uganda, (Le Pelley, 1959)
Blepharipa zebina (Walker), Tachinidae, Malaysia, (Thompson, 1943–1958), India, (Thompson, 1943–1958)
Exorista bombycis (Louis), Tachinidae, Malaysia, (Corbett and Miller, 1933)

Forcipomyia fuliginosa (Meigen), Ceratopogonidae, Malaysia, (Corbett and Miller, 1933)
Ooencyrtus papilionis Ashmead, Encyrtidae, Malaysia, (Corbett and Yusope, 1932)

Predators

Chrysocoris javanus Westwood, Scutelleridae, Democratic Republic of Congo, (Bredo, 1939)
Oecophylla smaragdina (Fabricius), Formicidae Formicinae, Malaysia, (Corbett, 1937)
Sycanus leucomesus Walker, Reduviidae, Malaysia, (Corbett and Yusope, 1932)

Epicampoptera andersonii (see p. 99)

Parasites

Brachymeria bottegi Masi, Chalcididae, Kenya, (Anderson, 1934)
Exorista bombycis (Louis), Tachinidae, Kenya, (Anderson, 1934)
Hemipimpla pulchra Morley, Ichneumonidae, Kenya, (Anderson, 1934)
Pimpla mahalensis Gribodo, Ichneumonidae, Kenya, (Evans, 1966a)
Xanthopimpla octopunctata (Kriechbaumer), Ichneumonidae, Kenya, (Anderson, 1934)

Predators

Eumenes maxillosa (De Geer), Vespidae, Uganda, (Hargreaves, H., 1932)
Glypsus conspicuus (Westwood), Pentatomidae, Kenya, (Le Pelley, 1959)
Glypsus vigil (Germar), Pentatomidae, Kenya, (Anderson, 1934)
Macrorhaphis acuta Dallas, Pentatomidae, Kenya, (Le Pelley, 1959), Malawi, (Le Pelley, 1968)

Epicampoptera marantica (see p. 99)

Parasite

Carcelia angulicornis Villeneuve, Tachinidae, Democratic Republic of Congo, (Leroy, 1936)

Epicampoptera stictigramma (see p. 99)

Parasites

Apanteles sp., Braconidae, Kenya, (Leeuwangh, 1965)
Brachymeria dunbrodyensis (Cameron), Chalcididae, Kenya, (Leeuwangh, 1965)
Casinaria sp., Ichneumonidae, Kenya, (Crowe and Leeuwangh, 1965)
Charops sp., Ichneumonidae, Kenya, (Crowe and Leeuwangh, 1965)
Euplectromorpha sp., Eulophidae, Kenya, (Crowe and Leeuwangh, 1965)
Exorista bombycis (Louis), Tachinidae, Kenya, (Leeuwangh, 1965)
Rasivalva lepelleyi (Wilkinson), Braconidae, Kenya, (Leeuwangh, 1965)

Epicampoptera strandi glauca (see p. 99)

Parasites

Telenomus narosus Nixon, Scelionidae, Côte d'Ivoire, (Griveaud, 1967)
Telenomus sp., Scelionidae, Nigeria, (Okelana, 1985)

Homona coffearia (see p. 100)

Parasites

Apanteles taragamae Viereck, Braconidae, Indonesia Java, (Wilkinson, 1928)
Camptotypus sellatus Kriechbaumer, Ichneumonidae, Papua New Guinea, (Szent-Ivany, 1958)
Macrocentrus homonae Nixon, Braconidae, Indonesia Java, (Gadd, 1946), Sri Lanka, (Gadd, 1946)
Theronia simillima Turner, Ichneumonidae, Papua New Guinea, (Szent-Ivany, 1958)

Lamprosema crocodora (see p. 102)

Parasites

Apanteles congoensis De Saeger, Braconidae, Democratic Republic of Congo, (Schmitz, 1949)
Brachymeria sp., Chalcididae, Democratic Republic of Congo, (Bredo, 1939)
Echthromorpha agrestoria occidentalis Krieger, Ichneumonidae, Democratic Republic of Congo, (Schmitz, 1949)
Hypomicrogaster vacillatrix (Wilkinson), Braconidae, Democratic Republic of Congo, (Schmitz, 1949)
Trichogrammatoidea lutea Girault, Trichogrammatidae, Democratic Republic of Congo, (Schmitz, 1949)

Hyperparasites

Aphanogmus reticulatus (Fouts), Ceraphronidae, Democratic Republic of Congo, (Schmitz, 1949)
Eurytoma syleptae Ferriere, Eurytomidae, Democratic Republic of Congo, (Schmitz, 1949)
Nesolynx phaeosoma (Waterston), Eulophidae, Kenya, (Evans, 1966a), Democratic Republic of Congo, (Schmitz, 1949)
Pediobius afronigripes Kerrich, Eulophidae, Democratic Republic of Congo, (Schmitz, 1949)

Latoia vivida (see p. 103)

Parasites

Carcelia vara Curran, Tachinidae, Zimbabwe, (Kutywayo, 1989)
Chaetexorista langi (Curran), Tachinidae, Zimbabwe, (Kutywayo, 1989)
Charops spinitarsis Cameron, Ichneumonidae, Zimbabwe, (Kutywayo, 1989)
Charops sp., Ichneumonidae, Kenya, (Le Pelley, 1959), Malawi, (Smee, 1930)
Chrysis sp., Chrysididae, Zimbabwe, (Kutywayo, 1989)
Euplectromorpha sp., Eulophidae, Malawi, (Smee, 1942)
Fornicia africana Wilkinson, Braconidae, Kenya, (Le Pelley, 1959), Zambia, (Wilkinson, 1930), Zimbabwe, (Kutywayo, 1989)
Gelis sp., Ichneumonidae, Zimbabwe, (Kutywayo, 1989)
Mesochorus sp., Ichneumonidae, Zimbabwe, (Kutywayo, 1989)
Palexorista gilvoides (Curran), Tachinidae, Kenya, (Le Pelley, 1959)
Palexorista patruelis (Mesnil), Tachinidae, Zimbabwe, (Kutywayo, 1989)
Platyplectrus sp., Eulophidae, Malawi, (Smee, 1942)

Predators

Macrorhaphis acuta Dallas, Pentatomidae, Zimbabwe, (Kutywayo, 1989)
Rhynocoris segmentarius (Germar), Reduviidae, Zimbabwe, (Kutywayo, 1989)

Leucoptera caffeina (see p. 91)

Parasites

Ageniaspis sp., Encyrtidae, Kenya, (Bess, 1964), Tanzania, (Tapley, 1960)
Amblymerus sp., Pteromalidae, Kenya, (Crowe and Greathead, 1970)
Apanteles bordagei Giard, Braconidae, Ethiopia, (Crowe and Greathead, 1970), East Africa, (Crowe and Greathead, 1970)
Apleurotropis lamellata (Kerrich), Eulophidae, Tanzania, (Kerrich, 1969)
Aprostocetus leucopterae (Ferriere), Eulophidae, Tanzania, (Notley, 1948)
Chrysocharis lepelleyi Ferriere, Eulophidae, Ethiopia, (Greathead, 1965), Tanzania, (Notley, 1948)
Cirrospilus cinctiventris Ferriere, Eulophidae, Tanzania, (Notley, 1948)

Cirrospilus variegatus Masi, Eulophidae, Tanzania, (Notley, 1948)
Closterocerus ritchiei (Ferriere), Eulophidae, Ethiopia, (Greathead, 1965), Kenya, (Le Pelley, 1959), Tanzania, (Notley, 1948)
Elasmus ?johnstoni Ferriere, Elasmidae, Ethiopia, (Crowe and Greathead, 1970)
Elasmus leucopterae Ferriere, Elasmidae, East Africa, (Crowe and Greathead, 1970), Tanzania, (Notley, 1948)
Notanisomorphella borborica (Giard), Eulophidae, Tanzania, (Notley, 1948)
Parahormius leucopterae Nixon, Braconidae, Tanzania, (Notley, 1948)
Parahormius sp., Braconidae, Kenya, (Crowe and Greathead, 1970)
Pediobius coffeicola (Ferriere), Eulophidae, Ethiopia, (Greathead, 1965), Tanzania, (Notley, 1948)
Platocharis coffeae (Ferriere), Eulophidae, Ethiopia, (Crowe and Greathead, 1970)
Schizocharis amaniensis Kerrich, Eulophidae, Tanzania, (Crowe and Greathead, 1970)
Sympiesis comosus Kerrich, Eulophidae, Tanzania, (Notley, 1948)

Hyperparasites

Apleurotropis lamellata (Kerrich), Eulophidae, East Africa, (Crowe and Greathead, 1970)
Aprostocetus leucopterae (Ferriere), Eulophidae, Uganda, (Crowe and Greathead, 1970)
Cirrospilus cinctiventris Ferriere, Eulophidae, Tanzania, (Notley, 1948)
Closterocerus africanus Waterston, Eulophidae, Tanzania, (Notley, 1948)
Closterocerus violaceus (Ferriere), Eulophidae, Tanzania, (Notley, 1948)
Pediobius coffeicola (Ferriere), Eulophidae, Tanzania, (Notley, 1948)
Sympiesis comosus Kerrich, Eulophidae, Tanzania, (Notley, 1948)

Leucoptera coma (see p. 91)

Parasites

Ageniaspis sp., Encyrtidae, Democratic Republic of Congo, (Decelle, 1962)
Apanteles bordagei Giard, Braconidae, Uganda, (Crowe and Greathead, 1970)
Aprostocetus leucopterae (Ferriere), Eulophidae, Uganda, (Crowe and Greathead, 1970)
Chrysocharis lepelleyi Ferriere, Eulophidae, Uganda, (Crowe and Greathead, 1970)
Cirrospilus cinctiventris Ferriere, Eulophidae, Uganda, (Crowe and Greathead, 1970)
Cirrospilus variegatus Masi, Eulophidae, Uganda, (Crowe and Greathead, 1970)
Closterocerus ritchiei (Ferriere), Eulophidae, Uganda, (Crowe and Greathead, 1970), Democratic Republic of Congo, (Decelle, 1962)

Elasmus leucopterae Ferriere, Elasmidae, Uganda, (Crowe and Greathead, 1970), Democratic Republic of Congo, (Decelle, 1962)
Eulophus sp., Eulophidae, Democratic Republic of Congo, (Decelle, 1962)
Mirax leucopterae Wilkinson, Braconidae, Democratic Republic of Congo, (Decelle, 1962)
Pediobius coffeicola (Ferriere), Eulophidae, Democratic Republic of Congo, (Decelle, 1962)
Sympiesis comosus Kerrich, Eulophidae, Uganda, (Crowe and Greathead, 1970)

Hyperparasites

Aprostocetus leucopterae (Ferriere), Eulophidae, Uganda, (Crowe and Greathead, 1970)
Closterocerus violaceus (Ferriere), Eulophidae, Uganda, (Crowe and Greathead, 1970)
Sympiesis comosus Kerrich, Eulophidae, Uganda, (Crowe and Greathead, 1970)

Leucoptera meyricki (see p. 91)

Parasites

Ageniaspis sp., Encyrtidae, Tanzania, (Le Pelley, 1968), Zimbabwe, (Kutywayo, 1989)
Anagyrus sp. near *kivuensis* Compere, Encyrtidae, Kenya, (Evans, 1966b)
Apanteles bordagei Giard, Braconidae, Kenya, (Bess, 1964), Tanzania, (Notley, 1948), Uganda, (Crowe and Greathead, 1970)
Apanteles sp., Braconidae, Zimbabwe, (Kutywayo, 1989)
Apleurotropis lamellata (Kerrich), Eulophidae, Tanzania, (Kerrich,1969)
Aprostocetus leucopterae (Ferriere), Eulophidae, Tanzania, (Notley, 1948)
Chrysocharis lepelleyi Ferriere, Eulophidae, Kenya, (Le Pelley, 1968), Tanzania, (Notley, 1948), Uganda, (Le Pelley, 1959)
Chrysonotomyia sp., Eulophidae, Kenya, (Kerrich, 1969)
Cirrospilus cinctiventris Ferriere, Eulophidae, Kenya, (Crowe and Greathead, 1970), Tanzania, (Notley, 1948), Uganda, (Le Pelley, 1959)
Cirrospilus crowei Kerrich, Eulophidae, Kenya, (Kerrich, 1969), Zimbabwe, (Kutywayo, 1989)
Cirrospilus longifasciatus Ferriere, Eulophidae, Tanzania, (Ferriere, 1936)
Cirrospilus variegatus Masi, Eulophidae, Kenya, (Bess, 1964), Malawi, (Crowe and Greathead, 1970), Tanzania, (Notley, 1948), Uganda, (Crowe and Greathead, 1970), Zimbabwe, (Kutywayo, 1989)
Cirrospilus sp., Eulophidae, Uganda, (Crowe and Greathead, 1970)
Closterocerus ritchiei (Ferriere), Eulophidae, Kenya, (Bess, 1964), Malawi, (Ferriere, 1936), Tanzania, (Notley, 1948), Zimbabwe, (Weaving, 1972)
Elasmus leucopterae Ferriere, Elasmidae, Kenya, (Le Pelley, 1968), Tanzania, (Notley, 1948)
Elasmus sp., Elasmidae, Zimbabwe, (Kutywayo, 1989)

Hemiteles sp., Ichneumonidae, Tanzania, (Notley, 1948)
Microterys sp., Encyrtidae, Zimbabwe, (Kutywayo, 1989)
Mirax leucopterae Wilkinson, Braconidae, Zimbabwe, (Kutywayo, 1989)
Notanisomorphella borborica (Giard), Eulophidae, Kenya, (Crowe and Greathead, 1970), Tanzania, (Notley, 1948), Zimbabwe, (Kutywayo, 1989)
Parahormius leucopterae Nixon, Braconidae, Kenya, (Bess, 1964), Tanzania, (Nixon, 1940), Zimbabwe, (Weaving, 1972)
Parahormius sp., Braconidae, Kenya, (Crowe and Greathead, 1970)
Pediobius coffeicola (Ferriere), Eulophidae, Kenya, (Bess, 1964), Tanzania, (Notley, 1948), Uganda, (Le Pelley, 1959), Zimbabwe, (Weaving, 1972)
Platocharis coffeae (Ferriere), Eulophidae, Tanzania, (Crowe and Greathead, 1970)
Platyplectrus sp., Eulophidae, Tanzania, (Crowe and Greathead, 1970)
Schizocharis amaniensis Kerrich, Eulophidae, Tanzania, (Crowe and Greathead, 1970)
Sympiesis comosus Kerrich, Eulophidae, Tanzania, (Notley, 1948)

Hyperparasites

Apleurotropis lamellata (Kerrich), Eulophidae, East Africa, (Crowe and Greathead, 1970)
Aprostocetus leucopterae (Ferriere), Eulophidae, Uganda, (Crowe and Greathead, 1970)
Chrysocharis lepelleyi Ferriere, Eulophidae, East Africa, (Crowe and Greathead, 1970)
Cirrospilus cinctiventris Ferriere, Eulophidae, Kenya, (Le Pelley, 1968), Uganda, (Le Pelley, 1968), Tanzania, (Notley, 1948)
Closterocerus africanus Waterston, Eulophidae, Kenya, (Le Pelley, 1968), Tanzania, (Notley, 1948)
Closterocerus ritchiei (Ferriere), Eulophidae, East Africa, (Crowe and Greathead, 1970)
Closterocerus violaceus (Ferriere), Eulophidae, Kenya, (Le Pelley, 1968), Malawi, (Ferriere, 1936), Tanzania, (Notley, 1948)
Pediobius coffeicola (Ferriere), Eulophidae, Kenya, (Le Pelley, 1968), Tanzania, (Notley, 1948), Uganda, (Le Pelley, 1968)
Pediobius sp. near *foveolatus* (Crawford), Eulophidae, Tanzania, (Crowe and Greathead, 1970)
Sympiesis comosus Kerrich, Eulophidae, Tanzania, (Notley, 1948)
Tetrastichus sp., Eulophidae, Tanzania, (Crowe and Greathead, 1970)

Parasa lepida (see p. 103)

Parasites

Apanteles parasae Rohwer, Braconidae, Indonesia Java, (Kalshoven, 1950–1951), Sri Lanka, (Le Pelley, 1968)

Chaetexorista javana Brauer & Bergenstamm, Tachinidae, Malaysia, (Le Pelley, 1968)
Eurytoma monemae Ruschka, Eurytomidae, Malaysia, (Thompson, 1943–1958)
Kriechbaumerella ayyari (Gahan), Chalcididae, India, (Gahan, 1919)
Praestochrysis shanghaiensis (F.Smith), Chrysididae, India, (Sevastopulo, 1937)

Perileucoptera coffeella (see p. 91)

Parasites

Achrysocharoides sp., Eulophidae, Puerto Rico, (Gallardo-Covas, 1988)
Allobracon ?primus (Muesebeck), Braconidae, Colombia, (Florez and De Hernandez, 1981)
Allobracon sp., Braconidae, Mexico, (Aranda, 1986)
Aprostocetus sp., Eulophidae, Colombia, (Cardenas and Posada, 2001)
Ceranisus sp., Eulophidae, Mexico, (Aranda, 1986)
Chrysocharis sp. near *nitetis* (Walker), Eulophidae, Mexico, (Aranda, 1986)
Chrysocharis sp., Eulophidae, Colombia, (Cardenas and Posada, 2001), Cuba, (Konnorova, 1987), Mexico, (Aranda, 1986)
Chrysonotomyia sp., Eulophidae, Colombia, (Roba, 1938), Mexico, (Aranda, 1986), Puerto Rico, (Gallardo-Covas, 1988)
Cirrospiloideus sp., Eulophidae, Puerto Rico, (Gallardo-Covas, 1988)
Cirrospilus sp., Eulophidae, Brazil, (Parra *et al.*, 1977), Cuba, (Konnorova, 1987), Mexico, (Aranda, 1986), Peru, (Enriquez *et al.*, 1975a)
Closterocerus cinctipennis Ashmead, Eulophidae, Cuba, (Bruner, 1929)
Closterocerus coffeellae Ihering, Eulophidae, Brazil, (Mendes, 1940), Colombia, (Roba, 1938)
Closterocerus flavicinctus De Santis, Eulophidae, Brazil, (De Santis, 1983)
Closterocerus leucopus Ashmead, Eulophidae, Puerto Rico, (Anon, 1940b)
Closterocerus lividus (Ashmead), Eulophidae, Puerto Rico, (Van Zwaluwenberg, 1917), Venezuela, (Box, 1927)
Closterocerus sp., Eulophidae, Guatemala, (Hamilton, 1967), Mexico, (Aranda, 1986)
Derostenus sp., Eulophidae, Puerto Rico, (Wolcott, 1935)
Elachertus sp., Eulophidae, Colombia, (Cardenas and Posada, 2001), Cuba, (Bruner, 1929), Mexico, (Aranda, 1986), Puerto Rico, (Wolcott, 1935)
Entedontini sp., Eulophidae, Guatemala, (Hamilton, 1967)
Eubazus punctatus (Ratzeburg), Braconidae, Brazil, (Perioto *et al.*, 2004)
Eubazus sp., Braconidae, Brazil, (Parra *et al.*, 1977)
Eulophus sp., Eulophidae, Brazil, (Mendes, 1940), Colombia, (Roba, 1938)
Gelis sp., Ichneumonidae, Mexico, (Aranda, 1986)
Horismenus aeneicollis Ashmead, Eulophidae, Brazil, (Mendes, 1940)
Horismenus cupreus (Ashmead), Eulophidae, Colombia, (Cardenas and Posada, 2001), Cuba, (Bruner, 1929), Puerto Rico, (Anon, 1940b)
Horismenus sp., Eulophidae, Brazil, (Parra *et al.*, 1977), Cuba, (Konnorova, 1987), Mexico, (Aranda, 1986), Puerto Rico, (Gallardo-Covas, 1988)

Microlycus sp., Eulophidae, Peru, (Enriquez *et al.*, 1975a)
Mirax insularis Muesebeck, Braconidae, Dominica, (Muesebeck, 1937), Guadeloupe, (Muesebeck, 1937), St Lucia, (Anon, 1939), Puerto Rico, (Anon, 1940a)
Mirax sp., Braconidae, Brazil, (Villacorta, 1980)
Neochrysocharis aratus (Walker), Eulophidae, Peru, (Enriquez *et al.*, 1975a)
Notanisomorphella borborica (Giard), Eulophidae, Reunion, (Bordage, 1914)
Pediobius sp., Eulophidae, Peru, (Enriquez *et al.*, 1975a)
Platocharis coffeae (Ferriere), Eulophidae, Reunion, (Kerrich, 1969)
Pnigalio coloni (Girault), Eulophidae, Mexico, (Aranda, 1986)
Pnigalio elongatus Yoshimoto, Eulophidae, Mexico, (Aranda, 1986)
Pnigalio sarasolai De Santis, Eulophidae, Colombia, (De Santis, 1983)
Proacrias sp., Eulophidae, Cuba, (Konnorova, 1987)
Stiropius letifer (Mann), Braconidae, Brazil, (Pickman Mann, 1872), Honduras, (Le Pelley, 1968), Nicaragua, (Le Pelley, 1968)
Stiropius sp., Braconidae, Guatemala, (Hamilton, 1967)
Sympiesis sp., Eulophidae, Mexico, (Aranda, 1986)
Tetrastichus sp., Eulophidae, Brazil, (Parra *et al.*, 1977), Colombia, (Florez and De Hernandez, 1981), Mexico, (Aranda, 1986), Puerto Rico, (Wolcott, 1935)
Zagrammosoma multilineatum (Ashmead), Eulophidae, Colombia, (De Santis, 1983), Cuba, (Bruner, 1929), Jamaica, (Gowdey, 1925), Mexico, (Aranda, 1986), Puerto Rico, (Van Zwaluwenberg, 1917), Venezuela, (Box, 1927)
Zagrammosoma sp., Eulophidae, Colombia, (Cardenas and Posada, 2001), Puerto Rico, (Gallardo-Covas, 1988), Mexico, (Aranda, 1986)

Hyperparasites

Proacrias coffeae Ihering, Eulophidae, Brazil, (Mendes, 1940), Colombia, (Roba, 1938)
Prochiloneurus sp., Encyrtidae, Puerto Rico, (Gallardo-Covas, 1988)

Predators

Apoica pallens Fabricius, Vespidae, Brazil, (Gusmao *et al.*, 2000)
Brachygastra lecheguana (Latreille), Vespidae, Brazil, (Parra *et al.*, 1977)
Calvolia sp., Winterschmidtiidae, Guatemala, (Hamilton, 1967)
Chrysopa sp., Chrysopidae, Colombia, (Florez and De Hernandez, 1981), Peru, (Enriquez *et al.*, 1975a)
Chrysoperla externa (Hagen), Chrysopidae, Brazil, (Ecole *et al.*, 2002)
Chrysoperla sp., Chrysopidae, Colombia, (Cardenas and Posada, 2001)
Eumenes sp., Vespidae, Brazil, (Carvalho *et al.*, 2005)
Polistes versicolor (Olivier), Vespidae, Brazil, (Gusmao *et al.*, 2000), Peru, (Enriquez *et al.*, 1975b)
Polistes sp., Vespidae, Colombia, (Cardenas and Posada, 2001)
Polybia juruana Ihering, Vespidae, Peru, (Enriquez *et al.*, 1975b)
Polybia paulista Ihering, Vespidae, Brazil, (Fragoso *et al.*, 2001)
Polybia rejecta forma belizensis Cameron, Vespidae, Peru, (Enriquez *et al.*, 1975b)

Polybia rufitarsis peruviana Bequaert, Vespidae, Peru, (Enriquez *et al.*, 1975b)
Polybia scutellaris (White), Vespidae, Brazil, (Parra *et al.*, 1977)
Polybia sp., Vespidae, Colombia, (Cardenas and Posada, 2001)
Protonectarina sylveirae (de Saussure), Vespidae, Brazil, (Parra *et al.*, 1977)
Protopolybia exigua (De Saussure), Vespidae, Brazil, (Fragoso *et al.*, 2001)
Pyemotes sp., Pyemotidae, Puerto Rico, (Anon, 1940b)
Synoeca surinama cyanea Fabricius., Vespidae, Brazil, (Carvalho *et al.*, 2005)
Tyrophagus putrescentiae Schrank, Tyroglyphidae, Puerto Rico, (Anon, 1940a)

Prophantis smaragdina (see p. 76)

Parasites

Apanteles coffea Wilkinson, Braconidae, Kenya, (Wilkinson, 1934), Democratic Republic of Congo, (De Saeger, 1941)
Cratocnema sp., Braconidae, Kenya, (Anon, 1992)
Macrocentrus sp., Braconidae, Kenya, (Anon, 1992)
Microbracon sp., Braconidae, Democratic Republic of Congo, (De Saeger, 1943)
Phanerotoma sp., Tachinidae, Kenya, (Ndungi, 1994), São Tomé and Príncipe, (Derron, 1977)
Pristomerus sp., Ichneumonidae, Kenya, (Ndungi, 1994)
Trichogrammatoidea sp., Trichogrammatidae, São Tomé and Príncipe, (Derron, 1977)

Tortrix dinota (see p. 100)

Parasites

Apanteles coffea Wilkinson, Braconidae, Zimbabwe, (Kutywayo, 1989)
Apanteles sp., Braconidae, Kenya, (Evans, 1970)
Ascogaster cava De Saeger, Braconidae, Kenya, (Evans, 1970)
Ascogaster sp. near *cava* De Saeger, Braconidae, Zimbabwe, (Kutywayo, 1989)
Ascogaster sp., Braconidae, Kenya, (Evans, 1970)
Brachymeria olethria (Waterston), Chalcididae, Kenya, (Evans *et al.*, 1968)
Distatrix ugandaensis (Gahan), Braconidae, Kenya, (Evans, 1970)
Euplectromorpha kiambuensis Ferriere, Eulophidae, Kenya, (Ferriere, 1941)
Glyptapanteles africanus (Cameron), Braconidae, Zimbabwe, (Kutywayo, 1989)
Trichogrammatoidea nana (Zehntner), Trichogrammatidae, Kenya, (Evans, 1970)

Hyperparasites

Nesolynx phaeosoma (Waterston), Eulophidae, Kenya, (Evans, 1966a)
Pediobius sp. near *foveolatus* (Crawford), Eulophidae, Kenya, (Evans, 1970)

Zeuzera coffeae (see p. 61)

Parasites

Bracon leuzerae Rohwer, Braconidae, India, (Gopinath, 1962), Indonesia Java, (Rohwer, 1918)
Bracon sp., Braconidae, Malaysia Sabah, (Khoo *et al.*, 1991)
Iphiaulax sp., Braconidae, Malaysia Sabah, (Khoo *et al.*, 1991)

Hemiptera

Antestiopsis intricata (see p. 79)

Parasites

Acroclisoides africanus Ferriere, Pteromalidae, Uganda, (Taylor, 1945), Democratic Republic of Congo, (Ferriere, 1940)
Anastatus antestiae Ferriere, Eupelmidae, Cameroon, (Carayon, 1954), Ethiopia, (Abebe, 1999), Uganda, (Greathead, 1966), Democratic Republic of Congo, (Leroy *et al.*, 1942)
Aridelus cameroni (Szepligeti), Braconidae, Uganda, (Greathead, 1966)
Aridelus rufus luteus (Szepligeti), Braconidae, Uganda, (Greathead, 1966)
Bogosia rubens (Villeneuve), Tachinidae, Ethiopia, (Greathead, 1966), Kenya, (Le Pelley, 1959), Uganda, (Greathead, 1966)
Corioxenos antestiae Blair, Stylopidae, Ethiopia, (Greathead, 1966), Kenya, (Greathead, 1966)
Gryon fulviventre (Crawford), Scelionidae, Ethiopia, (Abebe, 1999), Cameroon, (Carayon, 1954), Uganda, (Le Pelley, 1959)
Telenomus seychellensis Kieffer, Scelionidae, Cameroon, (Carayon, 1954), Kenya, (Greathead, 1966), Uganda, (Greathead, 1966)
Trissolcus mopsus (Nixon), Scelionidae, Cameroon, (Carayon, 1954), Ethiopia, (Nixon, 1935), Uganda, (Taylor, 1945)
Trissolcus suranus (Nixon), Scelionidae, Ethiopia, (Abebe, 1999), Uganda, (Taylor, 1945)

Predators

Hediocoris fasciatus Reuter, Reduviidae, Cameroon, (Carayon, 1954)
Nagusta punctaticollis Stal, Reduviidae, Cameroon, (Carayon, 1954)
Pseudophonoctonus formosus (Distant), Reduviidae, Cameroon, (Carayon, 1954)
Rhynocoris albopunctatus Stal, Reduviidae, Cameroon, (Carayon, 1954)

Antestiopsis orbitalis bechuana (see p. 79)

Parasites

Acroclisoides africanus Ferriere, Pteromalidae, Kenya, (Greathead, 1966)
Anastatus antestiae Ferriere, Eupelmidae, Kenya, (Le Pelley, 1968), Tanzania, (Greathead, 1966)
Aridelus rufus (Cameron), Braconidae, Tanzania, (Kirkpatrick, 1937), Zimbabwe, (Kutywayo, 1989)
Aridelus rufus luteus (Szepligeti), Braconidae, Kenya, (Greathead, 1966), Tanzania, (Greathead, 1966), Uganda, (Le Pelley, 1959), Zimbabwe, (Greathead, 1966)
Bogosia antinorii Rondani, Tachinidae, Kenya, (Le Pelley, 1959)
Corioxenos antestiae Blair, Stylopidae, Tanzania, (Kirkpatrick, 1937), Uganda, (Taylor, 1945)
Gryon fulviventre (Crawford), Scelionidae, Kenya, (Dry, 1921), Tanzania, (Ritchie, 1932)
Gryon sp., Scelionidae, Zimbabwe, (Kutywayo, 1989)
Telenomus seychellensis Kieffer, Scelionidae, Kenya, (Greathead, 1966), Malawi, (Smee, 1930), Tanzania, (Greathead, 1966)
Trissolcus sp., Scelionidae, Zimbabwe, (Kutywayo, 1989)

Antestiopsis orbitalis ghesquerei (see p. 79)

Parasites

Acroclisoides africanus Ferriere, Pteromalidae, Rwanda, (Foucart and Brion, 1962), Uganda, (Greathead, 1966)
Anastatus antestiae Ferriere, Eupelmidae, Rwanda, (Foucart and Brion, 1962), Tanzania, (Greathead, 1966), Uganda, (Greathead, 1966)
Anastatus antestiae var *hancocki* Ferriere, Eupelmidae, Uganda, (Ferriere, 1930)
Aridelus cameroni (Szepligeti), Braconidae, Uganda, (Taylor, 1945)
Aridelus rufus luteus (Szepligeti), Braconidae, Tanzania, (Greathead, 1966), Uganda, (Greathead, 1966)
Bogosia rubens (Villeneuve), Tachinidae, Rwanda, (Foucart and Brion, 1962), Tanzania, (Greathead, 1966), Uganda, (Greathead, 1966)
Corioxenos antestiae Blair, Stylopidae, Tanzania, (Greathead, 1966), Uganda, (Greathead, 1966)
Gryon fulviventre (Crawford), Scelionidae, Rwanda, (Greathead, 1966), Uganda, (Taylor, 1945), Democratic Republic of Congo, (Greathead, 1966)
Telenomus seychellensis Kieffer, Scelionidae, Tanzania, (Greathead, 1966), Uganda, (Greathead, 1966)
Trissolcus mopsus (Nixon), Scelionidae, Ethiopia, (Greathead, 1966), Rwanda, (Foucart and Brion, 1962), Uganda, (Taylor, 1945)
Trissolcus suranus (Nixon), Scelionidae, Tanzania, (Greathead, 1966), Uganda, (Greathead, 1966)

Antestiopsis orbitalis Southern African forms (see p. 79)

Parasites

Telenomus seychellensis Kieffer, Scelionidae, South Africa, (Nixon, 1935)
Telenomus ? seychellensis Kieffer, Scelionidae, South Africa, (Greathead, 1966)

Aspidiotus sp. (Fried egg scale) (see p. 127)

Predators

Chilocorus nigripes Mader, Coccinellidae, Kenya, (Mugo *et al*., 1997)
Cybocephalus sp., Othniidae, Kenya, (Crowe, 2004)
Exochomus flavipes Thunberg, Coccinellidae, Kenya, (Mugo *et al*., 1997)
Hyperaspis senegalensis Mulsant, Coccinellidae, Kenya, (Mugo *et al*., 1997)

Asterolecanium coffeae (see p. 123)

Parasites

Metaphycus hemilecanii Compere, Encyrtidae, Kenya (Compere, 1940)
Metaphycus lounsburyi (Howard), Encyrtidae, Kenya (James, 1932b) Tanzania (Ritchie, 1935)

Hyperparasites

Marietta buscki (Howard), Aphelinidae, Kenya (James, 1932b)
Marietta leopardina Motschulsky, Aphelinidae, Kenya (Le Pelley, 1959)

Predators

Chilocorus angolensis Crotch, Coccinellidae, Kenya (James, 1932b)
Chilocorus schioedtei Mulsant, Coccinellidae, Kenya (James, 1932b) Uganda (James, 1932b)
Exochomus melanocephalus (Zoubkoff), Coccinellidae, Kenya (James, 1932b)

Asterolecanium pustulans princeps (see p. 123)

Parasites

Encarsia citrina (Crawford), Encyrtidae, São Tomé and Príncipe (Castel Branco, 1971)

Metaphycus portoricensis (Dozier), Encyrtidae, São Tomé and Príncipe (Castel Branco, 1971)

Predator

Chilocorus cacti (Linnaeus), Coccinellidae, São Tomé and Príncipe (Castel Branco, 1971)

Ceroplastes brevicauda (see p. 129)

Parasites

Aloencyrtus umbrinus (Compere), Encyrtidae, Kenya, (Compere, 1939)
Anabrolepis sp., Encyrtidae, Kenya, (Crowe, 1962b)
Anasemion inutile (Compere), Encyrtidae, Kenya, (Crowe, 1962b)
Anicetus parvus Compere, Encyrtidae, Kenya, (Crowe, 1962b)
Bothriophryne purpurascens Compere, Encyrtidae, Kenya, (Crowe, 1962b)
Coccophagus sp. near *nigropleurum* Girault, Aphelinidae, Kenya, (Crowe, 1962b)
Diversinervus elegans Silvestri, Encyrtidae, Kenya, (Le Pelley, 1959)
Eupelmus saissetiae Silvestri, Eupelmidae, Kenya, (Crowe, 1962b)
Pachyneuron sp., Pteromalidae, Kenya, (Crowe, 1962b)
Scutellista caerulea (Boyer de Fonscolombe), Pteromalidae, Kenya, (Crowe, 1962b)
Scutellista sp., Pteromalidae, Kenya, (Crowe, 1962b)

Hyperparasites

Cheiloneurus obscurus Silvestri, Encyrtidae, Kenya, (Crowe, 1962b)
Marietta leopardina Motschulsky, Aphelinidae, Kenya, (Crowe, 1962b)
Tetrastichus injuriosus Compere, Eulophidae, Kenya, (Crowe, 1962b)
Tremblaya oleae (Silvestri), Encyrtidae, Kenya, (Crowe, 1962b)

Predators

Cheilomenes lunata (Fabricius), Coccinellidae, Kenya, (Crowe, 1962b)
Coccidiphaga scitula (Rambur), Noctuidae, Kenya, (Crowe, 1962b)
Eublemma costimacula (Saalmuller), Noctuidae, Kenya, (Crowe, 1962b)

Coccus alpinus (see p. 119)

Parasites

Adelencyrtus sp., Encyrtidae, Kenya, (Murphy, 1991)
Aloencyrtus saissetiae (Compere), Encyrtidae, Kenya, (Murphy, 1991)

Aloencyrtus ugandensis (Compere), Encyrtidae, Kenya, (Murphy, 1991)
Aphytis sp., Aphelinidae, Ethiopia, (Greathead, 1965)
Aprostocetus gravans (Silvestri), Eulophidae, Kenya, (Le Pelley, 1959)
Aprostocetus ?sicarius (Silvestri), Eulophidae, Kenya, (Murphy, 1991)
Coccidoxenoides perminutus Girault, Encyrtidae, Kenya, (Mugo *et al.*, 1997)
Coccophagus argocoxa Annecke & Insley, Aphelinidae, Kenya, (Murphy, 1991)
Coccophagus nubes Compere, Aphelinidae, Kenya, (Murphy, 1991), Tanzania, (Ritchie, 1932)
Coccophagus pulvinariae Compere, Aphelinidae, Kenya, (Murphy, 1991), Tanzania, (Ritchie, 1932)
Coccophagus rusti Compere, Aphelinidae, Kenya, (Murphy, 1991)
Coccophagus sp., Aphelinidae, Ethiopia, (Greathead, 1965), Kenya, (Murphy, 1991)
Diversinervus stramineus Compere, Encyrtidae, Kenya, (Murphy, 1991)
Encyrtus sp., Encyrtidae, Kenya, (Murphy, 1991)
Mahencyrtus sp., Encyrtidae, Kenya, (Murphy, 1991)
Metaphycus baruensis Noyes, Encyrtidae, Kenya, (Murphy, 1991)
Metaphycus stanleyi Compere, Encyrtidae, Kenya, (Murphy, 1991)
Metaphycus sp., Encyrtidae, Kenya, (Murphy, 1991)
Metaphycus sp. near *helvolus* (Compere), Encyrtidae, Ethiopia, (Greathead, 1965), Kenya, (Murphy, 1991)
Trichomasthus sp., Encyrtidae, Kenya, (Murphy, 1991)

Hyperparasites

Cheiloneurus cyanonotus Waterston, Encyrtidae, Kenya, (Murphy, 1991)
Marietta leopardina Motschulsky, Aphelinidae, Kenya, (Murphy, 1991)
Promuscidea ?comperella (Ghesquière), Aphelinidae, Kenya, (Murphy, 1991)

Predators

Anisochrysa sp., Chrysopidae, Kenya, (Murphy, 1991)
Chilocorus adustus Weise, Coccinellidae, Tanzania, (Le Pelley, 1968)
Chilocorus angolensis Crotch, Coccinellidae, Kenya, (Murphy, 1991), Tanzania, (Le Pelley, 1968)
Chilocorus nigripes Mader, Coccinellidae, Kenya, (Murphy, 1991)
Chilocorus quadrimaculatus (Weise), Coccinellidae, Kenya, (Mugo *et al.*, 1997)
Chilocorus schioedtei Mulsant, Coccinellidae, Kenya, (Le Pelley, 1968), Tanzania, (Ritchie, 1935), Uganda, (Gowdey, 1916)
Coccidiphaga scitula (Rambur), Noctuidae, Kenya, (Le Pelley, 1968), Tanzania, (Le Pelley, 1968)
Eublemma costimacula (Saalmuller), Noctuidae, Kenya, (Le Pelley, 1968), Tanzania, (Le Pelley, 1968)
Exochomus nigromaculatus (Goeze), Coccinellidae, Democratic Republic of Congo, (Mayné and Ghesquière, 1934)
Exochomus ventralis (Gerstaecker), Coccinellidae, Tanzania, (Ritchie, 1935)
Hippodamia variegata (Goeze), Coccinellidae, Kenya, (Mugo *et al.*, 1997)

Hyperaspis senegalensis Mulsant, Coccinellidae, Kenya, (Le Pelley, 1959), Tanzania, (Le Pelley, 1968)
Hyperaspis sp., Coccinellidae, Kenya, (Murphy, 1991)
Mallada boninensis (Okamoto), Chrysopidae, Kenya, (Murphy, 1991)

Ant attendant

Camponotus sp., Formicidae Formicinae, Democratic Republic of Congo, (Mayné and Ghesquière, 1934)

Coccus celatus (see p. 119)

Parasites

Aprostocetus ?sicarius (Silvestri), Eulophidae, Kenya, (Murphy, 1991)
Coccophagus argocoxa Annecke & Insley, Aphelinidae, Kenya, (Murphy, 1991)
Coccophagus nubes Compere, Aphelinidae, Kenya, (Murphy, 1991)
Diversinervus stramineus Compere, Encyrtidae, Kenya, (Murphy, 1991)
Mahencyrtus sp., Encyrtidae, Kenya, (Murphy, 1991)
Metaphycus baruensis Noyes, Encyrtidae, Kenya, (Murphy, 1991), Papua New Guinea, (Masamdu, 1989)
Metaphycus stanleyi Compere, Encyrtidae, Kenya, (Murphy, 1991)
Metaphycus sp. near *helvolus* (Compere), Encyrtidae, Kenya, (Murphy, 1991)
Metaphycus sp., Encyrtidae, Kenya, (Murphy, 1991)
Trichomasthus sp., Encyrtidae, Kenya, (Murphy, 1991)

Hyperparasites

Cheiloneurus sp., Papua New Guinea, (Apety, 1994)
Marietta leopardina Motschulsky, Aphelinidae, Kenya, (Murphy, 1991)
Promuscidea ?comperella (Ghesquière), Aphelinidae, Kenya, (Murphy, 1991)

Predators

Anisochrysa sp., Chrysopidae, Kenya, (Murphy, 1991)
Chilocorus adustus Weise, Coccinellidae, Tanzania, (Le Pelley, 1968)
Chilocorus angolensis Crotch, Coccinellidae, Kenya, (Murphy, 1991), Tanzania, (Le Pelley, 1968)
Chilocorus nigripes Mader, Coccinellidae, Kenya, (Murphy, 1991)
Chilocorus quadrimaculatus (Weise), Coccinellidae, Kenya, (Mugo *et al.*, 1997)
Chilocorus schioedtei Mulsant, Coccinellidae, Kenya, (Le Pelley, 1968), Tanzania, (Ritchie, 1935), Uganda, (Gowdey, 1916)
Coccidiphaga scitula (Rambur), Noctuidae, Kenya, (Le Pelley, 1968), Tanzania, (Le Pelley, 1968)
Eublemma costimacula (Saalmuller), Noctuidae, Kenya, (Le Pelley, 1968), Tanzania, (Le Pelley, 1968)

Exochomus nigromaculatus (Goeze), Coccinellidae, Democratic Republic of Congo, (Mayné and Ghesquière, 1934)
Exochomus ventralis (Gerstaecker), Coccinellidae, Tanzania, (Ritchie, 1935)
Hippodamia variegata (Goeze), Coccinellidae, Kenya, (Mugo *et al.*, 1997)
Hyperaspis senegalensis Mulsant, Coccinellidae, Kenya, (Le Pelley, 1959), Tanzania, (Le Pelley, 1968)
Hyperaspis sp., Coccinellidae, Kenya, (Murphy, 1991)
Mallada boninensis (Okamoto), Chrysopidae, Kenya, (Murphy, 1991)

Ant attendants

Solenopsis geminata (Fabricius), Formicidae Myrmecinae, Papua New Guinea, (Buckley and Gullan, 1991)
Tetramorium aculeatum (Mayr), Formicidae Myrmecinae, Uganda, (De Lotto, 1960)

Coccus hesperidum

Predators

Chilocorus bipustulatus (Linnaeus), Coccinellidae, Democratic Republic of Congo, (Mayné and Ghesquière, 1934)
Scymnus sp., Coccinellidae, Peru, (Wille, 1952)

Ant attendant

Oecophylla longinoda (Latreille), Formicidae Formicinae, Democratic Republic of Congo, (Mayné and Ghesquière, 1934)

Coccus viridis (see p. 119)

Parasites

Anicetus annulatus Timberlake, Encyrtidae, India, (Srinivasa, 1985)
Anicetus ceylonensis Howard, Encyrtidae, India, (Thompson, 1943–1958), Sri Lanka, (Thompson, 1943–1958)
Anicetus sp. near *annulatus* Timberlake, Encyrtidae, India, (Srinivasa, 1987)
Aprostocetus gravans (Silvestri), Eulophidae, Tanzania, (Ritchie, 1932)
Aprostocetus minutus (Howard), Eulophidae, Hawaii, (Reimer *et al.*, 1993), Indonesia Java, (Leefmans, 1929)
Aprostocetus purpureus (Cameron), Eulophidae, India, (Reddy *et al.*, 1990)
Aprostocetus sicarius (Silvestri), Eulophidae, Mauritius, (Waterston, 1916)
Cephaleta australiensis (Howard), Pteromalidae, Indonesia Java, (Leefmans, 1929)
Cephaleta australiensis var. *javensis* Girault, Pteromalidae, Indonesia Java, (Girault, 1916)
Cerapteroceroides sp. near *similis* (Ishii), Encyrtidae, India, (Srinivasa, 1987)

Cheiloneuromyia javensis Girault, Encyrtidae, India, (Sekhar, 1964), Indonesia Java, (Van der Goot, 1916)
Coccophagus bogoriensis (Koningsberger), Aphelinidae, India, (Sekhar, 1964), Indonesia Java, (Leefmans, 1929)
Coccophagus ceroplastae (Howard), Aphelinidae, Hawaii, (Fullaway, 1932), India, (Srinivasa, 1987), Java, (Thompson, 1943–1958), Malaysia, (Miller, 1931), Sri Lanka, (Bess, 1958)
Coccophagus cowperi Girault, Aphelinidae, India, (Anon, 1978)
Coccophagus hawaiiensis Timberlake, Aphelinidae, Hawaii, (Fullaway, 1932)
Coccophagus lycimnia (Walker), Aphelinidae, Hawaii, (Fullaway, 1920), India, (Sekhar, 1964)
Coccophagus ochraceus Howard, Aphelinidae, Hawaii, (Fullaway, 1932)
Coccophagus rusti Compere, Aphelinidae, Kenya, (Murphy, 1991)
Coccophagus sp., Aphelinidae, India, (Srinivasa, 1987), Malaysia, (Miller, 1931)
Diadiplosis sp., Cecidomyiidae, Cuba, (Bruner, 1929)
Diversinervus silvestrii Waterston, Encyrtidae, Mauritius, (Waterston, 1916)
Encyrtus aurantii (Geoffray), Encyrtidae, India, (Srinivasa, 1987)
Gahaniella saissetiae Timberlake, Encyrtidae, Hawaii, (Reimer *et al.*, 1993)
Marietta caridei (Brethes), Aphelinidae, Cuba, (Hernandez and Ceballos, 1991)
Metaphycus baruensis Noyes, Encyrtidae, Kenya, (Noyes, 1988), Papua New Guinea, (Apety, 1994)
Metaphycus helvolus (Compere), Encyrtidae, Cuba, (Kohler, 1980), India, (Srinivasa, 1987)
Metaphycus lichtensiae (Howard), Encyrtidae, India, (Srinivasa, 1987)
Metaphycus maculatus Agarwal, Encyrtidae, India, (Srinivasa, 1987)
Microterys nietneri (Motschulsky), Encyrtidae, Hawaii, (Fullaway, 1932), Malaysia, (Miller, 1931)
Myiocnema comperei Ashmead, Aphelinidae, Indonesia Java, (Leefmans, 1929)
Neobrachista javae Girault, Trichogrammatidae, Indonesia Java, (Leefmans, 1929)
Prochiloneurus sp. near *nigriflagellum* (Girault), Encyrtidae, India, (Reddy and Bhat, 1989)
Promuscidea unfasciativentris Girault, Aphelinidae, Indonesia Java, (Leefmans, 1929)
Tetrastichus ibseni (Girault), Eulophidae, Indonesia Java, (Girault, 1916)
Tetrastichus sp., Eulophidae, India, (Srinivasa, 1987)

Hyperparasites

Cheiloneurus latiscapus (Girault), Encyrtidae, Indonesia Java, (Van der Goot, 1916)
Cheiloneurus sp., Encyrtidae, Papua New Guinea, (Apety, 1994)
Coccidoctonus dubius Girault, Encyrtidae, Hawaii, (Fullaway, 1932)
Marietta caridei (Brethes), Aphelinidae, Cuba, (Hernandez and Ceballos, 1991)
Marietta leopardina Motschulsky, Aphelinidae, India, (Srinivasa, 1987), Kenya, (Murphy, 1991)
Walkerella dubia (Girault), Agaonidae, Indonesia Java, (Girault, 1916)

Predators

Azya elegans Gordon, Coccinellidae, Venezuela, (Hanks and Sadof, 1990)
Azya luteipes Mulsant, Coccinellidae, Brazil, (Da Costa Lima, 1936), Colombia, (Roba, 1938), Hawaii, (Reimer *et al.*, 1993), Peru, (Wille, 1952), Venezuela, (Box, 1927)
Cheilomenes sexmaculata (Fabricius), Coccinellidae, India, (Reddy *et al.*, 1990), Papua New Guinea, (Szent-Ivany, 1958)
Chilocorus adustus Weise, Coccinellidae, Tanzania, (Ritchie, 1935)
Chilocorus angolensis Crotch, Coccinellidae, Tanzania, (Ritchie, 1935)
Chilocorus cacti (Linnaeus), Coccinellidae, Cuba, (Bruner, 1929)
Chilocorus circumdatus (Gyllenhal), Coccinellidae, Hawaii, (Illingworth, 1929)
Chilocorus melanophthalmus Mulsant, Coccinellidae, Indonesia Java, (Keuchenius, 1915)
Chilocorus politus Mulsant, Coccinellidae, Malaysia, (Corbett and Miller, 1933)
Chilocorus schioedtei Mulsant, Coccinellidae, Kenya, (Le Pelley, 1968), Tanzania, (Ritchie, 1935), Uganda, (Gowdey, 1916)
Coccidiphaga scitula (Rambur), Noctuidae, Kenya, (Le Pelley, 1968), Tanzania, (Ritchie, 1935)
Cryptoblabes proleucella Hampson, Pyralidae, Sri Lanka, (Rutherford, 1914)
Cryptolaemus montrouzieri Mulsant, Coccinellidae, Hawaii, (Reimer *et al.*, 1993), New Caledonia, (Chazeau, 1981)
Curinus coeruleus Mulsant, Coccinellidae, Hawaii, (Reimer *et al.*, 1993)
Cybocephalus sp., Nitidulidae, Venezuela, (Hanks and Sadof, 1990)
Diomus sp., Coccinellidae, Venezuela, (Hanks and Sadof, 1990)
Eublemma costimacula (Saalmuller), Noctuidae, Kenya, (Le Pelley, 1968), Tanzania, (Ritchie, 1935)
Eublemma rubra Hampson, Noctuidae, Malaysia, (Corbett, 1937)
Exochomus ventralis (Gerstaecker), Coccinellidae, Tanzania, (Ritchie, 1935)
Halmus chalybeus (Boisduval), Coccinellidae, Hawaii, (Illingworth, 1929)
Harmonia sp. near *testudinaria* Mulsant, Coccinellidae, Papua New Guinea, (Szent-Ivany, 1958)
Hyperaspis senegalensis Mulsant, Coccinellidae, Tanzania, (Ritchie, 1935)
Hyperaspis silvestrii Weise, Coccinellidae, Hawaii, (Reimer *et al.*, 1993)
Jauravia pallidula Motschulsky, Coccinellidae, India, (Prakasan and Kumar, 1985)
Novius koebelei (Blackburn), Coccinellidae, Australia, (B. Pinese, Brisbane, Australia, 1999, personal communication)
Olla v-nigrum (Mulsant), Coccinellidae, Hawaii, (Reimer *et al.*, 1993)
Orcus janthinus Mulsant, Coccinellidae, Indonesia Java, (Keuchenius, 1915)
Orcus sp., Coccinellidae, Papua New Guinea, (Szent-Ivany, 1958)
Phrynocaria quadrivittata Fauvel, Coccinellidae, New Caledonia, (Chazeau, 1981)
Pseudocaecillius elutus africanus Badonell, Pseudocaecilidae, Democratic Republic of Congo, (Vayssière, 1955)
Rhyzobius ventralis Erichson, Coccinellidae, Hawaii, (Reimer *et al.*, 1993)
Scymnus sp., Coccinellidae, Peru, (Wille, 1952), Venezuela, (Hanks and Sadof, 1990)

Synona inaequalis (Fabricius), Coccinellidae, Hawaii, (Reimer *et al.*, 1993), New Caledonia, (Chazeau, 1981)
Telsimia sp., Coccinellidae, India, (Reddy *et al.*, 1990)

Ant attendants

Anoplolepis gracilipes (F.Smith), Formicidae Formicinae, Indonesia Java, (De Fluiter, 1939)
Aphaenogaster sp., Formicidae Myrmecinae, Australia, (B. Pinese, Brisbane, Australia, 1999, personal communication)
Brachymyrmex heeri Forel, Formicidae Formicinae, Puerto Rico, (Smith, 1942)
Brachymyrmex sp., Formicidae Formicinae, Colombia, (Roba, 1938)
Camponotus sp., Formicidae Formicinae, Brazil, (Bondar, 1928), Venezuela, (Hanks and Sadof, 1990)
Crematogaster brevispinosa var *minutior* Forel, Formicidae Myrmecinae, Jamaica, (Edwards, 1932)
Crematogaster steinheili Forel, Formicidae Myrmecinae, Puerto Rico, (Smith, 1942)
Crematogaster sp., Formicidae Myrmecinae, Colombia, (Roba, 1936), Venezuela, (Hanks and Sadof, 1990)
Dolichoderus thoracicus (F.Smith), Formicidae Dolichoderinae, India, (Coleman and Kannan, 1918), Indonesia Java, (Van der Goot, 1917)
Dolichoderus sp., Formicidae Dolichoderinae, Colombia, (Roba, 1938)
Iridomyrmex purpureus F. Smith, Formicidae Dolichoderinae, Australia, (B. Pinese, Brisbane, Australia, 1999, personal communication)
Linepithema melleum (Wheeler), Formicidae Dolichoderinae, Puerto Rico, (Smith, 1942)
Monomorium floricola (Jerdon), Formicidae Myrmecinae, Puerto Rico, (Smith, 1942)
Monomorium pharaonis (Linnaeus), Formicidae Myrmecinae, Australia, (B. Pinese, Brisbane, Australia, 1999, personal communication)
Monomorium sp. near *rothsteini* Forel, Formicidae Myrmecinae, Australia, (B. Pinese, Brisbane, Australia, 1999, personal communication)
Myrmelachista ramulorum Wheeler, Formicidae Formicinae, Puerto Rico, (Smith, 1942)
Myrmicaria arachnoides (F.Smith), Formicidae Myrmecinae, Indonesia Java, (Kalshoven, 1950–1951)
Notostigma sp., Formicidae Formicinae, Australia, (B. Pinese, Brisbane, Australia, 1999, personal communication)
Oecophylla longinoda (Latreille), Formicidae Formicinae, Tanzania Zanzibar, (Way, 1954a)
Oecophylla smaragdina (Fabricius), Formicidae Formicinae, Malaysia, (Corbett, 1937), Sri Lanka, (Bess, 1958)
Opisthopsis haddoni Emery, Formicidae Formicinae, Australia, (B. Pinese, Brisbane, Australia, 1999, personal communication)
Paratrechina fulva (Mayr), Formicidae Formicinae, Colombia, (Cardenas and Posada, 2001)

Paratrechina longicornis (Latreille), Formicidae Formicinae, Puerto Rico, (Smith, 1942)
Pheidole fallax antillensis Forel, Formicidae Myrmecinae, Puerto Rico, (Smith, 1942)
Pheidole punctulata Mayr, Formicidae Myrmecinae, No locality, (Le Pelley, 1968)
Pheidole speculifera Emery, Formicidae Myrmecinae, Kenya, (Anderson, 1927)
Pheidole subarmata Mayr, Formicidae Myrmecinae, Puerto Rico, (Smith, 1942)
Plagiolepis alluaudi Emery, Formicidae Formicinae, Australia, (B. Pinese, Brisbane, Australia, 1999, personal communication)
Solenopsis geminata (Fabricius), Formicidae Myrmecinae, Jamaica, (Edwards, 1935), Puerto Rico, (Smith, 1942)
Solenopsis saevissima (F.Smith), Formicidae Myrmecinae, Surinam, (Reyne, 1920)
Tapinoma melanocephalum (Fabricius), Formicidae Dolichoderinae, Puerto Rico, (Smith, 1942)
Technomyrmex detorquens (Walker), Formicidae Dolichoderinae, Sri Lanka, (Bess, 1958)
Wasmannia auropunctata (Roger), Formicidae Myrmecinae, Puerto Rico, (Smith, 1942)

Dulinus unicolor (see p. 116)

Predator

Stethoconus frappai Carayon, Miridae, Madagascar, (Carayon, 1960)

Dysmicoccus brevipes (see p. 148)

Predators

Cryptolaemus montrouzieri Mulsant, Coccinellidae, Hawaii, (Illingworth, 1929)
Rhyzobius ventralis Erichson, Coccinellidae, Hawaii, (Illingworth, 1929)

Ferrisia virgata (see p. 124)

Parasites

Aenasius advena Compere, Encyrtidae, India, (Balakrishnan *et al*., 1991)
Anagyrus qadrii (Hayat Alam & Agarwal), Encyrtidae, India, (Balakrishnan *et al*., 1991)
Anicetus annulatus Timberlake, Encyrtidae, India, (Balakrishnan *et al*., 1991)
Blepyrus insularis (Cameron), Encyrtidae, Hawaii, (Fullaway, 1932), India, (Balakrishnan *et al*., 1991)

Coccodiplosis coffeae (Barnes), Cecidomyiidae, Democratic Republic of Congo, (Barnes, 1939)
Gyranusoidea citrina (Compere), Encyrtidae, Kenya, (Compere, 1938)

Predators

Allograpta javana (Weidemann), Syrphidae, India, (Balakrishnan *et al.*, 1991)
Chilocorus angolensis Crotch, Coccinellidae, Kenya, (Kirkpatrick, 1927a)
Chrysopa flaveola Schneider, Chrysopidae, Indonesia Java, (Kalshoven, 1950–1951)
Chrysopa sp., Chrysopidae, Kenya, (Kirkpatrick, 1927b)
Cryptolaemus montrouzieri Mulsant, Coccinellidae, Indonesia Java, (Ultee, 1929)
Deraeocoris sp., Miridae, Uganda, (Le Pelley, 1959)
Diadiplosis coccidivora (Felt), Cecidomyiidae, India, (Balakrishnan *et al.*, 1991)
Eublemma costimacula (Saalmuller), Noctuidae, Tanzania, (Ritchie, 1930)
Exochomus melanocephalus (Zoubkoff), Coccinellidae, Uganda, (Le Pelley, 1959)
Exochomus nigromaculatus (Goeze), Coccinellidae, Democratic Republic of Congo, (Mayné and Ghesquière, 1934)
Hyperaspis delicatula Mulsant, Coccinellidae, Kenya, (Kirkpatrick, 1927a)
Hyperaspis senegalensis Mulsant, Coccinellidae, Tanzania, (Kirkpatrick, 1927a)
Leucopis africana Malloch, Chamaemyiidae, Kenya, (Le Pelley, 1959)
Leucopis sp., Chamaemyiidae, India, (Balakrishnan *et al.*, 1991)
Mallada sp., Chrysopidae, India, (Balakrishnan *et al.*, 1991)
Nephus c-luteum (Sicard), Coccinellidae, No locality, (Le Pelley, 1968)
Nephus peyerimhoffi Sicard, Coccinellidae, Uganda, (Le Pelley, 1959)
Scymnus sp., Coccinellidae, India, (Balakrishnan *et al.*, 1991)
Spalgis epeus (Westwood), Lycaenidae, India, (Chacko and Bhat, 1976)
Spalgis lemolea Druce, Lycaenidae, Sierra Leone, (Hargreaves, E., 1936), Uganda, (Le Pelley, 1968)

Habrochila ghesquerei (see p. 116)

Predators

Stethoconus distanti (Schouteden), Miridae, Rwanda, (Schouteden, 1946), Democratic Republic of Congo, (Schouteden, 1946)
Stethoconus sp., Miridae, Kenya, (Le Pelley, 1968), Uganda, (Hargreaves, H., 1936)

Habrochila placida (see p. 116)

Predator

Stethoconus sp., Miridae, Kenya, (Le Pelley, 1968)

Ischnaspis longirostris (see p. 128)

Parasite

Aphytis chrysomphali (Mercet), Aphelinidae, Kenya, (Lepesme, 1947), Tanzania, (Lepesme, 1947), Uganda, (Lepesme, 1947)

Predators

Chilocorus distigma Klug, Coccinellidae, Tanzania, (Le Pelley, 1959)
Chilocorus nigritus (Fabricius), Coccinellidae, India, (Coleman and Kannan, 1918)
Chilocorus wahlbergi Mulsant, Coccinellidae, Tanzania, (Le Pelley, 1959), Tanzania Zanzibar, (Way, 1954b)

Parasaissetia nigra (see p. 122)

Parasites

Aloencyrtus obscuratus (Waterston), Encyrtidae, Kenya, (Le Pelley, 1959)
Aloencyrtus saissetiae (Compere), Encyrtidae, Kenya, (Compere, 1939)
Arrhenophagus chionaspidis Aurivillius, Encyrtidae, Jamaica, (Gowdey, 1926)
Cephaleta brunniventris Motschulsky, Pteromalidae, Philippines, (Smith, 1944)
Coccophagus ceroplastae (Howard), Aphelinidae, Hawaii, (Fullaway, 1932), Puerto Rico, (Smith, 1944)
Coccophagus hawaiiensis Timberlake, Aphelinidae, Hawaii, (Fullaway, 1932)
Coccophagus longifasciatus Howard, Aphelinidae, Sri Lanka, (Ramakrishna Ayyar, 1925)
Coccophagus lycimnia (Walker), Aphelinidae, Uganda, (Le Pelley, 1959)
Coccophagus scutellaris (Dalman), Aphelinidae, Kenya, (Le Pelley, 1968), Puerto Rico, (Le Pelley, 1968)
Dicrodiplosis fulva (Felt), Cecidomyiidae, Sri Lanka, (De Silva, 1961)
Diversinervus elegans Silvestri, Encyrtidae, Kenya, (Le Pelley, 1959)
Encyrtus aurantii (Geoffray), Encyrtidae, Hawaii, (Smith, 1944)
Encyrtus infelix (Embleton), Encyrtidae, Hawaii, (Fullaway, 1932), Puerto Rico, (Smith, 1942)
Lecaniobius cockerelli Ashmead, Eupelmidae, Guyana, (Smith, 1944), Puerto Rico, (Smith, 1944)
Metaphycus stanleyi Compere, Encyrtidae, Kenya, (Compere, 1940)
Microterys nietneri (Motschulsky), Encyrtidae, Hawaii, (Fullaway, 1932), Sri Lanka, (Keuchenius, 1915)
Scutellista caerulea (Boyer de Fonscolombe), Pteromalidae, Hawaii, (Fullaway, 1932), Kenya, (Le Pelley, 1968)

Hyperparasites

Cheiloneurus cyanonotus Waterston, Encyrtidae, Kenya, (Compere, 1938)

Coccidoctonus dubius Girault, Encyrtidae, Hawaii, (Smith, 1945)
Eupelmus coccidivorus Gahan, Eupelmidae, Cuba, (Bruner, 1929)
Marietta sp., Aphelinidae, Puerto Rico, (Le Pelley, 1968)
Tremblaya oleae (Silvestri), Encyrtidae, Kenya, (Crowe, 1962b)

Ant attendants

Oecophylla longinoda (Latreille), Formicidae Formicinae, Tanzania Zanzibar, (Way, 1954a)
Oecophylla smaragdina (Fabricius), Formicidae Formicinae, India, (Coleman and Kannan, 1918), Indonesia Java, (Kalshoven, 1950–1951)
Tapinoma melanocephalum (Fabricius), Formicidae Dolichoderinae, Puerto Rico, (Smith, 1942)

Planococcus citri (see p. 124)

Parasites

Alamella flava Agarwal, Encyrtidae, India, (Reddy *et al.*, 1990)
Anagyrus agraensis Saraswat, Encyrtidae, India, (Reddy *et al.*, 1990)
Anagyrus aurantifrons Compere, Encyrtidae, Kenya, (Le Pelley, 1959), Uganda, (Le Pelley, 1959), Tanzania, (Ritchie, 1936), South Africa, (Ritchie, 1936)
Anagyrus greenii Howard, Encyrtidae, Indonesia Java, (De Fluiter, 1960)
Aprostocetus purpureus (Cameron), Eulophidae, India, (Reddy *et al.*, 1990)
Arhopoideus sp., Mymaridae, Zimbabwe, (Kutywayo, 1989)
Coccidoxenoides perminutus Girault, Encyrtidae, Hawaii, (Zimmerman, 1948), Kenya, (Le Pelley and Melville, 1939), Uganda, (Le Pelley and Melville, 1939)
Coccodiplosis smithi (Felt), Cecidomyiidae, Indonesia Java, (De Fluiter, 1936), Philippines, (Le Pelley, 1943b)
Holanusomyia pulchripennis Girault, Encyrtidae, Philippines, (Girault, 1915)
Leptomastidea abnormis (Girault), Encyrtidae, Hawaii, (Zimmerman, 1948)
Leptomastix abyssinica Compere, Encyrtidae, Eritrea, (Thompson, 1943–1958)
Leptomastix dactylopii Howard, Encyrtidae, Cuba, (Hernandez and Ceballos, 1993), India, (Prakasan and Bhat, 1985), Puerto Rico, (Dozier, 1927)
Leptomastix nigrocoxalis Compere, Encyrtidae, India, (Prakasan and Kumar, 1985)
Leptomastix trilongifasciatus Girault, Encyrtidae, Indonesia Java, (De Fluiter, 1960)
Metaphycus sp. near *lounsburyi* (Howard), Encyrtidae, Zimbabwe, (Kutywayo, 1989)
Tetracnemoidea indica (Ramakrishna Ayyar), Encyrtidae, India, (Pruthi and Mani, 1940)

Hyperparasites

Chartocerus niger (Ashmead), Signiphoridae, Puerto Rico, (Dozier, 1927)

Cheiloneurus gahani (Dozier), Encyrtidae, Puerto Rico, (Dozier, 1927)
Homalotylus sp., Encyrtidae, Indonesia Java, (De Fluiter, 1936)
Prochiloneurus io (Girault), Philippines, (Le Pelley, 1968)
Prochiloneurus rex (Girault), Philippines, (Le Pelley, 1968)
Prochiloneurus seini (Dozier), Puerto Rico, (Dozier, 1927)
Signiphora bifasciata Ashmead, Signiphoridae, Puerto Rico, (Dozier, 1927)
Signiphora giraulti Crawford, Signiphoridae, Trinidad and Tobago, (Crawford, 1913)

Predators

Brumoides suturalis (Fabricius), Coccinellidae, Indonesia Java, (De Fluiter, 1960)
Chrysoperla harrisii (Fitch), Chrysopidae, Guatemala, (Alvarado, 1935)
Coccinella transversalis Fabricius, Coccinellidae, Indonesia Java, (De Fluiter, 1960)
Cryptolaemus montrouzieri Mulsant, Coccinellidae, Indonesia Java, (Betrem, 1932)
Diadiplosis coccidivora (Felt), Cecidomyiidae, India, (Reddy *et al*., 1990)
Dicrodiplosis sp., Cecidomyiidae, India, (Reddy *et al*., 1990)
Domomyza perspicax (Knab), Drosophilidae, India, (Reddy *et al*., 1990)
Harmonia octomaculata (Fabricius), Coccinellidae, Indonesia Java, (De Fluiter, 1960)
Hyperaspis senegalensis Mulsant, Coccinellidae, Uganda, (Le Pelley, 1959)
Hyperaspis senegalensis hottentota Mulsant, Coccinellidae, Uganda, (Hancock, 1926)
Nephus roepkei de Fluiter, Coccinellidae, Indonesia Java, (De Fluiter, 1960)
Pseudoscymnus pallidicollis (Mulsant), Coccinellidae, India, (Reddy *et al*., 1990)
Scymnus apiciflavus Motschulsky, Coccinellidae, Indonesia Java, (De Fluiter, 1960)
Spalgis epeus (Westwood), Lycaenidae, India, (Reddy *et al*., 1997), Philippines, (Le Pelley, 1943b)
Spalgis lemolea Druce, Lycaenidae, Kenya, (James, 1932a), Uganda, (Hancock, 1926)

Ant attendants

Anoplolepis gracilipes (F.Smith), Formicidae Formicinae, India, (Coleman and Kannan, 1918), Indonesia Java, (Betrem, 1936)
Camponotus sp., Formicidae Formicinae, Brazil, (Bondar, 1928)
Lepisiota capensis (Mayr), Formicidae Formicinae, Kenya, (Anderson, 1927)
Monomorium pharaonis (Linnaeus), Formicidae Myrmecinae, Kenya, (Le Pelley, 1959)
Myrmelachista ramulorum Wheeler, Formicidae Formicinae, Puerto Rico, (Smith, 1942)
Myrmicaria natalensis eumenoides Gerstaeker, Formicidae Myrmecinae, Kenya, (Le Pelley, 1959)

Paratrechina jaegerskioeldi (Mayr), Formicidae Formicinae, Kenya, (Anderson, 1927)
Pseudolasius gowdei Wheeler, Formicidae Formicinae, Uganda, (Hargreaves, H., 1924)
Solenopsis punctaticeps Mayr, Formicidae Myrmecinae, Kenya, (James, 1933)

Planococcus kenyae (see p. 124)

Parasites

Anagyrus aurantifrons Compere, Encyrtidae, Kenya, (Le Pelley, 1959), Tanzania, (Ritchie, 1936), Uganda, (Le Pelley, 1959), South Africa, (Ritchie, 1936)
Anagyrus beneficians Compere, Encyrtidae, Kenya, (Le Pelley, 1943a), Uganda, (Compere, 1943)
Anagyrus bugandaensis Compere, Encyrtidae, Uganda, (Compere, 1939)
Anagyrus kivuensis Compere, Encyrtidae, Kenya, (Le Pelley, 1943a), Uganda, (Le Pelley, 1943a), Democratic Republic of Congo, (Compere, 1939)
Coccidoxenoides perminutus Girault, Encyrtidae, Hawaii, (Zimmerman, 1948), Kenya, (Le Pelley and Melville, 1939), Uganda, (Le Pelley and Melville, 1939)
Coccodiplosis coffeae (Barnes), Cecidomyiidae, Kenya, (Le Pelley, 1959), Tanzania, (Barnes, 1939), Uganda, (Barnes, 1939)
Coccophagus sp., Aphelinidae, Kenya, (Le Pelley and Melville, 1939), Uganda, (Le Pelley and Melville, 1939)
Gyranusoidea citrina (Compere), Encyrtidae, Kenya, (Compere, 1938)
Leptomastidea jeanneli (Mercet), Encyrtidae, Kenya, (Compere, 1939)
Leptomastix dactylopii Howard, Encyrtidae, Kenya, (Le Pelley and Melville, 1939), Tanzania, (Compere, 1938), Uganda, (Le Pelley and Melville, 1939)
Pseudaphycus sp., Encyrtidae, Kenya, (Le Pelley and Melville, 1939), Uganda, (Le Pelley and Melville, 1939)
Tetracnemoidea coffeicola (Kerrich), Encyrtidae, Kenya, (Kerrich, 1967), Uganda, (Kerrich, 1967)
Tetracnemus sp., Encyrtidae, Kenya, (Le Pelley and Melville, 1939), Uganda, (Le Pelley and Melville, 1939)

Hyperparasite

Prochiloneurus comperei Viaggiani, Encyrtidae, Kenya, (Compere, 1938), Tanzania, (Compere, 1938)

Predators

Allograpta calopus (Loew), Syrphidae, Kenya, (Kirkpatrick, 1927a)
Allograpta nasuta (Macquart), Syrphidae, Kenya, (Kirkpatrick, 1927a)
Aslauga purpurascens Holland, Lycaenidae, Kenya, (Le Pelley, 1959), Tanzania, (Ritchie, 1936)

Betasyrphus adligatus (Wiedemann), Syrphidae, Kenya, (Kirkpatrick, 1927a)
Caecillius sp., Caeciliidae, Kenya, (Pearman, 1932)
Ceratochrysa antica (Walker), Chrysopidae, Kenya, (Le Pelley, 1959)
Cheilomenes aurora (Gerstaecker), Coccinellidae, Kenya, (Kirkpatrick, 1927b)
Cheilomenes lunata (Fabricius), Coccinellidae, Kenya, (Kirkpatrick, 1927b)
Cheilomenes posticalis Fairmaire, Coccinellidae, Kenya, (Kirkpatrick, 1927b)
Cheilomenes propinqua vicina (Mulsant), Coccinellidae, Kenya, (Kirkpatrick, 1927b)
Chilocorus angolensis Crotch, Coccinellidae, Kenya, (Kirkpatrick, 1927a)
Chilocorus schioedtei Mulsant, Coccinellidae, Kenya, (Kirkpatrick, 1927b)
Chrysoperla congrua (Walker), Chrysopidae, Kenya, (Le Pelley, 1959)
Coccidiphaga scitula (Rambur), Noctuidae, Kenya, (Ritchie, 1936), Tanzania, (Ritchie, 1936)
Cryptolaemus montrouzieri Mulsant, Coccinellidae, Kenya, (James, 1932a), South Africa, (James, 1932a)
Deraeocoris oculatus (Reuter), Miridae, Kenya, (Anderson, 1931)
Deraeocoris ostentans (Stal), Miridae, Kenya, (Le Pelley, 1959)
Dysis quadrilineata Sicard, Coccinellidae, Kenya, (Kirkpatrick, 1927b)
Ectopsocus briggsi McLachan, Ectopsocidae, Kenya, (Kirkpatrick, 1927a)
Eublemma costimacula (Saalmuller), Noctuidae, Tanzania, (Kirkpatrick, 1927b)
Exochomus flavipes Thunberg, Coccinellidae, Kenya, (Kirkpatrick, 1927b)
Hyperaspis delicatula Mulsant, Coccinellidae, Kenya, (Kirkpatrick, 1927a)
Hyperaspis senegalensis Mulsant, Coccinellidae, Kenya, (Kirkpatrick, 1927a), Uganda, (Kirkpatrick, 1927a)
Leucopis africana Malloch, Chamaemyiidae, Kenya, (Kirkpatrick, 1927a)
Micromus sjostedti Weele, Hemerobiidae, Kenya, (Le Pelley, 1959)
Nephus peyerimhoffi Sicard, Coccinellidae, Kenya, (Le Pelley, 1959), Uganda, (Le Pelley, 1959), Tanzania, (Ritchie, 1936)
Platynaspis capicola Crotch, Coccinellidae, Kenya, (Kirkpatrick, 1927b)
Platynaspis kollari Mulsant, Coccinellidae, Kenya, (Kirkpatrick, 1927b)
Platynaspis marginata Sicard, Coccinellidae, Tanzania, (Ritchie, 1936)
Platynaspis salaamensis Weise, Coccinellidae, Tanzania, (Ritchie, 1936)
Scymnus guttigera (Korschefsky), Coccinellidae, Kenya, (Anderson, 1932)
Scymnus sp., Coccinellidae, Kenya, (Kirkpatrick, 1927b)
Sidis ancoralis (Sicard), Coccinellidae, Kenya, (Anderson, 1932)
Spalgis lemolea Druce, Lycaenidae, Kenya, (James, 1932a), Uganda, (Hancock, 1926)
Trichophthalmocapsus jamesi China, Miridae, Kenya, (China, 1932)
Ypsiloneura kirkpatricki Pearman, Entomobryidae, Kenya, (Pearman, 1932)

Ant attendants

Crematogaster sp., Formicidae Myrmecinae, Kenya, (Le Pelley, 1959)
Lepisiota capensis (Mayr), Formicidae Formicinae, Kenya, (Anderson, 1927)
Lepisiota incisa (Forel), Formicidae Formicinae, Kenya, (Le Pelley, 1932)
Monomorium pharaonis (Linnaeus), Formicidae Myrmecinae, Kenya, (Le Pelley, 1959)

Myrmicaria natalensis (F.Smith), Formicidae Myrmecinae, Tanzania, (Ritchie, 1936)
Pheidole punctulata Mayr, Formicidae Myrmecinae, Kenya, (Anderson, 1931), Tanzania, (Le Pelley, 1959), Uganda, (Le Pelley and Melville, 1939)
Pheidole speculifera Emery, Formicidae Myrmecinae, Kenya, (Anderson, 1927)
Technomyrmex albipes (F.Smith), Formicidae Dolichoderinae, Kenya, (Anderson, 1927)

Planococcus lilacinus (see p. 124)

Parasites

Anagyrus lilacini Ferriere, Encyrtidae, Philippines, (Le Pelley, 1943b)
Apanteles sp. near *sauros* Nixon, Braconidae, India, (Reddy *et al.*, 1990)
Blepyrus insularis (Cameron), Encyrtidae, India, (Reddy *et al.*, 1990)
Coccodiplosis smithi (Felt), Cecidomyiidae, Indonesia Lombok, (Le Pelley, 1943b), Indonesia Java, (Kalshoven, 1950–1951), Indonesia Bali, (Le Pelley, 1943b)
Gonatocerus sp., Mymaridae, India, (Reddy *et al.*, 1990)
Leptacis sp., Platygasteridae, India, (Reddy *et al.*, 1990)
Pseudaphycus orientalis Ferriere, Encyrtidae, Philippines, (Le Pelley, 1943b)

Predators

Brumoides suturalis (Fabricius), Coccinellidae, India, (Le Pelley, 1968)
Dicrodiplosis sp., Cecidomyiidae, India, (Reddy *et al.*, 1990)
Horniolus vietnamicus, Coccinellidae, India, (Irulandi *et al.*, 2001)
Pseudoscymnus pallidicollis (Mulsant), Coccinellidae, India, (Reddy *et al.*, 1990)
Scymnus apiciflavus Motschulsky, Coccinellidae, Indonesia Bali, (Le Pelley, 1943b), Indonesia Lombok, (Le Pelley, 1943b), Indonesia Java, (Le Pelley, 1943b)
Spalgis epeus (Westwood), Lycaenidae, India, (Le Pelley, 1943b), Indonesia, (Le Pelley, 1943b), Sri Lanka, (Green, 1904)

Planococcus longispinus

Predator

Telsimia rotundata (Motschulsky), Coccinellidae, Reunion, (Bordage, 1914), Sri Lanka, (Nietner, 1861)

Planococcus minor (see p. 124)

Parasite

Leptacis sp., Platygasteridae, India, (Reddy *et al.*, 1990)

Predators

Cryptolaemus affinis Crotch, Coccinellidae, Papua New Guinea, (Szent-Ivany, 1963)
Dicrodiplosis sp., Cecidomyiidae, India, (Reddy *et al.*, 1990)
Harmonia sp., Coccinellidae, Papua New Guinea, (Szent-Ivany, 1963)
Spalgis epeus (Westwood), Lycaenidae, India, (Reddy *et al.*, 1997)

Pulvinaria psidii (see p. 119)

Parasites

Megommata psidii Barnes, Cecidomyiidae, Democratic Republic of Congo, (Barnes, 1939)
Microterys nietneri (Motschulsky), Encyrtidae, Hawaii, (Fullaway, 1932)

Predators

Cheilomenes sexmaculata (Fabricius), Coccinellidae, Papua New Guinea, (Szent-Ivany, 1958)
Cryptolaemus montrouzieri Mulsant, Coccinellidae, Puerto Rico, (Wolcott, 1951)
Harmonia sp. near *testudinaria* Mulsant, Coccinellidae, Papua New Guinea, (Szent-Ivany, 1958)
Orcus sp., Coccinellidae, Papua New Guinea, (Szent-Ivany, 1958)

Ant attendant

Pheidole punctulata Mayr, Formicidae Myrmecinae, East Africa, (Le Pelley, 1968)

Ruspoliella coffea (see p. 117)

Predator

Geocoris ruficeps (Germar), Lygaeidae, Kenya, (Le Pelley, 1932)

Saissetia coffeae (see p. 122)

Parasites

Anicetus annulatus Timberlake, Encyrtidae, Hawaii, (Timberlake, 1919)
Cephaleta brunniventris Motschulsky, Pteromalidae, Sri Lanka, (Nietner, 1861)
Cephaleta fusciventris Motschulsky, Pteromalidae, Sri Lanka, (Nietner, 1861)
Coccophagus ceroplastae (Howard), Aphelinidae, Puerto Rico, (Smith, 1944)
Coccophagus flavescens Howard, Aphelinidae, Sri Lanka, (Ramakrishna Ayyar, 1925)
Coccophagus ochraceus Howard, Aphelinidae, Kenya, (Le Pelley, 1968)
Coccophagus tibialis Compere, Aphelinidae, Philippines, (Compere, 1931)
Diversinervus paradisicus (Motschulsky), Encyrtidae, Sri Lanka, (Nietner, 1861)
Encyrtus aurantii (Geoffray), Encyrtidae, Hawaii, (Smith, 1944)
Encyrtus infelix (Embleton), Encyrtidae, Cuba, (Bruner, 1929)
Gahaniella brasiliensis (Gomes), Encyrtidae, Brazil, (Gomes, 1941)
Gahaniella saissetiae Timberlake, Encyrtidae, Cuba, (Bruner, 1929)
Lecaniobius cockerelli Ashmead, Eupelmidae, Jamaica, (Gowdey, 1925)
Lecaniobius sp., Eupelmidae, Colombia, (Cardenas and Posada, 2001)
Microterys nietneri (Motschulsky), Encyrtidae, Sri Lanka, (Nietner, 1861)
Scutellista caerulea (Boyer de Fonscolombe), Pteromalidae, Colombia, (Cardenas and Posada, 2001), Hawaii, (Fullaway, 1932), Kenya, (Le Pelley, 1968), Sri Lanka, (Nietner, 1861)
Taftia saissetiae Gahan, Encyrtidae, Philippines, (Gahan, 1920)

Hyperparasites

Coccidoctonus dubius Girault, Encyrtidae, Hawaii, (Le Pelley, 1968)
Eupelmus coccidivorus Gahan, Eupelmidae, Cuba, (Bruner, 1929)

Predators

Azya luteipes Mulsant, Coccinellidae, Colombia, (Cardenas and Posada, 2001), Peru, (Wille, 1952)
Baccha bonleyi Curran, Syrphidae, Colombia, (Cardenas and Posada, 2001)
Chilocorus cacti (Linnaeus), Coccinellidae, Colombia, (Cardenas and Posada, 2001)
Chilocorus circumdatus (Gyllenhal), Coccinellidae, Sri Lanka, (Nietner, 1861)
Harmonia sp. near *testudinaria* Mulsant, Coccinellidae, Papua New Guinea, (Szent-Ivany, 1958)
Ocyptamus capitata (Loew), Syrphidae, Puerto Rico, (Smith, 1942)
Orcus janthinus Mulsant, Coccinellidae, Indonesia Java, (Vayssière, 1955)
Orcus sp., Coccinellidae, Papua New Guinea, (Szent-Ivany, 1958)

Ant attendants

Brachymyrmex heeri Forel, Formicidae Formicinae, Puerto Rico, (Smith, 1942)

Camponotus sp., Formicidae Formicinae, Brazil, (Bondar, 1928)
Crematogaster castanea F.Smith, Formicidae Myrmecinae, Kenya, (Le Pelley, 1959)
Crematogaster steinheili Forel, Formicidae Myrmecinae, Puerto Rico, (Smith, 1942)
Linepithema melleum (Wheeler), Formicidae Dolichoderinae, Puerto Rico, (Smith, 1942)
Monomorium floricola (Jerdon), Formicidae Myrmecinae, Puerto Rico, (Smith, 1942)
Myrmelachista ramulorum Wheeler, Formicidae Formicinae, Puerto Rico, (Smith, 1942)
Oecophylla longinoda (Latreille), Formicidae Formicinae, Côte d'Ivoire, (Mallamaire, 1937)
Paratrechina longicornis (Latreille), Formicidae Formicinae, Puerto Rico, (Smith, 1942)
Pheidole fallax antillensis Forel, Formicidae Myrmecinae, Puerto Rico, (Smith, 1942)
Pheidole punctulata Mayr, Formicidae Myrmecinae, East Africa, (Le Pelley, 1968)
Pheidole subarmata Mayr, Formicidae Myrmecinae, Puerto Rico, (Smith, 1942)
Solenopsis geminata (Fabricius), Formicidae Myrmecinae, Puerto Rico, (Smith, 1942)
Wasmannia auropunctata (Roger), Formicidae Myrmecinae, Puerto Rico, (Smith, 1942)

Saissetia oleae (see p. 122)

Parasites

Coccophagus ceroplastae (Howard), Aphelinidae, Cuba, (Bruner, 1929), Hawaii, (Fullaway, 1932), Indonesia Java, (Thompson, 1943–1958), Malaysia, (Miller, 1931), Puerto Rico, (Smith, 1944), Sri Lanka, (Le Pelley, 1968)
Eupelmus saissetiae Silvestri, Eupelmidae, Puerto Rico, (Dozier, 1927)
Lecaniobius cockerelli Ashmead, Eupelmidae, Cuba, (Bruner, 1929), Puerto Rico, (Dozier, 1927)
Plagiomerus cyaneus (Ashmead), Encyrtidae, Cuba, (Bruner, 1929)
Trichomasthus portoricensis (Crawford), Encyrtidae, Cuba, (Bruner, 1929)

Toxoptera aurantii (see p. 130)

Parasites

Aphelinus semiflavus Howard, Aphelinidae, Hawaii, (Fullaway, 1932)
Aphidius matricariae Haliday, Braconidae, Honduras, (Munoz, *c.*2000), Peru, (Wille, 1952)
Aphidius sp., Braconidae, Peru, (Wille, 1952)

Lysiphlebus testaceipes (Cresson), Braconidae, Honduras, (Munoz, *c.*2000), Puerto Rico, (Wolcott, 1951)
Lysiphlebus sp., Braconidae, Peru, (Wille, 1952)
Syrphophagus africanus (Gahan), Encyrtidae, Kenya, (Gahan, 1932)

Predators

Allograpta calopus (Loew), Syrphidae, Kenya, (Kirkpatrick, 1927a)
Allograpta nasuta (Macquart), Syrphidae, Kenya, (Kirkpatrick, 1927a)
Allograpta obliqua (Say), Syrphidae, Hawaii, (Illingworth, 1929)
Azya luteipes Mulsant, Coccinellidae, Honduras, (Munoz, *c.*2000), Peru, (Wille, 1952)
Baccha sp., Syrphidae, Puerto Rico, (Wolcott, 1933)
Betasyrphus adligatus (Wiedemann), Syrphidae, Kenya, (Kirkpatrick, 1927a)
Cheilomenes lunata (Fabricius), Coccinellidae, Kenya, (McDonald, 1937)
Chilocorus melanophthalmus Mulsant, Coccinellidae, Indonesia Java, (Keuchenius, 1915)
Chrysopa sp., Chrysopidae, Honduras, (Munoz, *c.*2000)
Coelophora novempunctata (Fabricius), Coccinellidae, Indonesia, (Vayssière, 1955)
Cycloneda sanguinea (Linnaeus), Coccinellidae, Peru, (Wille, 1952)
Dioprosopa clavata (Fabricius), Syrphidae, Puerto Rico, (Vayssière, 1955)
Hemerobius sp., Hemerobiidae, Honduras, (Munoz, *c.*2000)
Hippodamia sp., Coccinellidae, Honduras, (Munoz, *c.*2000)
Ischiodon scutellaris (Fabricius), Syrphidae, Sri Lanka, (Nietner, 1861)
Isora anceps Mulsant, Coccinellidae, Indonesia, (Vayssière, 1955), Democratic Republic of Congo, (Mayné and Ghesquière, 1934)
Mallada basalis (Walker), Chrysopidae, Hawaii, (Illingworth, 1929)
Micromus australis Hagen, Hemerobiidae, Sri Lanka, (Nietner, 1861)
Ocyptamus dimidiatus (Fabricius), Syrphidae, Puerto Rico, (Wolcott, 1924)
Ocyptamus fasciatus (Roeder), Syrphidae, Puerto Rico, (Wolcott, 1924)
Ornidia sp., Syrphidae, Honduras, (Munoz, *c.*2000)
Paragus borbonicus Macquart, Syrphidae, Puerto Rico, (Vayssière, 1955)
Paragus marshalli Bezzi, Syrphidae, Kenya, (Le Pelley, 1959), Tanzania, (Ritchie, 1935)
Platynaspis capicola Crotch, Coccinellidae, Tanzania, (Ritchie, 1935)
Platynaspis kollari Mulsant, Coccinellidae, Tanzania, (Ritchie, 1935)
Pseudaspidimerus flaviceps (Walker), Coccinellidae, Sri Lanka, (De Silva, 1961)
Scymnodes lividigaster (Mulsant), Coccinellidae, Hawaii, (Illingworth, 1929)
Scymnus sp., Coccinellidae, Honduras, (Munoz, *c.*2000), Peru, (Wille, 1952)
Synona inaequalis (Fabricius), Coccinellidae, Hawaii, (Illingworth, 1929), Indonesia, (Kalshoven, 1950–1951)
Syrphus nietneri Schiner, Syrphidae, Sri Lanka, (Nietner, 1861)

Ant attendant

Pheidole punctulata Mayr, Formicidae Myrmecinae, Kenya, (Kirkpatrick, 1927b)

Coleoptera

Araecerus fasciculatus (see p. 329)

Parasites

Anisopteromalus calandrae (Howard), Pteromalidae, Colombia, (Cabal-Concha, 1956)
Apanteles araeceri Wilkinson, Braconidae, Indonesia Java, (Wilkinson, 1928)
Cephalonomia gallicola (Ashmead), Bethylidae, Colombia, (Cabal-Concha, 1956)
Plastanoxus sp., Bethylidae, Colombia, (Cabal-Concha, 1956)

Predators

Cheyletus sp., Cheyletidae, Colombia, (Cabal-Concha, 1956)
Monieziella sp., Sarcoptidae, Colombia, (Cabal-Concha, 1956)

Bixadus sierricola (see p. 50)

Parasites

Gabunia ruficoxis Kreichbaumer, Ichneumonidae, Equatorial Guinea, (Baguena Corella, 1942)
Phorostoma sp., Tachinidae, Cameroon, (Lepesme and Paulian, 1943)

Dirphya nigricornis (see p. 53)

Parasites

Hybogaster varipalpis (Cameron), Braconidae, Kenya, (Crowe, 1962a), Uganda, (Le Pelley, 1959), Zimbabwe, (Kutywayo, 1989)
Zaglyptogastra pulchricauda (Szepligeti), Braconidae, Uganda, (Le Pelley, 1968)

Hypothenemus hampei (see p. 68)

Parasites

Cephalonomia hyalinipennis Ashmead, Bethylidae, Mexico, (Perez-Lachaud and Hardy, 1999)
Cephalonomia stephanoderis Betrem, Bethylidae, Brazil, (Benassi and Berti Filho, 1989), Ecuador, (Klein-Koch, 1990), El Salvador, (Galvez, 1992), Guatemala, (Galvez, 1992), Honduras, (Munoz, *c.*2000), Côte d'Ivoire, (Ticheler, 1961), Mexico, (Barrera *et al.*, 1990)
Cryptoxilos sp., Braconidae, Colombia, (Cardenas and Posada, 2001)

Heterospilus coffeicola Schmiedeknecht, Braconidae, Cameroon, (Pascalet, 1939), Honduras, (Munoz, *c.*2000), Indonesia Java, (Friederichs, 1925), Kenya, (Mugo *et al.*, 1997), Sri Lanka, (Hutson, 1939), Tanzania, (Harris, 1935), Uganda, (Hargreaves, H., 1926), Democratic Republic of Congo, (Sladden, 1934)
Phymastichus coffea La Salle, Eulophidae, Brazil, (Gutierrez *et al.*, 1998), El Salvador, (Galvez, 1992), Guatemala, Honduras, (Munoz, *c.*2000), Kenya, (La Salle, 1990), Mexico, (Galvez, 1992), Togo, (La Salle, 1990)
Prorops nasuta Waterston, Bethylidae, Brazil, (Da Fonseca, 1939), Cameroon, (Pascalet, 1939), Ecuador, (Klein-Koch, 1990), El Salvador, (Galvez, 1992), Guatemala, (Galvez, 1992), Honduras, (Munoz, *c.*2000), Indonesia Java, (Schweizer, 1932), Kenya, (Mugo *et al.*, 1997), Mexico, (Barrera *et al.*, 1990), Sri Lanka, (Rodrigo, 1941), Tanzania, (Harris, 1935), Uganda, (Hargreaves, H., 1935), Democratic Republic of Congo, (Bredo, 1934)

Predators

Brachymyrmex sp., Formicidae Formicinae, Colombia, (Bustillo *et al.*, 2002)
Calliodis sp., Anthocoridae, Colombia, (Cardenas and Posada, 2001)
Chrysoperla sp., Chrysopidae, Colombia, (Cardenas and Posada, 2001)
Crematogaster curvispinosa Mayr, Formicidae Myrmecinae, Brazil, (Da Fonseca and Araujo, 1939)
Crematogaster sp., Formicidae Myrmecinae, Colombia, (Bustillo *et al.*, 2002)
Dindymus rubiginosus Fabricius, Pyrrhocoridae, Indonesia Java, (Du Pasquier, 1932)
Leptophloeus sp. near *punctatus* Lefkovitch, Laemophloeiidae, Togo, (Vega *et al.*, 1999), Côte d'Ivoire, (Vega *et al.*, 1999)
Paratrechina sp., Formicidae Formicinae, Colombia, (Bustillo *et al.*, 2002)
Pheidole sp., Formicidae Myrmecinae, Colombia, (Bustillo *et al.*, 2002)
Prenolepis sp., Formicidae Formicinae, Colombia, (Bustillo *et al.*, 2002)
Scoloposcelis sp., Anthocoridae, Colombia, (Cardenas and Posada, 2001)
Solenopsis sp., Formicidae Myrmecinae, Colombia, (Bustillo *et al.*, 2002)
Wasmannia sp., Formicidae Myrmecinae, Colombia, (Bustillo *et al.*, 2002)

Monochamus leuconotus (see p. 43)

Parasites

Afrocoelichneumon didymatus (Morley), Ichneumonidae, Tanzania, (Tapley, 1960)
Aprostocetus sp., Eulophidae, Tanzania, (Tapley, 1960), South Africa, (Schoeman, 2005, personal communication)
Calliscelio sp., Scelionidae, Tanzania, (Tapley, 1960)
Cratichneumon sp., Ichneumonidae, Tanzania, (Tapley, 1960)
Hybogaster varipalpis (Cameron), Braconidae, Kenya, (Knight, 1939), Zimbabwe, (Kutywayo, 1989)

Iphiaulax sp., Braconidae, South Africa, (Schoeman, 2005, personal communication)
Nadia sp. near *ruficeps* Cameron, Ichneumonidae, Tanzania, (Tapley, 1960)
Oxysychus sp., Pteromalidae, South Africa, (Schoeman, 2005, personal communication)
Tetrastichus sp., Eulophidae, Tanzania, (Tapley, 1960)

Predators

Gyponyx retrocinctus (Chevrolat), Cleridae, South Africa, (Schoeman, 2005, personal communication)
Odontomachus haemotodus (Linnaeus), Formicidae Ponerinae, Democratic Republic of Congo, (Mayné, 1923)
Pheidole megacephala (Fabricius), Formicidae Myrmecinae, South Africa, (Schoeman, 2005, personal communication)
Plagiolepis sp., Formicidae Formicinae, South Africa, (Schoeman, 2005, personal communication)

Neonitocris princeps (see p. 53)

Parasite

Bathyaulax sp., Braconidae, Uganda, (Hargreaves, H., 1936)

Predator

Margasus afzelii (Stal), Reduviidae, Uganda, (Hancock, 1926)

Xylosandrus compactus (see p. 58)

Parasites

Mesobraconoides psolopterus (Wilkinson), Braconidae, Sierra Leone, (Wilkinson, 1931)
Tetrastichus xylebororum Domenichini, Eulophidae, Indonesia Java, (Domenichini, 1960)
Tetrastichus sp. near *xylebororum* Domenichini, Eulophidae, India, (Dhanam *et al.*, 1992)

Predators

Callimerus sp., Cleridae, India, (Sreedharan *et al.*, 1992)
Eupelmus sp., Eupelmidae, India, (Vinodkumar *et al.*, 1986)

Xylosandrus morigerus (see p. 58)

Parasite

Tetrastichus xylebororum Domenichini, Eulophidae, Indonesia Java, (Domenichini, 1960)

Xylotrechus quadripes (see p. 47)

Parasites

Apenesia sp., Bethylidae, India, (Seetharama *et al.*, 2002)
Diastephanus sp., Ichneumonidae, Thailand, (Visitpanich, 1994)
Doryctes tristriatus Kieffer, Braconidae, Vietnam, (Kieffer, 1921)
Epixorides caerulescens (Morley), Ichneumonidae, Sri Lanka, (Thompson, 1943–1958)
Eurytoma xylotrechi Ferriere, Eurytomidae, Vietnam, (Du Pasquier, 1932)
Gonophorus exquisitus (Tosquinet), Ichneumonidae, Vietnam, (Kieffer, 1921)
Metapelma sp., Eupelmidae, India, (Subramanian, 1934)
Mysepyris grandiceps Kieffer, Bethylidae, Vietnam, (Kieffer, 1921)
Ontsira bistriata (Kieffer), Braconidae, Vietnam, (Kieffer, 1921)
Ontsira brevipetiola (Kieffer), Braconidae, Vietnam, (Kieffer, 1921)
Ontsira palliata (Cameron), Braconidae, Vietnam, (Kieffer, 1921)
Parallorhogas pallidiceps (Perkins), Braconidae, India, (Prakasan *et al.*, 1986), Vietnam, (Kieffer, 1921)
Pristaulacus nigripes Kieffer, Evaniidae, Vietnam, (Kieffer, 1921)
Pristaulacus sp., Evaniidae, Thailand, (Visitpanich, 1994)
Promiscolus sesquistriatus Kieffer, Braconidae, Vietnam, (Kieffer, 1921)
Sclerodermus domesticus Klug, Bethylidae, Vietnam, (Kieffer, 1921)

Diptera

Anastrepha fraterculus (see p. 85)

Parasites

Aganaspis pelleranoi (Brethes), Eucoilidae, Brazil, (Raga *et al.*, 2002)
Doryctobracon areolatus (Szepligeti), Braconidae, Brazil, (Raga *et al.*, 2002)
Doryctobracon brasiliensis (Szepligeti), Braconidae, Brazil, (Raga *et al.*, 2002)
Doryctobracon crawfordi (Viereck), Braconidae, Colombia, (Cardenas and Posada, 2001)
Microcrasis sp., Braconidae, Colombia, (Cardenas and Posada, 2001)
Odontosema sp., Eucoilidae, Colombia, (Cardenas and Posada, 2001)
Opius sp., Braconidae, Brazil, (Raga *et al.*, 2002)

Utetes anastrephae (Viereck), Braconidae, Brazil, (Raga *et al.*, 2002), Colombia, (Cardenas and Posada, 2001)

Anastrepha obliqua (see p. 85)

Parasite

Utetes anastrephae (Viereck), Braconidae, Cuba, (Vasquez *et al.*, 1999)

Ceratitis capitata (see p. 85)

Parasites

Aceratoneuromyia indica (Silvestri), Eulophidae, Brazil, (Menezes-Marconi and Iba, 1955), Costa Rica, (Wharton *et al.*, 1981), Hawaii, (Menezes-Marconi and Iba, 1955)
Aganaspis nordlanderi Wharton, Eucoilidae, Costa Rica, (Wharton *et al.*, 1998)
Aganaspis pelleranoi (Brethes), Eucoilidae, Brazil, (Raga *et al.*, 2002) Eucoilidae, Costa Rica, (Wharton *et al.*, 1998)
Ceromya sp., Tachinidae, Zimbabwe, (Kutywayo, 1989)
Diachasmimorpha fullawayi (Silvestri), Braconidae, Hawaii, (Willard and Mason, 1937), Kenya, (Bianchi, 1936), Tanzania, (Bianchi, 1936)
Diachasmimorpha longicaudata (Ashmead), Braconidae, Guatemala, (Eskafi, 1990), Hawaii, (Vargas *et al.*, 1994)
Diachasmimorpha tryoni (Cameron), Braconidae, Hawaii, (Willard and Mason, 1937), Guatemala, (Sivinski *et al.*, 2000)
Dirhinus giffardii Silvestri, Chalcididae, Hawaii, (Vargas *et al.*, 2001)
Doryctobracon areolatus (Szepligeti), Braconidae, Brazil, (Raga *et al.*, 2002)
Doryctobracon brasiliensis (Szepligeti), Braconidae, Brazil, (Raga *et al.*, 2002)
Doryctobracon crawfordi (Viereck), Braconidae, Colombia, (Cardenas and Posada, 2001), Guatemala, (Eskafi, 1990)
Fopius arisanus (Sonan), Braconidae, Costa Rica, (Wharton *et al.*, 1981), Hawaii, (Vargas *et al.*, 1994)
Fopius caudatus (Szepligeti), Braconidae, Guatemala, (Lopez *et al.*, 2003)
Fopius ceratitivorus Wharton, Braconidae, Guatemala, (Lopez *et al.*, 2003), Kenya, (Wharton, 1999)
Ganaspis pelleranoi (Brenthes), Cynipidae, Costa Rica, (Wharton *et al.*, 1981)
Microcrasis sp., Braconidae, Colombia, (Cardenas and Posada, 2001)
Odontosema anastrephae Borgmeier, Eucoilidae, Costa Rica, (Wharton *et al.*, 1981)
Odontosema sp., Eucoilidae, Colombia, (Cardenas and Posada, 2001)
Opius bellus Gahan, Braconidae, Venezuela, (Guagliumi, 1963)
Opius sp., Braconidae, Brazil, (Raga *et al.*, 2002), Guatemala, (Eskafi, 1990)
Phaenocarpa anastrephae Muesebeck, Braconidae, Venezuela, (Guagliumi, 1963)

Psyttalia concolor (Szepligeti), Braconidae, Kenya, (Kimani-Njogu *et al.*, 2001)
Psyttalia humilis (Silvestri), Braconidae, Hawaii, (Willard and Mason, 1937), Kenya, (Bianchi, 1936)
Tetrastichus giffardianus Silvestri, Eulophidae, Brazil, (Menezes-Marconi and Iba, 1955), Hawaii, (Menezes-Marconi and Iba, 1955), Kenya, (Menezes-Marconi and Iba, 1955)
Utetes anastrephae (Viereck), Braconidae, Brazil, (Raga *et al.*, 2002), Colombia, (Cardenas and Posada, 2001)

Ceratitis rosa (see p. 84)

Parasites

Fopius ceratitivorus Wharton, Braconidae, Kenya, (Wharton, 1999)
Psyttalia concolor (Szepligeti), Braconidae, Kenya, (Kimani-Njogu *et al.*, 2001)

Trirhithrum coffeae (see p. 86)

Parasite

Diachasmimorpha fullawayi (Silvestri), Braconidae, Cameroon, (Steck *et al.*, 1986), Togo, (Steck *et al.*, 1986)
Fopius caudatus (Szepligeti), Braconidae, Cameroon, (Steck *et al.*, 1986)
Fopius ceratitivorus Wharton, Braconidae, Kenya, (Wharton, 1999)
Fopius desideratus (Bridwell), Braconidae, Togo, (Steck *et al.*, 1986)
Fopius near *desideratus* (Bridwell), Braconidae, Uganda, (Greathead, 1972)
Fopius silvestrii (Wharton), Braconidae, Cameroon, (Wharton, 1987), Togo, (Wharton, 1987)
Opius sp., Braconidae, Cameroon, (Steck *et al.*, 1986), Togo, (Steck *et al.*, 1986)
Psilus sp., Pteromalidae, Uganda, (Greathead, 1972)
Psyttalia concolor (Szepligeti), Braconidae, Kenya, (Kimani-Njogu *et al.*, 2001)
Psyttalia cosyrae (Wilkinson), Braconidae, Uganda, (Greathead, 1972)
Psyttalia perproxima (Silvestri), Braconidae, Cameroon, (Steck *et al.*, 1986), Togo, (Steck *et al.*, 1986)
Tetrastichus giffardianus Silvestri, Eulophidae, Uganda, (Greathead, 1972)
Tetrastichus sp., Eulophidae, Uganda, (Greathead, 1972)

References

Abasa, R.O. and Mathenge, W.M. (1972) Observations on the biology and host-searching behaviour of *Afromelanichneumon sporadicus* and *Cryptus nigropictus* (Hymenoptera Ichneumonidae) pupal parasites of *Ascotis selenaria reciprocaria* (Lepidoptera Geometridae). *Entomophaga* 17, 93–97.

Abebe, M. (1999) The role of parasites in the natural control of *Antestia* (*Antestiopsis intricata* (Ghesquiere and Carayon)). Possibilities for further control using exotic parasites. *Dix-huitieme Colloque Scientifique International sur le Café*, Helsinki, Finland, 2–8 August 1999, pp. 492–496.

Alvarado, J.A. (1935) Nuestros insectos auxiliares. Leon de los pulgones. *Revista Agricultura de Guatemala* 13, 227–230.

Anderson, T.J. (1927) Report of the Entomologist. *Report. Department of Agriculture, Kenya*, 1926, 135–147.

Anderson, T.J. (1931) Annual Report of the Senior Entomologist 1930. *Report. Department of Agriculture, Kenya*, 1930, 190–205.

Anderson, T.J. (1932) Entomolgy Section. *Report. Department of Agriculture, Kenya*, 1931, 99–117.

Anderson, T.J. (1934) Entomology Section. *Report. Department of Agriculture, Kenya*, 1933, 137–145.

Anon (1939) Entomological investigations. *Report. Agricultural Department, St. Lucia*, 1938, 33–38.

Anon (1940a) Report. *Porto Rico Insular Agricultural Experiment Station*, 1937–1938, 114 pp.

Anon (1940b) Report. *Porto Rico Insular Agricultural Experiment Station*, 1938–1939, 102 pp.

Anon (1978) Thirty-first annual report, 1977–1978. *Report. Coffee Board Research Department, India*, 53 pp.

Anon (1992) Control of berry moth (*Prophantis smaragdina* (Butler)) in coffee. *Kenya Coffee* 57, 1427–1428.

Apety, J. (1994) Imidan insecticide 500 for greenscale control in coffee. *Newsletter – Coffee Research Institute* No. 1, 7–8.

Aranda, D.E. (1986) *Control Naturel del Minador de la Hoja del Cafeto en Mexico, Leucoptera coffeella (Guer-Men. 1842.) (Lep-Lyonetiidae)*. Tesis profesional, Universidad Veracruzana. Xalapa Veracruz, Mexico, 217 pp.

Báguena Corella, L. (1942) Observaciones sobre *Longicornios perjudiciales* a varios cultivos arbóreos y especialmente a los cacao y café en los Territorios Españoles del Golfo de Guinea. *Publicación, Direcccion de Agricultura de los Territorios Españoles del Golfo de Guinea* 6, 39–91.

Balakrishnan, M.M., Sreedharan, K., Venkatesha, M.G., Krishnamoorthy, P. and Bhat, P.K. (1991) Observations on *Ferrisia virgata* (Ckll.) (Homoptera: Pseudococcidae) and its natural enemies on coffee, with new records of predators and host plants. *Journal of Coffee Research* 21, 11–19.

Barnes, H.F. (1939) Gall midges (Cecidomyiidae) associated with coffee. *Revue de Zoologie et de Botanique Africaines* 32, 324–336.

Barrera, J.F., Moore, D., Abraham, Y.J., Murphy, S.T. and Prior, C. (1990) Biological control of the coffee berry borer, *Hypothenemus hampei*, in Mexico and possibilities for further action. *Brighton Crop Protection Conference, Pests and Disease*, Brighton, UK, 1990, Vol. 1, pp. 391–396.

Benassi, V.L.R.M. and Berti Filho, E. (1989) Nota sobre a ocorrencia de *Cephalonomyia* sp. (Hymenoptera, Bethylidae) parasitando a broca-do-café, *Hypothenemus hampei* (Ferrari, 1867) (Coleoptera Scolytidae) no estado do Espirito Santo. *Revista de Agricultura Piracicaba* 64, 105–106.

Bess, H.A. (1958) The green scale *Coccus viridis* (Green) and ants. *Proceedings of the Hawaiian Entomological Society* 16, 349–355.

Bess, H.A. (1964) Populations of the leaf-miner *Leucoptera meyricki* and its parasites in sprayed and unsprayed coffee in Kenya. *Bulletin of Entomological Research* 55, 59–82.

Betrem, J.G. (1932) Verslag over de werkzaamheden van het Proefstation Malang in 1931. In: Ultee, A.J. (ed.) *Mededeelingen van Het Proefstation, Malang* 82, 55 pp.

Betrem, J.G. (1936) Gegevens omtrent de biologie van de dompolanluis (*Pseudococcus citri*) en de lamtoroluis (*Ferrisiana virgata*) *Archief voor de Koffiecult in Indonesie* 10, 43–84.

Bianchi, F.A. (1936) *Report on the U.S. Department of Agriculture Fruit Fly Expedition of 1935–1936*. US Department Agricultural Bureau, Entomology, 52 pp.

Bondar, G. (1928) Uma provacao da lavoura cafeeira na Bahia. *Correio Agricola* 6, 252–256.

Bordage, E. (1914) Notes biologiques recueilles à l'Ile de la Réunion. *Bulletin Scientifique de la France et de la Belgique* 47, 377–421.

Box, H.E. (1927) Notes sobre dos insectos perjudiciales a las matas de café en Venezuela. *Cámara Comercio*, 19 pp.

Bredo, H.J. (1934) La lutte biologique et son importance économique au Congo Belge. *Bulletin Agricole du Congo Belge* 25, 3–20.

Bredo, H.J. (1939) Catalogue des pricipaux insectes et nematodes parasites des caféiers au Congo Belge. *Bulletin Agricole du Congo Belge* 30, 266–307.

Bruner, S.C. (1929) Reseña de las plagas del cafeto en Cuba. *Estación Experimental Agronomica, Circular No.* 68, 38 pp.

Buckley, R.C. and Gullan, P.J. (1991) More aggressive ant species (Hymenoptera: Formicidae) provide better protection for soft scales and mealybugs (Homoptera: Coccidae, Pseudococcidae). *Biotropica* 23, 282–286.

Bustillo, A.E., Cardenas, R. and Posada, F.J. (2002) Natural enemies and competitors of *Hypothenemus hampei* (Ferrari) (Coleoptera: Scolytidae) in Colombia. *Neotropical Entomology* 31, 635–639.

Cabal-Concha, A. (1956) Biologia y control del gorgojo del café: *Araecerus fasciculatus* De Geer, en Barranquilla, Colombia. *Revista de la Facultad Nacional de Agronomia* 18, 30–31, 49–72.

Carayon, J. (1954) Les *Antestiopsis* du caféier en Afrique tropicale française. *Bulletin Scientifique. Ministere des Colonies de la France d'Outre Mer. Section Technique d'Agriculture Tropicale* 5, 363–373.

Carayon, J. (1960) *Stethoconus frappai* n.sp. miride predateur du tingide du caféier *Dulinus unicolor* Sign. à Madagascar. *Journal d'Agriculture Tropicale et de Botanique Appliquee* 7, 110–120.

Cardenas, M.R. and Posada, F.J. (2001) *Los Insectos y Otres Habitantes de Cafetales y Platanales*. Centro Nacional de Investigaciones de Café, Colombia, 250 pp.

Carvalho, G.A., Miranda, J.C., Moura, A.P., Rocha, L.C.D., Reis, P.R. and Vilela, F.Z. (2005) Controle de *Leucoptera coffeella* (Guérin-Mèneville & Perrottet, 1842) (Lepidoptera: Lyonetiidae) com insecticidas granulados e seus efeitos sobre vespas predadoras e parasitóides. *Arquivos do Instituto Biologico, São Paulo* 72, 63–72.

Castel Branco, A.J.F. (1971) Cochonilhas de S. Tomé e Príncipe *Asterolecanium pustulans* (Cockerell) e *A. pustulans princeps* nov. var. (Hemiptera Homoptera). *Eos* 46, 27–36.

Chacko, M.J. and Bhat, P.K. (1976) Record of *Ferrisia virgata* and its natural enemy *Spalgis epius* on coffee in India. *Journal of Coffee Research* 6, 56–57.

Chazeau, J. (1981) Donneés sur la biologie de *Coelophora quadrivittata* (Coleoptera Coccinellidae) predateur de *Coccus viridis* (Homoptera Coccidae) en Nouvelle-Caledonie. *Entomophaga* 26, 301–311.

China, W.E. (1932) A new species of *Trichophthalmocapsus* Popp. from Kenya. *Annals and Magazine of Natural History* 10, 594–597.

Coleman, L.C. and Kannan, K.K. (1918) Some scale insect pests of coffee in South India. *Mysore Department of Agriculture Entomological Series Bulletin* 4, 66 pp.

Compere, H. (1931) A revision of the species of *Coccophagus*. *Proceedings of the United States National Museum* 78, 132 pp.

Compere, H. (1938) A report on some miscellaneous African Encyrtidae in the British Museum. *Bulletin of Entomological Research* 29, 315–337.

Compere, H. (1939) A second report on some miscellaneous Encyrtidae in the British Museum. *Bulletin of Entomological Research* 30, 1–26.

Compere, H. (1940) The African species of *Metaphycus* Mercet. *Bulletin of Entomological Research* 31, 7–33.

Compere, H. (1943) A new and economically important species of *Anagyrus* from Africa. *Bulletin of Entomological Research* 34, 129–130.

Corbett, G.H. (1937) Report of the Division of Entomology, 1936. *General Series. Department of Agriculture, Straits Settlements and Federated Malay States* 26, 29–48.

Corbett, G.H. and Miller, N.C.E. (1933) A list of insects with their parasites and predators in Malaya. *Scientific Series. Department of Agriculture, Straits Settlements and Federated Malay States* 13, 15 pp.

Corbett, G.H. and Yusope, M. (1932) The coffee clear-wing hawk moth (*Cephanodes hylas* L.) *Malayan Agricultural Journal* 20, 508–517.

Crawford, J.C. (1913) Descriptions of new Hymenoptera No. 8. *Proceedings of the United States National Museum* 46, 343–352.

Crosskey, R.W. (1970) The identity of *Palexorista quadrizonula* (Thomson) (Diptera), a Tachinid parasite of Lepidopterous pests in Africa. *Bulletin of Entomological Research* 59, 579–583.

Crowe, T.J. (1962a) The biology and control of *Dirpha nigricornis* (Olivier), a pest of coffee in Kenya. *Journal of the Entomological Society of Southern Africa* 25, 304–312.

Crowe, T.J. (1962b) The white waxy scale. *Kenya Coffee* 27, 93–95.

Crowe, T.J. (2004) Coffee pests in Africa. In: Wintgens, J.N. (ed.) *Coffee: Growing, Processing, Sustainable Production. A Guidebook for Growers, Processors, Traders and Researchers.* Wiley, Chichester, UK, 1022 pp.

Crowe, T.J. and Greathead, D.J. (1970) Parasites of *Leucoptera* spp. on coffee in East Africa. An annotated list. *East African Agricultural and Forestry Journal* 35, 364–371.

Crowe, T.J. and Leeuwangh, J. (1965) The green looper. *Kenya Coffee* 30, 3 pp.

Da Costa Lima, A. (1936) *Terceiro Catalogo dos Insectos que Viviem nas Plantas do Brasil.* Ministry da Agricultura, Rio de Janeiro, 460 pp.

Da Fonseca, J.P. (1939) A 'broca' e o sombreamento dos cafezias. *Biológico* 5, 133–136.

Da Fonseca, J.P. and Araujo, R.L. (1939) Insetos inimigos do *Hypothenemus hampei* (Ferr.) *Boletim Biologico* 4, 486–504.

Decelle, J. (1962) La chenille des feuilles du caféier Robusta, *Leucoptera coma* Ghesq. *Parasitica* 18, 177–198.

De Fluiter, H.J. (1936) Onderzoekingen omtrent de witte luizen van de koffie. *Bergcultures* 10, 1630–1639.

De Fluiter, H.J. (1939) Het witte luis-vraagstuk bij de Koffie. *Bergcultures* 13, 760–765.

De Fluiter, H.J. (1960) Krankheiten und Schädlinge In: Von Bernegg, S. (ed.) *Tropische und Subtropische Weltwirtschafts-pflanzen III, Tiel Band 2 Kaffe*. Stuttgart, Germany, pp. 96–102.

De Lotto, G. (1960) The green scales of coffee in Africa south of the Sahara. *Bulletin of Entomological Research* 51, 389–403.

Derron, M. (1977) *Prophantis smaragdina* Butler and *Cryptophlebia colivora* Meyrick, two important pests on *Coffea arabica* L. on the island of São Tomé. *Mitteilungen der Schweizerischen Entomologischen Gesellschaft* 50, 149–151.

De Saeger, H. (1941) Le genre *Apanteles* au Congo Belge. Contribution à l'étude du genre. *Revue de Zoologie et de Botanique Africaines* 34, 322–347.

De Saeger, H. (1943) Quelques braconides nouveaux du genre *Microbracon*. *Revue de Zoologie et de Botanique Africaines* 36, 361–389.

De Santis, L. (1983) Eulofidos (Hymenoptera) de Colombia y Brasil parasitos de *Leucoptera coffeella* (Guerin-Meneville). *Revista Colombiana de Entomologia* 9, 9–12.

De Silva, M.D. (1961) A preliminary list of the native parasites and predators of insect pests in

Ceylon. *Tropical Agriculturalist and Magazine of the Ceylon Agricultural Society* 117, 115–141.
Dhanam, M., Raju, T. and Bhat, P.K. (1992) *Tetrastichus* sp. nr. *xylebororum* Domenichini (Hymenoptera: Eulophidae), a natural enemy of *Xylosandrus compactus* (Eichhoff). *Journal of Coffee Research* 22, Suppl., 127–128.
Domenichini, G. (1960) Descrizione di un nuovo *Tetrastichus* malese parassita di *Xyleborus* spp. *Entomologische Berichten Amsterdam* 20, 256–258.
Dozier, H.L. (1927) Notes on Porto Rican scale parasites. *Journal of the Department of Agriculture, Puerto Rico* 10, 267–277.
Dry, F.W. (1921) The egg parasites of the coffee bug in Kenya Colony. *Bulletin of Entomological Research* 12, 191–201.
Du Pasquier, R. (1932) Principales maladies parasitaires du théier et du caféier en Extrême-Orient. *Bulletin Économique de l'Indochine* 35, 223B–253B, 367B–415B, 589B–618B, 689B–720B.
Ecole, C.C., Silva, R.A., Louzada, J.N.C., Moraes, J.C., Barbosa, L.R. and Ambrogi, B.G. (2002) Predacao de ovos, larvas e pupas do bicho-mineiro-do-cafeeiro, *Leucoptera coffeella* (Guerin-Meneville & Perrottet, 1842) (Lepidoptera: Lyonetiidae) por *Chrysoperla externa* (Hagen, 1861) (Neuroptera: Chrysopidae). *Ciencia e Agrotecnologia* 26, 318–324.
Edwards, W.H. (1932) Report of the Government Entomologist. *Annual Report of the Department of Agriculture, Jamaica* 1931, 22–24.
Edwards, W.H. (1935) Report of the Government Entomologist. *Report on the Department of Agriculture, Jamaica*, 1934, 24–28.
Enriquez, E., Bejarano, S. and Vila, V. (1975a) Morfologia ciclo biologico y comportamiento de *Leucoptera coffeella* Guer-Men. *Revista Peruana de Entomologia* 18, 79–81.
Enriquez, E., Bejarano, S. and Vila, V. (1975b) Observaciones sobre avispas predatores de *Leucoptera coffeella* Guer-Men. en le centro y sur del Peru. *Revista Peruana de Entomologia* 18, 82–83.
Eskafi, F.M. (1990) Parasitism of fruit flies *Ceratatis capitata* and *Anastrepha* spp. (Diptera: Tephritidae) in Guatemala. *Entomophaga* 35, 355–362.
Evans, D.E. (1966a) Entomology. *Report 1965–1966 Coffee Research Foundation, Kenya.*
Evans, D.E. (1966b) Field evaluation of Bidrin against the coffee leaf miner (*Leucoptera meyricki* Ghesq.) in Kenya. *Turrialba* 16, 125–129.
Evans, D.E. (1970) The parasites of three Kenya coffee Tortricids. *Journal of the Entomological Society of Southern Africa* 33, 349–350.
Evans, D.E., Andrade, O. and Mathenge, W.M. (1968) The biology and control of *Archips occidentalis* (Wals.) and *Tortrix dinota* Meyr. (Lepidoptera Tortricidae) on coffee in Kenya. *Journal of the Entomological Society of Southern Africa* 31, 133–140.
Ferrière, C. (1930) On some egg parasites from Africa. *Bulletin of Entomological Research* 21, 34–44.
Ferrière, C. (1936) The parasites of the coffee leaf-miners in Africa. *Bulletin of Entomological Research* 27, 477–491.
Ferrière, C. (1940) Notes on two egg parasites of *Antestia lineaticollis* Stal. *Bulletin of Entomological Research* 31, 145–146.
Ferrière, C. (1941) New species of Euplectrini from Europe, Africa and Asia. *Bulletin of Entomological Research* 32, 17–48.
Florez, D.E. and De Hernandez, M.R. (1981) Fluctuacion de la poblacion del minador de las hojas del cafeto, *Leucoptera coffeella* y de sus enemigos naturales en el Valle del Cauca. *Revista Colombiana de Entomologia* 7, 29–38.
Foucart, G. and Brion, L. (1962) Control measures against pests of arabica coffee trees in Ruanda. *Document I.A.P.S.C.* (62) 6, 10 pp.
Fragoso, D.B., Jusselino-Filho, P., Guedes, R.N.C. and Proque, R. (2001) Seletividade de inseticidas

a vespas predadoras de *Leucoptera coffeella* (Guer.-Menev.) (Lepidoptera: Lyonetiidae). *Neotropical Entomology* 30, 139–143.

Friederichs, K. (1925) Verslag over het kweeken en de eerst verspreiding der Sluipwesp, *Prorops nasuta* Waterst. *Mededeelingen van het Koffiebessenboeboek-Fonds, Malang* 13, 1–11.

Fullaway, D.T. (1920) Natural control of scale insects in Hawaii. *Proceedings of the Hawaiian Entomological Society* 4, 237–246.

Fullaway, D.T. (1932) Hymenopterous parasites of the Coccidae etc. in Hawaii. *Proceedings of the Hawaiian Entomological Society* 8, 111–120.

Gadd, C.H. (1946) *Macrocentrus homonae*, a polyembryonic parasite of tea tortrix (*Homona coffearia*). *Ceylon Journal of Science* 23, 67–79.

Gahan, A.B. (1919) Report on a small collection of Indian parasitic hymenoptera. *Proceedings of the United States National Museum* 56, 513–524.

Gahan, A.B. (1920) New reared parasitic hymenoptera from the Philippines. *Philippine Journal of Science* 17, 343–351.

Gahan, A.B. (1932) Two new species of Encyrtidae from Kenya Colony. *Stylops* 1, 220–222.

Gallardo-Covas, F. (1988) Faunal survey of the coffee leaf miner, *Leucoptera coffeella*, parasitoids of Puerto Rico. *Journal of Agriculture of the University of Puerto Rico* 72, 255–263.

Galvez, G.C. (1992) El control biologico de la broca como una alternativa viable para Centroamerica. *Boletin de Promecafé* 57, 6–11.

Girault, A.A. (1915) New genera of Chalcidoid Hymenoptera. *Journal of the New York Entomological Society* 23, 165–173.

Girault, A.A. (1916) New Javanese Chalcidoid Hymenoptera. *Proceedings of the United States National Museum* 51, 479–485.

Gomes, J.G. (1941) Um novo parasito de Coccidae. *Boletim da Sociedade Brasileira de Entomologia* 4, 401–404.

Gopinath, K. (1962) Studies on the Braconid *Amyosoma zeuzerae*, an ectoparasite of the coffee red borer, *Zeuzera coffeae*. *Current Science* 31, 5–6.

Gowdey, C.C. (1916) Report of the Government Entomologist. Report of the *Department of Agriculture, Uganda 1915–1916*, 48–53.

Gowdey, C.C. (1925) Report of the Government Entomologist. *Report of the Department of Science and Agriculture, Jamaica 1924*, 17–20.

Gowdey, C.C. (1926) Catalogus Insectorum Jamaicensis. *Entomological Bulletin. Department of Agriculture, Jamaica* 4, 114 pp.

Greathead, D.J. (1965) Observation on coffee pests in Ethiopia. *FAO Coffee Mission to Ethiopia 1964–1965*, FAO Rome, 198 pp.

Greathead, D.J. (1966) The parasites of *Antestiopsis* spp. in East Africa and a discussion of the possibilities of biological control. *Technical Bulletin of the Commonwealth Institute of Biological Control* 7, 113–137.

Greathead, D.J. (1972) Notes on coffee fruit flies and their parasites at Kawanda (Uganda) *Technical Bulletin of the Commonwealth Institute of Biological Control* 15, 18 pp.

Green, E.E. (1904) *Coccidae of Ceylon*. London (five vols), Dulau & Co., London.

Griveaud, P. (1967) Lépidoptères défoliateurs des rubiacées en Côte d'Ivoire. *Cahiers ORSTOM Serie Biologie* 3, 3–41.

Guagliumi, P. (1963) La mosca del Mediterraneo (*Ceratitis capitata* Wied.: Dipt., Trypetidae) in Venezuela. *Rivista di Agricoltura Subtropicale e Tropicale* 57, 364–369.

Gusmao, M.R., Picanco, M., Gonring, A.H.R. and Moura, M.F. (2000) Seletividade fisiologica de inseticidas a vespidae predadores do bicho-mineiro-do-cafeeiro. *Pesquisa Agropecuaria Brasileira* 35, 681–686.

Gutierrez, A.P., Villacorta, A., Cure, J.R. and Ellis, C.K. (1998) Analise tritrofica do café (*Coffea arabica*) – sistema broca-do-café, (*Hypothenemus hampei* (Ferrari)) – parasitoide. *Anais da Sociedade Entomologica do Brasil* 27, 357–385.

Hamilton, D.W. (1967) Injurious and beneficial insects in coffee plantations of Costa Rica and Guatemala 1964. *Journal of Economic Entomology* 60, 1409–1413.

Hancock, G.L.R. (1926) Report of the Assistant Entomologist. *Report of the Department of Agriculture, Uganda* 1925, 25–28.

Hanks, L.M. and Sadof, C.S. (1990) The effect of ants on nymphal survival of *Coccus viridis* (Homoptera: Coccidae). *Biotropica* 22, 210–213.

Hargreaves, E. (1936) Entomological Work. *Report of the Department of Agriculture, Sierra Leone*, 1934, 16–18.

Hargreaves, H. (1924) Report of the Government Entomologist. *Report of the Department of Agriculture, Uganda*, 1922, 29–32.

Hargreaves, H. (1926) Notes on the coffee berry borer. *Bulletin of Entomological Research* 16, 346–354.

Hargreaves, H. (1932) Report of the Government Entomologist. *Report of the Department of Agriculture, Uganda*, 1931, 43–47.

Hargreaves, H. (1935) *Stephanoderes hampei* Ferr. coffee-borer, in Uganda. *East African Agricultural Journal* 1, 218–224.

Hargreaves, H. (1936) Report of the Government Entomologist. *Report of the Department of Agriculture, Uganda*, 1935–1936 2, 8–11.

Harris, W.V. (1935) Report of the Assistant Entomologist. *Report. Department of Agriculture, Tanganyika Territory* 1934, 84–89.

Hernandez, M. and Ceballos, M. (1991) *Marietta caridei* (Brethes) (Hymenoptera Chalcidoidea: Aphelinidae) on *Coccus viridis* (Green) (Homoptera: Coccidae) in Cuba. *Revista de Proteccion Vegetal* 6, 77–78.

Hernandez, M. and Ceballos, M. (1993) Especies de *Leptomastix* y *Leptomastid*ea presentes en cafetales Cubanos. Datos preliminares de comportamiento y biologia de *Leptomastix dactylopii*. *Revista de Proteccion Vegetal* 8, 203–205.

Hutson, J.C. (1939) Report of the Entomology Division. *Administrative Report of the Director of Agriculture, Ceylon* 1938, D36–D41.

Illingworth, J.F. (1929) Preliminary notes on pests of agricultural crops of Kona 1928. *Proceedings of the Hawaiian Entomological Society* 12, 248–254.

Irulandi, S., Kumar, P.K.V., Seetharama, H.G. and Sreedharan, K. (2001). Bioefficacy of neem formulations alone and in combination with synthetic insecticide against mealybug, *Planococcus citri* (Risso) on coffee. *Journal of Coffee Research* 29, 56–60.

James, H.C. (1932a) Coffee mealybug research. *Bulletin. Department of Agriculture, Kenya* 18, 18 pp.

James, H.C. (1932b) The control of *Asterolecanium*. *Bulletin. Department of Agriculture, Kenya* 23, 4 pp.

James, H.C. (1933) Taxonomic notes on the coffee mealy bugs of Kenya. *Bulletin of Entomological Research* 24, 429–436.

Kalshoven, L.G.E. (1950–1951) *De Plagen van de Cultuurgewassen in Indonesië*. Van Hoeve, Gravenhage/Bandoeng, Netherlands/Indonesia, 1065 pp.

Kerrich, G.J. (1967) On the classification of the Anagyrine Encyrtidae. *Bulletin of the British Museum of Natural History, Entomology* 20, 143–250.

Kerrich, G.J. (1969) Systematic studies on Eulophid parasites (Hymenoptera: Chalcidoidea), mostly on coffee leaf miner in Africa. *Bulletin of Entomological Research* 59, 195–228.

Keuchenius, P.E. (1915) Onderzoekingen en beschouwingen over eenige schadelijke Schildluizen van de Koffiekultur op Java. *Mededeelingen van het Besoekisch Proefstation, Djember* 16, 63.

Khoo, K.C., Ooi, P.A.C. and Ho, C.T. (1991) *Crop pests and their management in Malaysia*. Tropical Press, Kuala Lumpur, 242 pp.

Kieffer, J.J. (1921) Sur divers Hymenopteres destructeurs des Cerambycides nuisibles au Caféier et au Bambou. *Bulletin Agricole de l' Insitute Scientifique* 3, 129–140.

Kimani-Njogu, S.W., Trostle, M.K., Wharton, R.A., Woolley, J.B. and Raspi, A. (2001) Biosystematics of the *Psyttalia concolor* species complex (Hymenoptera: Braconidae: Opiinae): the identity of populations attacking *Ceratitis capitata* (Diptera: Tephritidae) in coffee in Kenya. *Biological Control* 20, 167–174.

Kirkpatrick, T.W. (1927a) The common coffee mealybug (*Pseudococcus lilacinus* Ckll.) in Kenya Colony. *Bulletin. Department of Agriculture, Kenya* 18, 110 pp.

Kirkpatrick, T.W. (1927b) Biological control of insect pests, with particular reference to the control of the common coffee mealybug in Kenya Colony. *Proceedings of the South and East Africa Agricultural Conference, 1926*, pp. 184–196.

Kirkpatrick, T.W. (1937) Studies on the ecology of coffee plantations in East Africa II. The autecology of *Antestia* species with a particular account of a strepsipterous parasite. *Transactions of the Royal Entomological Society of London* 86, 247–343.

Klein-Koch, C. (1990) Natural regulation factors and classical biological control of the coffee berry borer (*Hypothenemus hampei* Ferrari) in Ecuador. *Integrated Pest Management in Tropical and Subtropical Cropping Systems*, 8–15 February 1989, Bad Durkheim, Germany, pp. 331–344.

Knight, C.D. (1939) Observations on the life history and control of the white borer of coffee in Kenya. *East African Agricultural Journal* 5, 61–67.

Kohler, G. (1980) Los parasitos y episitos de la guagua verde del cafeto (*Coccus viridis* Green) (Hemiptera Coccidae) en cafetales de Cuba. *Centro Agricola* 7, 75–105.

Konnorova, E. (1987) Complejo parasitico de *Leucoptera coffeella* (Guer) en distintas localidades. *Centro Agricola* 14, 81–87.

Kutywayo, D. (1989) An annotated list of parasitoids and predators of coffee insect pests in Zimbabwe. *Zimbabwe Journal of Agricultural Research* 27, 11–20.

La Salle, J. (1990) A new genus and species of *Tetrastichinae* (Hymenoptera: Eulophidae) parasitic on the coffee berry borer *Hypothenemus hampei* (Ferrari) (Coleoptera Scolytidae). *Bulletin of Entomological Research* 80, 7–10.

Leefmans, S. (1929) Preliminary list of parasites and predators of some important insect pests in the Netherlands Indies. *Proceedings of the 4th Pacific Scientific Congress*, Java, 1929, pp. 561–568.

Leeuwangh, J. (1965) The biology of *Epigynopteryx stictigramma* (Hampson), a pest of coffee in Kenya. *Journal of the Entomological Society of Southern Africa* 28, 21–31.

Le Pelley, R.H. (1932) Lygus, a pest in coffee in Kenya Colony. *Bulletin of Entomological Research* 23, 85–99.

Le Pelley, R.H. (1943a) The establishment of a new species of *Anagyrus* in Kenya. *Bulletin of Entomological Research* 34, 131–133.

Le Pelley, R.H. (1943b) An oriental mealybug (*Pseudococcus lilacinus*) and its insect enemies. *Transactions of the Royal Entomological Society of London* 93, 73–93.

Le Pelley, R.H. (1959) *Agricultural Insects of East Africa*. East African High Commission, Nairobi, 307 pp.

Le Pelley, R.H. (1968) *Pests of Coffee*. Longmans, London, 590 pp.

Le Pelley, R.H. and Melville, A.R. (1939) Entomological work on coffee. *Report. Department of Agriculture, Kenya* 1938 2, 34–41.

Lepesme, P. (1947) *Les insectes des palmiers*. Lechevalier, Paris, 903 pp.

Lepesme, P. and Paulian, R. (1943) Une Tachinaire parasite des larves du monohamme du caféier. *Revue de Zoologie et de Botanique Africaines* 37, 89–92.

Leroy, J.V. (1936) Observations relatives à quelques insectes attaquant le caféier. *Publications de l'Institut National pour l'Étude Agronomique du Congo Belge* 8, 30 pp.

Leroy, J.V., Hendrickx, F.L. and Lefevre, P.C. (1942) Les *Antestia* spp. au Kivu. *Publications de l'Institut National pour l'Étude Agronomique du Congo Belge* 26, 69 pp.

Lopez, M., Sivinski, J., Rendon, P., Holler, T., Bloem, K., Copeland, R., Trostle, M. and Aluja, M.

(2003) Colonization of *Fopius ceratitivorus*, a newly discovered African egg-pupal parasitoid (Hymenoptera: Braconidae) of *Ceratitis capitata* (Diptera: Tephritidae). *Florida Entomologist* 86, 53–60.

Mallamaire, A. (1937) Les principaux nématodes, myriapodes et insectes parasites des caféiers cultiveés dans l'Ouest Africain francais. *Annales Agricoles de l'Afrique Occidentale Francais et Étrangère* 1, 1–45.

Masamdu, R. (1989) Factors influencing the effectiveness of *Metaphycus baruensis* Noyes (Hymenoptera: Encyrtidae) in parasitising *Coccus* spp. (Hemiptera: Coccidae) on coffee in Papua and New Guinea. *PNG Coffee* 8, 89–91.

Mayné, R. (1923) Principaux ennemis des caféiers au Congo Belge. *Annales de Gembloux* 29, 377–384

Mayné, R. and Ghesquière, J. (1934) Hémiptères nuisibles aux végétaux du Congo Belge. *Annales de Gembloux* 40, 41 pp.

McDonald, J. (ed.) (1937) *Coffee in Kenya*. Kenya Department of Agriculture, Nairobi, 210 pp.

Mendes, L.O.T. (1940) Os parasitas do 'bicho mineiro das folhas de café'. *Leucoptera coffeella. Revista del Instituto de Café do Estado de São Paulo* 15, 6–12.

Menezes-Mariconi, F.A. and Iba, S. (1955) A mosca do Mediterraneo. *Biologico* 21, 17–32.

Miller, N.C.E. (1931) *Coccus viridis* Green. The 'green scale' of coffee. *Scientific Series. Department of Agriculture, Straits Settlements and Federated Malay States* 7, 17–29.

Muesebeck, C.F.W. (1937) A new West Indian species of *Mirax* Haliday parasitic on the coffee leaf miner. *Proceedings of the Entomological Society of Washington* 39, 139–141.

Mugo, H.M., Omondi, G.O. and Ndugo, S.M. (1997) Biological control of coffee insect pests in Kenya. *Dix-septième Colloque Scientifique International sur le Café*. 20–25 July 1997, Nairobi, pp. 646–652.

Munoz, R.H. (*c*.2000) *Plagas Insectiles del Cafeto*. Honduras, 32 pp.

Murphy, S.T. (1991) Insect natural enemies of coffee green scales (Hemiptera:Coccidae) in Kenya and their potential for biological control of *Coccus celatus* and *C. viridis* in Papua and New Guinea. *Entomophaga* 36, 519–529.

Ndungi, A.W. (1994) New natural enemies of the coffee berry moth, *Prophantis smaragdina* (Butler) in Kenya. *Kenya Coffee* 59, 688, 1725–1726.

Nietner, J. (1861) *Observations on the Enemies of the Coffee Tree in Ceylon. Ceylon Times*, Colombo, 31 pp.

Nixon, G.E.J. (1935) A revision of the African Telenominae. *Transactions of the Royal Entomological Society of London* 83, 73–103.

Nixon, G.E.J. (1940) New genera and species of Hormiinae with a note on *Hormiopterus* Giraud. *Annals and Magazine of Natural History* 5, 473–493.

Nixon, G.E.J. (1971) Two new species of *Apanteles* from Europe and Africa with a redescription of the African *Apanteles rutilans* Nixon. *Journal of Natural History* 5, 361–365.

Notley, F.B. (1948) The *Leucoptera* leaf miners of coffee on Kilimanjaro I. *Leucoptera coffeella* Guer. *Bulletin of Entomological Research* 39, 399–416.

Noyes, J.S. (1988) *Metaphycus baruensis* sp. n. (Hymenoptera: Encyrtidae) parasitic in green scales, *Coccus* spp. (Hemiptera: Coccidae), on coffee in Kenya. *Bulletin of Entomological Research* 78, 131–134.

Okelana, F.A. (1985) Oviposition pattern, hatching and parasitism of eggs of *Epicampoptera strandi* Byrk. ssp *glauca* Hmps. (Lepidoptera: Drepanidae) on coffee in Nigeria. *Café-Cacao-Thé* 29, 273–276.

Parra, J.R.P., Goncalves, W., Gravena, S. and Marconato, A.R. (1977) Parasitos e predadores do bicho-mineiro do cafeeiro, *Perileucoptera coffeella* (Guerin-Meneville) im São Paulo. *Anais de Sociedade Entomologica do Brazil* 6, 138–143.

Pascalet, P. (1939) La lutte biologique contre *Stephanoderes hampei* ou scolyte du caféier au Cameroun. *Revue de Botanique Appliquée et de l'Agriculture Tropicale* 19, 753–764.

Pearman, J.V. (1932) Some coccophagus psocids (Psocoptera) from East Africa. *Stylops* 1, 90–96.

Perez-Lachaud, G. and Hardy, I.C.W. (1999) Reproductive biology of *Cephalonomia hyalinipennis* (Hymenoptera: Bethylidae), a native parasitoid of the coffee berry borer, *Hypothenemus hampei* (Coleoptera: Scolytidae), in Chiapas, Mexico. *Biological Control* 14, 152–158.

Perioto, N.W., Lara, R.I.R., Selegatto, A. and Luciano, E.S. (2004) Himenópteres parasitóides (Insecta, Hymenoptera) coletados em cultura de café *Coffea arabica* L. (Rubiaceae) em Ribeirão Preto, SP, Brasil. *Arquivos do Instituto Biológico, São Paulo* 71, 41–44.

Pickman Mann, B. (1872) The white coffee leaf-miner. *American Naturalist* 6, 332–341, 596–607.

Prakasan, C.B. and Bhat, P.K. (1985) Interference of the predator *Cryptolaemus montrouzieri* with the performance of a newly introduced parasite, *Leptomastix dactylopii*. *Journal of Coffee Research* 15, 29–32.

Prakasan, C.B. and Kumar, M.G. (1985) New record of natural enemies on coffee mealybugs and green scale. *Journal of Coffee Research* 15, 53–54.

Prakasan, C.B., Sreedharan, K. and Bhat, P.K. (1986) New record of a parasite of coffee white stem borer *Xylotrechus quadripes* Chevr. from India. *Journal of Coffee Research* 16, 38–40.

Pruthi, H.S. and Mani, M.S. (1940) Biological notes on Indian parasitic Chalcidoidea. *Miscellaneous Bulletins. Indian Council of Agricultural Research* 30, 44 pp.

Raga, A., Oliveira Preste, D.A. de, Souza Filho, M.E., Sato, M.E., Siloto, R.C. and Zucchi, R.A. (2002) Occurrence of fruit flies in coffee varieties in the State of São Paulo, Brazil. *Boletin de Sanidad Vegetal, Plagas* 28, 519–524.

Ramakrishna Ayyar, T.V. (1925) A checklist of Indo-Ceylonese Chalcid flies. *Spolia Zeylanika* 13, 235–254.

Reddy, K.B. and Bhat, P.K. (1989) Effect of relative humidity and temperature on the biotic agents of green scale Coccus viridis (Green). *Journal of Coffee Research* 19, 82–87.

Reddy, K.B., Sreedharan, K., Prakasan, C.B. and Bhat, P.K. (1990) New records of natural enemies of *Planococcus* spp., *Coccus viridis* (Green) and *Ferrisia virgata* (Cockerell) on coffee in India. *Journal of Coffee Research* 20, 153–156.

Reddy, K.B., Bhat, P.K. and Naidu, R. (1997) Suppression of mealybugs and green scale infesting coffee with natural enemies in Karnataka. *Pest Management and Economic Zoology* 5, 119–121.

Reimer, N.J., Cope, M. and Yasuda, G. (1993) Interference of *Pheidole megacephala* (Hymenoptera: Formicidae) with biological control of *Coccus viridis* (Homoptera: Coccidae) in coffee. *Environmental Entomology* 22, 483–488.

Reyne, A. (1920) Verslag van den Entomoloog. *Verslagen. Departement van den Landbouw, Suriname 1919*, 20–24.

Ritchie, A.H. (1930) Entomologist's work. *Report. Department of Agriculture, Tanganyika Territory 1928–1929*, 29–34.

Ritchie, A.H. (1932) Report of the Entomologist. *Report. Department of Agriculture, Tanganyika Territory 1931*, 83–86.

Ritchie, A.H. (1935) Report of the Entomologist. *Report. Department of Agriculture, Tanganyika Territory 1934*, 73–83.

Ritchie, A.H. (1936) Report of the Entomologist. *Report. Department of Agriculture, Tanganyika Territory 1935*, 95–103.

Roba, R.P. (1936) El estado sanitario de los cafetales del Norte de Santander el punto de vista entomológico. *Revista Cafetera de Colombia* 6, 2091–2095.

Roba, R.P. (1938) Notes sur quelques insectes s'attaquant au caféier en Colombie. *Bulletin et Annales de la Societe Royale Entomologique de Belgique* 78, 235–239.

Rodrigo, E. (1941) *Administrative Report of the Acting Director of Agriculture, Ceylon* 1940, 18 pp.

Rohwer, S.A. (1918) Descriptions and notes on some Ichneumon flies from Indonesia. *Proceedings of the United States National Museum* 54, 563–570.

Rutherford, A. (1914) Report of the entomologist. *Report of the Department of Agriculture, Ceylon* 1912–1913, 9–12.

Schmitz, G. (1949) La pyrale du caféier robusta *Dichocrocis crocodora* Meyrick. Biologie et moyens de lutte. *Publications de l'Institut National pour l'Étude Agronomique du Congo Belge* 41, 132 pp.

Schouteden, H. (1946) Trois Mirides africains nouveaux du genre *Apollodotus* Dist. *Revue de Zoologie et de Botanique Africaines* 39, 327–330.

Schweizer, J. (1932) Verslag over het jaar 1931. *Mededeelingen van het Besoekisch Proefstation, Djember* 48, 87 pp.

Seetharama, H.G., Kumar, P.K.V. and Sreedharan, K. (2002) *Apenesia* sp.: a potential natural enemy for the biological control of coffee white stem borer. *Indian Coffee* 66, 24–25.

Sekhar, P.S. (1964) Pests of Coffee. In: Pant, N.C. (ed.) *Entomology in India 1938–1963: Indian Journal of Entomology Jubilee Issue*, 99–109.

Sevastopulo, D.G. (1937) *Chrysis shanghaiensis* Smith, a parasite of *Parasa lepida* Cramer. *Proceedings of the Royal Entomological Society of London (A)* 12, 11.

Sivinski, J., Jeronimo, F. and Holler, T. (2000) Development of aerial releases of *Diachasmimorpha tryoni* (Cameron) (Hymenoptera: Braconidae), a parasitoid that attacks the Mediterranean fruit fly, *Ceratitis capitata* (Weidemann) (Diptera: Tephritidae), in the Guatemalan highlands. *Biocontrol Science and Technology* 10, 15–25.

Sladden, G.E. (1934) Le *Stephanoderes hampei* Ferr. *Bulletin Agricole du Congo Belge* 25, 26–77.

Smee, C. (1930) Report of the Government Entomologist. *Report. Department of Agriculture, Nyasaland, 1929*, 13–21.

Smee, C. (1942) Report of the Entomologist 1941. *Nyasaland, Department of Agriculture*, 14 pp.

Smith, H.D. (1945) La 'mosca prieta' de los citricos en la costa occidental de México y la importación y colonización de *Eretmocerus serius* para su control. *Fitófilo* 4, 67–103.

Smith, M.R. (1942) The relationship of ants and other organisms to certain scale insects on coffee in Puerto Rico. *Journal of Agriculture of the University of Puerto Rico* 26, 21–27.

Smith, R.H. (1944) Bionomics and control of the Nigra scale, *Saissetia nigra*. *Hilgardia* 16, 255–288.

Sreedharan, K., Balakrishnan, M.M. and Bhat, P.K. (1992) *Callimerus* sp. (Coleoptera: Cleridae), a predator of the shot-hole borer, *Xylosandrus compactus* (Eichh.). *Journal of Coffee Research* 22, 139–142.

Srinivasa, M.V. (1985) A new record of *Anicetus annulatus* Timberlake (Encyrtidae Hymenoptera) as a parasite of *Coccus viridis* (Green) (Coccidae Homoptera) attacking coffee. *Plant Protection Bulletin, India* 37, 37.

Srinivasa, M.V. (1987) New parasites and host plants of coffee green scale (*Coccus viridis* (Green)) Homoptera: Coccidae in South India. *Journal of Coffee Research* 17, 122–123.

Steck, G.J., Gilstrap, F.E., Wharton, R.A. and Hart, W.G. (1986) Braconid parasitoids of Tephritidae (Diptera) infesting coffee and other fruits in West–Central Africa. *Entomophaga* 31, 59–67.

Subramanian, T.V. (1934) The coffee stem borer. *Bulletin of the Department of Agriculture, Mysore State* 11, 18 pp.

Szent-Ivany, J.J.H. (1958) Insects of cultivated plants in the Central Highlands of New Guinea. *Proceedings of the 10th International Congress of Entomology* 3, Montreal, Canada, 1956, pp. 427–437.

Szent-Ivany, J.J.H. (1963) A biological control experiment against coffee mealybug in the Wau valley. *Highlands Quarterly Bulletin and Journal of the Highland Farmers Settlers Association, Goroka, Papua New Guinea, April 1963*, 8–9.

Tapley, R.G. (1960) The white coffee borer, *Anthores leuconotus* Pasc. and its control. *Bulletin of Entomological Research* 51, 279–301.
Taylor, T.C.H. (1945) Recent investigations of *Antestia* species in Uganda. *East African Agricultural Journal* 10, 223–233; 11, 47–55.
Thompson, W.R. (1943–1958) *A Catalogue of the Parasites and Predators of Insect Pests*. Belville, Ontario, Canada.
Ticheler, J.H.G. (1961) An analytical study of the epidemiology of the coffee berry borer in the Ivory Coast. *Mededeelingen van de Landbouwhoogeschool te Wageningen* 61, 1–49.
Timberlake, P.H. (1919) Observations on the sources of Hawaiian Encyrtidae. *Proceedings of the Hawaiian Entomological Society* 4, 183–196.
Ultee, A.J. (1929) Verslag over de werkzaamheden van het Proefstation Malang in 1928. *Mededeelingen van het Proefstation, Malang* 69, 64 pp.
Van der Goot, P. (1916) De oeconomische beteekenis der gramang-mier. *Mededeelingen van het Proefstation, Midden-Java, Salatiga* 22, 122 pp.
Van der Goot, P. (1917) De zwarte cacao-meir (*Dolichoderus bituberculatus* Mayr) en haar beteekenis voor de cacaocultuur op Java. *Mededeelingen van het proefstation, Midden-Java, Salatiga* 25, 142 pp.
Van Zwaluwenberg, R.H. (1917) Insects affecting coffee in Porto Rico. *Journal of Economic Entomology* 10, 513–517.
Vargas, R.I., Walsh, W.A., Chiou-Ling Hsu, Spencer, J., Mackey, B. and Whitehand, L. (1994) Effects of sterile Mediterranean fruit fly (Diptera: Tephritidae) releases on the target species, a nontarget tephritid, and a braconid parasitoid (Hymenoptera: Braconidae) in commercial coffee fields. *Journal of Economic Entomology* 87, 653–660.
Vargas, R.I., Peck, S.L., McQuate, G.T., Jackson, C.G., Stark, J.D. and Armstrong, J.W. (2001) Potential for areawide integrated management of Mediterranean fruit fly (Diptera: Tephritidae) with a braconid parasitoid and a novel bait spray. *Journal of Economic Entomology* 94, 817–825.
Vazquez, L.L., Perez, I., Navarro, A. and Casin, J.C. (1999) Occurrence and management of fruit flies in Cuba. *Bulletin OEPP* 29, 163–166.
Vayssière, P. (1955) Les animaux parasites du caféier. In: Coste, R. (ed.) *Les Caféiers et les Cafés dans le Monde*. Larose, Paris, pp. 233–318.
Vega, F.E., Mercadier, G., Damon, A. and Kirk, A. (1999) Natural enemies of the coffee berry borer, *Hypothenemus hampei* (Ferrari) (Coleoptera: Scolytidae) in Togo and Côte d'Ivoire, and other insects associated with coffee beans. *African Entomology* 7, 243–248.
Villacorta, A. (1980) Alguns fatores que afetam a populacao estacional de *Perileucoptera coffeella* Guerin-Meneville no norte do Parana Londrina P.R. *Anais da Sociedade Entomologica do Brasil* 9, 23–32.
Vinodkumar, P.K., Balakrishnan, M.M. and Govindarajan, T.S. (1986) *Eupelmus* sp. (Hymenoptera Eupelmidae) – a new predator of *Xylosandrus compactus*. *Journal of Coffee Research* 16, 35–37.
Visitpanich, J. (1994) The parasitoid wasps of the coffee stem borer, *Xylotrechus quadripes* Chevrolat (Coleoptera, Cerambycidae) in northern Thailand. *Japanese Journal of Entomology* 62, 597–606.
Waterston, J. (1916) Notes on Coccid-feeding Chalcidoidea I. *Bulletin of Entomological Research* 7, 137–144.
Way, M.J. (1954a) Studies on the life history and ecology of the ant *Oecophylla longinoda*. *Bulletin of Entomological Research* 45, 93–112.
Way, M.J. (1954b) Studies on the association of the ant *Oecophylla longinoda* with the scale insect *Saissetia zanzibarensis* Williams. *Bulletin of Entomological Research* 45, 113–134.
Weaving, A.J.S. (1972) A preliminary screening trial of insecticides for the control of the coffee

leaf-miner, *Leucoptera meyricki* Ghesq. (Lepidoptera Lyonetiidae) in Rhodesia. *Rhodesian Journal of Agricultural Research* 10, 79–89.
Wharton, R.A. (1987) Changes in nomenclature and classification of some Opiine Braconidae (Hymenoptera). *Proceedings of the Entomological Society of Washington* 89, 61–73.
Wharton, R.A. (1999) A review of the Old World genus *Fopius* Wharton (Hymenoptera: Braconidae Opiinae), with description of two new species reared from fruit-infesting Tephritidae (Diptera). *Journal of Hymenoptera Research* 8, 48–64.
Wharton, R.A., Gilstrap, F.E., Rhode, R.H., Fischel, M. and Hart, W.G. (1981) Hymenopterous egg-pupal and larval-pupal parasitoids of *Ceratitis capitata* and *Anastrepha* spp. (Diptera Tephritidae) in Costa Rica. *Entomophaga* 26, 285–290.
Wharton, R.A., Ovruski, S.M. and Gilstrap, F.E. (1998) Neotropical Eucoilidae (Cynipoidea) associated with fruit-infesting Tephritidae, with new records from Argentina, Bolivia and Costa Rica. *Journal of Hymenoptera Research* 7, 102–115.
Wheatley, P.E. (1963) The giant coffee looper *Ascotis selenaria reciprocaria*. *East African Agricultural and Forestry Journal* 29, 143–146.
Wilkinson, D.S. (1928) A revision of the Indo-Australian species of the genus *Apanteles* Pts I & II. *Bulletin of Entomological Research* 19, 79–105, 109–146.
Wilkinson, D.S. (1930) New Braconidae and other notes. *Bulletin of Entomological Research* 21, 275–285.
Wilkinson, D.S. (1931) Four new species of Ichneumonoidea. *Bulletin of Entomological Research* 22, 393–397.
Wilkinson, D.S. (1934) On some Microgasterinae. *Stylops* 3, 118–120.
Willard, H.F. and Mason, A.C. (1937) Parasitization of the Mediterranean fruit fly in Hawaii 1914–1933. *Circular of the U.S. Department of Agriculture* 439, 18 pp.
Wille, J.E. (1952) *Entomologia agricola del Peru. Manual para Entomólogos, Ingenieros, Agronómos, Agricultores y Estudientes de Agricultura*. Junta de Sanidad Vegetal, Ministerio de Agricultura, Lima, 544 pp.
Wolcott, G.N. (1924) Insectae Portoricensis. A preliminary annotated check-list of the insects of Porto Rico with descriptions of some new species. *Journal of the Department of Agriculture Puerto Rico* 7, 313 pp.
Wolcott, G.N. (1933) *An Economic Entomology of the West Indies*. Porto Rico Entomological Society, 688 pp.
Wolcott, G.N. (1935) Report of the Entomologist. *Report. Porto Rico Insular Agricultural Experimental Station 1933–1934*, 142–144.
Wolcott, G.N. (1951) The insects of Puerto Rico. *Journal of Agriculture of the University of Puerto Rico* 32, 1–975.
Zimmerman, E.C. (1948) *Insects of Hawaii Vol. 5, Homoptera Sternorhyncha*. University Press, Honolulu, 464 pp.

Appendix B Pollination of Coffee

The majority of *Coffea* species are diploid and self-infertile and therefore have to be cross-pollinated, whereas *C. arabica* is tetraploid, self-fertile and at times cleistogamic, and so relies less on cross-pollination. The rate of natural cross-pollination of *C. arabica* seems to vary from place to place and is probably influenced by environmental factors.

In Brazil, Krug (1935) found self-fertilization to be the rule and Castillo-Zatapa (1976) showed cross-pollination rates in Colombia to be < 10%. Carvalho and Krug (1949) later quoted an average of around 12% in Brazil. Taschdijian (1932), by contrast, found in Brazil that very few fruits were set when the flowers were only wind-pollinated and when the possibility of insect and self-pollination was excluded. Using a fasciated form and a red seedling form as indicators, he showed experimentally that cross-pollination could be up to 40–90%. In wild *C. arabica* in Ethiopia Meyer (1965) found 40–60% self-fertility, so the rate of cross-pollination could be higher in the wild than in the cultivated crop.

Both species flower after a period of dry weather and the duration of flowering is short, lasting only a few days. A study of the phenology of flowering of *C. arabica* in Brazil showed that the entire reproductive cycle lasted 2 years, beginning in September of each year (Camargo and Camargo, 2001). After a vegetative phase, the reproductive buds are fully formed by July or August of the following year, when they enter a state of semi-dormancy.

Maturation follows the accumulation of around 350 mm potential evapotranspiration, beginning early in April. Flowering occurs between September and December, 8–15 d after suitable rain. Hacquart (1941) studied the flowering of *C. canephora* in the Democratic Republic of Congo and found that a combination of a drought of 10–30 d, followed by at least 10 mm of rain, induced flowering 9 days later.

In the past, little attention seems to have been paid to the role which insects might play in the pollination of coffee flowers, and the assumption has been made that the crop is primarily wind-pollinated, with some assistance from insects (Free, 1970). Franssen (1932) states that coffee is only rarely visited by bees and that cross-pollination is almost non-existent. Le Pelley (1968), in the standard work on coffee insects, makes no mention of pollinating insects. Although he does have a section on pollination in his review of 1973 (Le Pelley, 1973), he reiterates the view held by Free (1970) that wind is the chief method of cross-pollination but that insects do play a part; however, he mentions neither specific insects nor the degree of their involvement.

Even as early as 1911, however, coffee-planters in India were sufficiently alarmed at the effect which the destruction of wild bees might be having on their crop to ask for an entomological investigation. Fletcher (1915) found that wild bees did act as pollinators, *Apis dorsata* Fabricius being the most important followed by *A. cerana* Fabricius and *A. florea* Fabricius. In a number of countries, beekeeping has been advocated as being beneficial to coffee production, including Tanzania (Smith, 1958), Kenya (McDonald, 1937), Puerto Rico (Lower, 1911; Sein, 1923; Montealegre, 1946) and Costa Rica (Rudin, 1942).

In Jamaica, Raw and Free (1977) enclosed small plots of *C. arabica* in mesh cages, two of which had beehives enclosed within them and two without bees. These were compared with uncaged plots. Plots with bees yielded more berries and a greater weight of berries than plots caged without bees.

Willmer and Stone (1989) made a study of the pollination of *C. canephora* in Papua New Guinea. They bagged flowers with cotton sheets which allowed only self-pollination, with gauze which allowed wind pollination but excluded flying insects or with gauze bags on branches which had been greased to prevent ant incursion. The mean number of pollen grains per stigma was eight times greater on unbagged flowers than for any of the other treatments, and there was no significant difference between the bagged treatments.

The most frequent insect visitors were *Megachile* (*Creightonella*) *frontalis* (Fabricius), *Amegilla sapiens*, the social bees, *Apis* sp and *Trigona* sp., and Syrphid flies. Willmer and Stone were of the opinion that *M. frontalis* was responsible for much of the pollination. This species cuts leaf discs to line its underground nest and has been recorded as a minor pest of coffee by Michener and Szent-Ivany (1960).

Klein *et al.* (2003a, b, c) studied pollination in an area of Sulawesi where fields of *C. arabica* and *C. canephora* grew in close proximity. Using a similar bagging system to that of Willmer and Stone, they showed that open pollination and hand cross-pollination of both species gave significantly better fruit set than when insects were excluded, but with *C. canephora* hand pollination was superior to open pollination (Klein *et al.*, 2003b). They found that bees were the most frequent visitors and recorded a number of species. The percentage fruit set of *C. arabica* increased with the number of bee species present, from around 60% with three species present to 90% when 20 species were present, but not with the number of bee individuals (Klein *et al.*, 2003c).

Individuals of the seven species of social bees, *Apis* and *Trigona*, visited flowers twice as frequently as the eight species of solitary bees but solitary bees effected a significantly higher rate of fruit set than social bees, so it was inferred that the latter were the more important pollinators. The social bees tended to nest in forest, so there was a negative linear relationship between number of social species visiting and distance from forest. However, with the solitary bees, which favour open conditions for nesting, there was a positive linear correlation between number of species visiting and light intensity but not with distance from forest (Klein *et al.*, 2003a).

Interest in insect pollination has increased in South and Central America in recent years following the spread through the continents of African bees. It was thought that *Apis mellifera* Linnaeus (the domesticated honeybee) stock of tropical origin would be better adapted to the climate of Brazil than bees of European origin previously used in Brazil and, during the 1950s, new stock was sought in Africa. A queen of the aggressive *A. mellifera scutellata* Lepeletier of Tanzanian origin escaped from quarantine in 1957, and her offspring colonized the surrounding countryside, rapidly spreading throughout South and Central America. They had spread as far as Texas by 1990 and have since colonized > 13 million km^2. The majority of hive bees in South and Central America are now African bees.

Malerbo-Souza *et al.* (2003) investigated the role played by African bees in the pollination of *C. arabica* in Brazil. They found that 98% of the bees collected nectar from the flowers and only 2% pollen but, nevertheless, the rate of pollination was 38.8% greater when bees had access to flowers than when they were excluded. Apart from *A. mellifera scutellata* – which was recorded at 89% of flowers – much smaller numbers of flowers (2–3%) were visited by other bees, including *Chloralictus* sp., *Trigona spinipes*, *Xylocopa* sp. and *Tetragonisca angustula* (Latreille).

Amaral (1972), on the other hand, found a smaller increase in pollen grains (13.6%) when bees had access to the flowers than when they were excluded, but he did show an increase in visitation the closer hives were placed to the crop.

In Panama, Roubik (2002a, b) found visiting bees to be most abundant in the vicinity of nearby forest and recorded several species, although the African bee was by far the most frequent visitor. There appeared to be an increase in fruit retention in both Caturra and Catimor as a result of bee pollination, and also an increase in seed size.

Ricketts (2004) also showed increased bee activity at sites close to forest in Costa Rica, there being a significantly greater number of visits to flowers over a 20-min period and number of pollen grains per stigma at sites 50 m from forest compared to sites 800 or 1600 m distant. He found a number of bee species to be involved, but *A. mellifera* was the most common.

Although coffee flowers are attractive to a number of insects searching for nectar or pollen, such as thrips, Syrphid and other flies, moths, butterflies and ants it is evident that bees, both social and solitary, are by far the most important as potential pollinators. If coffee pollination is to be enhanced, it is these species which must be encouraged. As different families and subfamilies have different nesting requirements, these must be taken into account in manipulating the coffee environment.

The species recorded as visiting coffee flowers belong to three families, the Apidae, Halticidae and Megachilidae. All feed their larvae on pollen.

Apidae

Apidae Anthophorinae (Digger bees)

These are relatively large solitary bees that make burrows in flat ground or in banks. Their cells are lined with wax and the egg is laid on a semi-liquid substrate. Individuals tend to be gregarious, and several nests are constructed close together in a suitable patch of soil. These and other ground-living bees could be encouraged by patches of bare ground or soil banks close to the coffee plantation.

Apidae Apinae (Social bees)

The subfamily Apinae contains all the social bees and is by far the largest group. *Apis* live in large colonies and produce wax combs consisting of a large number of cells in which the young stages are nurtured or in which honey is stored. Some species live in cavities in trees or between rocks, etc. whereas with others such as *A. dorsata* the comb is fully exposed and hung from overhanging cliffs or the branches of large trees.

By and large they are aggressive bees, although the domesticated honeybee has been tamed through centuries of breeding. *Trigona* by contrast, although living in quite large colonies, mainly in hollow trees, store less honey and are stingless. *Bombus* (bumble bees) nest mainly underground. Before the introduction of hive bees into the Americas, stingless bees such as *Melipona* species were cultivated in Central America – and still are to some extent, but are less popular because their capacity for honey production is much less than that of honeybees. They could be encouraged as pollinators and would be more 'user friendly' to coffee workers than the aggressive African bee, which has already claimed around 1000 lives during its short history on the continent.

Apidae Xylocopinae (Carpenter bees)

Burrows are made in sound wood or in pithy stems in some species. There is no lining to the cell and the egg is laid on top of a firm pellet of pollen.

Halticidae (Sweat bees)

Small bees that generally nest in the soil but occasionally use rotten wood. The cell is lined with wax and the egg deposited on top of a dry pellet of pollen. They are well known in tropical countries for their annoying habit of feeding on sweat, particularly around the eyes.

Megachilidae (Leaf-cutter bees)

These are generally large bees, some of which nest underground and others in hollow stems or holes in wood. The cells of some species are lined with discs of leaf, whereas others use pellets, mud or other substances. Some members, such as *Coelioxys*, are cleptoparasites, living within nests of other bee species.

It is hoped that now it has been acknowledged that insects do have a significant effect on coffee pollination that research in the field will increase. A comprehensive list of those insects visiting coffee flowers is given below.

Potential Hymenopterous Pollinators: Social Bees

Apidae

Apinae Apini

Apis cerana Fabricius	Sulawesi (Klein *et al.*, 2003a, b, c), India (Fletcher, 1915)
Apis dorsata Fabricius	India (Fletcher, 1915)
Apis dorsata binghami Cockerell	Sulawesi (Klein *et al.*, 2003a, b, c)
Apis florea Fabricius	India (Fletcher, 1915)
Apis mellifera Linnaeus	Jamaica (Raw and Free, 1977), Panama (Roubik, 2002a, b), Costa Rica (Ricketts, 2004), Venezuela (Manrique and Thimann, 2002), Brazil (Malerbo-Souza *et al.*, 2003; Favero, 2002)
Apis mellifera ligustica Spinola	Costa Rica (Badilla and Ramirez, 1991)
Apis nigrocincta Smith	Sulawesi (Klein *et al.*, 2003a, b, c)
Apis sp.	Papua New Guinea (Willmer and Stone, 1989)

Apinae Bombini

Bombus ephippiatus Say	Panama (Roubik, 2002a, b)
Bombus pullatus Franklin	Panama (Roubik, 2002a, b)
Bombus volucelloides Gribodo	Panama (Roubik, 2002a, b)

Apinae Euglossini

Eulaema polychroma (Mocsary)	Panama (Roubik, 2002a, b)

Apinae Meliponini

Melipona fasciata Latreille	Costa Rica (Ricketts, 2004)
Melipona panamica Cockerell	Panama (Roubik, 2002a, b)
Nannotrigona mellaria (Smith)	Costa Rica (Ricketts, 2004)

Nannotrigona perilampoides (Cresson)	Panama (Roubik, 2002a, b)
Paratrigona ornaticeps (Schwarz)	Panama (Roubik, 2002a, b)
Partamona bilineata (Say)	Panama (Roubik, 2002a, b)
Partamona cupira (Smith)	Costa Rica (Ricketts, 2004)
Plebeia frontalis (Friese)	Costa Rica (Ricketts, 2004)
Plebeia jatiformis (Cockerell)	Costa Rica (Ricketts, 2004)
Scaptotrigona subobscuripennis (Schwarz)	Panama (Roubik, 2002a, b)
Tetragonisca angustula (Latreille)	Brazil (Malerbo-Souza *et al.*, 2003)
Trigona amalthea Olivier	Panama (Roubik, 2002a, b)
Trigona angustula (Latreille)	Panama (Roubik, 2002a, b), Costa Rica (Ricketts, 2004)
Trigona bocandei Spinola	Ghana (Forsyth, 1966)
Trigona braunsi Kohl	Uganda (Le Pelley, 1959)
Trigona clavipes (Fabricius)	Costa Rica (Ricketts, 2004)
Trigona corvina Cockerell	Panama (Roubik, 2002a, b), Costa Rica (Ricketts, 2004)
Trigona dorsalis Friese	Panama (Roubik, 2002a, b), Costa Rica (Ricketts, 2004)
Trigona fulviventris Guerin-Meneville	Panama (Roubik, 2002a, b), Costa Rica (Ricketts, 2004)
Trigona fuscipennis Friese	Costa Rica (Ricketts, 2004)
Trigona nigerrima Cresson	Panama (Roubik, 2002a, b)
Trigona spinipes (Fabricius)	Brazil (Malerbo-Souza *et al.*, 2003)
Trigona terminata Smith	Sulawesi (Klein *et al.*, 2003a, b, c)
Trigona sp.	Sulawesi (Klein *et al.*, 2003a, b, c), Papua New Guinea (Willmer and Stone, 1989)
Trigonisca sp.	Costa Rica (Ricketts, 2004)

Potential Hymenopterous Pollinators: Solitary Bees

Apidae

Apinae Anthophorini

Amegilla sapiens (Cockerell)	Papua New Guinea (Willmer and Stone, 1989)
Amegilla whiteheadi (Cockerell)	Sulawesi (Klein *et al.*, 2003b)
Amegilla sp.	Sulawesi (Klein *et al.*, 2003a, b, c)
Amegilla sp. near *samarensis* (Cockerell)	Sulawesi (Klein *et al.*, 2003b)

Apinae Centridini

Centris dirrhoda Moure	Jamaica (Raw and Free, 1977)
Centris festiva F. Smith	Panama (Roubik, 2002a, b)
Centris sp.	Panama (Roubik, 2002a, b)
Epicharis rustica (Olivier)	Panama (Roubik, 2002a, b)

Apinae Exomalopsini

Exomalopsis sp.	Jamaica (Raw and Free, 1977)

Xylocopinae Ceratinini

Ceratina rugifrons Smith	Sulawesi (Klein *et al.*, 2003b)
Ceratina sp.	Sulawesi (Klein *et al.*, 2003a, c)

Xylocopinae Xylocopini

Xylocopa aestuans (Linnaeus)	Sulawesi (Klein *et al.*, 2003a, b, c)
Xylocopa dejeanii Lepeletier	Sulawesi (Klein *et al.*, 2003b)
Xylocopa dejeanii nigrocerulea Smith	Sulawesi (Klein *et al.*, 2003a, c)
Xylocopa smithii Ritsema	Sulawesi (Klein *et al.*, 2003b)
Xylocopa sp.	Brazil (Favero, 2002)

Halticidae

Halticinae Augochlorini

Augochlora sp.	Panama (Roubik, 2002a, b)

Halticinae Halticini

Dialictus spp.	Jamaica (Raw and Free, 1977), Brazil (Malerbo-Souza *et al.*, 2003)
Lasioglossum sp.	Panama (Roubik, 2002a, b)
Patellapis sp.	Sulawesi (Klein *et al.*, 2003b)

Nomiinae

Lipotriches sp.	Sulawesi (Klein *et al.*, 2003b)
Nomia thoracica Smith	Sulawesi (Klein *et al.*, 2003b)

Megachilidae

Megachilinae Megachilini

Megachile (Callochile) sp. near *bakeri* Cockerell	Sulawesi (Klein *et al.*, 2003b)
Megachile (Callomegachile) *clotho* Smith	Sulawesi (Klein *et al.*, 2003c)
Megachile (Callomegachile) *incisa* Smith	Sulawesi (Klein *et al.*, 2003b)
Megachile (Callomegachile) *terminalis* Smith	Sulawesi (Klein *et al.*, 2003b)
Megachile (Creightonella) *frontalis* (Fabricius)	Sulawesi (Klein *et al.*, 2003a, c), Papua New Guinea (Willmer and Stone, 1989)
Megachile (Creightonella) *frontalis atrata* Smith	Sulawesi (Klein *et al.*, 2003b)
Megachile (Eumegachilana) *tuberculata* Smith	Sulawesi (Klein *et al.*, 2003b)
Megachile (Paracella) sp.	Sulawesi (Klein *et al.*, 2003b)
Megachile sp.	Sulawesi (Klein *et al.*, 2003a)

Megachilinae Osmiinii

Heriades spp.	Sulawesi (Klein *et al.*, 2003a, b, c)

Other Recorded Insects

Apidae

Apinae Melectini

Thyreus nitidus quartina (Cockerell)[a]	Sulawesi (Klein *et al.*, 2003b)

Heliconiidae

Dryas iulia (Fabricius)	Jamaica (Raw and Free, 1977)
Heliconius charithonia (Linnaeus)	Jamaica (Raw and Free, 1977)

Hesperiidae

Astraptes jaira (Butler)	Jamaica (Raw and Free, 1977)
Cymaenes tripunctus (Herrich-Schaffer)	Jamaica (Raw and Free, 1977)
Panoquina sylvicola (Herrich-Schaffer)	Jamaica (Raw and Free, 1977)
Polygonus leo (Gmelin)	Jamaica (Raw and Free, 1977)
Proteides mercurius (Fabricius)	Jamaica (Raw and Free, 1977)
Urbanus proteus (Linnaeus)	Jamaica (Raw and Free, 1977)

Megachilidae

Megachilinae Megachilini

Coelioxys smithii Dalla Torre[a]	Sulawesi (Klein *et al.*, 2003b)
Coelioxys(Torridapis) ducalis Smith[a]	Sulawesi (Klein *et al.*, 2003b)

Nymphalidae

Metamorpha stelenes (Linnaeus)	Jamaica (Raw and Free, 1977)

Papilionidae

Battus polydamas (Linnaeus)	Jamaica (Raw and Free, 1977)

Phlaeothripidae

Haplothrips tenuipennis Bagnall	India (Sekhar and Sekhar, 1964)

Pieridae

Ascia monuste (Linnaeus)	Jamaica (Raw and Free, 1977)
Eurema sp.	Jamaica (Raw and Free, 1977)
Phoebis sennae (Linnaeus)	Jamaica (Raw and Free, 1977)
Polistes crinitus (Felton)	Jamaica (Raw and Free, 1977)

Thripidae

Frankliniella parvula Hood	Trinidad and Tobago (Billes, 1941)
Thrips florum Schmutz	India (Sekhar and Sekhar, 1964)
Thrips parvispinus Karny	Indonesia (Kalshoven, 1950–1951)

Vespidae

Brachygastra sp.	Panama (Roubik, 2002a, b)

[a] These species are cleptoparasites in nests of other bees.

References

Amaral, E. (1972) *Polinização Entomófila de* Coffea arabica *L., Raio de Ação e Coleta de Pólen Pela* Apis mellifera *Linnaeus 1758 (Hymenoptera: Apidae), em Cafezal Florido.* Tese – Escola Superior de Agronomia 'Luiz de Queroz', Universidade de São Paulo, Piracicaba, 82 pp.

Badilla, E. and Ramirez, B. (1991) Polinizacion de cafe por Apis mellifera L. y otros insectos en Costa Rica. *Turrialba* 41, 285–288.

Billes, D.J. (1941) Pollination of *Theobroma cacao* L. in Trinidad, B.W.I. *Tropical Agriculture* 18, 151–156.

Camargo, A.P. and Camargo, M.B.P. (2001) Definição e esquematização das fases fenológicas do cafeeiro arábica nas condições tropicais do Brasil. *Bragantia* 60, 65–68.

Carvalho, A. and Krug, C.A. (1949) Genetica de *Coffea* XII. Hetereditariedade da cor amarela da semente. *Bragantia* 9, 193–202.

Castillo-Zapata, J. (1976) Tasa de polinization cruzada del cafe arabigo en la region de Chinchina. *Cenicafe* 27, 78–88.

Fávero, A.C. (2002) *Polinização Entomófila em Soja* (Glycine max *L. Merrill, var. FT 2000) e Café* (Coffea arabica *L., variedades Catuai Vermelho – IAC 144 e Mundo Novo*). Monografia – Faculdade de Ciências Agrárias e Veterinárias, Univesidade Estadual Paulista, Jaboticabal, Brazil, 44 pp.

Fletcher, T.B. (1915) Bees and the fertilization of coffee. *Madras Department of Agriculture Bulletin* 4, 38 pp.

Forsyth, J. (1966) *Agricultural Insects of Ghana*. Ghana Universities Press, 163 pp.

Franssen, C.J.H. (1932) De beteekenis van *Apis indica* als bloembestuivend insect. *Bergcultures* 6, 1417–1423.

Free, J. (1970) *Insect Pollination of Crops*. Academic Press, London, 684 pp.

Hacquart, H. (1941) Periodicité de la floraison et de la fructification du caféier Robusta a l'equateur. *Bulletin Agricole du Congo Belge*, 32, 496–536.

Kalshoven, L.G.E. (1950–1951) *De Plagen van de Cultuurgewassen in Indonesië*. Van Hoeve, Gravenhage/Bandoeng, Netherlands/Indonesia, 1065 pp.

Klein, A.-M., Steffan-Dewenter, I. and Tscharntke, T. (2003a) Pollination of *Coffea canephora* in relation to local and regional agroforestry management. *Journal of Applied Ecology* 40, 837–845.

Klein, A.-M., Steffan-Dewenter, I. and Tscharntke, T. (2003b) Bee pollination and fruit set of *Coffea arabica* and *C. canephora* (Rubiaceae). *American Journal of Botany* 90, 153–157.

Klein, A.-M., Steffan-Dewenter, I. and Tscharntke, T. (2003c) Fruit set of highland coffee increases with the diversity of pollinating bees. *Proceedings of the Royal Society* (B) 270, 955–961.

Krug, C.A. (1935) Controle da pollinizacao nas flores do cafeeiro. *Boletim Technico Instituto Agronomico, Sao Paulo* No. 15, 12 pp.
Le Pelley, R.H. (1959) *Agricultural Insects of East Africa*. East African High Commission, Nairobi, 307 pp.
Le Pelley, R.H. (1968) *Pests of Coffee*. Longmans, London, 590 pp.
Le Pelley, R.H. (1973) Coffee Insects. *Annual Review of Entomology* 18, 121–142.
Lower, W.V. (1911) Beekeeping in Puerto Rico. *Puerto Rico Agricultural Experimental Station Circular* No. 13, 31 pp.
Malerbo-Souza, D.T., Nogueira-Couto, R.H., Couto, L.A. and De Souza, J.C. (2003) Atrativo para as abelhas *Apis mellifera* e polinizacao em cafe (*Coffea arabica* L.). *Brazilian Journal of Veterinary Research and Animal Science* 40, 272–278.
Manrique, A.J. and Thimann, R.E. (2002) Coffee (*Coffea arabica*) pollination with Africanized honeybees in Venezuela. *Interciencia* 27, 414–416,
Mc Donald, J.H. (1937) *Coffee in Kenya*. Kenya Department of Agriculture, Nairobi, 210 pp.
Meyer, F. (1965) Notes on wild *Coffea arabica* from southwestern Ethiopia, with some historical considerations. *Economic Botany* 19, 136–151.
Michener, C.D. and Szent-Ivany, J.J.H. (1960) Observations on the biology of a leaf-cutter bee *Megachile frontalis* in New Guinea. *Papua and New Guinea Agricultural Journal* 13, 22–35.
Montealegre, M.R. (1946) The fertilization of coffee flowers. *Revista del Instituto de Defensa del Cafe de Costa Rica* 15, 337–340.
Raw, A. and Free, J.B. (1977) The pollination of coffee (*Coffea arabica*) by honeybees. *Tropical Agriculture* 54, 365–370.
Ricketts, T.H. (2004) Tropical forest fragments enhance pollinator activity in nearby coffee crops. *Conservation Biology* 18, 1262–1271.
Roubik, D.W. (2002a) Feral African bees augment Neotropical coffee yield. In: Kevan, P. and Imperatiz Fonseca, V.L. (eds) *Pollinating Bees – the Conservation Link between Agriculture and Nature*. Ministry of Environment, Brasilia, Brazil, pp. 255–266.
Roubik, D.W. (2002b) The value of bees to the coffee harvest. *Nature* 417, 708.
Rudin, J. (1942) Bees as pollination agents in fruit trees. *Revista del Instituto de Defensa del Cafe de Costa Rica* 12, 490–491.
Sein, F. (1923) Las abejas en los cafetales. *Puerto Rico Agricultural Experimental Station Circular* No. 79, 6 pp.
Sekhar, P.S. and Sekhar, S. (1964) Investigations on thrips occurring on arabica coffee. I. Bionomics. *Indian Coffee* 28, 173–179.
Smith, F.G. (1958) Beekeeping operations in Tanganyika, 1949–1957. *Bee World* 39, 29–36.
Taschdijian, E. (1932) Beobachrungen über variabilität, dominanz und vizinismus bei *Coffea arabica*. *Zeitung Züchtung* 17, 341–354.
Willmer, P.G. and Stone, G.N. (1989) Incidence of entomophilous pollination of lowland coffee (*Coffea canephora*); the role of leaf-cutter bees in Papua New Guinea. *Entomologia Experimentalis et Applicata* 50, 113–124.

Index